Limit Cycles and Homoclinic Networks in Two-Dimensional Polynomial Systems

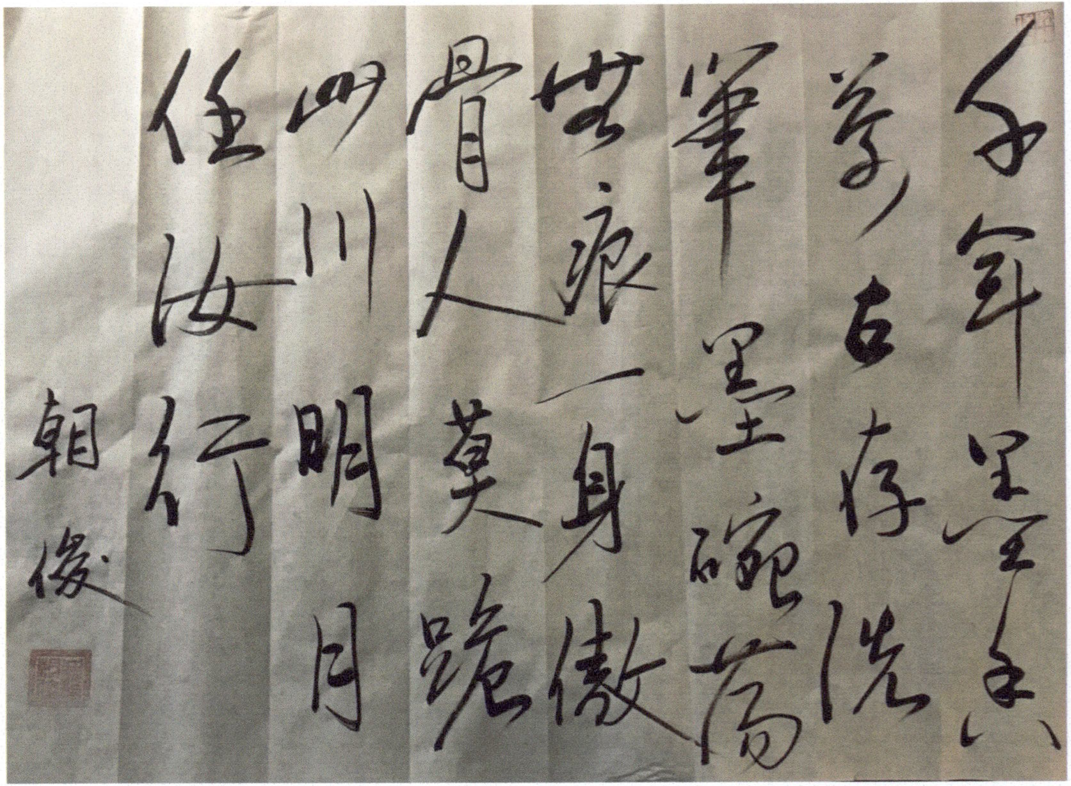

千年墨香万古存，洗笔墨碗荡无痕，一身傲骨人莫跪，山川明月任汝行。

—— 朝俊 《评院士赋》

Albert C. J. Luo

Limit Cycles and Homoclinic Networks in Two-Dimensional Polynomial Systems

With 30 figures

高等教育出版社
HIGHER EDUCATION PRESS

Springer

Albert C. J. Luo
Department of Mechanical
and Mechatronics Engineering
Southern Illinois University Edwardsville
Edwardsville, IL, USA

ISBN 978-981-97-2616-5 ISBN 978-981-97-2617-2 (eBook)
https://doi.org/10.1007/978-981-97-2617-2

Jointly published with Higher Education Press Limited Company
The print edition is not for sale in the mainland of China. Customers from the mainland of China please order the print book from: Higher Education Press Limited Company.

This Springer imprint is published by the registered company Springer Nature Singapore Pte Ltd.
The registered company address is: 152 Beach Road, #21-01/04 Gateway East, Singapore 189721, Singapore

Preface

This book is a monograph about limit cycles and homoclinic networks in polynomial systems. The study of dynamical behaviors of polynomial dynamical systems has been stimulated by the Hilbert's 16th problem in 1900. Too many scientists have tried to work on the Hilbert sixteenth problem. Until now, no significant results have been achieved yet.

In this book, the properties of equilibriums in planar polynomial dynamical systems are studied. The homoclinic networks of sources, sinks, and saddles in self-univariate polynomial systems are discussed, and the corresponding bifurcation theory is developed. The corresponding first integral manifolds are determined analytically, and networks of source, sinks, and saddles are illustrated. The homoclinic networks of saddles and centers (or limit cycles) in crossing-univariate polynomial systems are discussed, and the corresponding bifurcation theory is developed. The corresponding first integral manifolds are polynomial functions. The homoclinic networks of saddles and centers are illustrated, which are without any sources and sinks. Since the maximum numbers of equilibriums for such two types of planar polynomial systems with the same degrees are discussed, the maximum centers and saddles in homoclinic networks are obtained, and the maximum numbers of sinks, sources, and saddles in homoclinic networks without centers are obtained as well. Such studies are to achieve global dynamics of planar polynomial dynamical systems, which can help one study global behaviors in nonlinear dynamical systems in physics, chemical reaction dynamics, engineering dynamics, and so on.

In this book, five chapters are included. Chapter 1 is for instruction to polynomial dynamical systems. Chapter 2 discusses homoclinic networks without centers. The corresponding bifurcations for homoclinic networks without centers are presented in Chap. 3. Chapter 4 discusses homoclinic networks with centers and saddles. The corresponding bifurcations for homoclinic with centers are presented in Chap. 5.

Finally, the author hopes the materials presented herein can provide a better understanding of nonlinear dynamics in science and engineering.

Edwardsville, IL, USA Albert C. J. Luo

Contents

Chapter 1
Introduction

1.1 Hilbert's 16th Problem

Consider a dynamical system with a differential equation as

$$\dot{x}_1 \equiv \frac{dx_1}{dt} = P(x_1, x_2),\ \dot{x}_2 \equiv \frac{dx_2}{dt} = Q(x_1, x_2) \tag{1.1}$$

where $P(x_1, x_2)$ and $Q(x_1, x_2)$ are real polynomials of degree n. The second part of Hilbert's 16th problem is to decide an upper bound for the number of limit cycles in polynomial vector fields of degree n and, similar to the first part, investigate their relative positions. The original problem can be found in [1]. For the first part, it is about much many branches of the algebraic curves determined by

$$\frac{dx_1}{dx_2} = \frac{P(x_1, x_2)}{Q(x_1, x_2)}. \tag{1.2}$$

If the curve manifolds are polynomial functions, the first and second parts are similar. If the first integral manifolds are not polynomial functions, the first and second parts are different. For the limit cycles, the first integral manifolds will be considered as polynomial functions. For other cases, they will not be discussed in this book.

In fact, the limit cycle or the branches of curves can be separated by homoclinic orbits relative to the equilibriums. To look for the maximum limit cycles, the maximum equilibriums (or singular points) should be solved. Consider

$$\begin{aligned} P(x_1, x_2) &= \sum_{k=0}^{n_1} \alpha_{12k}(x_2) x_1^k = \sum_{k=0}^{n_2} \alpha_{11k}(x_1) x_2^k, \\ Q(x_1, x_2) &= \sum_{k=0}^{n_1} \alpha_{22k}(x_2) x_1^k = \sum_{k=0}^{n_2} \alpha_{21k}(x_1) x_2^k; \end{aligned} \tag{1.3}$$

A. C. J. Luo, *Limit Cycles and Homoclinic Networks in Two-Dimensional Polynomial Systems*, https://doi.org/10.1007/978-981-97-2617-2_1

where

$$\alpha_{12k}(x_2) = \sum_{l=0}^{n-k} \beta_{12l} x_2^l, \alpha_{11k}(x_1) = \sum_{l=0}^{n-k} \beta_{11l} x_1^l,$$
$$\alpha_{22k}(x_2) = \sum_{l=0}^{n-k} \beta_{22l} x_2^l, \alpha_{21k}(x_1) = \sum_{l=0}^{n-k} \beta_{21l} x_1^l. \tag{1.4}$$

Thus, if

$$P(x_1, x_2) = 0,\ Q(x_1, x_2) = 0, \tag{1.5}$$

then four cases are given by

$$\sum_{k=0}^{n_1} \alpha_{12k}(x_2) x_1^k = 0, \sum_{k=0}^{n_1} \alpha_{22k}(x_2) x_1^k = 0;$$
$$\sum_{k=0}^{n_1} \alpha_{12k}(x_2) x_1^k = 0, \sum_{k=0}^{n_2} \alpha_{21k}(x_1) x_2^k = 0;$$
$$\sum_{k=0}^{n_2} \alpha_{11k}(x_1) x_2^k = 0, \sum_{k=0}^{n_1} \alpha_{22k}(x_2) x_1^k = 0;$$
$$\sum_{k=0}^{n_2} \alpha_{11k}(x_1) x_2^k = 0, \sum_{k=0}^{n_2} \alpha_{21k}(x_1) x_2^k = 0. \tag{1.6}$$

From the above four cases, the algebraic basic theorem gives the foregoing algebraic equations having $n \times n$ roots and only $n \times n$ roots. The real roots are equilibriums (singular points). If all of $n \times n$ roots are real, then the maximum equilibriums (singular points) of Eq. (1.1) are $n \times n$ and cannot be more than $n \times n$. Suppose equilibriums are obtained by

$$x_1^* = a_{11k}\ k \in \{1, 2, \cdots, n_1\},$$
$$x_1^* = a_{21k}\ k \in \{1, 2, \cdots, n - n_1\};$$
$$x_2^* = a_{12k}\ k \in \{1, 2, \cdots, n - n_1\},$$
$$x_2^* = a_{22k}\ k \in \{1, 2, \cdots, n_1\}. \tag{1.7}$$

Thus the deformation of Eq. (1.1) gives

$$\dot{x}_1 = a_{110} \prod_{s_1=1}^{n_1} (x_1 - a_{11s_1}) \prod_{l_1=1}^{n-n_1} (x_2 - a_{12l_1}),$$
$$\dot{x}_2 = a_{220} \prod_{s_2=1}^{n-n_1} (x_1 - a_{21s_2}) \prod_{l_2=1}^{n_1} (x_2 - a_{22l_2}). \tag{1.8}$$

Without lose of generality, consider two extreme cases. For $n_1 = n$,

$$\dot{x}_1 = a_{110} \prod_{s_1=1}^{n} (x_1 - a_{11s_1}), \dot{x}_2 = a_{220} \prod_{l_2=1}^{n} (x_2 - a_{22l_2}), \quad (1.9)$$

and for $n_1 = 0$,

$$\dot{x}_1 = a_{120} \prod_{l_1=1}^{n} (x_2 - a_{12l_1}), \ \dot{x}_2 = a_{210} \prod_{s_2=1}^{n} (x_1 - a_{21s_2}). \quad (1.10)$$

In this book, from the above two cases, the properties of $n \times n$ equilibriums will be discussed. Further, the limit cycles and homoclinic networks of planar dynamical systems for the two cases will be studied. Such studies will help one better understand the nonlinear dynamics in mathematics, physics, and engineering.

1.2 A Brief History

Too many scientists have worked on the Hilbert sixteenth problem, which cannot be discussed one by one herein. For a brief history, in 1923, Dulac [2] discussed the proof of the finiteness theorem, which was not considered to be true until 1970s. In 1982, Ilyashenko [3] discussed singular points and limit cycles of differential equations in real and complex planes, which are based on normal form analysis. The phase portraits of normal forms are based on the topological imagination rather than studying the depth of the corresponding dynamical behaviors. In 1984, Ilyashenko [4] also presented limit cycle of polynomial vector fields with non-degenerate singular points on the real plane. The review articles on limit cycles were discussed through the local theory of differential equations in Ilyashenko [5]. In 1991, Ilyashenko [6] showed Dulac's proof contains an essential gap and presented the theorems for finiteness of limit cycles.

The proof is based on Poincare maps without differential equations. The Dulac formal form series was adopted. In such series, polynomial functions are based on linear cases to be imagined. For higher-order singularity, such polynomial logarithmic function cannot exist. The centennial research history of Hilbert's sixteenth problem was given by Ilyashenko [7]. In 2003, Li [8] also gave a systematical review on Hilbert's 16th problem, the bifurcations of planar polynomial vector fields are based on the traditional analysis. The simple versions of the Hilbert 16th problem include the Abel equation, the Lienard equation and van del Pol equation, the infinitesimal Hilbert problem, and the Hilbert-Arnold problem. For the Abel equation, the equation is on the cylinder polynomial in y and periodic in x. Such a problem can be referenced to Shahshahani [9], Lins Neto [10] and Panov [11]. Smale [12] listed the Lienard equation as one of the 21th century mathematical problems.

The early work originated from van del Pol equation [13, 14] and Lienard equation [15]. The experiments proved the existence of limit cycles in the Poincare sense. In 1939, Bautin [16] announced the number of limit cycles $N(2) \geq 3$ in a quadratic system, and the full proof was given by Bautin [17] in 1952. In 1955, Petrovski and Landis [18, 19] claimed the quadratic system has three limit cycles and the general discussion can be found as well. However, in 1979, Shi [20] gave a quadratic system having four limit cycles, and Chen and Wang [21] discussed the relative position, and the number of, the limit cycles in quadratic differential equation. In 2001, Ilyashenko and Panov [22] obtained the uniform upper bound for the number of limit cycles for the general Lienard equation. In 1977, Arnold [23, 24] proposed the weakened, infinitesimal (or tangential) Hilbert 16th problems. In fact, such a deformed problem should not belong to the original Hilbert's sixteen problem. The possible bifurcations of limit cycles for Hilbert's sixteenth problem were mentioned in Li [8], and the corresponding discussion can also be found from Ye [25] and Zhang et al. [26].

The limit cycles are separated by homoclinic networks. To determine possible maximum limit cycles, the maximum equilibriums should be determined. In 2022, Luo [27] presented an alternative theory of singularity and stability of two-dimensional linear systems. To understand singularities in two-dimensional quadratic systems, a theory of two-dimensional quadratic systems with single-variable quadratic vector fields was presented in Luo [28]. The theories for planar dynamical systems with self-univariate and crossing-univariate quadratic vector fields were developed in Luo [29, 30]. In 2023, Luo [31] discussed the nonlinear dynamics of two-dimensional systems with product quadratic vector fields. If one is interested in bifurcations for such polynomial systems, the detailed discussion for quadratic and cubic polynomial systems are in Luo [32–34].

1.3 Book Layout

In this book, five chapters are included. The first chapter is for instruction to polynomial dynamical systems. Chapter 2 will discuss homoclinic networks without centers. The corresponding bifurcations for homoclinic networks without centers will be discussed in Chap. 3. Chapter 4 will discuss homoclinic networks with centers and saddles. The corresponding bifurcations for homoclinic with centers will be presented.

In Chap. 2, the homoclinic networks of sources, sinks, and saddles in self-univariate polynomial systems are discussed, and the numbers of sources, sinks and saddles are determined through a theorem, and the first integral manifolds are developed. The corresponding proof of the theorem is completed and a few illustrations of networks for source, sinks and saddles are presented for a better understanding of the homoclinic networks. Such homoclinic networks are without any centers even if the networks are separated by the homoclinic orbits.

In Chap. 3, the appearing and switching bifurcations are discussed for the homoclinic networks of non-singular and singular sources, sinks, saddles with singular saddle-sources, saddle-sinks, and double-saddles in self-univariate polynomial systems. The first integral manifolds for non-singular and singular equilibrium networks are determined. The illustrations of singular equilibriums to networks of non-singular sources, sinks and saddles are given.

In Chap. 4, the homoclinic networks of positive and negative saddles with clockwise and counter-clockwise limit cycles in crossing-univariate polynomial systems are studied secondly, and the numbers of saddles and centers are determined through a theorem and the first integral manifolds are determined through polynomial functions. The corresponding proof of the theorem is given, and a few illustrations of networks of saddles and centers are given to show the corresponding geometric structures. Such homoclinic networks of saddles and centers are without any sources and sinks.

In Chap. 5, the appearing and switching bifurcations are studied for homoclinic networks of singular and non-singular saddles and centers with singular parabola-saddles and double-inflection saddles in crossing-univariate polynomial systems, and the first integral manifolds of such homoclinic networks are determined through polynomial functions. The illustrations of singular equilibriums to networks of non-singular saddles and centers are given.

References

1. Hilbert, D. (1900). Mathematische problem. II ICM Paris 1900. *Nachrichten von der Koniglichen Gesellschaft der Wissenschaften zu Gottingen*. Kl. 253–297.
2. Dulac, H. (1923). Sur les cycles limites. *Bulletin de la Société Mathématique de France, 51*, 45–188.
3. Ilyashenko, Y. S. (1982). Singular points and limit cycles of differential equations in the real and complex plane. *Preprint, Pushchino 39: Science Research Computer Centre, Academy Science* USSR (in Russian).
4. Ilyashenko, Yu. S. (1984). Limit cycles of polynomial vector fields with nondegenerate singular points on the real plane. *Functional Analysis and Its Applications, 18*, 199–209.
5. Ilyashenko, Yu. S. (1985). Dulac's memoir "On limit cycles" and related problems of the local theory of differential equations. *Russian Mathematical Surveys, 40*, 1–49.
6. Ilyashenko, Y. S. (1991). Finiteness theorems for limit cycles, AMS. *Providence, RI*.
7. Ilyashenko, Yu. S. (2002). Centennial history of Hilbert's 16th problem. *Bulletin of the American Mathematical Society, 39*(3), 301–354.
8. Li, J. (2003). Hilbert's 16th problem and bifurcations of planar polynomial vector fields. *International Journal of Bifurcation and Chaos, 13*, 47–106.
9. Shahshahani, S. (1981). Periodic solutions of polynomial first-order differential equations. *Nonlinear Analysis, 5*(2), 157–165.
10. Lins Neto, A. (1980). On the number of solutions of the equation (), for. *Inventiones Mathematicae, 59*(1), 67–76.
11. Panov, A. (1999). Variety of Poincare Mapping for cubic equations with variable coefficients. *Functional Analysis and Its Applications, 33*, 84–88.
12. Smale, S. (1998). Mathematical problems for the next century. *Mathematics Intelligencer, 20*(2), 7–15.

13. van der Pol, B. (1920). A theory of the amplitude of free and forced triode vibrations". *Radio Review, 1*, 701–710.
14. van der Pol, B. (1926). On "relaxation-oscillations." *The London, Edinburgh, and Dublin Philosophical Magazine and Journal of Science Ser, 7*(2), 978–992.
15. Liénard, A. (1928). Etude des oscillations entretenues, *Revue générale de l'électricité, 23*, 901–912 and 946–954.
16. Bautin, N. N. (1939). Du nombre de cycles limites naissant en cas de variation des coefficients d'un état d'équilibre du type foyer ou centre. *Doklady Akademii Nauk, 24*, 669–672.
17. Bautin, N. N. (1952). On the number of limit cycles appearing with variation of the coefficients from an equilibrium state of the type of a focus or a center. *Matematicheskii Sbornik (N.S.), 30*(72) 1, 181–196.
18. Petrovskii, I. G., Landis, E. M. (1955). On the number of limit cycles of the equation , where P and Q are polynomials of the second degree. *Matematicheskii Sbornik (N.S.), 37*(79) 2, 209–250 (in Russian).
19. Petrovskii, I. G., Landis, E. M. (1957). On the number of limit cycles of the equation , where P and Q are polynomials. *Matematicheskii Sbornik (N.S.), 43*(85) 2, 149–168 (in Russian).
20. Shi, S. (1979). A concrete example of the existence of four limit cycles for plane quadratic systems. *Science China Ser. A, 11*, 1051–1056 (in Chinese).
21. Chen, L. S., & Wang, M. S. (1979). The relative position, and the number of limit cycles of a quadratic differential system. *Acta Math Sinica, 22*, 51–758.
22. Ilyashenko, Yu. S., & Panov, A. (2001). Some upper estimates of the number of limit cycles of planar vector fields with applications to Lienard equations. *Moscow Mathematics Journal, 1*, 583–599.
23. Arnold, V. I. (1977). Loss of stability of self-oscillations lose to resonance and versal deformations of equivalent vector fields. *Functional Analysis and Its Applications, 11*, 85–92.
24. Arnold, V. I. (1983). *Geometric methods in theory of ordinary differential equation.* Springer.
25. Ye, Y. (1986). Theory of limit cycles. Translations of Mathematics Monographs, 66 American Mathematics Society.
26. Zhang, Z., Ding, T., Juang, W., Dong, Z. (1992). Qualitative theory of differential equations. Translations of Mathematics Monographs, 101 American Mathematics Society.
27. Luo, A. C. J. (2022). A theory for singularity and stability in two-dimensional linear systems. *Journal of Vibration Testing and System Dynamics, 6*(1), 63–105.
28. Luo, A. C. J. (2022). Singularity and 1-dimensional flows in 2-D single-variable quadratic systems. *Journal of Vibration Testing and System Dynamics, 6*(2), 107–194.
29. Luo, A. C. J. (2023). Bifurcations and saddle-sink-source networks in variable-independent quadratic systems. *Journal of Vibration Testing and System Dynamics, 7*(1), 59–112.
30. Luo, A. C. J. (2023). Bifurcations and saddle-limit cycle networks in crossing-variable quadratic systems. *Journal of Vibration Testing and System Dynamics, 7*(2), 187–252.
31. Luo, A. C. J. (2023). Dynamics and bifurcations in a quadratic nonlinear system with univariate product vector fields. *Journal of Vibration Testing and System Dynamics, 7*(3), 327–397.
32. Luo, A. C. J. (2023). *Quadratic dynamical systems* (Vol. I). Univariate Vector Fields, HEP/Springer.
33. Luo, A. C. J. (2024). *Self-univariate cubic systems.* Springer.
34. Luo, A. C. J. (2024). *Crossing-univariate cubic systems.* Springer.

Chapter 2
Homoclinic Networks Without Centers

In this Chapter, the homoclinic networks of sources, sinks, and saddles in self-univariate polynomial systems are discussed, and the numbers of sources, sinks and saddles are determined through a theorem, and the first integral manifolds are developed. The corresponding proof of the theorem is completed and a few illustrations of networks for source, sinks and saddles are presented for a better understanding of the homoclinic networks. Such homoclinic networks are without any centers even if the networks are separated by the homoclinic orbits.

2.1 Sources, Sinks, and Saddles

In this section, as in Luo [1], consider homoclinic networks of maximized sinks, sources, and saddles in two-dimensional polynomial nonlinear systems. A polynomial system with a self-univariate polynomial vector field is considered, and the corresponding dynamical behaviors will be presented through the following theorem.

Theorem 2.1 *Without of lose of generality, consider a self-univariate polynomial dynamical system as*

$$
\begin{aligned}
\dot{x}_{j_1} &= a_{j_1 j_1 0} \prod_{s_1=1}^{m} (x_{j_1} - a_{j_1 j_1 s_1}), \\
\dot{x}_{j_2} &= a_{j_2 j_2 0} \prod_{l_1=1}^{n} (x_{j_2} - a_{j_2 j_2 l_1}), \\
& j_1, j_2 \in \{1, 2\}; j_1 \neq j_2.
\end{aligned}
\tag{2.1}
$$

A. C. J. Luo, *Limit Cycles and Homoclinic Networks in Two-Dimensional Polynomial Systems*, https://doi.org/10.1007/978-981-97-2617-2_2

The first integral manifold is

$$\sum_{l_1=1}^{n} \frac{1}{\prod_{l_2=1,l_2\neq l_1}^{3}(a_{j_2j_2l_1}-a_{j_2j_2l_2})} \ln\frac{|x_{j_2}-a_{j_2j_2l_1}|}{|x_{j_20}-a_{j_2j_2l_1}|}$$
$$=\frac{a_{j_2j_20}}{a_{j_1j_10}}\sum_{s_1=1}^{m}\frac{1}{\prod_{s_2=1,s_2\neq s_1}^{3}(a_{j_1j_1s_1}-a_{j_1j_1s_2})}\ln\frac{|x_{j_1}-a_{j_1j_1s_1}|}{|x_{j_10}-a_{j_1j_1s_1}|}. \tag{2.2}$$

The equilibrium of $(x_{j_1}^*, x_{j_2}^*)=(a_{j_1j_1s_1}, a_{j_2j_2l_1})$ $(s_1, s_2\in\{1,2,\cdots,m\}, s_1\neq s_2; l_1, l_2\in\{1,2,3,\cdots,n\}, l_1\neq l_2)$ *possesses the following properties.*

- *For* $a_{j_1j_10}\prod_{s_2=1,s_2\neq s_1}^{m}(a_{j_1j_1s_1}-a_{j_1j_1s_2})>0$ *and* $a_{j_2j_20}\prod_{l_2=1,l_2\neq l_1}^{n}(a_{j_2j_2l_1}-a_{j_2j_2l_2})>0$,
$$(a_{j_1j_1s_1}, a_{j_2j_2l_1})=\underbrace{(\mathrm{SO},\mathrm{SO})}_{\text{source}}. \tag{2.3}$$
The *equilibrium of* $(x_{j_1}^*, x_{j_2}^*)=(a_{j_1j_1s_1}, a_{j_2j_2l_1})$ *is an* (SO,SO)-*source.*

- *For* $a_{j_1j_10}\prod_{s_2=1,s_2\neq s_1}^{m}(a_{j_1j_1s_1}-a_{j_1j_1s_2})<0$ *and* $a_{j_2j_20}\prod_{l_2=1,l_2\neq l_1}^{n}(a_{j_2j_2l_1}-a_{j_2j_2l_2})>0$,
$$(a_{j_1j_1s_1}, a_{j_2j_2l_1})=\underbrace{(\mathrm{SI},\mathrm{SO})}_{\text{saddle}}. \tag{2.4}$$
The *equilibrium of* $(x_{j_1}^*, x_{j_2}^*)=(a_{j_1j_1s_1}, a_{j_2j_2l_1})$ *is an* (SI,SO)-*saddle.*

- *For* $a_{j_1j_10}\prod_{s_2=1,s_2\neq s_1}^{m}(a_{j_1j_1s_1}-a_{j_1j_1s_2})>0$ *and* $a_{j_2j_20}\prod_{l_2=1,l_2\neq l_1}^{n}(a_{j_2j_2l_1}-a_{j_2j_2l_2})<0$,
$$(a_{j_1j_1s_1}, a_{j_2j_2l_1})=\underbrace{(\mathrm{SO},\mathrm{SI})}_{\text{saddle}}. \tag{2.5}$$
The *equilibrium of* $(x_{j_1}^*, x_{j_2}^*)=(a_{j_1j_1s_1}, a_{j_2j_2l_1})$ *is called an* (SO,SI)-*saddle.*

- *For* $a_{j_1j_10}\prod_{s_2=1,s_2\neq s_1}^{m}(a_{j_1j_1s_1}-a_{j_1j_1s_2})<0$ *and* $a_{j_2j_20}\prod_{l_2=1,l_2\neq l_1}^{n}(a_{j_2j_2l_1}-a_{j_2j_2l_2})<0$,
$$(a_{j_1j_1s_1}, a_{j_2j_2l_1})=\underbrace{(\mathrm{SI},\mathrm{SI})}_{\text{sink}}. \tag{2.6}$$
The *equilibrium of* $(x_{j_1}^*, x_{j_2}^*)=(a_{j_1j_1s_1}, a_{j_2j_2l_1})$ *is an* (SI,SI)-*sink.*

Define a notation for homoclinic networks as

$$
\bigcup_{s=1}^{m}\bigcup_{l=1}^{n}\underbrace{(a_{j_1j_1s}, a_{j_2j_2l})}_{XX}
= \begin{Bmatrix}
(a_{j_1j_11}, a_{j_2j_2n}) & (a_{j_1j_12, j_2j_2n}) & \cdots & (a_{j_1j_1m, j_2j_2n}) \\
(a_{j_1j_11}, a_{j_2j_2(n-1)}) & (a_{j_1j_12}, a_{j_2j_2(n-1)}) & \cdots & (a_{j_1j_1m}, a_{j_2j_2(n-1)}) \\
\vdots & \vdots & \cdots & \vdots \\
(a_{j_1j_11}, a_{j_2j_21}) & (a_{j_1j_12}, a_{j_2j_21}) & \cdots & (a_{j_1j_1m}, a_{j_2j_21})
\end{Bmatrix}_{n\times m}. \tag{2.7}
$$

(i$_1$) For $m = 2m_1 + 1$ and $n = 2n_1 + 1$, the homoclinic networks of $(2m_1 + 1) \times (2n_1 + 1)$ equilibriums have the following properties.

- For $a_{j_1j_10} > 0$ *and* $a_{j_2j_20} > 0$,

$$
\bigcup_{s=1}^{2m_1+1}\bigcup_{l=1}^{2n_1+1}\underbrace{(a_{j_1j_1s}, a_{j_2j_2l})}_{XX} =
\begin{Bmatrix}
\underbrace{(SO,SO)}_{\text{source}} & \underbrace{(SI,SO)}_{\text{saddle}} & \cdots & \underbrace{(SO,SO)}_{\text{source}} \\
\underbrace{(SO,SI)}_{\text{saddle}} & \underbrace{(SI,SI)}_{\text{sink}} & \cdots & \underbrace{(SO,SI)}_{\text{saddle}} \\
\vdots & \vdots & \vdots & \vdots \\
\underbrace{(SO,SO)}_{\text{source}} & \underbrace{(SI,SO)}_{\text{saddle}} & \cdots & \underbrace{(SO,SO)}_{\text{source}}
\end{Bmatrix}_{(2n_1+1)\times(2m_1+1)}. \tag{2.8}
$$

In the network, the number of (SO,SI)-saddles is $n_1 \times (m_1 + 1)$, the number of (SI,SO)-saddles is $(n_1 + 1) \times m_1$;the number of (SO,SO)-sources is $(n_1 + 1) \times (m_1 + 1)$; and the number of (SI,SI)-sink is $n_1 \times m_1$.

- For $a_{j_1j_10} < 0$ *and* $a_{j_2j_20} > 0$,

$$
\bigcup_{s=1}^{2m_1+1}\bigcup_{l=1}^{2n_1+1}\underbrace{(a_{j_1j_1s}, a_{j_2j_2l})}_{XX} =
\begin{Bmatrix}
\underbrace{(SI,SO)}_{\text{saddle}} & \underbrace{(SO,SO)}_{\text{source}} & \cdots & \underbrace{(SI,SO)}_{\text{saddle}} \\
\underbrace{(SI,SI)}_{\text{sink}} & \underbrace{(SO,SI)}_{\text{saddle}} & \cdots & \underbrace{(SI,SI)}_{\text{sink}} \\
\vdots & \vdots & \vdots & \vdots \\
\underbrace{(SI,SO)}_{\text{saddle}} & \underbrace{(SO,SO)}_{\text{source}} & \cdots & \underbrace{(SI,SO)}_{\text{saddle}}
\end{Bmatrix}_{(2n_1+1)\times(2m_1+1)}. \tag{2.9}
$$

In the network, the number of (SO,SI)-saddles is $n_1 \times m_1$, the number of (SI,SO)-saddles is $(n_1 + 1) \times (m_1 + 1)$; the number of (SO,SO)-sources is $(n_1 + 1) \times m_1$; and the number of (SI,SI)-sink is $n_1 \times (m_1 + 1)$.

- For $a_{j_1j_10} > 0$ *and* $a_{j_2j_20} < 0$,

$$\bigcup_{s=1}^{2m_1+1}\bigcup_{l=1}^{2n_1+1}\underbrace{(a_{j_1j_1s}, a_{j_2j_2l})}_{\mathrm{XX}} = \left\{\begin{matrix} \underbrace{(\mathrm{SO,SI})}_{\text{saddle}} & \underbrace{(\mathrm{SI,SI})}_{\text{sink}} & \cdots & \underbrace{(\mathrm{SO,SI})}_{\text{saddle}} \\ \underbrace{(\mathrm{SO,SO})}_{\text{source}} & \underbrace{(\mathrm{SI,SO})}_{\text{saddle}} & \cdots & \underbrace{(\mathrm{SO,SO})}_{\text{source}} \\ \vdots & \vdots & \vdots & \vdots \\ \underbrace{(\mathrm{SO,SI})}_{\text{saddle}} & \underbrace{(\mathrm{SI,SI})}_{\text{sink}} & \cdots & \underbrace{(\mathrm{SO,SI})}_{\text{saddle}} \end{matrix}\right\}_{(2n_1+1)\times(2m_1+1)}. \tag{2.10}$$

In the network, the number of (SO,SI)-saddles is $(n_1+1)\times(m_1+1)$, the number of (SI,SO)-saddles is $n_1 \times m_1$; the number of (SO,SO)-sources is $n_1 \times (m_1+1)$; and the number of (SI,SI)-sink is $(n_1+1) \times m_1$.

- For $a_{j_1j_10} < 0$ *and* $a_{j_2j_20} < 0$,

$$\bigcup_{s=1}^{2m_1+1}\bigcup_{l=1}^{2n_1+1}\underbrace{(a_{j_1j_1s}, a_{j_2j_2l})}_{\mathrm{XX}} = \left\{\begin{matrix} \underbrace{(\mathrm{SI,SI})}_{\text{sink}} & \underbrace{(\mathrm{SO,SI})}_{\text{saddle}} & \cdots & \underbrace{(\mathrm{SI,SI})}_{\text{sink}} \\ \underbrace{(\mathrm{SI,SO})}_{\text{saddle}} & \underbrace{(\mathrm{SO,SO})}_{\text{source}} & \cdots & \underbrace{(\mathrm{SI,SO})}_{\text{saddle}} \\ \vdots & \vdots & \vdots & \vdots \\ \underbrace{(\mathrm{SI,SI})}_{\text{sink}} & \underbrace{(\mathrm{SO,SI})}_{\text{saddle}} & \cdots & \underbrace{(\mathrm{SI,SI})}_{\text{sink}} \end{matrix}\right\}_{(2n_1+1)\times(2m_1+1)}. \tag{2.11}$$

In the network, the number of (SO,SI)-saddles is $(n_1+1) \times m_1$, the number of (SI,SO)-saddles is $n_1 \times (m_1+1)$; the number of (SO,SO)-sources is $n_1 \times m_1$; and the number of (SI,SI)-sink is $(n_1+1) \times (m_1+1)$.

(i_2) The numbers of saddles, sink and sources for $(2m_1+1)\times(2n_1+1)$-equilibriums are summarized in Table 2.1.

(ii_1) For $m = 2m_1$ and $n = 2n_1+1$, the equilibrium networks with $(2m_1)\times(2n_1+1)$ equilibriums have the following properties.

Table 2.1 Numbers of saddles, sinks, and sources in homoclinic networks ($m = 2m_1+1, n = 2n_1+1$)

$(a_{j_1j_10}, a_{j_2j_20})$	(SO,SI)-saddle	(SI,SO)-saddle	(SO,SO)-source	(SI,SI)-sink
$(+,+)$	$n_1 \times (m_1+1)$	$(n_1+1)\times m_1$	$(n_1+1)\times(m_1+1)$	$n_1 \times m_1$
$(-,+)$	$n_1 \times m_1$	$(n_1+1)\times(m_1+1)$	$(n_1+1)\times m_1$	$n_1 \times (m_1+1)$
$(+,-)$	$(n_1+1)\times(m_1+1)$	$n_1 \times m_1$	$n_1 \times (m_1+1)$	$(n_1+1)\times m_1$
$(-,-)$	$(n_1+1)\times m_1$	$n_1 \times (m_1+1)$	$n_1 \times m_1$	$(n_1+1)\times(m_1+1)$

- For $a_{j_1j_10} > 0$ and $a_{j_2j_20} > 0$,

$$\bigcup_{s=1}^{2m_1}\bigcup_{l=1}^{2n_1+1}\underbrace{(a_{j_1j_1s}, a_{j_2j_2l})}_{\mathrm{XX}} = \begin{Bmatrix} \underbrace{(\mathrm{SI,SO})}_{\text{saddle}} & \underbrace{(\mathrm{SO,SO})}_{\text{source}} & \cdots & \underbrace{(\mathrm{SO,SO})}_{\text{source}} \\ \underbrace{(\mathrm{SI,SI})}_{\text{sink}} & \underbrace{(\mathrm{SO,SI})}_{\text{saddle}} & \cdots & \underbrace{(\mathrm{SO,SI})}_{\text{saddle}} \\ \vdots & \vdots & \vdots & \vdots \\ \underbrace{(\mathrm{SI,SO})}_{\text{saddle}} & \underbrace{(\mathrm{SO,SO})}_{\text{source}} & \cdots & \underbrace{(\mathrm{SO,SO})}_{\text{source}} \end{Bmatrix}_{(2n_1+1)\times(2m_1)}. \tag{2.12}$$

In the network, the number of (SO,SI)-saddles is $n_1 \times m_1$, the number of (SI,SO)-saddles is $(n_1+1)\times m_1$; the number of (SO,SO)-sources is $(n_1+1)\times m_1$; and the number of (SI,SI)-sinks is $n_1 \times m_1$.

- For $a_{j_1j_10} > 0$ and $a_{j_2j_20} > 0$,

$$\bigcup_{s=1}^{2m_1}\bigcup_{l=1}^{2n_1+1}\underbrace{(a_{j_1j_1s}, a_{j_2j_2l})}_{\mathrm{XX}} = \begin{Bmatrix} \underbrace{(\mathrm{SO,SO})}_{\text{source}} & \underbrace{(\mathrm{SI,SO})}_{\text{saddle}} & \cdots & \underbrace{(\mathrm{SI,SO})}_{\text{saddle}} \\ \underbrace{(\mathrm{SO,SI})}_{\text{saddle}} & \underbrace{(\mathrm{SI,SI})}_{\text{sink}} & \cdots & \underbrace{(\mathrm{SI,SI})}_{\text{sink}} \\ \vdots & \vdots & \vdots & \vdots \\ \underbrace{(\mathrm{SO,SO})}_{\text{source}} & \underbrace{(\mathrm{SI,SO})}_{\text{saddle}} & \cdots & \underbrace{(\mathrm{SI,SO})}_{\text{saddle}} \end{Bmatrix}_{(2n_1+1)\times(2m_1)}. \tag{2.13}$$

In the network, the number of (SO,SI)-saddles is $n_1 \times m_1$, the number of (SI,SO)-saddles is $(n_1+1)\times m_1$; the number of (SO,SO)-sources is $(n_1+1)\times m_1$; and the number of (SI,SI)-sinks is $n_1 \times m_1$.

- For $a_{j_1j_10} > 0$ and $a_{j_2j_20} > 0$,

$$\bigcup_{s=1}^{2m_1}\bigcup_{l=1}^{2n_1+1}\underbrace{(a_{j_1j_1s}, a_{j_2j_2l})}_{\mathrm{XX}} = \begin{Bmatrix} \underbrace{(\mathrm{SI,SI})}_{\text{sink}} & \underbrace{(\mathrm{SO,SI})}_{\text{saddle}} & \cdots & \underbrace{(\mathrm{SO,SI})}_{\text{saddle}} \\ \underbrace{(\mathrm{SI,SO})}_{\text{saddle}} & \underbrace{(\mathrm{SO,SO})}_{\text{source}} & \cdots & \underbrace{(\mathrm{SO,SO})}_{\text{source}} \\ \vdots & \vdots & \vdots & \vdots \\ \underbrace{(\mathrm{SI,SI})}_{\text{sink}} & \underbrace{(\mathrm{SO,SI})}_{\text{saddle}} & \cdots & \underbrace{(\mathrm{SO,SI})}_{\text{saddle}} \end{Bmatrix}_{(2n_1+1)\times(2m_1)}. \tag{2.14}$$

In the network, the number of (SO,SI)-saddles is $(n_1+1)\times m_1$, the number of (SI,SO)-saddles is $n_1 \times m_1$; the number of (SO,SO)-sources is $n_1\times m_1$; and the number of (SI,SI)-sinks is $(n_1+1)\times m_1$.

- For $a_{j_1j_10} > 0$ and $a_{j_2j_20} > 0$,

$$\bigcup_{s=1}^{2m_1}\bigcup_{l=1}^{2n_1+1}\underbrace{(a_{j_1j_1s}, a_{j_2j_2l})}_{XX} = \left\{\begin{array}{cccc} \underbrace{\text{(SO,SI)}}_{\text{saddle}} & \underbrace{\text{(SI,SI)}}_{\text{sink}} & \cdots & \underbrace{\text{(SI,SI)}}_{\text{sink}} \\ \underbrace{\text{(SO,SO)}}_{\text{source}} & \underbrace{\text{(SI,SO)}}_{\text{saddle}} & \cdots & \underbrace{\text{(SI,SO)}}_{\text{saddle}} \\ \vdots & \vdots & \vdots & \vdots \\ \underbrace{\text{(SO,SI)}}_{\text{saddle}} & \underbrace{\text{(SI,SI)}}_{\text{sink}} & \cdots & \underbrace{\text{(SI,SI)}}_{\text{sink}} \end{array}\right\}_{(2n_1+1)\times(2m_1)}. \tag{2.15}$$

In the network, the number of (SO,SI)-saddles is $(n_1+1)\times m_1$, the number of (SI,SO)-saddles is $n_1\times m_1$; the number of (SO,SO)-sources is $n_1\times m_1$; and the number of (SI,SI)-sinks is $(n_1+1)\times m_1$.

(ii_2) The numbers of saddles, sinks and sources for $(2m_1)\times(2n_1+1)$-equilibriums are summarized in Table 2.2.

(iii_1) For $m = 2m_1+1$ and $n = 2n_1$, the equilibrium networks with $(2m_1+1)\times(2n_1)$ equilibriums have the following properties.

- For $a_{j_1j_10} > 0$ *and* $a_{j_2j_20} > 0$,

$$\bigcup_{s=1}^{2m_1+1}\bigcup_{l=1}^{2n_1}\underbrace{(a_{j_1j_1s}, a_{j_2j_2l})}_{XX} = \left\{\begin{array}{cccc} \underbrace{\text{(SO,SO)}}_{\text{source}} & \underbrace{\text{(SI,SO)}}_{\text{saddle}} & \cdots & \underbrace{\text{(SO,SO)}}_{\text{source}} \\ \underbrace{\text{(SO,SI)}}_{\text{saddle}} & \underbrace{\text{(SI,SI)}}_{\text{sink}} & \cdots & \underbrace{\text{(SO,SI)}}_{\text{saddle}} \\ \vdots & \vdots & \vdots & \vdots \\ \underbrace{\text{(SO,SI)}}_{\text{saddle}} & \underbrace{\text{(SI,SI)}}_{\text{sink}} & \cdots & \underbrace{\text{(SO,SI)}}_{\text{saddle}} \end{array}\right\}_{(2n_1)\times(2m_1+1)}. \tag{2.16}$$

Table 2.2 Numbers of saddles, sinks, and sources in homoclinic networks ($m = 2m_1, n = 2n_1+1$)

$(a_{j_1j_10}, a_{j_2j_20})$	(SO,SI)-saddle	(SI,SO)-saddle	(SO,SO)-source	(SI,SI)-sink
$(+,+)$	$n_1\times m_1$	$(n_1+1)\times m_1$	$(n_1+1)\times m_1$	$n_1\times m_1$
$(-,+)$	$n_1\times m_1$	$(n_1+1)\times m_1$	$(n_1+1)\times m_1$	$n_1\times m_1$
$(+,-)$	$(n_1+1)\times m_1$	$n_1\times m_1$	$n_1\times m_1$	$(n_1+1)\times m_1$
$(-,-)$	$(n_1+1)\times m_1$	$n_1\times m_1$	$n_1\times m_1$	$(n_1+1)\times m_1$

In the network, the number of (SO,SI)-saddles is $n_1 \times (m_1 + 1)$, the number of (SI,SO)-saddles is $n_1 \times m_1$; the number of (SO,SO)-sources is $n_1 \times (m_1 + 1)$; and the number of (SI,SI)-sinks is $n_1 \times m_1$.

- For $a_{j_1j_10} < 0$ *and* $a_{j_2j_20} > 0$,

$$\bigcup_{s=1}^{2m_1+1}\bigcup_{l=1}^{2n_1}\underbrace{(a_{j_1j_1s}, a_{j_2j_2l})}_{\text{XX}} = \left\{\begin{matrix} \underbrace{\text{(SI,SO)}}_{\text{saddle}} & \underbrace{\text{(SO,SO)}}_{\text{source}} & \cdots & \underbrace{\text{(SI,SO)}}_{\text{saddle}} \\ \underbrace{\text{(SI,SI)}}_{\text{sink}} & \underbrace{\text{(SO,SI)}}_{\text{saddle}} & \cdots & \underbrace{\text{(SI,SI)}}_{\text{sink}} \\ \vdots & \vdots & \vdots & \vdots \\ \underbrace{\text{(SI,SI)}}_{\text{sink}} & \underbrace{\text{(SO,SI)}}_{\text{saddle}} & \cdots & \underbrace{\text{(SI,SI)}}_{\text{sink}} \end{matrix}\right\}_{(2n_1)\times(2m_1+1)}. \tag{2.17}$$

In the network, the number of (SO,SI)-saddles is $n_1 \times m_1$, the number of (SI,SO)-saddles is $n_1 \times (m_1 + 1)$; the number of (SO,SO)-sources is $n_1 \times m_1$; and the number of (SI,SI)-sink is $n_1 \times (m_1 + 1)$.

- For $a_{j_1j_10} > 0$ *and* $a_{j_2j_20} < 0$,

$$\bigcup_{s=1}^{2m_1+1}\bigcup_{l=1}^{2n_1}\underbrace{(a_{j_1j_1s}, a_{j_2j_2l})}_{\text{XX}} = \left\{\begin{matrix} \underbrace{\text{(SO,SI)}}_{\text{saddle}} & \underbrace{\text{(SI,SI)}}_{\text{sink}} & \cdots & \underbrace{\text{(SO,SI)}}_{\text{saddle}} \\ \underbrace{\text{(SO,SO)}}_{\text{source}} & \underbrace{\text{(SI,SO)}}_{\text{saddle}} & \cdots & \underbrace{\text{(SO,SO)}}_{\text{source}} \\ \vdots & \vdots & \vdots & \vdots \\ \underbrace{\text{(SO,SO)}}_{\text{source}} & \underbrace{\text{(SI,SO)}}_{\text{saddle}} & \cdots & \underbrace{\text{(SO,SO)}}_{\text{source}} \end{matrix}\right\}_{(2n_1)\times(2m_1+1)}. \tag{2.18}$$

In the network, the number of (SO,SI)-saddles is $n_1 \times (m_1 + 1)$, the number of (SI,SO)-saddles is $n_1 \times m_1$; the number of (SO,SO)-sources is $n_1 \times (m_1 + 1)$; and the number of (SI,SI)-sink is $n_1 \times m_1$.

- For $a_{j_1j_10} < 0$ *and* $a_{j_2j_20} < 0$,

$$\bigcup_{s=1}^{2m_1+1}\bigcup_{l=1}^{2n_1}\underbrace{(a_{j_1j_1s}, a_{j_2j_2l})}_{\text{XX}} = \left\{\begin{matrix} \underbrace{\text{(SI,SI)}}_{\text{sink}} & \underbrace{\text{(SO,SI)}}_{\text{saddle}} & \cdots & \underbrace{\text{(SI,SI)}}_{\text{sink}} \\ \underbrace{\text{(SI,SO)}}_{\text{saddle}} & \underbrace{\text{(SO,SO)}}_{\text{source}} & \cdots & \underbrace{\text{(SI,SO)}}_{\text{saddle}} \\ \vdots & \vdots & \vdots & \vdots \\ \underbrace{\text{(SI,SO)}}_{\text{saddle}} & \underbrace{\text{(SO,SO)}}_{\text{source}} & \cdots & \underbrace{\text{(SI,SO)}}_{\text{saddle}} \end{matrix}\right\}_{(2n_1)\times(2m_1+1)}. \tag{2.19}$$

In the network, the number of (SO,SI)-saddles is $n_1 \times m_1$, the number of (SI,SO)-saddles is $n_1 \times (m_1 + 1)$; the number of (SO,SO)-sources is $n_1 \times m_1$; and the number of (SI,SI)-sink is $n_1 \times (m_1 + 1)$.

(iii$_2$) The numbers of saddles, sink and sources for $(2m_1 + 1) \times (2n_1)$-equilibriums are summarized in Table 2.3.

(iv$_1$) For $m = 2m_1$ and $n = 2n_1$, the equilibrium networks with $(2m_1) \times (2n_1)$ equilibriums have the following properties.

- For $a_{j_1 j_1 0} > 0$ *and* $a_{j_2 j_2 0} > 0$,

$$\bigcup_{s=1}^{2m_1}\bigcup_{l=1}^{2n_1}\underbrace{(a_{j_1 j_1 s}, a_{j_2 j_2 l})}_{XX} = \left\{\begin{array}{cccc} \underbrace{(SI,SO)}_{\text{saddle}} & \underbrace{(SO,SO)}_{\text{source}} & \cdots & \underbrace{(SO,SO)}_{\text{source}} \\ \underbrace{(SI,SI)}_{\text{sink}} & \underbrace{(SO,SI)}_{\text{saddle}} & \cdots & \underbrace{(SO,SI)}_{\text{saddle}} \\ \vdots & \vdots & \vdots & \vdots \\ \underbrace{(SI,SI)}_{\text{sink}} & \underbrace{(SO,SI)}_{\text{saddle}} & \cdots & \underbrace{(SO,SI)}_{\text{saddle}} \end{array}\right\}_{(2n_1)\times(2m_1)}. \quad (2.20)$$

In the network, the number of (SO,SI)-saddles is $n_1 \times m_1$, the number of (SI,SO)-saddles is $n_1 \times m_1$; the number of (SO,SO)-sources is $n_1 \times m_1$; and the number of (SI,SI)-sinks is $n_1 \times m_1$.

- For $a_{j_1 j_1 0} < 0$ *and* $a_{j_2 j_2 0} > 0$,

$$\bigcup_{s=1}^{2m_1}\bigcup_{l=1}^{2n_1}\underbrace{(a_{j_1 j_1 s}, a_{j_2 j_2 l})}_{XX} = \left\{\begin{array}{cccc} \underbrace{(SO,SO)}_{\text{source}} & \underbrace{(SI,SO)}_{\text{saddle}} & \cdots & \underbrace{(SI,SO)}_{\text{saddle}} \\ \underbrace{(SO,SI)}_{\text{saddle}} & \underbrace{(SI,SI)}_{\text{sink}} & \cdots & \underbrace{(SI,SI)}_{\text{sink}} \\ \vdots & \vdots & \vdots & \vdots \\ \underbrace{(SO,SI)}_{\text{saddle}} & \underbrace{(SI,SI)}_{\text{sink}} & \cdots & \underbrace{(SI,SI)}_{\text{sink}} \end{array}\right\}_{(2n_1)\times(2m_1)}. \quad (2.21)$$

Table 2.3 Numbers of saddles, sinks, and sources in homoclinic networks ($m = 2m_1 + 1$, $n = 2n_1$)

$(a_{j_1 j_1 0}, a_{j_2 j_2 0})$	(SO,SI)-saddle	(SI,SO)-saddle	(SO,SO)-source	(SI,SI)-sink
$(+, +)$	$n_1 \times (m_1 + 1)$	$n_1 \times m_1$	$n_1 \times (m_1 + 1)$	$n_1 \times m_1$
$(-, +)$	$n_1 \times m_1$	$n_1 \times (m_1 + 1)$	$n_1 \times m_1$	$n_1 \times (m_1 + 1)$
$(+, -)$	$n_1 \times (m_1 + 1)$	$n_1 \times m_1$	$n_1 \times (m_1 + 1)$	$n_1 \times m_1$
$(-, -)$	$n_1 \times m_1$	$n_1 \times (m_1 + 1)$	$n_1 \times m_1$	$n_1 \times (m_1 + 1)$

In the network, the number of (SO,SI)-saddles is $n_1 \times m_1$, the number of (SI,SO)-saddles is $n_1 \times m_1$; the number of (SO,SO)-sources is $n_1 \times m_1$; and the number of (SI,SI)-sink is $n_1 \times m_1$.

- For $a_{j_1 j_1 0} > 0$ *and* $a_{j_2 j_2 0} < 0$,

$$\bigcup_{s=1}^{2m_1}\bigcup_{l=1}^{2n_1}\underbrace{(a_{j_1j_1s}, a_{j_2j_2l})}_{\text{XX}} = \begin{Bmatrix} \underbrace{\text{(SI,SI)}}_{\text{sink}} & \underbrace{\text{(SO,SI)}}_{\text{saddle}} & \cdots & \underbrace{\text{(SO,SI)}}_{\text{saddle}} \\ \underbrace{\text{(SI,SO)}}_{\text{saddle}} & \underbrace{\text{(SO,SO)}}_{\text{source}} & \cdots & \underbrace{\text{(SO,SO)}}_{\text{source}} \\ \vdots & \vdots & \vdots & \vdots \\ \underbrace{\text{(SI,SO)}}_{\text{saddle}} & \underbrace{\text{(SO,SO)}}_{\text{source}} & \cdots & \underbrace{\text{(SO,SO)}}_{\text{source}} \end{Bmatrix}_{(2n_1)\times(2m_1)} . \quad (2.22)$$

In the network, the number of (SO,SI)-saddles is $n_1 \times m_1$, the number of (SI,SO)-saddles is $n_1 \times m_1$; the number of (SO,SO)-sources is $n_1 \times m_1$; and the number of (SI,SI)-sink is $n_1 \times m_1$.

- For $a_{j_1 j_1 0} < 0$ *and* $a_{j_2 j_2 0} < 0$,

$$\bigcup_{s=1}^{2m_1}\bigcup_{l=1}^{2n_1}\underbrace{(a_{j_1j_1s}, a_{j_2j_2l})}_{\text{XX}} = \begin{Bmatrix} \underbrace{\text{(SO,SI)}}_{\text{saddle}} & \underbrace{\text{(SI,SI)}}_{\text{sink}} & \cdots & \underbrace{\text{(SI,SI)}}_{\text{sink}} \\ \underbrace{\text{(SO,SO)}}_{\text{source}} & \underbrace{\text{(SI,SO)}}_{\text{saddle}} & \cdots & \underbrace{\text{(SI,SO)}}_{\text{saddle}} \\ \vdots & \vdots & \vdots & \vdots \\ \underbrace{\text{(SO,SO)}}_{\text{source}} & \underbrace{\text{(SI,SO)}}_{\text{saddle}} & \cdots & \underbrace{\text{(SI,SO)}}_{\text{saddle}} \end{Bmatrix}_{(2n_1)\times(2m_1)} . \quad (2.23)$$

In the network, the number of (SO,SI)-saddles is $n_1 \times m_1$, the number of (SI,SO)-saddles is $n_1 \times m_1$; the number of (SO,SO)-sources is $n_1 \times m_1$; and the number of (SI,SI)-sink is $n_1 \times m_1$.

(iv$_2$) The numbers of saddles, sink and sources for $(2m_1) \times (2n_1)$-equilibriums are summarized in Table 2.4.

Table 2.4 Numbers of saddles, sinks and sources in homoclinic networks ($m = 2m_1, n = 2n_1$)

$(a_{j_1j_10}, a_{j_2j_20})$	(SO,SI)-saddle	(SI,SO)-saddle	(SO,SO)-source	(SI,SI)-sink
$(+,+)$	$n_1 \times m_1$	$n_1 \times m_1$	$n_1 \times m_1$	$n_1 \times m_1$
$(-,+)$	$n_1 \times m_1$	$n_1 \times m_1$	$n_1 \times m_1$	$n_1 \times m_1$
$(+,-)$	$n_1 \times m_1$	$n_1 \times m_1$	$n_1 \times m_1$	$n_1 \times m_1$
$(-,-)$	$n_1 \times m_1$	$n_1 \times m_1$	$n_1 \times m_1$	$n_1 \times m_1$

2.2 Proof of Theorem 2.1

Without lose of generality, consider a self-univariate polynomial dynamical system as

$$\dot{x}_{j_1} = a_{j_1 j_1 0} \prod_{s_1=1}^{m} (x_{j_1} - a_{j_1 j_1 s_1}),$$
$$\dot{x}_{j_2} = a_{j_2 j_2 0} \prod_{l_1=1}^{n} (x_{j_2} - a_{j_2 j_2 l_1}),$$
$$j_1, j_2 \in \{1, 2\}; j_1 \neq j_2.$$

In phase space,

$$\frac{dx_{j_1}}{dx_{j_2}} = \frac{a_{j_1 j_1 0} \prod_{s_1=1}^{m} (x_{j_1} - a_{j_1 j_1 s_1})}{a_{j_2 j_2 0} \prod_{l_1=1}^{n} (x_{j_2} - a_{j_2 j_2 l_1})},$$

and

$$\frac{dx_{j_1}}{(x_{j_1} - a_{j_1 j_1 s_1}) \prod_{s_2=1, s_2 \neq s_1}^{m} (x_{j_1} - a_{j_1 j_1 s_2})}$$
$$= \frac{a_{j_1 j_1 0}}{a_{j_2 j_2 0}} \frac{dx_{j_2}}{(x_{j_2} - a_{j_2 j_2 l_1}) \prod_{l_2=1, l_2 \neq l_1}^{n} (x_{j_2} - a_{j_2 j_2 l_2})},$$

With an initial condition $(x_{j_1 0}, x_{j_2 0})$, the first integral manifold is

$$\sum_{l_1=1}^{n} \frac{1}{\prod_{l_2=1, l_2 \neq l_1}^{3} (a_{j_2 j_2 l_1} - a_{j_2 j_2 l_2})} \ln \frac{|x_{j_2} - a_{j_2 j_2 l_1}|}{|x_{j_2 0} - a_{j_2 j_2 l_1}|}$$
$$= \frac{a_{j_2 j_2 0}}{a_{j_1 j_1 0}} \sum_{s_1=1}^{m} \frac{1}{\prod_{s_2=1, s_2 \neq s_1}^{3} (a_{j_1 j_1 s_1} - a_{j_1 j_1 s_2})} \ln \frac{|x_{j_1} - a_{j_1 j_1 s_1}|}{|x_{j_1 0} - a_{j_1 j_1 s_1}|}.$$

For $(x^*_{j_1}, x^*_{j_2}) = (a_{j_1 j_1 s_1}, a_{j_2 j_2 l_1})(s_1, s_2 \in \{1, 2, \cdots, m\}, s_1 \neq s_2; l_1, l_2 \in \{1, 2, \cdots, n\}, l_1 \neq l_2)$, *the variational equations* are

$$\Delta \dot{x}_{j_1} = [a_{j_1 j_1 0} \prod_{s_2=1, s_2 \neq s_1}^{m} (a_{j_1 j_1 s_1} - a_{j_1 j_1 s_2})] \Delta x_{j_1}$$
$$\Delta \dot{x}_{j_2} = [a_{j_2 j_2 0} \prod_{l_1=1}^{n} (a_{j_2 j_2 l_2} - a_{j_2 j_2 l_1})] \Delta x_{j_2}.$$

For $a_{j_1j_10}\prod_{s_2=1,s_2\neq s_1}^{m}(a_{j_1j_1s_1}-a_{j_1j_1s_2})>0$ and $a_{j_1j_10}\prod_{s_2=1,s_2\neq s_1}^{m}(a_{j_1j_1s_1}-a_{j_1j_1s_2})<0$, the flows at $(x_{j_1}^*,x_{j_2}^*)=(a_{j_1j_1s_1},a_{j_2j_2l_1})$ are source and sink flows in the x_{j_1}-direction, and the flows at $(x_{j_1}^*,x_{j_2}^*)=(a_{j_1j_1s_1},a_{j_2j_2l_1})$ are source and sink flows for $a_{j_2j_20}\prod_{l_2=1,l_2\neq l_1}^{n}(a_{j_2j_2l_1}-a_{j_2j_2l_2})>0$ and $a_{j_2j_20}\prod_{l_2=1,l_2\neq l_1}^{n}(a_{j_2j_2l_1}-a_{j_2j_2l_2})<0$ in the x_{j_2}-direction.

Therefore, the equilibrium of $(x_{j_1}^*,x_{j_2}^*)=(a_{j_1j_1s_1},a_{j_2j_2l_1})$ $(s_1,s_2\in\{1,2,\cdots,m\}$, $s_1\neq s_2$; $l_1,l_2\in\{1,2,3,\cdots,n\}$, $l_1\neq l_2)$ possesses the following properties in Eqs. (2.3)–(2.6).

- For $a_{j_1j_10}\prod_{s_2=1,s_2\neq s_1}^{m}(a_{j_1j_1s_1}-a_{j_1j_1s_2})>0$ and $a_{j_2j_20}\prod_{l_2=1,l_2\neq l_1}^{n}(a_{j_2j_2l_1}-a_{j_2j_2l_2})>0$,

$$(a_{j_1j_1s_1},a_{j_2j_2l_1})=\underbrace{(\mathrm{SO},\mathrm{SO})}_{\text{source}}.$$

  The equilibrium of $(x_{j_1}^*,x_{j_2}^*)=(a_{j_1j_1s_1},a_{j_2j_2l_1})$ is an (SO,SO)-source.

- For $a_{j_1j_10}\prod_{s_2=1,s_2\neq s_1}^{m}(a_{j_1j_1s_1}-a_{j_1j_1s_2})<0$ and $a_{j_2j_20}\prod_{l_2=1,l_2\neq l_1}^{n}(a_{j_2j_2l_1}-a_{j_2j_2l_2})>0$,

$$(a_{j_1j_1s_1},a_{j_2j_2l_1})=\underbrace{(\mathrm{SI},\mathrm{SO})}_{\text{saddle}}.$$

  The equilibrium of $(x_{j_1}^*,x_{j_2}^*)=(a_{j_1j_1s_1},a_{j_2j_2l_1})$ is an (SI,SO)-saddle.

- For $a_{j_1j_10}\prod_{s_2=1,s_2\neq s_1}^{m}(a_{j_1j_1s_1}-a_{j_1j_1s_2})>0$ and $a_{j_2j_20}\prod_{l_2=1,l_2\neq l_1}^{n}(a_{j_2j_2l_1}-a_{j_2j_2l_2})<0$,

$$(a_{j_1j_1s_1},a_{j_2j_2l_1})=\underbrace{(\mathrm{SO},\mathrm{SI})}_{\text{saddle}}.$$

  The equilibrium of $(x_{j_1}^*,x_{j_2}^*)=(a_{j_1j_1s_1},a_{j_2j_2l_1})$ is called an (SO,SI)-saddle.

- For $a_{j_1j_10}\prod_{s_2=1,s_2\neq s_1}^{m}(a_{j_1j_1s_1}-a_{j_1j_1s_2})<0$ and $a_{j_2j_20}\prod_{l_2=1,l_2\neq l_1}^{n}(a_{j_2j_2l_1}-a_{j_2j_2l_2})<0$,

$$(a_{j_1j_1s_1},a_{j_2j_2l_1})=\underbrace{(\mathrm{SI},\mathrm{SI})}_{\text{sink}}.$$

  The equilibrium of $(x_{j_1}^*,x_{j_2}^*)=(a_{j_1j_1s_1},a_{j_2j_2l_1})$ is an (SI,SI)-sink.

From the above equations, the equilibrium networks can be developed. Consider four cases:

- For $m = 2m_1 + 1$ and $n = 2n_1 + 1$, the corresponding equilibrium networks can be matrixed.
- For $m = 2m_1$ and $n = 2n_1 + 1$, the corresponding equilibrium networks can be matrixed.
- For $m = 2m_1 + 1$ and $n = 2n_1$, the corresponding equilibrium networks can be matrixed.
- For $m = 2m_1$ and $n = 2n_1$, the corresponding equilibrium networks can be matrixed.

The (SO,SI)-saddles, (SI,SO)-saddles, (SO,SO)-sources, (SI,SI)-sinks are determined and counted for each row and column. Summations of numbers yield the total numbers of saddles, sources, and sinks.

For simplicity, define a notation for homoclinic networks as

$$\bigcup_{s=1}^{m}\bigcup_{l=1}^{n}\underbrace{(a_{j_1j_1s}, a_{j_2j_2l})}_{\text{XX}} = \left\{\begin{matrix} (a_{j_1j_1}1, a_{j_2j_2n}) & (a_{j_1j_1}2, {}_{j_2j_2n}) & \cdots & (a_{j_1j_1}m, {}_{j_2j_2n}) \\ (a_{j_1j_1}1, a_{j_2j_2(n-1)}) & (a_{j_1j_1}2, a_{j_2j_2(n-1)}) & \cdots & (a_{j_1j_1}m, a_{j_2j_2(n-1)}) \\ \vdots & \vdots & \cdots & \vdots \\ (a_{j_1j_1}1, a_{j_2j_2}1) & (a_{j_1j_1}2, a_{j_2j_2}1) & \cdots & (a_{j_1j_1}m, a_{j_2j_2}1) \end{matrix}\right\}_{n\times m}.$$

(i_1) For $m = 2m_1 + 1$ and $n = 2n_1 + 1$,, the equilibrium networks with $(2m_1 + 1) \times (2n_1 + 1)$ equilibriums have the following properties as in Eqs. (2.8)–(2.11).

- For $a_{j_1j_1 0} > 0$ *and* $a_{j_2j_2 0} > 0$,

$$\bigcup_{s=1}^{2m_1+1}\bigcup_{l=1}^{2n_1+1}\underbrace{(a_{j_1j_1s}, a_{j_2j_2l})}_{\text{XX}} = \left\{\begin{matrix} \underbrace{(\text{SO,SO})}_{\text{source}} & \underbrace{(\text{SI,SO})}_{\text{saddle}} & \cdots & \underbrace{(\text{SO,SO})}_{\text{source}} \\ \underbrace{(\text{SO,SI})}_{\text{saddle}} & \underbrace{(\text{SI,SI})}_{\text{sink}} & \cdots & \underbrace{(\text{SO,SI})}_{\text{saddle}} \\ \vdots & \vdots & \vdots & \vdots \\ \underbrace{(\text{SO,SO})}_{\text{source}} & \underbrace{(\text{SI,SO})}_{\text{saddle}} & \cdots & \underbrace{(\text{SO,SO})}_{\text{source}} \end{matrix}\right\}_{(2n_1+1)\times(2m_1+1)}.$$

In the network, the number of (SO,SI)-saddles is $n_1 \times (m_1 + 1)$, the number of (SI,SO)-saddles is $(n_1 + 1) \times m_1$; the number of (SO,SO)-sources is $(n_1 + 1) \times (m_1 + 1)$; and the number of (SI,SI)-sink is $n_1 \times m_1$.

- For $a_{j_1j_1 0} < 0$ *and* $a_{j_2j_2 0} > 0$,

$$\bigcup_{s=1}^{2m_1+1}\bigcup_{l=1}^{2n_1+1}\underbrace{(a_{j_1j_1s},a_{j_2j_2l})}_{XX}=\left\{\begin{array}{cccc}\underbrace{(SI,SO)}_{saddle} & \underbrace{(SO,SO)}_{source} & \cdots & \underbrace{(SI,SO)}_{saddle}\\ \underbrace{(SI,SI)}_{sink} & \underbrace{(SO,SI)}_{saddle} & \cdots & \underbrace{(SI,SI)}_{sink}\\ \vdots & \vdots & \vdots & \vdots\\ \underbrace{(SI,SO)}_{saddle} & \underbrace{(SO,SO)}_{source} & \cdots & \underbrace{(SI,SO)}_{saddle}\end{array}\right\}_{(2n_1+1)\times(2m_1+1)}.$$

In the network, the number of (SO,SI)-saddles is $n_1 \times m_1$, the number of (SI,SO)-saddles is $(n_1 + 1) \times (m_1 + 1)$; the number of (SO,SO)-sources is $(n_1 + 1) \times m_1$; and the number of (SI,SI)-sink is $n_1 \times (m_1 + 1)$.

- For $a_{j_1j_10} > 0$ *and* $a_{j_2j_20} < 0$,

$$\bigcup_{s=1}^{2m_1+1}\bigcup_{l=1}^{2n_1+1}\underbrace{(a_{j_1j_1s},a_{j_2j_2l})}_{XX}=\left\{\begin{array}{cccc}\underbrace{(SO,SI)}_{saddle} & \underbrace{(SI,SI)}_{sink} & \cdots & \underbrace{(SO,SI)}_{saddle}\\ \underbrace{(SO,SO)}_{source} & \underbrace{(SI,SO)}_{saddle} & \cdots & \underbrace{(SO,SO)}_{source}\\ \vdots & \vdots & \vdots & \vdots\\ \underbrace{(SO,SI)}_{saddle} & \underbrace{(SI,SI)}_{sink} & \cdots & \underbrace{(SO,SI)}_{saddle}\end{array}\right\}_{(2n_1+1)\times(2m_1+1)}.$$

In the network, the number of (SO,SI)-saddles is $(n_1+1)\times(m_1+1)$, the number of (SI,SO)-saddles is $n_1 \times m_1$; the number of (SO,SO)-sources is $n_1 \times (m_1+1)$; and the number of (SI,SI)-sink is $(n_1 + 1) \times m_1$.

- For $a_{j_1j_10} < 0$ *and* $a_{j_2j_20} < 0$,

$$\bigcup_{s=1}^{2m_1+1}\bigcup_{l=1}^{2n_1+1}\underbrace{(a_{j_1j_1s},a_{j_2j_2l})}_{XX}=\left\{\begin{array}{cccc}\underbrace{(SI,SI)}_{sink} & \underbrace{(SO,SI)}_{saddle} & \cdots & \underbrace{(SI,SI)}_{sink}\\ \underbrace{(SI,SO)}_{saddle} & \underbrace{(SO,SO)}_{source} & \cdots & \underbrace{(SI,SO)}_{saddle}\\ \vdots & \vdots & \vdots & \vdots\\ \underbrace{(SI,SI)}_{sink} & \underbrace{(SO,SI)}_{saddle} & \cdots & \underbrace{(SI,SI)}_{sink}\end{array}\right\}_{(2n_1+1)\times(2m_1+1)}.$$

In the network, the number of (SO,SI)-saddles is $(n_1 + 1) \times m_1$, the number of (SI,SO)-saddles is $n_1 \times (m_1 + 1)$; the number of (SO,SO)-sources is $n_1 \times m_1$; and the number of (SI,SI)-sink is $(n_1 + 1) \times (m_1 + 1)$.

(i_2) From case (i_1), the numbers of saddles, sink and sources for $(2m_1+1) \times (2n_1+1)$ -equilibriums are summarized as in Table 2.1.

(ii_1) For $m = 2m_1$ and $n = 2n_1 + 1$, the equilibrium networks with $(2m_1) \times (2n_1 + 1)$ equilibriums have the following properties as in Eqs. (2.12)–(2.15).

- For $a_{j_1 j_1 0} > 0$ *and* $a_{j_2 j_2 0} > 0$,

$$\bigcup_{s=1}^{2m_1}\bigcup_{l=1}^{2n_1+1}\underbrace{(a_{j_1j_1s},a_{j_2j_2l})}_{XX}=\left\{\begin{matrix}\underbrace{(SI,SO)}_{saddle} & \underbrace{(SO,SO)}_{source} & \cdots & \underbrace{(SO,SO)}_{source}\\ \underbrace{(SI,SI)}_{sink} & \underbrace{(SO,SI)}_{saddle} & \cdots & \underbrace{(SO,SI)}_{saddle}\\ \vdots & \vdots & \vdots & \vdots\\ \underbrace{(SI,SO)}_{saddle} & \underbrace{(SO,SO)}_{source} & \cdots & \underbrace{(SO,SO)}_{source}\end{matrix}\right\}_{(2n_1+1)\times(2m_1)}.$$

In the network, the number of (SO,SI)-saddles is $n_1 \times m_1$, the number of (SI,SO)-saddles is $(n_1 + 1) \times m_1$; the number of (SO,SO)-sources is $(n_1 + 1) \times m_1$; and the number of (SI,SI)-sinks is $n_1 \times m_1$.

- For $a_{j_1 j_1 0} < 0$ *and* $a_{j_2 j_2 0} > 0$,

$$\bigcup_{s=1}^{2m_1}\bigcup_{l=1}^{2n_1+1}\underbrace{(a_{j_1j_1s},a_{j_2j_2l})}_{XX}=\left\{\begin{matrix}\underbrace{(SO,SO)}_{source} & \underbrace{(SI,SO)}_{saddle} & \cdots & \underbrace{(SI,SO)}_{saddle}\\ \underbrace{(SO,SI)}_{saddle} & \underbrace{(SI,SI)}_{sink} & \cdots & \underbrace{(SI,SI)}_{sink}\\ \vdots & \vdots & \vdots & \vdots\\ \underbrace{(SO,SO)}_{source} & \underbrace{(SI,SO)}_{saddle} & \cdots & \underbrace{(SI,SO)}_{saddle}\end{matrix}\right\}_{(2n_1+1)\times(2m_1)}.$$

In the network, the number of (SO,SI)-saddles is $n_1 \times m_1$, the number of (SI,SO)-saddles is $(n_1 + 1) \times m_1$; the number of (SO,SO)-sources is $(n_1 + 1) \times m_1$; and the number of (SI,SI)-sinks is $n_1 \times m_1$.

- For $a_{j_1 j_1 0} > 0$ *and* $a_{j_2 j_2 0} < 0$,

$$\bigcup_{s=1}^{2m_1}\bigcup_{l=1}^{2n_1+1}\underbrace{(a_{j_1j_1s},a_{j_2j_2l})}_{XX}=\left\{\begin{matrix}\underbrace{(SI,SI)}_{sink} & \underbrace{(SO,SI)}_{saddle} & \cdots & \underbrace{(SO,SI)}_{saddle}\\ \underbrace{(SI,SO)}_{saddle} & \underbrace{(SO,SO)}_{source} & \cdots & \underbrace{(SO,SO)}_{source}\\ \vdots & \vdots & \vdots & \vdots\\ \underbrace{(SI,SI)}_{sink} & \underbrace{(SO,SI)}_{saddle} & \cdots & \underbrace{(SO,SI)}_{saddle}\end{matrix}\right\}_{(2n_1+1)\times(2m_1)}.$$

In the network, the number of (SO,SI)-saddles is $(n_1 + 1) \times m_1$, the number of (SI,SO)-saddles is $n_1 \times m_1$; the number of (SO,SO)-sources is $n_1 \times m_1$; and the number of (SI,SI)-sinks is $(n_1 + 1) \times m_1$.

- For $a_{j_1 j_1 0} < 0$ *and* $a_{j_2 j_2 0} < 0$,

$$\bigcup_{s=1}^{2m_1} \bigcup_{l=1}^{2n_1+1} \underbrace{(a_{j_1 j_1 s}, a_{j_2 j_2 l})}_{\text{XX}} = \left\{ \begin{array}{cccc} \underbrace{\text{(SO,SI)}}_{\text{saddle}} & \underbrace{\text{(SI,SI)}}_{\text{sink}} & \cdots & \underbrace{\text{(SI,SI)}}_{\text{sink}} \\ \underbrace{\text{(SO,SO)}}_{\text{source}} & \underbrace{\text{(SI,SO)}}_{\text{saddle}} & \cdots & \underbrace{\text{(SI,SO)}}_{\text{saddle}} \\ \vdots & \vdots & \vdots & \vdots \\ \underbrace{\text{(SO,SI)}}_{\text{saddle}} & \underbrace{\text{(SI,SI)}}_{\text{sink}} & \cdots & \underbrace{\text{(SI,SI)}}_{\text{sink}} \end{array} \right\}_{(2n_1+1)\times(2m_1)}.$$

In the network, the number of (SO,SI)-saddles is $(n_1 + 1) \times m_1$, the number of (SI,SO)-saddles is $n_1 \times m_1$; the number of (SO,SO)-sources is $n_1 \times m_1$; and the number of (SI,SI)-sinks is $(n_1 + 1) \times m_1$.

(ii_2) Form case (ii_2), the numbers of saddles, sinks and sources for $(2m_1) \times (2n_1 + 1)$-equilibriums are summarized in Table 2.2.

(iii_1) For $m = 2m_1 + 1$ and $n = 2n_1$, the equilibrium networks with $(2m_1 + 1) \times (2n_1)$ equilibriums have the following properties as in Eqs. (2.16)–(2.19).

- For $a_{j_1 j_1 0} > 0$ *and* $a_{j_2 j_2 0} > 0$,

$$\bigcup_{s=1}^{2m_1+1} \bigcup_{l=1}^{2n_1} \underbrace{(a_{j_1 j_1 s}, a_{j_2 j_2 l})}_{\text{XX}} = \left\{ \begin{array}{cccc} \underbrace{\text{(SO,SO)}}_{\text{source}} & \underbrace{\text{(SI,SO)}}_{\text{saddle}} & \cdots & \underbrace{\text{(SO,SO)}}_{\text{source}} \\ \underbrace{\text{(SO,SI)}}_{\text{saddle}} & \underbrace{\text{(SI,SI)}}_{\text{sink}} & \cdots & \underbrace{\text{(SO,SI)}}_{\text{saddle}} \\ \vdots & \vdots & \vdots & \vdots \\ \underbrace{\text{(SO,SI)}}_{\text{saddle}} & \underbrace{\text{(SI,SI)}}_{\text{sink}} & \cdots & \underbrace{\text{(SO,SI)}}_{\text{saddle}} \end{array} \right\}_{(2n_1)\times(2m_1+1)}.$$

In the network, the number of (SO,SI)-saddles is $n_1 \times (m_1 + 1)$, the number of (SI,SO)-saddles is $n_1 \times m_1$; the number of (SO,SO)-sources is $n_1 \times (m_1 + 1)$; and the number of (SI,SI)-sinks is $n_1 \times m_1$.

- For $a_{j_1 j_1 0} < 0$ *and* $a_{j_2 j_2 0} > 0$,

$$\bigcup_{s=1}^{2m_1+1}\bigcup_{l=1}^{2n_1}\underbrace{(a_{j_1j_1s},a_{j_2j_2l})}_{\text{XX}}=\left\{\begin{matrix}\underbrace{\text{(SI,SO)}}_{\text{saddle}} & \underbrace{\text{(SO,SO)}}_{\text{source}} & \cdots & \underbrace{\text{(SI,SO)}}_{\text{saddle}}\\ \underbrace{\text{(SI,SI)}}_{\text{sink}} & \underbrace{\text{(SO,SI)}}_{\text{saddle}} & \cdots & \underbrace{\text{(SI,SI)}}_{\text{sink}}\\ \vdots & \vdots & \vdots & \vdots\\ \underbrace{\text{(SI,SI)}}_{\text{sink}} & \underbrace{\text{(SO,SI)}}_{\text{saddle}} & \cdots & \underbrace{\text{(SI,SI)}}_{\text{sink}}\end{matrix}\right\}_{(2n_1)\times(2m_1+1)}.$$

In the network, the number of (SO,SI)-saddles is $n_1 \times m_1$, the number of (SI,SO)-saddles is $n_1 \times (m_1 + 1)$; the number of (SO,SO)-sources is $n_1 \times m_1$; and the number of (SI,SI)-sink is $n_1 \times (m_1 + 1)$.

- For $a_{j_1j_10} > 0$ *and* $a_{j_2j_20} < 0$,

$$\bigcup_{s=1}^{2m_1+1}\bigcup_{l=1}^{2n_1}\underbrace{(a_{j_1j_1s},a_{j_2j_2l})}_{\text{XX}}=\left\{\begin{matrix}\underbrace{\text{(SO,SI)}}_{\text{saddle}} & \underbrace{\text{(SI,SI)}}_{\text{sink}} & \cdots & \underbrace{\text{(SO,SI)}}_{\text{saddle}}\\ \underbrace{\text{(SO,SO)}}_{\text{source}} & \underbrace{\text{(SI,SO)}}_{\text{saddle}} & \cdots & \underbrace{\text{(SO,SO)}}_{\text{source}}\\ \vdots & \vdots & \vdots & \vdots\\ \underbrace{\text{(SO,SO)}}_{\text{source}} & \underbrace{\text{(SI,SO)}}_{\text{saddle}} & \cdots & \underbrace{\text{(SO,SO)}}_{\text{source}}\end{matrix}\right\}_{(2n_1)\times(2m_1+1)}.$$

In the network, the number of (SO,SI)-saddles is $n_1 \times (m_1 + 1)$, the number of (SI,SO)-saddles is $n_1 \times m_1$; the number of (SO,SO)-sources is $n_1 \times (m_1 + 1)$; and the number of (SI,SI)-sink is $n_1 \times m_1$.

- For $a_{j_1j_10} < 0$ *and* $a_{j_2j_20} < 0$,

$$\bigcup_{s=1}^{2m_1+1}\bigcup_{l=1}^{2n_1}\underbrace{(a_{j_1j_1s},a_{j_2j_2l})}_{\text{XX}}=\left\{\begin{matrix}\underbrace{\text{(SI,SI)}}_{\text{sink}} & \underbrace{\text{(SO,SI)}}_{\text{saddle}} & \cdots & \underbrace{\text{(SI,SI)}}_{\text{sink}}\\ \underbrace{\text{(SI,SO)}}_{\text{saddle}} & \underbrace{\text{(SO,SO)}}_{\text{source}} & \cdots & \underbrace{\text{(SI,SO)}}_{\text{saddle}}\\ \vdots & \vdots & \vdots & \vdots\\ \underbrace{\text{(SI,SO)}}_{\text{saddle}} & \underbrace{\text{(SO,SO)}}_{\text{source}} & \cdots & \underbrace{\text{(SI,SO)}}_{\text{saddle}}\end{matrix}\right\}_{(2n_1)\times(2m_1+1)}.$$

In the network, the number of (SO,SI)-saddles is $n_1 \times m_1$, the number of (SI,SO)-saddles is $n_1 \times (m_1 + 1)$; the number of (SO,SO)-sources is $n_1 \times m_1$; and the number of (SI,SI)-sink is $n_1 \times (m_1 + 1)$.

(iii$_2$) From case (iii$_1$), the numbers of saddles, sink and sources for $(2m_1 + 1) \times (2n_1)$-equilibriums are summarized in Table 2.3.

(iv$_1$) For $m = 2m_1$ and $n = 2n_1$, the equilibrium networks with $(2m_1) \times (2n_1)$ equilibriums have the following properties as in Eqs. (2.20)–(2.23).

- For $a_{j_1j_10} > 0$ *and* $a_{j_2j_20} > 0$,

$$\bigcup_{s=1}^{2m_1}\bigcup_{l=1}^{2n_1}\underbrace{(a_{j_1j_1s}, a_{j_2j_2l})}_{XX} = \left\{\begin{array}{cccc} \underbrace{(SI,SO)}_{saddle} & \underbrace{(SO,SO)}_{source} & \cdots & \underbrace{(SO,SO)}_{source} \\ \underbrace{(SI,SI)}_{sink} & \underbrace{(SO,SI)}_{saddle} & \cdots & \underbrace{(SO,SI)}_{saddle} \\ \vdots & \vdots & \vdots & \vdots \\ \underbrace{(SI,SI)}_{sink} & \underbrace{(SO,SI)}_{saddle} & \cdots & \underbrace{(SO,SI)}_{saddle} \end{array}\right\}_{(2n_1)\times(2m_1)}.$$

In the network, the number of (SO,SI)-saddles is $n_1 \times m_1$, the number of (SI,SO)-saddles is $n_1 \times m_1$; the number of (SO,SO)-sources is $n_1 \times m_1$; and the number of (SI,SI)-sinks is $n_1 \times m_1$.

- For $a_{j_1j_10} < 0$ *and* $a_{j_2j_20} > 0$,

$$\bigcup_{s=1}^{2m_1}\bigcup_{l=1}^{2n_1}\underbrace{(a_{j_1j_1s}, a_{j_2j_2l})}_{XX} = \left\{\begin{array}{cccc} \underbrace{(SO,SO)}_{source} & \underbrace{(SI,SO)}_{saddle} & \cdots & \underbrace{(SI,SO)}_{saddle} \\ \underbrace{(SO,SI)}_{saddle} & \underbrace{(SI,SI)}_{sink} & \cdots & \underbrace{(SI,SI)}_{sink} \\ \vdots & \vdots & \vdots & \vdots \\ \underbrace{(SO,SI)}_{saddle} & \underbrace{(SI,SI)}_{sink} & \cdots & \underbrace{(SI,SI)}_{sink} \end{array}\right\}_{(2n_1)\times(2m_1)}.$$

In the network, the number of (SO,SI)-saddles is $n_1 \times m_1$, the number of (SI,SO)-saddles is $n_1 \times m_1$; the number of (SO,SO)-sources is $n_1 \times m_1$; and the number of (SI,SI)-sink is $n_1 \times m_1$.

- For $a_{j_1j_10} > 0$ *and* $a_{j_2j_20} < 0$,

$$\bigcup_{s=1}^{2m_1}\bigcup_{l=1}^{2n_1}\underbrace{(a_{j_1j_1s}, a_{j_2j_2l})}_{XX} = \left\{\begin{array}{cccc} \underbrace{(SI,SI)}_{sink} & \underbrace{(SO,SI)}_{saddle} & \cdots & \underbrace{(SO,SI)}_{saddle} \\ \underbrace{(SI,SO)}_{saddle} & \underbrace{(SO,SO)}_{source} & \cdots & \underbrace{(SO,SO)}_{source} \\ \vdots & \vdots & \vdots & \vdots \\ \underbrace{(SI,SO)}_{saddle} & \underbrace{(SO,SO)}_{source} & \cdots & \underbrace{(SO,SO)}_{source} \end{array}\right\}_{(2n_1)\times(2m_1)}.$$

In the network, the number of (SO,SI)-saddles is $n_1 \times m_1$, the number of (SI,SO)-saddles is $n_1 \times m_1$; the number of (SO,SO)-sources is $n_1 \times m_1$; and the number of (SI,SI)-sink is $n_1 \times m_1$.

- For $a_{j_1j_10} < 0$ *and* $a_{j_2j_20} < 0$,

$$\bigcup_{s=1}^{2m_1}\bigcup_{l=1}^{2n_1}\underbrace{(a_{j_1j_1s}, a_{j_2j_2l})}_{\text{XX}} = \left\{\begin{matrix} \underbrace{\text{(SO,SI)}}_{\text{saddle}} & \underbrace{\text{(SI,SI)}}_{\text{sink}} & \cdots & \underbrace{\text{(SI,SI)}}_{\text{sink}} \\ \underbrace{\text{(SO,SO)}}_{\text{source}} & \underbrace{\text{(SI,SO)}}_{\text{saddle}} & \cdots & \underbrace{\text{(SI,SO)}}_{\text{saddle}} \\ \vdots & \vdots & \vdots & \vdots \\ \underbrace{\text{(SO,SO)}}_{\text{source}} & \underbrace{\text{(SI,SO)}}_{\text{saddle}} & \cdots & \underbrace{\text{(SI,SO)}}_{\text{saddle}} \end{matrix}\right\}_{(2n_1)\times(2m_1)}.$$

In the network, the number of (SO,SI)-saddles is $n_1 \times m_1$, the number of (SI,SO)-saddles is $n_1 \times m_1$; the number of (SO,SO)-sources is $n_1 \times m_1$; and the number of (SI,SI)-sink is $n_1 \times m_1$.

(iv$_2$) From case (iv$_1$), the numbers of saddles, sink and sources for $(2m_1) \times (2n_1)$-equilibriums are summarized in Table 2.4.

In the end, Theorem 2.1 is proved. ■

2.3 Homoclinic Networks Without Centers

(A) Consider a 2-dimensional system with $(2m_1) \times (2n_1) = 2 \times 2$ as

$$\begin{aligned} \dot{x}_1 &= a_{110}(x_1 - a_{111})(x_1 - a_{112}), \\ \dot{x}_2 &= a_{220}(x_2 - a_{221})(x_2 - a_{222}). \end{aligned} \tag{2.24}$$

The first integral manifold is

$$\begin{aligned} &\frac{a_{110}}{a_{222} - a_{221}}\Big(\ln\frac{|x_2 - a_{222}|}{|x_{20} - a_{222}|} - \ln\frac{|x_2 - a_{221}|}{|x_{20} - a_{221}|}\Big) \\ &= \frac{a_{220}}{a_{112} - a_{111}}\Big(\ln\frac{|x_1 - a_{112}|}{|x_{10} - a_{112}|} - \ln\frac{|x_1 - a_{111}|}{|x_{10} - a_{111}|}\Big). \end{aligned} \tag{2.25}$$

Phase portraits for the 2-dimensional systems near the simple equilibriums of the saddles, sinks and sources presented in Fig. 2.1a–d for $(a_{110} > 0, a_{220} > 0)$, $(a_{110} < 0, a_{220} > 0)$, $(a_{110} > 0, a_{220} < 0)$ and $(a_{110} < 0, a_{220} < 0)$. For all cases, one sink equilibrium, one source equilibrium, and two opposite saddles are for the four simple equilibriums of $(x_1^*, x_2^*) = (a_{111}, a_{221}), (a_{112}, a_{221}), (a_{111}, a_{222}), (a_{112}, a_{222})$. The four simple equilibriums are based on the bifurcation of double-saddle bifurcations.

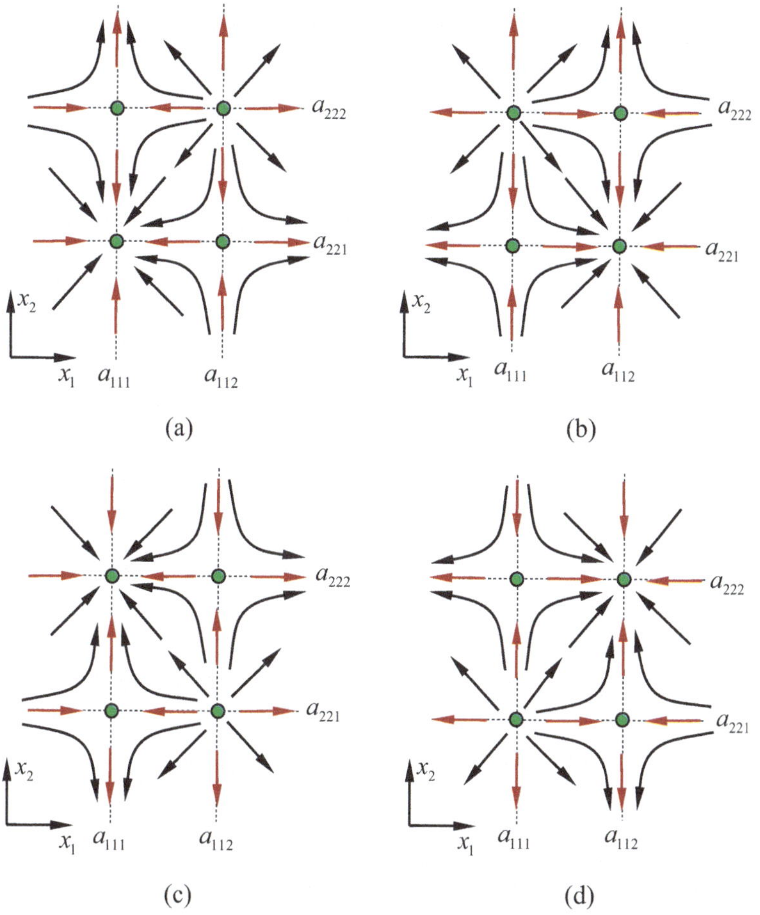

Fig. 2.1 Phase portraits for 2-dimensional systems on the x_1-direction with $x_1^* = a_{111}, a_{112}$ and on the x_2-direction with $x_2^* = a_{221}, a_{222}$. The four sets of four simple equilibriums: **a** ($a_{110} > 0, a_{220} > 0$), **b** ($a_{110} < 0, a_{220} > 0$), **c** ($a_{110} > 0, a_{220} < 0$), **d** ($a_{110} < 0, a_{220} < 0$)

Table 2.5 Numbers of saddles, sinks and sources in networks for 2×2-equilibriums

(a_{110}, a_{220})	(SO,SI)-saddle	(SI,SO)-saddle	(SO,SO)-source	(SI,SI)-sink
$(+, +)$	$m_1 \times n_1 = 1$	$m_1 \times n_1 = 1$	$m_1 \times n_1 = 1$	$m_1 \times n_1 = 1$
$(-, +)$	$m_1 \times n_1 = 1$	$m_1 \times n_1 = 1$	$m_1 \times n_1 = 1$	$m_1 \times n_1 = 1$
$(+, -)$	$m_1 \times n_1 = 1$	$m_1 \times n_1 = 1$	$m_1 \times n_1 = 1$	$m_1 \times n_1 = 1$
$(-, -)$	$m_1 \times n_1 = 1$	$m_1 \times n_1 = 1$	$m_1 \times n_1 = 1$	$m_1 \times n_1 = 1$

$$\left\{\begin{array}{l}(a_{111}, a_{222})\ (a_{112}, a_{222})\\(a_{111}, a_{221})\ (a_{112}, a_{221})\end{array}\right\} = \left\{\begin{array}{l}\underbrace{\text{(SI,SO)}}_{\text{saddle}}\ \underbrace{\text{(SO,SO)}}_{\text{source}}\\\underbrace{\text{(SI,SI)}}_{\text{sink}}\ \underbrace{\text{(SO,SI)}}_{\text{saddle}}\end{array}\right\}$$

$$\text{for } a_{110} > 0, a_{220} > 0; \tag{2.26}$$

$$\left\{\begin{array}{l}(a_{111}, a_{222})\ (a_{112}, a_{222})\\(a_{111}, a_{221})\ (a_{112}, a_{221})\end{array}\right\} = \left\{\begin{array}{l}\underbrace{\text{(SO,SO)}}_{\text{source}}\ \underbrace{\text{(SI,SO)}}_{\text{saddle}}\\\underbrace{\text{(SO,SI)}}_{\text{saddle}}\ \underbrace{\text{(SI,SI)}}_{\text{sink}}\end{array}\right\}$$

$$\text{for } a_{110} < 0, a_{220} > 0; \tag{2.27}$$

$$\left\{\begin{array}{l}(a_{111}, a_{222})\ (a_{112}, a_{222})\\(a_{111}, a_{221})\ (a_{112}, a_{221})\end{array}\right\} = \left\{\begin{array}{l}\underbrace{\text{(SI,SI)}}_{\text{sink}}\ \underbrace{\text{(SO,SI)}}_{\text{saddle}}\\\underbrace{\text{(SI,SO)}}_{\text{saddle}}\ \underbrace{\text{(SO,SO)}}_{\text{source}}\end{array}\right\}$$

$$\text{for } a_{110} > 0, a_{220} < 0; \tag{2.28}$$

$$\left\{\begin{array}{l}(a_{111}, a_{222})\ (a_{112}, a_{222})\\(a_{111}, a_{221})\ (a_{112}, a_{221})\end{array}\right\} = \left\{\begin{array}{l}\underbrace{\text{(SO,SI)}}_{\text{saddle}}\ \underbrace{\text{(SI,SI)}}_{\text{sink}}\\\underbrace{\text{(SO,SO)}}_{\text{source}}\ \underbrace{\text{(SI,SO)}}_{\text{saddle}}\end{array}\right\}$$

$$\text{for } a_{110} < 0, a_{220} < 0. \tag{2.29}$$

The numbers of saddles, sinks, and saddles in networks for 2×2-equilibriums are listed in Table 2.5.

(B) Consider a 2-dimensional system with $(2m_1 + 1) \times (2n_1) = 3 \times 2$ equilibriums as

$$\begin{aligned}\dot{x}_1 &= a_{110}(x_1 - a_{111})(x_1 - a_{112})(x_1 - a_{113}),\\\dot{x}_2 &= a_{220}(x_2 - a_{221})(x_2 - a_{222}).\end{aligned} \tag{2.30}$$

The first integral manifold is given by

$$\begin{aligned}&\frac{1}{a_{222} - a_{221}}(\ln \frac{|x_2 - a_{222}|}{|x_{20} - a_{222}|} - \ln \frac{|x_2 - a_{221}|}{|x_{20} - a_{221}|})\\&= \frac{a_{220}}{a_{110}} \sum\nolimits_{s_1=1}^{3} \frac{1}{\prod_{s_2=1, s_2 \neq s_1}^{3} (a_{11s_1} - a_{11s_2})} \ln \frac{|x_1 - a_{11s_1}|}{|x_{10} - a_{11s_1}|}.\end{aligned} \tag{2.31}$$

Phase portraits for the 2-dimensional systems near the simple equilibriums of the saddles, sinks and sources presented in Fig. 2.2a–d for ($a_{110} > 0, a_{220} > 0$),

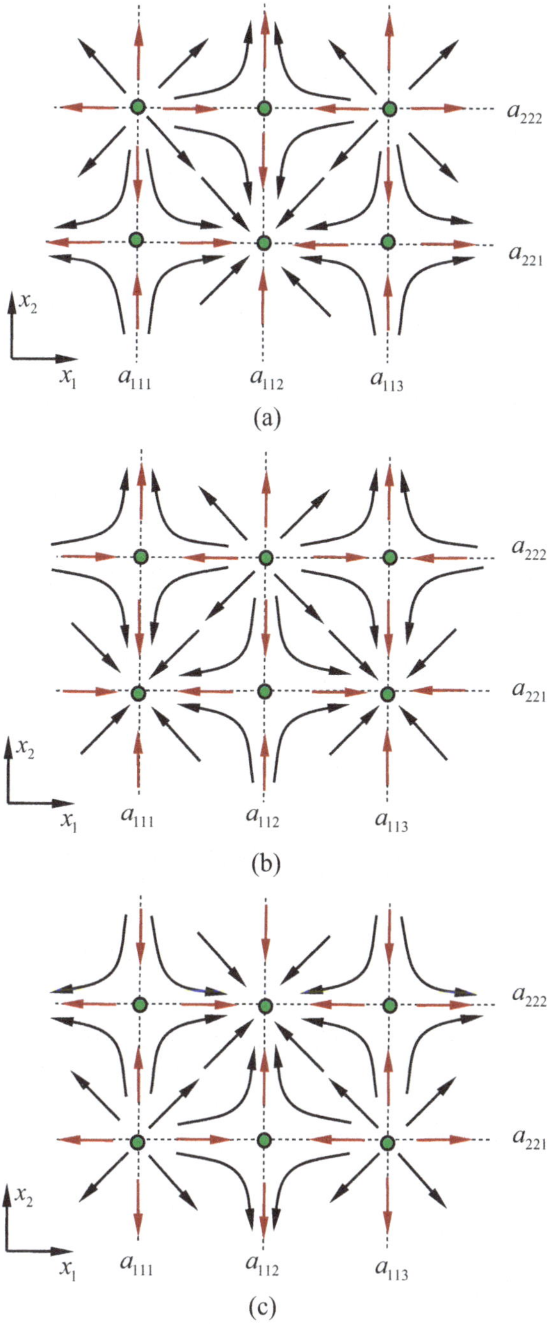

Fig. 2.2 Phase portraits for 2-dimensional systems on the x_1-direction with $x_1^* = a_{111}, a_{112}, a_{113}$ and on the x_2-direction with $x_2^* = a_{221}, a_{222}$. The four networks of sink, source and saddle: **a** ($a_{110} > 0, a_{220} > 0$), **b** ($a_{110} < 0, a_{220} > 0$), **c** ($a_{110} > 0, a_{220} < 0$), **d** ($a_{110} < 0, a_{220} < 0$)

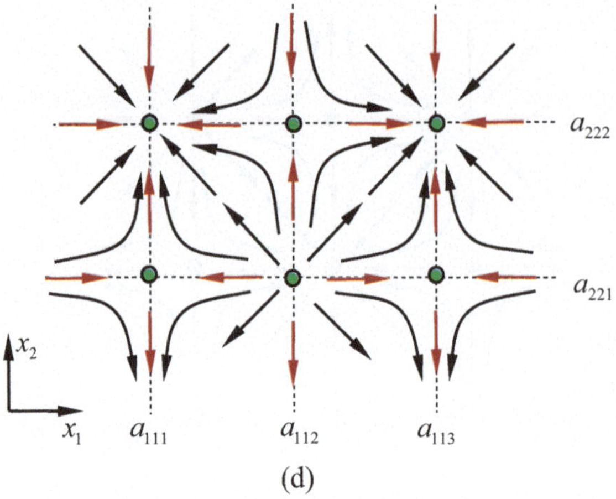

Fig. 2.2 (continued)

Table 2.6 Numbers of saddles, sinks and sources in networks for 3×2-equilibriums

(a_{110}, a_{220})	(SO,SI)-saddle	(SI,SO)-saddle	(SO,SO)-source	(SI,SI)-sink
$(+,+)$	$n_1 \times (m_1+1) = 2$	$n_1 \times m_1 = 1$	$n_1 \times (m_1+1) = 2$	$m_1 \times n_1 = 1$
$(-,+)$	$n_1 \times m_1 = 1$	$n_1 \times (m_1+1) = 2$	$n_1 \times m_1 = 1$	$n_1 \times (m_1+1) = 2$
$(+,-)$	$n_1 \times (m_1+1) = 2$	$n_1 \times m_1 = 1$	$n_1 \times (m_1+1) = 2$	$m_1 \times n_1 = 1$
$(-,-)$	$n_1 \times m_1 = 1$	$n_1 \times (m_1+1) = 2$	$n_1 \times m_1 = 1$	$n_1 \times (m_1+1) = 2$

$(a_{110} < 0, a_{220} > 0)$, $(a_{110} > 0, a_{220} < 0)$ and $(a_{110} < 0, a_{220} < 0)$. For all cases, the six simple equilibriums are generated through the (3,2)-saddle-source (sink) appearing bifurcations, and the (3,2) homoclinic network without centers are given as follows.

$$\left\{ \begin{matrix} (a_{111}, a_{222}) & (a_{112}, a_{222}) & (a_{113}, a_{222}) \\ (a_{111}, a_{221}) & (a_{112}, a_{221}) & (a_{113}, a_{221}) \end{matrix} \right\} = \left\{ \begin{matrix} \underbrace{(\text{SO,SO})}_{\text{source}} & \underbrace{(\text{SI,SO})}_{\text{saddle}} & \underbrace{(\text{SO,SO})}_{\text{source}} \\ \underbrace{(\text{SO,SI})}_{\text{saddle}} & \underbrace{(\text{SI,SI})}_{\text{sink}} & \underbrace{(\text{SO,SI})}_{\text{saddle}} \end{matrix} \right\}$$

$$\text{for } a_{110} > 0, a_{220} > 0; \tag{2.32}$$

$$\left\{ \begin{matrix} (a_{111}, a_{222}) & (a_{112}, a_{222}) & (a_{113}, a_{222}) \\ (a_{111}, a_{221}) & (a_{112}, a_{221}) & (a_{113}, a_{221}) \end{matrix} \right\} = \left\{ \begin{matrix} \underbrace{(\text{SI,SO})}_{\text{saddle}} & \underbrace{(\text{SO,SO})}_{\text{source}} & \underbrace{(\text{SI,SO})}_{\text{saddle}} \\ \underbrace{(\text{SI,SI})}_{\text{sink}} & \underbrace{(\text{SO,SI})}_{\text{saddle}} & \underbrace{(\text{SI,SI})}_{\text{sink}} \end{matrix} \right\}$$

$$\text{for } a_{110} < 0, a_{220} > 0; \tag{2.33}$$

$$\left\{\begin{matrix} (a_{111},a_{222}) & (a_{112},a_{222}) & (a_{113},a_{222}) \\ (a_{111},a_{221}) & (a_{112},a_{221}) & (a_{113},a_{221}) \end{matrix}\right\} = \left\{\begin{matrix} \underbrace{(\mathrm{SO,SI})}_{\text{saddle}} & \underbrace{(\mathrm{SI,SI})}_{\text{sink}} & \underbrace{(\mathrm{SO,SI})}_{\text{saddle}} \\ \underbrace{(\mathrm{SO,SO})}_{\text{source}} & \underbrace{(\mathrm{SI,SO})}_{\text{saddle}} & \underbrace{(\mathrm{SO,SO})}_{\text{source}} \end{matrix}\right\}$$

for $a_{110} > 0, a_{220} < 0$; (2.34)

$$\left\{\begin{matrix} (a_{111},a_{222}) & (a_{112},a_{222}) & (a_{113},a_{222}) \\ (a_{111},a_{221}) & (a_{112},a_{221}) & (a_{113},a_{221}) \end{matrix}\right\} = \left\{\begin{matrix} \underbrace{(\mathrm{SI,SI})}_{\text{sink}} & \underbrace{(\mathrm{SO,SI})}_{\text{saddle}} & \underbrace{(\mathrm{SI,SI})}_{\text{sink}} \\ \underbrace{(\mathrm{SI,SO})}_{\text{saddle}} & \underbrace{(\mathrm{SO,SO})}_{\text{source}} & \underbrace{(\mathrm{SI,SO})}_{\text{saddle}} \end{matrix}\right\} \quad (2.35)$$

for $a_{110} < 0, a_{220} < 0$.

The numbers of saddles, sinks, and sources in the networks of 3×2-equilibriums are listed in Table 2.6.

(C) Consider a 2-dimensional system with $(2m_1) \times (2n_1 + 1) = 2 \times 3$ equilibriums as

$$\begin{aligned} \dot{x}_1 &= a_{110}(x_1 - a_{111})(x_1 - a_{112}), \\ \dot{x}_2 &= a_{220}(x_2 - a_{221})(x_2 - a_{222})(x_2 - a_{223}). \end{aligned} \quad (2.36)$$

The first integral manifold is given by

$$\begin{aligned} &\sum_{l_1=1}^{3} \frac{1}{\prod_{l_2=1,l_2\neq l_1}^{3}(a_{22l_1} - a_{22l_2})} \ln \frac{|x_2 - a_{22l_1}|}{|x_{20} - a_{22l_1}|} \\ &= \frac{a_{220}}{a_{110}} \sum_{s_1=1}^{2} \frac{1}{\prod_{s_2=1,s_2\neq s_1}^{2}(a_{11s_1} - a_{11s_2})} \ln \frac{|x_1 - a_{11s_1}|}{|x_{10} - a_{11s_1}|}. \end{aligned} \quad (2.37)$$

Phase portraits for the 2-dimensional systems near the simple equilibriums of the saddles, sinks and sources presented in Fig. 2.3a–d for $(a_{110} > 0, a_{220} > 0)$, $(a_{110} < 0, a_{220} > 0)$, $(a_{110} > 0, a_{220} < 0)$ and $(a_{110} < 0, a_{220} < 0)$. For all cases, the six simple equilibriums are based on the third-order sink, source and saddle bifurcations. The numbers of saddles, sinks, and sources in the networks of 2×3-equilibriums are listed in Table 2.7.

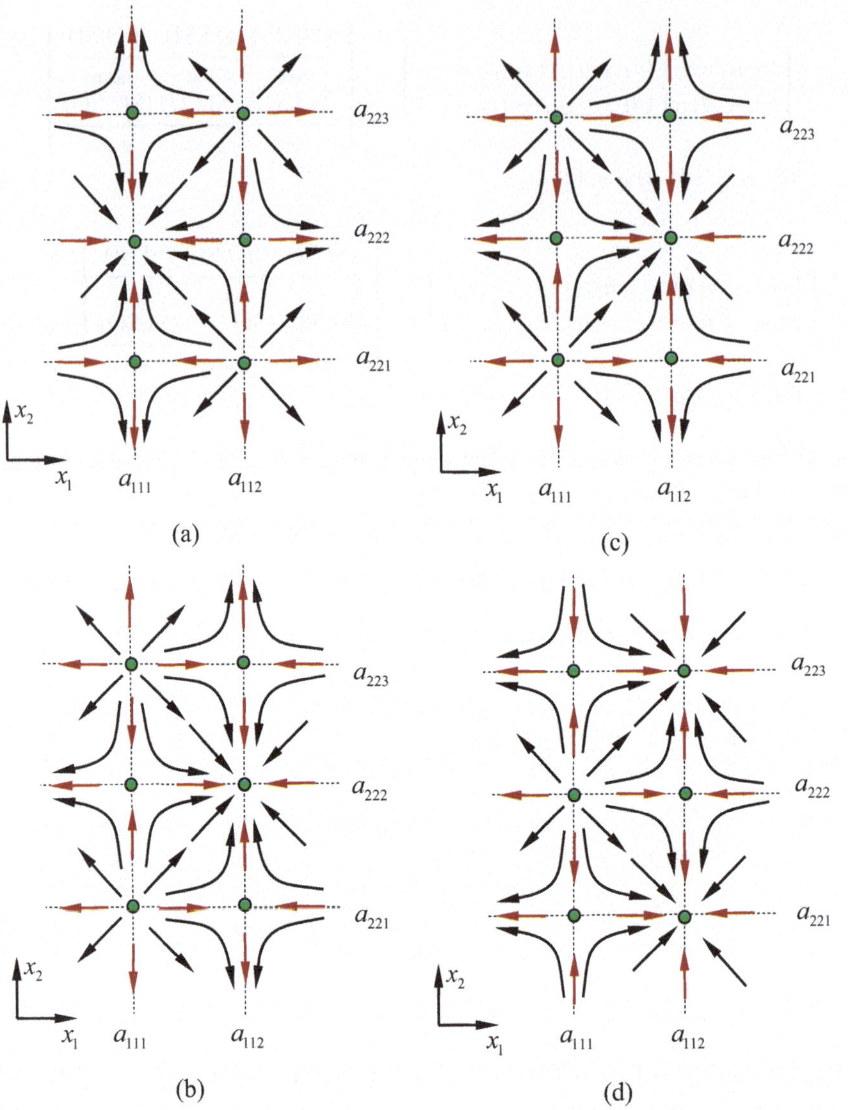

Fig. 2.3 Phase portraits for 2-dimensional systems on the x_1-direction with $x_1^* = a_{111}, a_{112}$ and on the x_2-direction with $x_2^* = a_{221}, a_{222}, a_{223}$. The four networks of sink, source and saddle: **a** ($a_{110} > 0, a_{220} > 0$), **b** ($a_{110} < 0, a_{220} > 0$), **c** ($a_{110} > 0, a_{220} < 0$), **d** ($a_{110} < 0, a_{220} < 0$)

Table 2.7 Numbers of saddles, sinks and sources in networks for 2 × 3-equilibriums

(a_{110}, a_{220})	(SO,SI)-saddle	(SI,SO)-saddle	(SO,SO)-source	(SI,SI)-sink
$(+,+)$	$n_1 \times m_1 = 1$	$(n_1+1) \times m_1 = 2$	$(n_1+1) \times m_1 = 2$	$n_1 \times m_1 = 1$
$(-,+)$	$n_1 \times m_1 = 1$	$(n_1+1) \times m_1 = 2$	$(n_1+1) \times m_1 = 2$	$n_1 \times m_1 = 1$
$(+,-)$	$(n_1+1) \times m_1 = 2$	$n_1 \times m_1 = 1$	$n_1 \times m_1 = 1$	$(n_1+1) \times m_1 = 2$
$(-,-)$	$(n_1+1) \times m_1 = 2$	$n_1 \times m_1 = 1$	$n_1 \times m_1 = 1$	$(n_1+1) \times m_1 = 2$

$$\left\{\begin{array}{cc} (a_{111}, a_{223}) & (a_{112}, a_{223}) \\ (a_{111}, a_{222}) & (a_{112}, a_{222}) \\ (a_{111}, a_{221}) & (a_{112}, a_{221}) \end{array}\right\} = \left\{\begin{array}{cc} \underbrace{(\mathrm{SI,SO})}_{\text{saddle}} & \underbrace{(\mathrm{SO,SO})}_{\text{source}} \\ \underbrace{(\mathrm{SI,SI})}_{\text{sink}} & \underbrace{(\mathrm{SO,SI})}_{\text{saddle}} \\ \underbrace{(\mathrm{SI,SO})}_{\text{saddle}} & \underbrace{(\mathrm{SO,SO})}_{\text{source}} \end{array}\right\}$$

$$\text{for } a_{110} > 0, a_{220} > 0; \tag{2.38}$$

$$\left\{\begin{array}{cc} (a_{111}, a_{223}) & (a_{112}, a_{223}) \\ (a_{111}, a_{222}) & (a_{112}, a_{222}) \\ (a_{111}, a_{221}) & (a_{112}, a_{221}) \end{array}\right\} = \left\{\begin{array}{cc} \underbrace{(\mathrm{SO,SO})}_{\text{source}} & \underbrace{(\mathrm{SI,SO})}_{\text{saddle}} \\ \underbrace{(\mathrm{SO,SI})}_{\text{saddle}} & \underbrace{(\mathrm{SI,SI})}_{\text{sink}} \\ \underbrace{(\mathrm{SO,SO})}_{\text{source}} & \underbrace{(\mathrm{SI,SO})}_{\text{saddle}} \end{array}\right\}$$

$$\text{for } a_{110} < 0, a_{220} > 0; \tag{2.39}$$

$$\left\{\begin{array}{cc} (a_{111}, a_{223}) & (a_{112}, a_{223}) \\ (a_{111}, a_{222}) & (a_{112}, a_{222}) \\ (a_{111}, a_{221}) & (a_{112}, a_{221}) \end{array}\right\} = \left\{\begin{array}{cc} \underbrace{(\mathrm{SI,SI})}_{\text{sink}} & \underbrace{(\mathrm{SO,SI})}_{\text{saddle}} \\ \underbrace{(\mathrm{SI,SO})}_{\text{saddle}} & \underbrace{(\mathrm{SO,SO})}_{\text{source}} \\ \underbrace{(\mathrm{SI,SI})}_{\text{sink}} & \underbrace{(\mathrm{SO,SI})}_{\text{saddle}} \end{array}\right\}$$

$$\text{for } a_{110} > 0, a_{220} < 0; \tag{2.40}$$

$$\left\{\begin{array}{cc} (a_{111}, a_{223}) & (a_{112}, a_{223}) \\ (a_{111}, a_{222}) & (a_{112}, a_{222}) \\ (a_{111}, a_{221}) & (a_{112}, a_{221}) \end{array}\right\} = \left\{\begin{array}{cc} \underbrace{(\mathrm{SO,SI})}_{\text{saddle}} & \underbrace{(\mathrm{SI,SI})}_{\text{sink}} \\ \underbrace{(\mathrm{SO,SO})}_{\text{source}} & \underbrace{(\mathrm{SI,SO})}_{\text{saddle}} \\ \underbrace{(\mathrm{SO,SI})}_{\text{saddle}} & \underbrace{(\mathrm{SI,SI})}_{\text{sink}} \end{array}\right\}$$

$$\text{for } a_{110} < 0, a_{220} < 0. \tag{2.41}$$

(D) Consider a 2-dimensional system with $(2m_1+1)\times(2n_1+1)=3\times 3$ equilibriums as

$$\begin{aligned}\dot{x}_1 &= a_{110}(x_1-a_{111})(x_1-a_{112})(x_1-a_{113}),\\ \dot{x}_2 &= a_{220}(x_2-a_{221})(x_2-a_{222})(x_2-a_{223}).\end{aligned} \tag{2.42}$$

The first integral manifold is given by

$$\begin{aligned}&\sum_{l_1=1}^{3}\frac{1}{\prod_{l_2=1,l_2\neq l_1}^{3}(a_{22l_1}-a_{22l_2})}\ln\frac{|x_2-a_{22l_1}|}{|x_{20}-a_{22l_1}|}\\ &=\frac{a_{220}}{a_{110}}\sum_{s_1=1}^{3}\frac{1}{\prod_{s_2=1,s_2\neq s_1}^{3}(a_{11s_1}-a_{11s_2})}\ln\frac{|x_1-a_{11s_1}|}{|x_{10}-a_{11s_1}|}.\end{aligned} \tag{2.43}$$

Phase portraits for the 2-dimensional systems near the simple equilibriums of the saddles, sinks and sources presented in Fig. 2.4a–d for $(a_{110}>0, a_{220}>0)$, $(a_{110}<0, a_{220}>0)$, $(a_{110}>0, a_{220}<0)$ and $(a_{110}<0, a_{220}<0)$. For all cases, the nine simple equilibriums are based on the third-order sink, source and saddle bifurcations. The numbers of saddles, sinks, and sources in the networks of 3×3-equilibriums are listed in Table 2.8.

$$\left\{\begin{matrix}(a_{111},a_{223}) & (a_{112},a_{223}) & (a_{113},a_{223})\\ (a_{111},a_{222}) & (a_{112},a_{222}) & (a_{113},a_{222})\\ (a_{111},a_{221}) & (a_{112},a_{221}) & (a_{113},a_{221})\end{matrix}\right\}=\left\{\begin{matrix}\underbrace{(\text{SO,SO})}_{\text{source}} & \underbrace{(\text{SI,SO})}_{\text{saddle}} & \underbrace{(\text{SO,SO})}_{\text{source}}\\ \underbrace{(\text{SO,SI})}_{\text{saddle}} & \underbrace{(\text{SI,SI})}_{\text{sink}} & \underbrace{(\text{SO,SI})}_{\text{saddle}}\\ \underbrace{(\text{SO,SO})}_{\text{source}} & \underbrace{(\text{SI,SO})}_{\text{saddle}} & \underbrace{(\text{SO,SO})}_{\text{source}}\end{matrix}\right\}$$

$$\text{for } a_{110}>0,\ a_{220}>0; \tag{2.44}$$

$$\left\{\begin{matrix}(a_{111},a_{223}) & (a_{112},a_{223}) & (a_{113},a_{223})\\ (a_{111},a_{222}) & (a_{112},a_{222}) & (a_{113},a_{222})\\ (a_{111},a_{221}) & (a_{112},a_{221}) & (a_{113},a_{221})\end{matrix}\right\}=\left\{\begin{matrix}\underbrace{(\text{SI,SO})}_{\text{saddle}} & \underbrace{(\text{SO,SO})}_{\text{source}} & \underbrace{(\text{SI,SO})}_{\text{saddle}}\\ \underbrace{(\text{SI,SI})}_{\text{sink}} & \underbrace{(\text{SO,SI})}_{\text{saddle}} & \underbrace{(\text{SI,SI})}_{\text{sink}}\\ \underbrace{(\text{SI,SO})}_{\text{saddle}} & \underbrace{(\text{SO,SO})}_{\text{source}} & \underbrace{(\text{SI,SO})}_{\text{saddle}}\end{matrix}\right\}$$

$$\text{for } a_{110}<0,\ a_{220}>0; \tag{2.45}$$

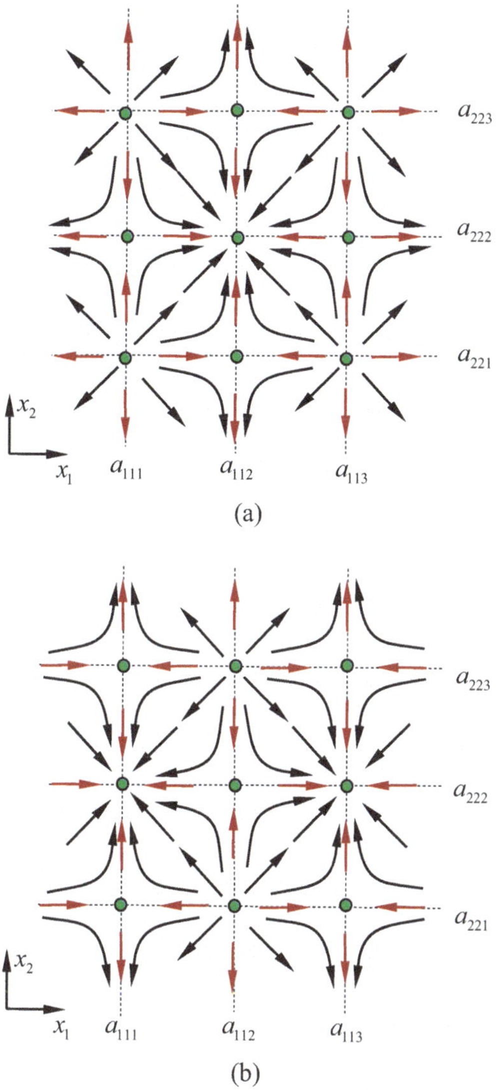

Fig. 2.4 Phase portraits for 2-dimensional systems on the x_1-direction with $x_1^* = a_{111}, a_{112}, a_{113}$ and on the x_2-direction with $x_2^* = a_{221}, a_{222}, a_{223}$. The four networks of sink, source and saddle: **a** ($a_{110} > 0, a_{220} > 0$), **b** ($a_{110} < 0, a_{220} > 0$), **c** ($a_{110} > 0, a_{220} < 0$), **d** ($a_{110} < 0, a_{220} < 0$)

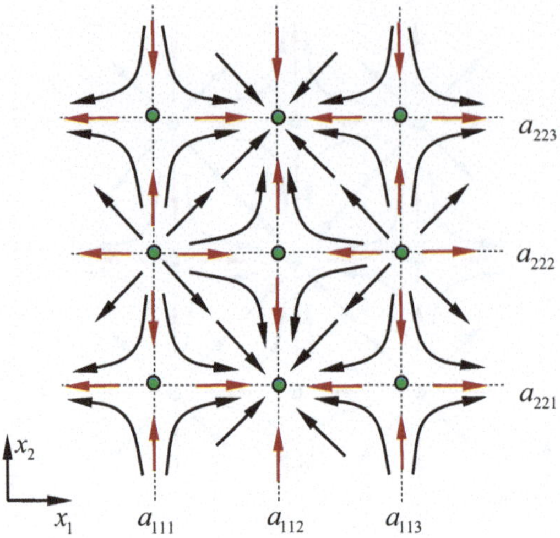

(c)

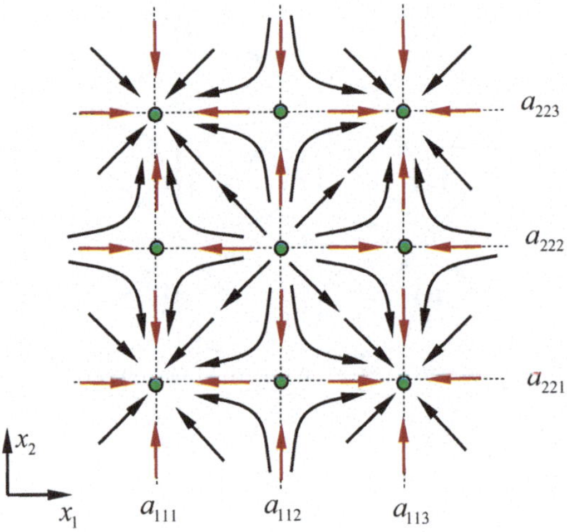

(d)

Fig. 2.4 (continued)

$$\left\{\begin{array}{ccc} (a_{111}, a_{223}) & (a_{112}, a_{223}) & (a_{113}, a_{223}) \\ (a_{111}, a_{222}) & (a_{112}, a_{222}) & (a_{113}, a_{222}) \\ (a_{111}, a_{221}) & (a_{112}, a_{221}) & (a_{113}, a_{221}) \end{array}\right\} = \left\{\begin{array}{ccc} \underbrace{(\mathrm{SO},\mathrm{SI})}_{\text{saddle}} & \underbrace{(\mathrm{SI},\mathrm{SI})}_{\text{sink}} & \underbrace{(\mathrm{SO},\mathrm{SI})}_{\text{saddle}} \\ \underbrace{(\mathrm{SO},\mathrm{SO})}_{\text{source}} & \underbrace{(\mathrm{SI},\mathrm{SO})}_{\text{saddle}} & \underbrace{(\mathrm{SO},\mathrm{SO})}_{\text{source}} \\ \underbrace{(\mathrm{SO},\mathrm{SI})}_{\text{saddle}} & \underbrace{(\mathrm{SI},\mathrm{SI})}_{\text{sink}} & \underbrace{(\mathrm{SO},\mathrm{SI})}_{\text{saddle}} \end{array}\right\}$$

for $a_{110} > 0, a_{220} < 0;$ (2.46)

Table 2.8 Numbers of saddles, sinks and sources in networks for 3 × 3-equilibriums

(a_{110}, a_{220})	(SO,SI)-saddle	(SI,SO)-saddle	(SO,SO)-source	(SI,SI)-sink
$(+,+)$	$n_1 \times (m_1+1) = 2$	$(n_1+1) \times m_1 = 2$	$(n_1+1) \times (m_1+1) = 4$	$n_1 \times m_1 = 1$
$(-,+)$	$n_1 \times m_1 = 1$	$(n_1+1) \times (m_1+1) = 4$	$(n_1+1) \times m_1 = 2$	$n_1 \times (m_1+1) = 2$
$(+,-)$	$(n_1+1) \times (m_1+1) = 4$	$n_1 \times m_1 = 1$	$n_1 \times (m_1+1) = 2$	$(n_1+1) \times m_1 = 2$
$(-,-)$	$(n_1+1) \times m_1 = 2$	$n_1 \times (m_1+1) = 2$	$n_1 \times m_1 = 1$	$(n_1+1) \times (m_1+1) = 4$

$$\left\{\begin{matrix} (a_{111}, a_{223}) & (a_{112}, a_{223}) & (a_{113}, a_{223}) \\ (a_{111}, a_{222}) & (a_{112}, a_{222}) & (a_{113}, a_{222}) \\ (a_{111}, a_{221}) & (a_{112}, a_{221}) & (a_{113}, a_{221}) \end{matrix}\right\} = \left\{\begin{matrix} \underbrace{\text{(SI,SI)}}_{\text{sink}} & \underbrace{\text{(SO,SI)}}_{\text{saddle}} & \underbrace{\text{(SI,SI)}}_{\text{sink}} \\ \underbrace{\text{(SI,SO)}}_{\text{saddle}} & \underbrace{\text{(SO,SO)}}_{\text{source}} & \underbrace{\text{(SI,SO)}}_{\text{saddle}} \\ \underbrace{\text{(SI,SI)}}_{\text{sink}} & \underbrace{\text{(SO,SI)}}_{\text{saddle}} & \underbrace{\text{(SI,SI)}}_{\text{sink}} \end{matrix}\right\}$$

for $a_{110} < 0, a_{220} < 0$. (2.47)

(E) Consider a 2-dimensional system with $(2m_1 + 1) \times (2n_1) = 3 \times 4$ equilibriums as

$$\begin{aligned} \dot{x}_1 &= a_{110}(x_1 - a_{111})(x_1 - a_{112})(x_1 - a_{113}), \\ \dot{x}_2 &= a_{220}(x_2 - a_{221})(x_2 - a_{222})(x_2 - a_{223})(x_2 - a_{224}). \end{aligned} \tag{2.48}$$

The first integral manifold is given by

$$\begin{aligned} &\sum_{l_1=1}^{4} \frac{1}{\prod_{l_2=1, l_2 \neq l_1}^{4} (a_{22l_1} - a_{22l_2})} \ln \frac{|x_2 - a_{22l_1}|}{|x_{20} - a_{22l_1}|} \\ &= \frac{a_{220}}{a_{110}} \sum_{s_1=1}^{3} \frac{1}{\prod_{s_2=1, s_2 \neq s_1}^{3} (a_{11s_1} - a_{11s_2})} \ln \frac{|x_1 - a_{11s_1}|}{|x_{10} - a_{11s_1}|}. \end{aligned} \tag{2.49}$$

Phase portraits for the 2-dimensional systems near the simple equilibriums of the saddles, sinks and sources presented in Fig. 2.5a–d for $(a_{110} > 0, a_{220} > 0)$, $(a_{110} < 0, a_{220} > 0)$, $(a_{110} > 0, a_{220} < 0)$ and $(a_{110} < 0, a_{220} < 0)$. For all cases, the twelve (12) simple equilibriums are based on the (3,4)-saddle-node bifurcations. The numbers of saddles, sinks and sources in the networks of 4 × 3-equilibriums are listed in Table 2.9.

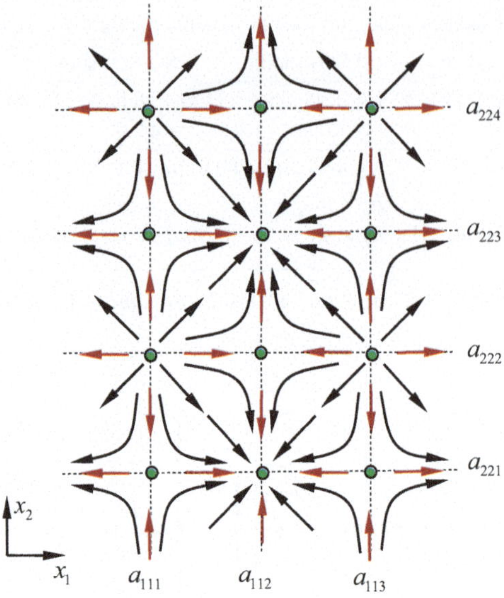

(a)

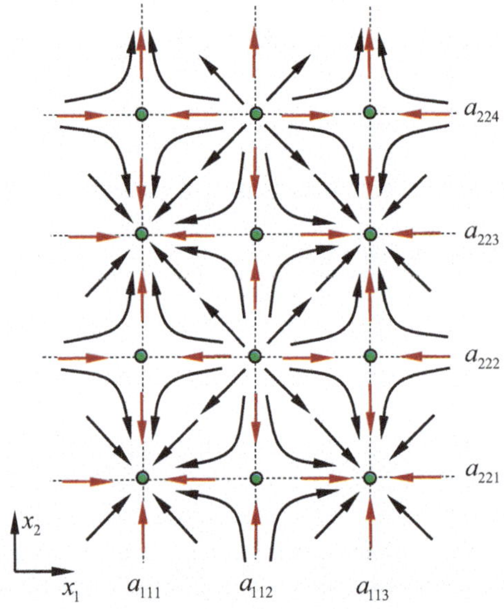

(b)

Fig. 2.5 Phase portraits for 2-dimensional systems on the x_1-direction with $x_1^* = a_{111}, a_{112}, a_{113}$ and on the x_2-direction with $x_2^* = a_{221}, a_{222}, a_{223}, a_{224}$. The four networks of sink, source and saddle: **a** ($a_{110} > 0$, $a_{220} > 0$), **b** ($a_{110} < 0$, $a_{220} > 0$), **c** ($a_{110} > 0$, $a_{220} < 0$), **d** ($a_{110} < 0$, $a_{220} < 0$)

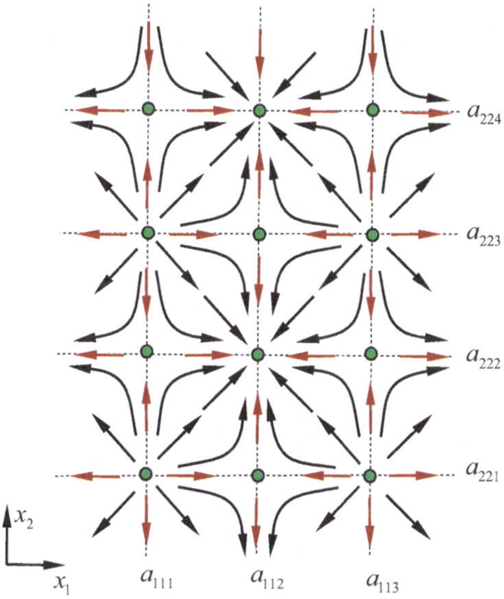

(c)

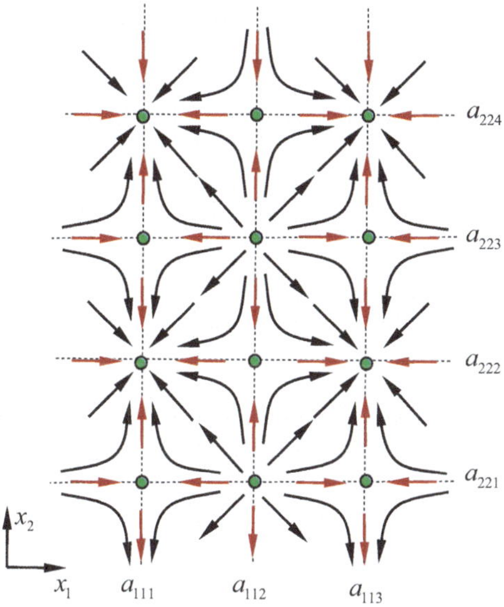

(d)

Fig. 2.5 (continued)

Table 2.9 Numbers of saddles, sinks and sources in networks for 3 × 4-equilibriums

(a_{110}, a_{220})	(SO,SI)-saddle	(SI,SO)-saddle	(SO,SO)-source	(SI,SI)-sink
$(+,+)$	$n_1 \times (m_1+1) = 4$	$n_1 \times m_1 = 2$	$n_1 \times (m_1+1) = 4$	$n_1 \times m_1 = 2$
$(-,+)$	$n_1 \times m_1 = 2$	$n_1 \times (m_1+1) = 4$	$n_1 \times m_1 = 2$	$n_1 \times (m_1+1) = 4$
$(+,-)$	$n_1 \times (m_1+1) = 4$	$n_1 \times m_1 = 2$	$n_1 \times (m_1+1) = 4$	$n_1 \times m_1 = 2$
$(-,-)$	$n_1 \times m_1 = 2$	$n_1 \times (m_1+1) = 4$	$n_1 \times m_1 = 2$	$n_1 \times (m_1+1) = 4$

$$\bigcup_{s=1}^{3}\bigcup_{l=1}^{4}\underbrace{(a_{11s}, a_{22l})}_{\text{XX}} = \begin{Bmatrix} \underbrace{\text{(SO,SO)}}_{\text{source}} & \underbrace{\text{(SI,SO)}}_{\text{saddle}} & \underbrace{\text{(SO,SO)}}_{\text{source}} \\ \underbrace{\text{(SO,SI)}}_{\text{saddle}} & \underbrace{\text{(SI,SI)}}_{\text{sink}} & \underbrace{\text{(SO,SI)}}_{\text{saddle}} \\ \underbrace{\text{(SO,SO)}}_{\text{source}} & \underbrace{\text{(SI,SO)}}_{\text{saddle}} & \underbrace{\text{(SO,SO)}}_{\text{source}} \\ \underbrace{\text{(SO,SI)}}_{\text{saddle}} & \underbrace{\text{(SI,SI)}}_{\text{sink}} & \underbrace{\text{(SO,SI)}}_{\text{saddle}} \end{Bmatrix}$$
$$\text{for } a_{110} > 0,\, a_{220} > 0; \tag{2.50}$$

$$\bigcup_{s=1}^{3}\bigcup_{l=1}^{4}\underbrace{(a_{11s}, a_{22l})}_{\text{XX}} = \begin{Bmatrix} \underbrace{\text{(SI,SO)}}_{\text{saddle}} & \underbrace{\text{(SO,SO)}}_{\text{source}} & \underbrace{\text{(SI,SO)}}_{\text{saddle}} \\ \underbrace{\text{(SI,SI)}}_{\text{sink}} & \underbrace{\text{(SO,SI)}}_{\text{saddle}} & \underbrace{\text{(SI,SI)}}_{\text{sink}} \\ \underbrace{\text{(SI,SO)}}_{\text{saddle}} & \underbrace{\text{(SO,SO)}}_{\text{source}} & \underbrace{\text{(SI,SO)}}_{\text{saddle}} \\ \underbrace{\text{(SI,SI)}}_{\text{sink}} & \underbrace{\text{(SO,SI)}}_{\text{saddle}} & \underbrace{\text{(SI,SI)}}_{\text{sink}} \end{Bmatrix}$$
$$\text{for } a_{110} < 0,\, a_{220} > 0; \tag{2.51}$$

$$\bigcup_{s=1}^{3}\bigcup_{l=1}^{4}\underbrace{(a_{11s}, a_{22l})}_{\text{XX}} = \begin{Bmatrix} \underbrace{\text{(SO,SI)}}_{\text{saddle}} & \underbrace{\text{(SI,SI)}}_{\text{sink}} & \underbrace{\text{(SO,SI)}}_{\text{saddle}} \\ \underbrace{\text{(SO,SO)}}_{\text{source}} & \underbrace{\text{(SI,SO)}}_{\text{saddle}} & \underbrace{\text{(SO,SO)}}_{\text{source}} \\ \underbrace{\text{(SO,SI)}}_{\text{saddle}} & \underbrace{\text{(SI,SI)}}_{\text{sink}} & \underbrace{\text{(SO,SI)}}_{\text{saddle}} \\ \underbrace{\text{(SO,SO)}}_{\text{source}} & \underbrace{\text{(SI,SO)}}_{\text{saddle}} & \underbrace{\text{(SO,SO)}}_{\text{source}} \end{Bmatrix}$$
$$\text{for } a_{110} > 0,\, a_{220} < 0; \tag{2.52}$$

$$\bigcup_{s=1}^{3}\bigcup_{l=1}^{4}\underbrace{(a_{11s},a_{22l})}_{\text{XX}}=\left\{\begin{matrix}\underbrace{(\text{SI,SI})}_{\text{sink}} & \underbrace{(\text{SO,SI})}_{\text{saddle}} & \underbrace{(\text{SI,SI})}_{\text{sink}}\\ \underbrace{(\text{SI,SO})}_{\text{saddle}} & \underbrace{(\text{SO,SO})}_{\text{source}} & \underbrace{(\text{SI,SO})}_{\text{saddle}}\\ \underbrace{(\text{SI,SI})}_{\text{sink}} & \underbrace{(\text{SO,SI})}_{\text{saddle}} & \underbrace{(\text{SI,SI})}_{\text{sink}}\\ \underbrace{(\text{SI,SO})}_{\text{saddle}} & \underbrace{(\text{SO,SO})}_{\text{source}} & \underbrace{(\text{SI,SO})}_{\text{saddle}}\end{matrix}\right\}$$

for $a_{110} < 0, a_{220} < 0.$ (2.53)

(F) Consider a 2-dimensional system with $(2m_1) \times (2n_1 + 1) = 4 \times 3$ equilibriums as

$$\begin{aligned}\dot{x}_1 &= a_{110}(x_1 - a_{111})(x_1 - a_{112})(x_1 - a_{113})(x_1 - a_{114}),\\ \dot{x}_2 &= a_{220}(x_2 - a_{221})(x_2 - a_{222})(x_2 - a_{223}).\end{aligned} \tag{2.54}$$

The first integral manifold is given by

$$\begin{aligned}&\sum_{l_1=1}^{3}\frac{1}{\prod_{l_2=1,l_2\neq l_1}^{3}(a_{22l_1}-a_{22l_2})}\ln\frac{|x_2-a_{22l_1}|}{|x_{20}-a_{22l_1}|}\\ &=\frac{a_{220}}{a_{110}}\sum_{s_1=1}^{4}\frac{1}{\prod_{s_2=1,s_2\neq s_1}^{4}(a_{11s_1}-a_{11s_2})}\ln\frac{|x_1-a_{11s_1}|}{|x_{10}-a_{11s_1}|}.\end{aligned} \tag{2.55}$$

Phase portraits for the 2-dimensional systems near the simple equilibriums of the saddles, sinks and sources presented in Fig. 2.6a–d for $(a_{110} > 0, a_{220} > 0)$, $(a_{110} < 0, a_{220} > 0)$, $(a_{110} > 0, a_{220} < 0)$ and $(a_{110} < 0, a_{220} < 0)$. For all cases, the twelve simple equilibriums are based on the (4,3)-saddle-node bifurcations. The numbers of saddles, sinks and sources in the networks of 4×3-equilibriums are listed in Table 2.10.

$$\bigcup_{s=1}^{4}\bigcup_{l=1}^{3}\underbrace{(a_{11s},a_{22l})}_{\text{XX}}=\left\{\begin{matrix}\underbrace{(\text{SI,SO})}_{\text{saddle}} & \underbrace{(\text{SO,SO})}_{\text{source}} & \underbrace{(\text{SI,SO})}_{\text{saddle}} & \underbrace{(\text{SO,SO})}_{\text{source}}\\ \underbrace{(\text{SI,SI})}_{\text{sink}} & \underbrace{(\text{SO,SI})}_{\text{saddle}} & \underbrace{(\text{SI,SI})}_{\text{sink}} & \underbrace{(\text{SO,SI})}_{\text{saddle}}\\ \underbrace{(\text{SI,SO})}_{\text{saddle}} & \underbrace{(\text{SO,SO})}_{\text{source}} & \underbrace{(\text{SI,SO})}_{\text{saddle}} & \underbrace{(\text{SO,SO})}_{\text{source}}\end{matrix}\right\}$$

for $a_{110} > 0, a_{220} > 0;$ (2.56)

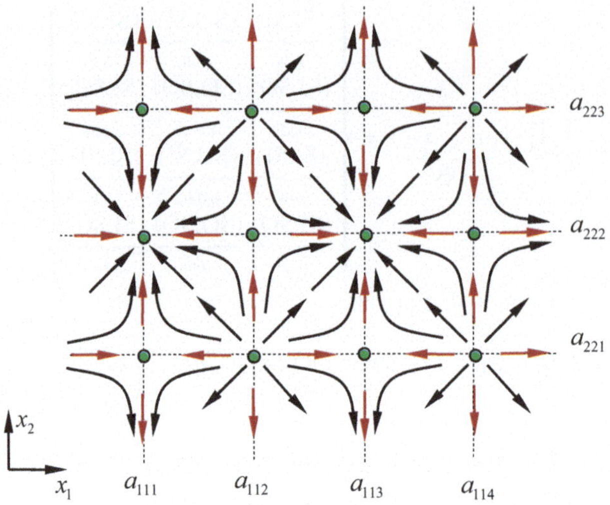

(a)

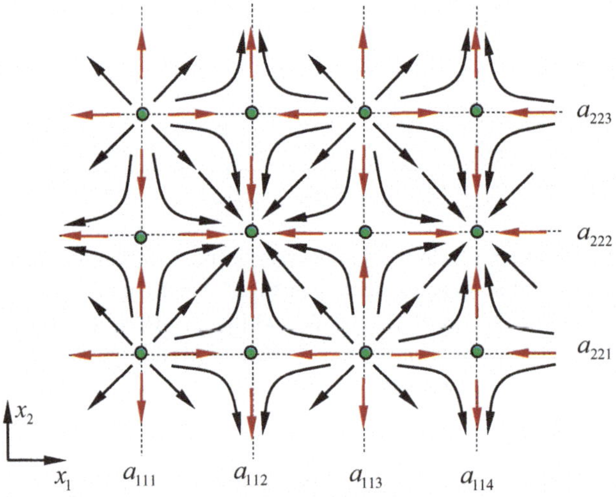

(b)

Fig. 2.6 Phase portraits for 2-dimensional systems on the x_1-direction with $x_1^* = a_{111}, a_{112}, a_{113}, a_{114}$ and on the x_2-direction with $x_2^* = a_{221}, a_{222}, a_{223}$. The four networks of sink, source and saddle: **a** ($a_{110} > 0, a_{220} > 0$), **b** ($a_{110} < 0, a_{220} > 0$), **c** ($a_{110} > 0, a_{220} < 0$), **d** ($a_{110} < 0, a_{220} < 0$)

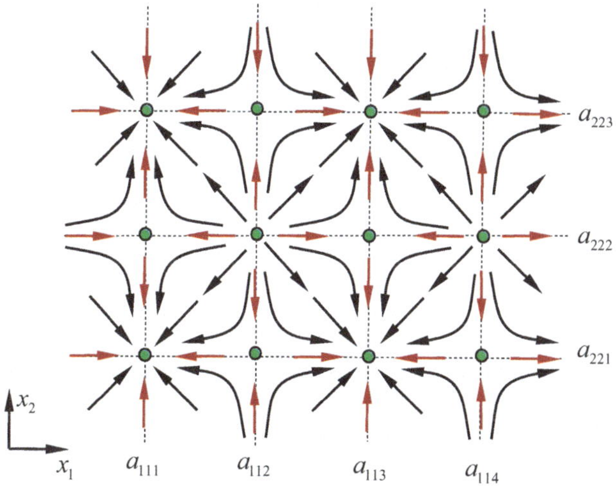

(c)

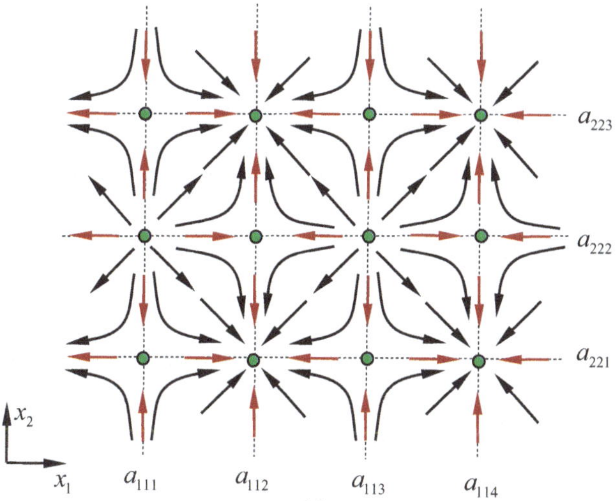

(d)

Fig. 2.6 (continued)

Table 2.10 Numbers of saddles, sinks and sources in networks for 4×3-equilibriums

(a_{110}, a_{220})	(SO,SI)-saddle	(SI,SO)-saddle	(SO,SO)-source	(SI,SI)-sink
$(+, +)$	$n_1 \times m_1 = 2$	$(n_1 + 1) \times m_1 = 4$	$(n_1 + 1) \times m_1 = 4$	$n_1 \times m_1 = 2$
$(-, +)$	$n_1 \times m_1 = 2$	$(n_1 + 1) \times m_1 = 4$	$(n_1 + 1) \times m_1 = 4$	$n_1 \times m_1 = 2$
$(+, -)$	$(n_1 + 1) \times m_1 = 4$	$n_1 \times m_1 = 2$	$n_1 \times m_1 = 2$	$(n_1 + 1) \times m_1 = 4$
$(-, -)$	$(n_1 + 1) \times m_1 = 4$	$n_1 \times m_1 = 2$	$n_1 \times m_1 = 2$	$(n_1 + 1) \times m_1 = 4$

$$\bigcup_{s=1}^{4}\bigcup_{l=1}^{3}\underbrace{(a_{11s},a_{22l})}_{\text{XX}}=\left\{\begin{array}{cccc}\underbrace{(\text{SO,SO})}_{\text{source}} & \underbrace{(\text{SI,SO})}_{\text{saddle}} & \underbrace{(\text{SO,SO})}_{\text{source}} & \underbrace{(\text{SI,SO})}_{\text{saddle}}\\ \underbrace{(\text{SO,SI})}_{\text{saddle}} & \underbrace{(\text{SI,SI})}_{\text{sink}} & \underbrace{(\text{SO,SI})}_{\text{saddle}} & \underbrace{(\text{SI,SI})}_{\text{sink}}\\ \underbrace{(\text{SO,SO})}_{\text{source}} & \underbrace{(\text{SI,SO})}_{\text{saddle}} & \underbrace{(\text{SO,SO})}_{\text{source}} & \underbrace{(\text{SI,SO})}_{\text{saddle}}\end{array}\right\}$$

for $a_{110} < 0, a_{220} > 0$; (2.57)

$$\bigcup_{s=1}^{4}\bigcup_{l=1}^{3}\underbrace{(a_{11s},a_{22l})}_{\text{XX}}=\left\{\begin{array}{cccc}\underbrace{(\text{SI,SI})}_{\text{sink}} & \underbrace{(\text{SO,SI})}_{\text{saddle}} & \underbrace{(\text{SI,SI})}_{\text{sink}} & \underbrace{(\text{SO,SI})}_{\text{saddle}}\\ \underbrace{(\text{SI,SO})}_{\text{saddle}} & \underbrace{(\text{SO,SO})}_{\text{source}} & \underbrace{(\text{SI,SO})}_{\text{saddle}} & \underbrace{(\text{SO,SO})}_{\text{source}}\\ \underbrace{(\text{SI,SI})}_{\text{sink}} & \underbrace{(\text{SO,SI})}_{\text{saddle}} & \underbrace{(\text{SI,SI})}_{\text{sink}} & \underbrace{(\text{SO,SI})}_{\text{saddle}}\end{array}\right\}$$

for $a_{110} > 0, a_{220} < 0$; (2.58)

$$\bigcup_{s=1}^{4}\bigcup_{l=1}^{3}\underbrace{(a_{11s},a_{22l})}_{\text{XX}}=\left\{\begin{array}{cccc}\underbrace{(\text{SO,SI})}_{\text{saddle}} & \underbrace{(\text{SI,SI})}_{\text{sink}} & \underbrace{(\text{SO,SI})}_{\text{saddle}} & \underbrace{(\text{SI,SI})}_{\text{sink}}\\ \underbrace{(\text{SO,SO})}_{\text{source}} & \underbrace{(\text{SI,SO})}_{\text{saddle}} & \underbrace{(\text{SO,SO})}_{\text{source}} & \underbrace{(\text{SI,SO})}_{\text{saddle}}\\ \underbrace{(\text{SO,SI})}_{\text{saddle}} & \underbrace{(\text{SI,SI})}_{\text{sink}} & \underbrace{(\text{SO,SI})}_{\text{saddle}} & \underbrace{(\text{SI,SI})}_{\text{sink}}\end{array}\right\}$$

for $a_{110} < 0, a_{220} < 0$. (2.59)

(G) Consider a 2-dimensional system with $(2m_1) \times (2n_1) = 4 \times 4$ equilibriums as

$$\begin{aligned}\dot{x}_1 &= a_{110}(x_1 - a_{111})(x_1 - a_{112})(x_1 - a_{113})(x_1 - a_{114}),\\ \dot{x}_2 &= a_{220}(x_2 - a_{221})(x_2 - a_{222})(x_2 - a_{223})(x_2 - a_{224}).\end{aligned} \tag{2.60}$$

The first integral manifold is given by

$$\begin{aligned}&\sum_{l_1=1}^{4}\frac{1}{\prod\limits_{l_2=1,l_2\neq l_1}^{4}(a_{22l_1}-a_{22l_2})}\ln\frac{|x_2-a_{22l_1}|}{|x_{20}-a_{22l_1}|}\\ &=\frac{a_{220}}{a_{110}}\sum_{s_1=1}^{4}\frac{1}{\prod_{s_2=1,s_2\neq s_1}^{4}(a_{11s_1}-a_{11s_2})}\ln\frac{|x_1-a_{11s_1}|}{|x_{10}-a_{11s_1}|}.\end{aligned} \tag{2.61}$$

Phase portraits for the 2-dimensional systems near the simple equilibriums of the saddles, sinks and sources presented in Fig. 2.7a–d for $(a_{110} > 0, a_{220} > 0)$, $(a_{110} < 0, a_{220} > 0)$, $(a_{110} > 0, a_{220} < 0)$ and $(a_{110} < 0, a_{220} < 0)$. For all cases, the sixteen simple equilibriums are based on the fourth-order double-saddle bifurcations. The numbers of saddles, sinks, and sources in the networks of 4×4-equilibriums are listed in Table 2.11.

$$\bigcup_{s=1}^{4}\bigcup_{l=1}^{4}\underbrace{(a_{11s}, a_{22l})}_{\text{XX}} = \left\{\begin{matrix} \underbrace{(\text{SI,SO})}_{\text{saddle}} & \underbrace{(\text{SO,SO})}_{\text{source}} & \underbrace{(\text{SI,SO})}_{\text{saddle}} & \underbrace{(\text{SO,SO})}_{\text{source}} \\ \underbrace{(\text{SI,SI})}_{\text{sink}} & \underbrace{(\text{SO,SI})}_{\text{saddle}} & \underbrace{(\text{SI,SI})}_{\text{sink}} & \underbrace{(\text{SO,SI})}_{\text{saddle}} \\ \underbrace{(\text{SI,SO})}_{\text{saddle}} & \underbrace{(\text{SO,SO})}_{\text{source}} & \underbrace{(\text{SI,SO})}_{\text{saddle}} & \underbrace{(\text{SO,SO})}_{\text{source}} \\ \underbrace{(\text{SI,SI})}_{\text{sink}} & \underbrace{(\text{SO,SI})}_{\text{saddle}} & \underbrace{(\text{SI,SI})}_{\text{sink}} & \underbrace{(\text{SO,SI})}_{\text{saddle}} \end{matrix}\right\}$$

$$\text{for } a_{110} > 0,\ a_{220} > 0; \tag{2.62}$$

$$\bigcup_{s=1}^{4}\bigcup_{l=1}^{4}\underbrace{(a_{11s}, a_{22l})}_{\text{XX}} = \left\{\begin{matrix} \underbrace{(\text{SO,SO})}_{\text{source}} & \underbrace{(\text{SI,SO})}_{\text{saddle}} & \underbrace{(\text{SO,SO})}_{\text{source}} & \underbrace{(\text{SI,SO})}_{\text{saddle}} \\ \underbrace{(\text{SO,SI})}_{\text{saddle}} & \underbrace{(\text{SI,SI})}_{\text{sink}} & \underbrace{(\text{SO,SI})}_{\text{saddle}} & \underbrace{(\text{SI,SI})}_{\text{sink}} \\ \underbrace{(\text{SO,SO})}_{\text{source}} & \underbrace{(\text{SI,SO})}_{\text{saddle}} & \underbrace{(\text{SO,SO})}_{\text{source}} & \underbrace{(\text{SI,SO})}_{\text{saddle}} \\ \underbrace{(\text{SO,SI})}_{\text{saddle}} & \underbrace{(\text{SI,SI})}_{\text{sink}} & \underbrace{(\text{SO,SI})}_{\text{saddle}} & \underbrace{(\text{SI,SI})}_{\text{sink}} \end{matrix}\right\}$$

$$\text{for } a_{110} < 0,\ a_{220} > 0; \tag{2.63}$$

$$\bigcup_{s=1}^{4}\bigcup_{l=1}^{4}\underbrace{(a_{11s}, a_{22l})}_{\text{XX}} == \left\{\begin{matrix} \underbrace{(\text{SI,SI})}_{\text{sink}} & \underbrace{(\text{SO,SI})}_{\text{saddle}} & \underbrace{(\text{SI,SI})}_{\text{sink}} & \underbrace{(\text{SO,SI})}_{\text{saddle}} \\ \underbrace{(\text{SI,SO})}_{\text{saddle}} & \underbrace{(\text{SO,SO})}_{\text{source}} & \underbrace{(\text{SI,SO})}_{\text{saddle}} & \underbrace{(\text{SO,SO})}_{\text{source}} \\ \underbrace{(\text{SI,SI})}_{\text{sink}} & \underbrace{(\text{SO,SI})}_{\text{saddle}} & \underbrace{(\text{SI,SI})}_{\text{sink}} & \underbrace{(\text{SO,SI})}_{\text{saddle}} \\ \underbrace{(\text{SI,SO})}_{\text{saddle}} & \underbrace{(\text{SO,SO})}_{\text{source}} & \underbrace{(\text{SI,SO})}_{\text{saddle}} & \underbrace{(\text{SO,SO})}_{\text{source}} \end{matrix}\right\}$$

$$\text{for } a_{110} > 0,\ a_{220} < 0; \tag{2.64}$$

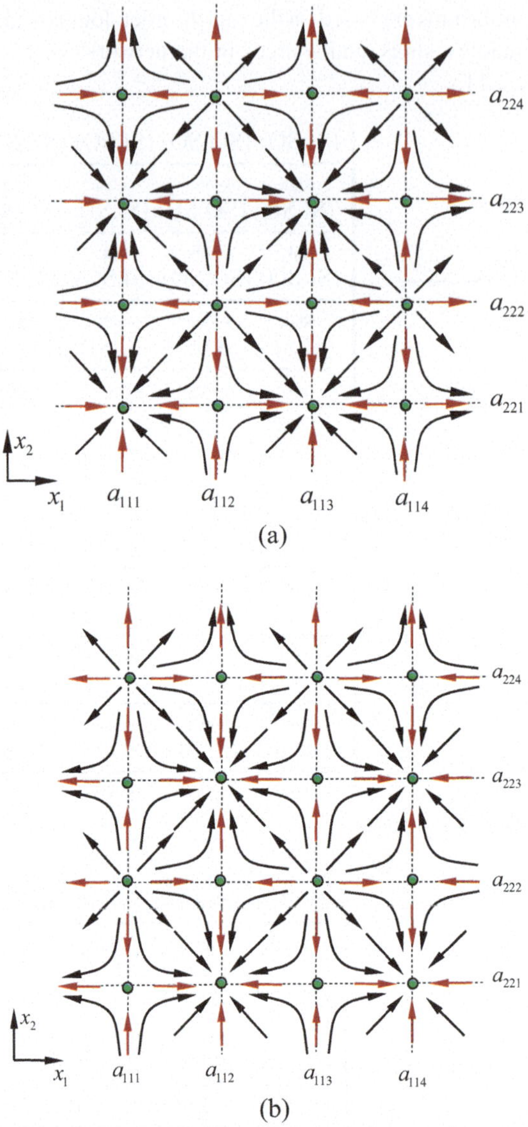

Fig. 2.7 Phase portraits for 2-dimensional systems on the x_1-direction with $x_1^* = a_{111}, a_{112}, a_{113}, a_{114}$ and on the x_2-direction with $x_2^* = a_{221}, a_{222}, a_{223}, a_{224}$. The four networks of sink, source and saddle: **a** ($a_{110} > 0, a_{220} > 0$), **b** ($a_{110} < 0, a_{220} > 0$), **c** ($a_{110} > 0, a_{220} < 0$), **d** ($a_{110} < 0, a_{220} < 0$)

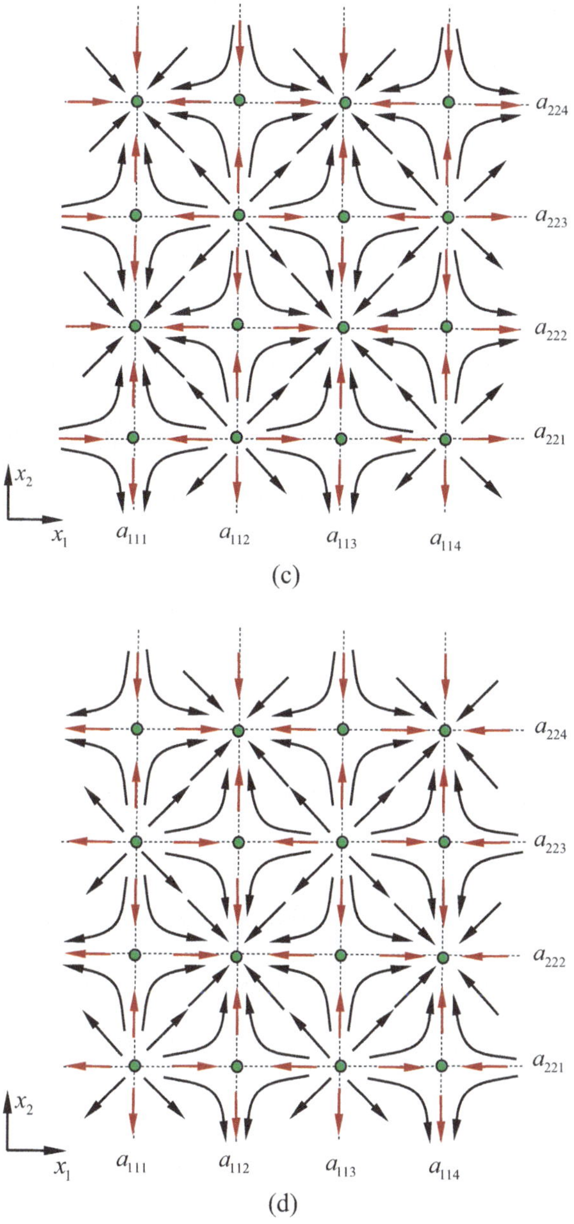

Fig. 2.7 (continued)

Table 2.11 Numbers of saddles, sinks and sources in networks for 4 × 4-equilibriums

(a_{110}, a_{220})	(SO,SI)-saddle	(SI,SO)-saddle	(SO,SO)-source	(SI,SI)-sink
$(+, +)$	$n_1 \times m_1 = 4$	$n_1 \times m_1 = 4$	$n_1 \times m_1 = 4$	$n_1 \times m_1 = 4$
$(-, +)$	$n_1 \times m_1 = 4$	$n_1 \times m_1 = 4$	$n_1 \times m_1 = 4$	$n_1 \times m_1 = 4$
$(+, -)$	$n_1 \times m_1 = 4$	$n_1 \times m_1 = 4$	$n_1 \times m_1 = 4$	$n_1 \times m_1 = 4$
$(-, -)$	$n_1 \times m_1 = 4$	$n_1 \times m_1 = 4$	$n_1 \times m_1 = 4$	$n_1 \times m_1 = 4$

$$\bigcup_{s=1}^{4}\bigcup_{l=1}^{4}\underbrace{(a_{11s}, a_{22l})}_{\text{XX}} = \left\{\begin{array}{cccc} \underbrace{\text{(SO,SI)}}_{\text{saddle}} & \underbrace{\text{(SI,SI)}}_{\text{sink}} & \underbrace{\text{(SO,SI)}}_{\text{saddle}} & \underbrace{\text{(SI,SI)}}_{\text{sink}} \\ \underbrace{\text{(SO,SO)}}_{\text{source}} & \underbrace{\text{(SI,SO)}}_{\text{saddle}} & \underbrace{\text{(SO,SO)}}_{\text{source}} & \underbrace{\text{(SI,SO)}}_{\text{saddle}} \\ \underbrace{\text{(SO,SI)}}_{\text{saddle}} & \underbrace{\text{(SI,SI)}}_{\text{sink}} & \underbrace{\text{(SO,SI)}}_{\text{saddle}} & \underbrace{\text{(SI,SI)}}_{\text{sink}} \\ \underbrace{\text{(SO,SO)}}_{\text{source}} & \underbrace{\text{(SI,SO)}}_{\text{saddle}} & \underbrace{\text{(SO,SO)}}_{\text{source}} & \underbrace{\text{(SI,SO)}}_{\text{saddle}} \end{array}\right\}$$

$$\text{for } a_{110} < 0,\ a_{220} < 0. \tag{2.65}$$

Reference

1. Luo, A. C. J. (2024). Limit cycles and homoclinic networks in 2-dimensional Polynomial system. *AIP Chaos*, 34, 022104 (66 pages).

Chapter 3
Bifurcations for Homoclinic Networks Without Centers

In this chapter, the appearing and switching bifurcations are discussed for the homoclinic networks of non-singular and singular sources, sinks, saddles with singular saddle-sources, saddle-sinks, and double-saddles in self-univariate polynomial systems. The first integral manifolds for non-singular and singular equilibrium networks are determined. The illustrations of singular equilibriums to networks of non-singular sources, sinks and saddles are given.

3.1 Higher-Order Singularity and Bifurcations

To discuss singular equilibriums and bifurcations in polynomial systems, as in Luo [1], consider a polynomial system with self-univariate polynomial vector fields, and the corresponding dynamical behaviors will be presented through the following theorem.

Theorem 3.1

(i) *Consider a self-univariate polynomial dynamical system as*

$$\begin{aligned}&\dot{x}_{j_1} = a_{j_1j_10}(x_{j_1} - a_{j_1j_11})^m,\ \dot{x}_{j_2} = a_{j_2j_20}(x_{j_2} - a_{j_2j_21})^n,\\&j_1, j_2 \in \{1, 2\}; j_1 \neq j_2.\end{aligned} \tag{3.1}$$

The first integral manifold is

$$\begin{aligned}&\frac{1}{n-1}\Big[\frac{1}{(x_{j_2} - a_{j_2j_21})^{n-1}} - \frac{1}{(x_{j_20} - a_{j_2j_21})^{n-1}}\Big]\\&= \frac{a_{j_2j_20}}{a_{j_1j_10}}\frac{1}{m-1}\Big[\frac{1}{(x_{j_1} - a_{j_1j_11})^{m-1}} - \frac{1}{(x_{j_20} - a_{j_1j_11})^{m-1}}\Big],\end{aligned}$$

A. C. J. Luo, *Limit Cycles and Homoclinic Networks in Two-Dimensional Polynomial Systems*, https://doi.org/10.1007/978-981-97-2617-2_3

for $m, n \neq 1$;

$$-\ln \frac{|x_{j_2} - a_{j_2 j_2 1}|}{|x_{j_2 0} - a_{j_2 j_2 1}|} = \frac{a_{j_2 j_2 0}}{a_{j_1 j_1 0}} \frac{1}{m-1} \left[\frac{1}{(x_{j_1} - a_{j_1 j_1 1})^{m-1}} - \frac{1}{(x_{j_2 0} - a_{j_1 j_1 1})^{m-1}} \right],$$

for $n = 1$ and $m \neq 1$;

$$\frac{1}{n-1} \left[\frac{1}{(x_{j_2} - a_{j_2 j_2 1})^{n-1}} - \frac{1}{(x_{j_2 0} - a_{j_2 j_2 1})^{n-1}} \right] = -\frac{a_{j_2 j_2 0}}{a_{j_1 j_1 0}} \ln \frac{|x_{j_1} - a_{j_1 j_1 1}|}{|x_{j_2 0} - a_{j_1 j_1 1}|},$$

for $m = 1$ and $n \neq 1$;

$$\ln \frac{|x_{j_2} - a_{j_2 j_2 1}|}{|x_{j_2 0} - a_{j_2 j_2 1}|} = \frac{a_{j_2 j_2 0}}{a_{j_1 j_1 0}} \ln \frac{|x_{j_1} - a_{j_1 j_1 1}|}{|x_{j_2 0} - a_{j_1 j_1 1}|},$$

for $m, n = 1$. (3.2)

(i_1) For $m = 2m_1 + 1$ and $n = 2n_1 + 1$, the equilibrium of $(x_{j_1}^*, x_{j_2}^*) = (a_{j_1 j_1 1}, a_{j_2 j_2 1})$ has the following properties.

- For $a_{j_1 j_1 0} > 0$ and $a_{j_2 j_2 0} > 0$,

$$(a_{j_1 j_1 1}, a_{j_2 j_2 1}) = \underbrace{((2m_1{+}1)^{\text{th}}\text{SO},(2n_1{+}1)^{\text{th}}\text{SO})}_{((2m_1+1),(2n_1+1))\text{-source}} \tag{3.3}$$

- For $a_{j_1 j_1 0} < 0$ and $a_{j_2 j_2 0} > 0$,

$$(a_{j_1 j_1 1}, a_{j_2 j_2 1}) = \underbrace{((2m_1{+}1)^{\text{th}}\text{SI},(2n_1{+}1)^{\text{th}}\text{SO})}_{((2m_1+1),(2n_1+1))\text{-saddle}} \tag{3.4}$$

- For $a_{j_1 j_1 0} > 0$ and $a_{j_2 j_2 0} < 0$,

$$(a_{j_1 j_1 1}, a_{j_2 j_2 1}) = \underbrace{((2m_1{+}1)^{\text{th}}\text{SO},(2n_1{+}1)^{\text{th}}\text{SI})}_{((2m_1+1),(2n_1+1))\text{-saddle}} \tag{3.5}$$

- For $a_{j_1 j_1 0} < 0$ and $a_{j_2 j_2 0} < 0$,

$$(a_{j_1 j_1 1}, a_{j_2 j_2 1}) = \underbrace{((2m_1{+}1)^{\text{th}}\text{SI},(2n_1{+}1)^{\text{th}}\text{SI})}_{((2m_1+1),(2n_1+1))\text{-sink}} \tag{3.6}$$

(i_2) For $m = 2m_1$ and $n = 2n_1 + 1$, the equilibrium of $(x_{j_1}^*, x_{j_2}^*) = (a_{j_1 j_1 1}, a_{j_2 j_2 1})$ has the following properties.

- For $a_{j_1 j_1 0} > 0$ and $a_{j_2 j_2 0} > 0$,

$$(a_{j_1 j_1 1}, a_{j_2 j_2 1}) = \underbrace{((2m_1)^{\text{th}}\text{US},(2n_1{+}1)^{\text{th}}\text{SO})}_{((2m_1),(2n_1+1))\text{-upper-saddle source}} \tag{3.7}$$

- For $a_{j_1 j_1 0} < 0$ and $a_{j_2 j_2 0} > 0$,

$$(a_{j_1 j_1 1}, a_{j_2 j_2 1}) = \underbrace{((2m_1)^{\text{th}}\text{LS},(2n_1+1)^{\text{th}}\text{SO})}_{((2m_1),(2n_1+1))\text{-lower-saddle source}} \tag{3.8}$$

- For $a_{j_1 j_1 0} > 0$ and $a_{j_2 j_2 0} < 0$,

$$(a_{j_1 j_1 1}, a_{j_2 j_2 1}) = \underbrace{((2m_1)^{\text{th}}\text{US},(2n_1+1)^{\text{th}}\text{SI})}_{((2m_1),(2n_1+1))\text{-upper-saddle sink}} \tag{3.9}$$

- For $a_{j_1 j_1 0} < 0$ and $a_{j_2 j_2 0} < 0$,

$$(a_{j_1 j_1 1}, a_{j_2 j_2 1}) = \underbrace{((2m_1)^{\text{th}}\text{LS},(2n_1+1)^{\text{th}}\text{SI})}_{((2m_1),(2n_1+1))\text{-lower-saddle sink}} \tag{3.10}$$

(i_3) For $m = 2m_1 + 1$ and $n = 2n_1$, the equilibrium of $(x^*_{j_1}, x^*_{j_2}) = (a_{j_1 j_1 1}, a_{j_2 j_2 1})$ has the following properties.

- For $a_{j_1 j_1 0} > 0$ and $a_{j_2 j_2 0} > 0$,

$$(a_{j_1 j_1 1}, a_{j_2 j_2 1}) = \underbrace{((2m_1+1)^{\text{th}}\text{SO},(2n_1)^{\text{th}}\text{US})}_{((2m_1+1),(2n_1))\text{-upper-saddle source}}. \tag{3.11}$$

- For $a_{j_1 j_1 0} < 0$ and $a_{j_2 j_2 0} > 0$,

$$(a_{j_1 j_1 1}, a_{j_2 j_2 1}) = \underbrace{((2m_1+1)^{\text{th}}\text{SI},(2n_1)^{\text{th}}\text{US})}_{((2m_1+1),(2n_1))\text{-upper-saddle sink}}. \tag{3.12}$$

- For $a_{j_1 j_1 0} > 0$ and $a_{j_2 j_2 0} < 0$,

$$(a_{j_1 j_1 1}, a_{j_2 j_2 1}) = \underbrace{((2m_1+1)^{\text{th}}\text{SO},(2n_1)^{\text{th}}\text{LS})}_{((2m_1+1),(2n_1))\text{-lower-saddle source}}. \tag{3.13}$$

- For $a_{j_1 j_1 0} < 0$ and $a_{j_2 j_2 0} < 0$,

$$(a_{j_1 j_1 1}, a_{j_2 j_2 1}) = \underbrace{((2m_1+1)^{\text{th}}\text{SI},(2n_1)^{\text{th}}\text{LS})}_{((2m_1+1),(2n_1))\text{-lower-saddle sink}}. \tag{3.14}$$

(i_4) For $m = 2m_1$ and $n = 2n_1$, the equilibrium of $(x^*_{j_1}, x^*_{j_2}) = (a_{j_1 j_1 1}, a_{j_2 j_2 1})$ has the following properties.

- For $a_{j_1 j_1 0} > 0$ and $a_{j_2 j_2 0} > 0$,

$$(a_{j_1j_11}, a_{j_2j_21}) = \underbrace{((2m_1)^{\text{th}}\text{US},(2n_1)^{\text{th}}\text{US})}_{((2m_1),(2n_1))\text{-double saddle}} \tag{3.15}$$

- For $a_{j_1j_10} < 0$ and $a_{j_2j_20} > 0$,

$$(a_{j_1j_11}, a_{j_2j_21}) = \underbrace{((2m_1)^{\text{th}}\text{LS},(2n_1)^{\text{th}}\text{US})}_{((2m_1),(2n_1))\text{-double saddle}} \tag{3.16}$$

- For $a_{j_1j_10} > 0$ and $a_{j_2j_20} < 0$,

$$(a_{j_1j_11}, a_{j_2j_21}) = \underbrace{((2m_1)^{\text{th}}\text{US},(2\text{n}_1)^{\text{th}}\text{LS})}_{((2m_1),(2n_1))\text{-upper-saddle sink}} \tag{3.17}$$

- For $a_{j_1j_10} < 0$ and $a_{j_2j_20} < 0$,

$$(a_{j_1j_11}, a_{j_2j_21}) = \underbrace{((2m_1)^{\text{th}}\text{LS},(2n_1)^{\text{th}}\text{LS})}_{((2m_1),(2n_1))\text{-double-saddle}} \tag{3.18}$$

(ii) *Consider a self-univariate polynomial dynamical system as*

$$\begin{aligned}
\dot{x}_{j_1} &= a_{j_1j_10} \prod_{r=1,\,\Sigma_{r=1}^{p} m_r = m}^{p} (x_{j_1} - a_{j_1j_1r})^{m_r},\\
\dot{x}_{j_2} &= a_{j_2j_20} \prod_{s=1,\,\Sigma_{s=1}^{q} n_s = n}^{q} (x_{j_2} - a_{j_2j_2s})^{n_s},\\
& j_1, j_2 \in \{1, 2\};\, j_1 \neq j_2.
\end{aligned} \tag{3.19}$$

The first integral manifold is

$$\begin{aligned}
&\sum_{s_1=1}^{q}\sum_{l_2=1}^{n_{s_1}} \frac{B_{l_2n_{s_1}}}{l_2-1}\Big[\frac{1}{(x_{j_2}-a_{j_2j_2s_1})^{l_2-1}} - \frac{1}{(x_{j_20}-a_{j_2j_2s_1})^{l_2-1}}\Big]\\
&= \frac{a_{j_2j_20}}{a_{j_1j_10}}\sum_{r_1=1}^{p}\sum_{l_1=1}^{m_{r_1}} \frac{A_{l_1m_{r_1}}}{l_1-1}\Big[\frac{1}{(x_{j_1}-a_{j_1j_1r_1})^{l_1-1}} - \frac{1}{(x_{j_10}-a_{j_1j_1r_1})^{l_1-1}}\Big],\\
&\text{for } l_1, l_2 \neq 1;\\
&\lim_{l_2\to 1}\frac{1}{(l_2-1)}\Big[\frac{1}{(x_{j_2}-a_{j_2j_2s_1})^{l_2-1}} - \frac{1}{(x_{j_20}-a_{j_2j_2s_1})^{l_2-1}}\Big] = -\ln\frac{|x_{j_2}-a_{j_2j_2s_1}|}{|x_{j_20}-a_{j_2j_2s_1}|},\\
&\text{for } l_2 = 1;\\
&\lim_{l_1\to 1}\frac{1}{l_1-1}\Big[\frac{1}{(x_{j_1}-a_{j_1j_1r_1})^{l_1-1}} - \frac{1}{(x_{j_10}-a_{j_1j_1r_1})^{l_1-1}}\Big] = -\ln\frac{|x_{j_1}-a_{j_1j_1r_1}|}{|x_{j_10}-a_{j_1j_1r_1}|},
\end{aligned}$$

for $l_1 = 1$. (3.20)

where

$$\begin{aligned}
A_{l_1 m_{r_1}} &= \frac{1}{(m_{r_1}-l_1)!}\frac{d^{m_{r_1}-l_1}}{dx_{j_1}^{m_{r_1}-l_1}}\frac{1}{\prod_{r_2=1,r_2\neq r_1}^{p}(x_{j_1}-a_{j_1j_1r_2})^{m_{r_2}}}\Big|_{x_{j_1}^*=a_{j_1j_1r_1}},\\
A_{m_{r_1} m_{r_1}} &= \frac{1}{\prod_{r_2=1,r_2\neq r,}^{p}(a_{j_1j_1r_1}-a_{j_1j_1r_2})^{m_{r_2}}};\\
B_{l_2 n_{s_1}} &= \frac{1}{(n_{s_1}-l_2)!}\frac{d^{n_{s_1}-l_2}}{dx_{j_2}^{n_{s_1}-l_2}}\frac{1}{\prod_{s_2=1,s_2\neq s_1}^{q}(x_{j_2}-a_{j_2j_2s_2})^{n_{s_2}}}\Big|_{x_{j_2}^*=a_{j_2j_2s_1}},\\
B_{n_{s_1} n_{s_1}} &= \frac{1}{\prod_{s_2=1,s_2\neq s_1}^{q}(a_{j_2j_2s_1}-a_{j_2j_2s_2})^{n_{s_2}}}.
\end{aligned}\tag{3.21}$$

The singular equilibrium network with $\Sigma_{r_i=1}^{p} m_{r_i} = m$ and $\Sigma_{r_i=1}^{q} n_{s_i} = n$ is defined as

$$\begin{aligned}
&\bigcup_{r=1}^{p}\bigcup_{s=1}^{q}\underbrace{(a_{j_1j_1r},a_{j_2j_2s})}_{(m_r,n_s)\text{-XX}} \equiv \begin{Bmatrix} (a_{j_1j_11},a_{j_2j_2q}) & \cdots & (a_{j_1j_1p},a_{j_2j_2q})\\ \vdots & \cdots & \vdots\\ (a_{j_1j_11},a_{j_2j_21}) & \cdots & (a_{j_1j_1p},a_{j_2j_21})\end{Bmatrix}_{q\times p}\\
&= \begin{bmatrix} \underbrace{((m_1)^{\text{th}}\text{XX},(n_q)^{\text{th}}\text{XX})}_{(m_1,n_q)\text{-XX}} & \cdots & \underbrace{((m_p)^{\text{th}}\text{XX},(n_q)^{\text{th}}\text{XX})}_{(m_{p_i}^{(i)},n_{q_i}^{(i)})\text{-XX}}\\ \vdots & \vdots & \vdots\\ \underbrace{((m_1)^{\text{th}}\text{XX},(n_1)^{\text{th}}\text{XX})}_{(m_1,n_1)\text{-XX}} & \cdots & \underbrace{((m_p)^{\text{th}}\text{XX},(n_1)^{\text{th}}\text{XX})}_{(m_p,n_1)\text{-XX}}\end{bmatrix}_{q\times p}.
\end{aligned}\tag{3.22}$$

(ii_1) For $m_{r_1} = 2m_{r_1 1}+1$ and $n_{s_1} = 2n_{s_1 1}+1$, the equilibrium of $(x_{j_1}^*, x_{j_2}^*) = (a_{j_1j_1r_1}, a_{j_2j_2s_1})$ has the following properties.

- For $a_{j_1j_10}\prod_{r_2=1,r_2\neq s_1}^{p}(a_{j_1j_1r_1}-a_{j_1j_1r_2})^{m_{r_2}} > 0$ and $a_{j_2j_20}\prod_{s_2=1,s_2\neq s_1}^{q}(a_{j_2j_2s_1}-a_{j_2j_2s_2})^{n_{s_2}} > 0$,

$$(a_{j_1j_1r_1},a_{j_2j_2s_1}) = \underbrace{((2m_{r_1 1}+1)^{\text{th}}\text{SO},(2n_{s_1 1}+1)^{\text{th}}\text{SO})}_{((2m_{r_1 1}+1),(2n_{s_1 1}+1))\text{-source}}.\tag{3.23}$$

- For $a_{j_1j_10}\prod_{r_2=1,r_2\neq s_1}^{p}(a_{j_1j_1r_1}-a_{j_1j_1r_2})^{m_{r_2}} < 0$ and $a_{j_2j_20}\prod_{s_2=1,s_2\neq s_1}^{q}(a_{j_2j_2s_1}-a_{j_2j_2s_2})^{n_{s_2}} > 0$,

$$(a_{j_1j_1r_1},a_{j_2j_2s_1}) = \underbrace{((2m_{r_1 1}+1)^{\text{th}}\text{SI},(2n_{s_1 1}+1)^{\text{th}}\text{SO})}_{((2m_{r_1 1}+1),(2n_{s_1 1}+1))\text{-saddle}}.\tag{3.24}$$

- For $a_{j_1j_10}\prod_{r_2=1,r_2\neq s_1}^{p}(a_{j_1j_1r_1}-a_{j_1j_1r_2})^{m_{r_2}}>0$ and $a_{j_2j_20}\prod_{s_2=1,s_2\neq s_1}^{q}(a_{j_2j_2s_1}-a_{j_2j_2s_2})^{n_{s_2}}<0$,

$$(a_{j_1j_1r_1},a_{j_2j_2s_1})=\underbrace{((2m_{r_1 1}+1)^{\text{th}}\text{SO},(2n_{s_1 1}+1)^{\text{th}}\text{SI})}_{((2m_{r_1 1}+1),(2n_{s_1 1}+1))\text{-saddle}}. \tag{3.25}$$

- For $a_{j_1j_10}\prod_{r_2=1,r_2\neq s_1}^{p}(a_{j_1j_1r_1}-a_{j_1j_1r_2})^{m_{r_2}}<0$ and $a_{j_2j_20}\prod_{s_2=1,s_2\neq s_1}^{q}(a_{j_2j_2s_1}-a_{j_2j_2s_2})^{n_{s_2}}<0$,

$$(a_{j_1j_1r_1},a_{j_2j_2s_1})=\underbrace{((2m_{r_1 1}+1)^{\text{th}}\text{SI},(2n_{s_1 1}+1)^{\text{th}}\text{SI})}_{((2m_{r_1 1}+1),(2n_{s_1 1}+1))\text{-sink}}. \tag{3.26}$$

(ii$_2$) For $m_{r_1}=2m_{r_1 1}$ and $n_{s_1}=2n_{s_1 1}+1$, the equilibrium of $(x_{j_1}^*,x_{j_2}^*)=(a_{j_1j_1r_1},a_{j_2j_2s_1})$ has the following properties.

- For $a_{j_1j_10}\prod_{r_2=1,r_2\neq s_1}^{p}(a_{j_1j_1r_1}-a_{j_1j_1r_2})^{m_{r_2}}>0$ and $a_{j_2j_20}\prod_{s_2=1,s_2\neq s_1}^{q}(a_{j_2j_2s_1}-a_{j_2j_2s_2})^{n_{s_2}}>0$,

$$(a_{j_1j_1r_1},a_{j_2j_2s_1})=\underbrace{((2m_{r_1 1})^{\text{th}}\text{US},(2n_{s_1 1}+1)^{\text{th}}\text{SO})}_{((2m_{r_1 1}),(2n_{s_1 1}+1))\text{-upper-saddle source}}. \tag{3.27}$$

- For $a_{j_1j_10}\prod_{r_2=1,r_2\neq s_1}^{p}(a_{j_1j_1r_1}-a_{j_1j_1r_2})^{m_{r_2}}<0$ and $a_{j_2j_20}\prod_{s_2=1,s_2\neq s_1}^{q}(a_{j_2j_2s_1}-a_{j_2j_2s_2})^{n_{s_2}}>0$,

$$(a_{j_1j_1r_1},a_{j_2j_2s_1})=\underbrace{((2m_{r_1 1})^{\text{th}}\text{LS},(2n_{s_1 1}+1)^{\text{th}}\text{SO})}_{((2m_{r_1 1}),(2n_{s_1 1}+1))\text{-lower-saddle source}}. \tag{3.28}$$

- For $a_{j_1j_10}\prod_{r_2=1,r_2\neq s_1}^{p}(a_{j_1j_1r_1}-a_{j_1j_1r_2})^{m_{r_2}}>0$ and $a_{j_2j_20}\prod_{s_2=1,s_2\neq s_1}^{q}(a_{j_2j_2s_1}-a_{j_2j_2s_2})^{n_{s_2}}<0$,

$$(a_{j_1j_1r_1},a_{j_2j_2s_1})=\underbrace{((2m_{r_1 1})^{\text{th}}\text{US},(2n_{s_1 1}+1)^{\text{th}}\text{SI})}_{((2m_{r_1 1}),(2n_{s_1 1}+1))\text{-upper-saddle sink}}. \tag{3.29}$$

- For $a_{j_1j_10}\prod_{r_2=1,r_2\neq s_1}^{p}(a_{j_1j_1r_1}-a_{j_1j_1r_2})^{m_{r_2}}<0$ and $a_{j_2j_20}\prod_{s_2=1,s_2\neq s_1}^{q}(a_{j_2j_2s_1}-a_{j_2j_2s_2})^{n_{s_2}}<0$,

$$(a_{j_1j_1r_1},a_{j_2j_2s_1})=\underbrace{((2m_{r_1 1})^{\text{th}}\text{LS},(2n_{s_1 1}+1)^{\text{th}}\text{SI})}_{((2m_{r_1 1}),(2n_{s_1 1}+1))\text{-lower-saddlesink}}. \tag{3.30}$$

(ii$_3$) For $m_{r_1} = 2m_{r_1 1} + 1$ and $n_{s_1} = 2n_{s_1 1}$, the equilibrium of $(x^*_{j_1}, x^*_{j_2}) = (a_{j_1 j_1 r_1}, a_{j_2 j_2 s_1})$ has the following properties.

- For $a_{j_1 j_1 0} \prod_{r_2=1, r_2 \neq s_1}^{p} (a_{j_1 j_1 r_1} - a_{j_1 j_1 r_2})^{m_{r_2}} > 0$ and $a_{j_2 j_2 0} \prod_{s_2=1, s_2 \neq s_1}^{q} (a_{j_2 j_2 s_1} - a_{j_2 j_2 s_2})^{n_{s_2}} > 0$,

$$(a_{j_1 j_1 r_1}, a_{j_2 j_2 s_1}) = \underbrace{((2m_{r_1 1}+1)^{\text{th}}\text{SO},(2n_{s_1 1})^{\text{th}}\text{US})}_{((2m_{r_1 1}+1),(2n_{s_1 1}))\text{-upper-saddle source}}. \tag{3.31}$$

- For $a_{j_1 j_1 0} \prod_{r_2=1, r_2 \neq s_1}^{p} (a_{j_1 j_1 r_1} - a_{j_1 j_1 r_2})^{m_{r_2}} < 0$ and $a_{j_2 j_2 0} \prod_{s_2=1, s_2 \neq s_1}^{q} (a_{j_2 j_2 s_1} - a_{j_2 j_2 s_2})^{n_{s_2}} > 0$,

$$(a_{j_1 j_1 r_1}, a_{j_2 j_2 s_1}) = \underbrace{((2m_{r_1 1}+1)^{\text{th}}\text{SI},(2n_{s_1 1})^{\text{th}}\text{US})}_{((2m_{r_1 1}+1),(2n_{s_1 1}))\text{-upper-saddle sink}}. \tag{3.32}$$

- For $a_{j_1 j_1 0} \prod_{r_2=1, r_2 \neq s_1}^{p} (a_{j_1 j_1 r_1} - a_{j_1 j_1 r_2})^{m_{r_2}} > 0$ and $a_{j_2 j_2 0} \prod_{s_2=1, s_2 \neq s_1}^{q} (a_{j_2 j_2 s_1} - a_{j_2 j_2 s_2})^{n_{s_2}} < 0$,

$$(a_{j_1 j_1 r_1}, a_{j_2 j_2 s_1}) = \underbrace{((2m_{r_1 1}+1)^{\text{th}}\text{SO},(2n_{s_1 1})^{\text{th}}\text{LS})}_{((2m_{r_1 1}+1),(2n_{s_1 1}))\text{-lower-saddle source}}. \tag{3.33}$$

- For $a_{j_1 j_1 0} \prod_{r_2=1, r_2 \neq s_1}^{p} (a_{j_1 j_1 r_1} - a_{j_1 j_1 r_2})^{m_{r_2}} < 0$ and $a_{j_2 j_2 0} \prod_{s_2=1, s_2 \neq s_1}^{q} (a_{j_2 j_2 s_1} - a_{j_2 j_2 s_2})^{n_{s_2}} < 0$,

$$(a_{j_1 j_1 r_1}, a_{j_2 j_2 s_1}) = \underbrace{((2m_{r_1 1}+1)^{\text{th}}\text{SI},(2n_{s_1 1})^{\text{th}}\text{LS})}_{((2m_{r_1 1}+1),(2n_{s_1 1}))\text{-lower-saddle sink}}. \tag{3.34}$$

(ii$_4$) For $m_{r_1} = 2m_{r_1 1}$ and $n_{s_1} = 2n_{s_1 1}$, the equilibrium of $(x^*_{j_1}, x^*_{j_2}) = (a_{j_1 j_1 r_1}, a_{j_2 j_2 s_1})$ has the following properties.

- For $a_{j_1 j_1 0} \prod_{r_2=1, r_2 \neq s_1}^{p} (a_{j_1 j_1 r_1} - a_{j_1 j_1 r_2})^{m_{r_2}} > 0$ and $a_{j_2 j_2 0} \prod_{s_2=1, s_2 \neq s_1}^{q} (a_{j_2 j_2 s_1} - a_{j_2 j_2 s_2})^{n_{s_2}} > 0$,

$$(a_{j_1 j_1 r_1}, a_{j_2 j_2 s_1}) = \underbrace{((2m_{r_1 1})^{\text{th}}\text{US},(2n_{s_1 1})^{\text{th}}\text{US})}_{((2m_{r_1 1}),(2n_{s_1 1}))\text{-double-saddle}}. \tag{3.35}$$

- For $a_{j_1 j_1 0} \prod_{r_2=1, r_2 \neq s_1}^{p} (a_{j_1 j_1 r_1} - a_{j_1 j_1 r_2})^{m_{r_2}} < 0$ and $a_{j_2 j_2 0} \prod_{s_2=1, s_2 \neq s_1}^{q} (a_{j_2 j_2 s_1} - a_{j_2 j_2 s_2})^{n_{s_2}} > 0$,

$$(a_{j_1 j_1 r_1}, a_{j_2 j_2 s_1}) = \underbrace{((2m_{r_1 1})^{\text{th}}\text{LS},(2n_{s_1 1})^{\text{th}}\text{US})}_{((2m_{r_1 1}),(2n_{s_1 1}))\text{-double-saddle}}. \tag{3.36}$$

- For $a_{j_1j_10}\prod_{r_2=1,r_2\neq s_1}^{p}(a_{j_1j_1r_1}-a_{j_1j_1r_2})^{m_{r_2}}>0$ and $a_{j_2j_20}\prod_{s_2=1,s_2\neq s_1}^{q}(a_{j_2j_2s_1}-a_{j_2j_2s_2})^{n_{s_2}}<0$,

$$(a_{j_1j_1r_1},a_{j_2j_2s_1})=\underbrace{((2m_{r_1 1})^{\text{th}}\text{US},(2n_{s_1 1})^{\text{th}}\text{LS})}_{((2m_{r_1 1}),(2n_{s_1 1}))\text{-doubel-saddle}}. \tag{3.37}$$

- For $a_{j_1j_10}\prod_{r_2=1,r_2\neq s_1}^{p}(a_{j_1j_1r_1}-a_{j_1j_1r_2})^{m_{r_2}}<0$ and $a_{j_2j_20}\prod_{s_2=1,s_2\neq s_1}^{q}(a_{j_2j_2s_1}-a_{j_2j_2s_2})^{n_{s_2}}<0$,

$$(a_{j_1j_1r_1},a_{j_2j_2s_1})=\underbrace{((2m_{r_1 1})^{\text{th}}\text{LS},(2n_{s_1 1})^{\text{th}}\text{LS})}_{((2m_{r_1 1}),(2n_{s_1 1}))\text{-double-saddle}}. \tag{3.38}$$

(iii) *Consider a self-univariate polynomial dynamical system as*

$$\dot{x}_{j_1}=a_{j_1j_10}\prod_{s_1=1}^{m}(x_{j_1}-a_{j_1j_1s_1}),\ \dot{x}_{j_2}=a_{j_2j_20}\prod_{l_1=1}^{n}(x_{j_2}-a_{j_2j_2l_1}), \tag{3.39}$$
$$j_1,j_2\in\{1,2\};\ j_1\neq j_2.$$

The first integral manifold is

$$\begin{aligned}&\sum_{l_1=1}^{n}\frac{1}{\prod_{l_2=1,l_2\neq l_1}^{n}(a_{j_2j_2l_1}-a_{j_2j_2l_2})}\ln\frac{|x_{j_2}-a_{j_2j_2l_1}|}{|x_{j_20}-a_{j_2j_2l_1}|}\\&=\frac{a_{j_2j_20}}{a_{j_1j_10}}\sum_{s_1=1}^{m}\frac{1}{\prod_{s_2=1,s_2\neq s_1}^{m}(a_{j_1j_1s_1}-a_{j_1j_1s_2})}\ln\frac{|x_{j_1}-a_{j_1j_1s_1}|}{|x_{j_10}-a_{j_1j_1s_1}|}.\end{aligned} \tag{3.40}$$

The simple-equilibrium network is defined as

$$\begin{aligned}&\bigcup_{s=1}^{m}\bigcup_{l=1}^{n}\underbrace{(a_{j_1j_1s},a_{j_2j_2l})}_{XX}\\&\equiv\left\{\begin{array}{cccc}(a_{j_1j_11},a_{j_2j_2n}) & (a_{j_1j_12},a_{j_2j_2n}) & \cdots & (a_{j_1j_1m},a_{j_2j_2n})\\ (a_{j_1j_11},a_{j_2j_2(n-1)}) & (a_{j_1j_12},a_{j_2j_2(n-1)}) & \cdots & (a_{j_1j_1m},a_{j_2j_2(n-1)})\\ \vdots & \vdots & \cdots & \vdots\\ (a_{j_1j_11},a_{j_2j_21}) & (a_{j_1j_12},a_{j_2j_21}) & \cdots & (a_{j_1j_1m},a_{j_2j_21})\end{array}\right\}_{n\times m}.\end{aligned} \tag{3.41}$$

The equilibrium of $(x_{j_1}^{*},x_{j_2}^{*})=(a_{j_1j_1s_1},a_{j_2j_2l_1})$ $(s_1,s_2\in\{1,2,\cdots,m\},s_1\neq s_2;$ $l_1,l_2\in\{1,2,\cdots,n\},l_1\neq l_2)$ *possesses the following properties.*

- *For* $a_{j_1j_10}\prod_{s_2=1,s_2\neq s_1}^{m}(a_{j_1j_1s_1}-a_{j_1j_1s_2})>0$ *and* $a_{j_2j_20}\prod_{l_2=1,l_2\neq l_1}^{n}(a_{j_2j_2l_1}-a_{j_2j_2l_2})>0$,

$$(a_{j_1j_1s_1},a_{j_2j_2l_1})=\underbrace{(\text{SO,SO})}_{\text{source}}. \tag{3.42}$$

The *equilibrium of* $(x_{j_1}^*,x_{j_2}^*)=(a_{j_1j_1s_1},a_{j_2j_2l_1})$ *is an* (SO,SO)-*source.*

- *For* $a_{j_1j_10}\prod_{s_2=1,s_2\neq s_1}^{m}(a_{j_1j_1s_1}-a_{j_1j_1s_2})<0$ *and* $a_{j_2j_20}\prod_{l_2=1,l_2\neq l_1}^{n}(a_{j_2j_2l_1}-a_{j_2j_2l_2})>0$,

$$(a_{j_1j_1s_1},a_{j_2j_2l_1})=\underbrace{(\text{SI,SO})}_{\text{saddle}}. \tag{3.43}$$

The *equilibrium of* $(x_{j_1}^*,x_{j_2}^*)=(a_{j_1j_1s_1},a_{j_2j_2l_1})$ *is an* (SI,SO)-*saddle.*

- *For* $a_{j_1j_10}\prod_{s_2=1,s_2\neq s_1}^{m}(a_{j_1j_1s_1}-a_{j_1j_1s_2})>0$ *and* $a_{j_2j_20}\prod_{l_2=1,l_2\neq l_1}^{n}(a_{j_2j_2l_1}-a_{j_2j_2l_2})<0$,

$$(a_{j_1j_1s_1},a_{j_2j_2l_1})=\underbrace{(\text{SO,SI})}_{\text{saddle}}. \tag{3.44}$$

The *equilibrium of* $(x_{j_1}^*,x_{j_2}^*)=(a_{j_1j_1s_1},a_{j_2j_2l_1})$ *is an* (SO,SI)-*saddle.*

- *For* $a_{j_1j_10}\prod_{s_2=1,s_2\neq s_1}^{m}(a_{j_1j_1s_1}-a_{j_1j_1s_2})<0$ *and* $a_{j_2j_20}\prod_{l_2=1,l_2\neq l_1}^{n}(a_{j_2j_2l_1}-a_{j_2j_2l_2})<0$,

$$(a_{j_1j_1s_1},a_{j_2j_2l_1})=\underbrace{(\text{SI,SI})}_{\text{sink}}. \tag{3.45}$$

The *equilibrium of* $(x_{j_1}^*,x_{j_2}^*)=(a_{j_1j_1s_1},a_{j_2j_2l_1})$ *is an* (SI,SI)-*sink.*

(iv) For $m=2m_1+1$ and $n=2n_1+1$, the equilibrium of$(x_{j_1}^*,x_{j_2}^*)=(a_{j_1j_11},a_{j_2j_21})$ has the following bifurcation properties.

(iv$_1$) For $a_{j_1j_10}>0$ and $a_{j_2j_20}>0$, there is a $((2m_1{+}1)^{\text{th}}\text{SO},(2n_1{+}1)^{\text{th}}\text{SO})$-source equilibrium as

$$(a_{j_1j_11},a_{j_2j_21})=\underbrace{((2m_1{+}1)^{\text{th}}\text{SO},(2n_1{+}1)^{\text{th}}\text{SO})}_{((2m_1+1),(2n_1+1))\text{-source}}. \tag{3.46}$$

There are three following $((2m_1{+}1)^{\text{th}}\text{SO},(2n_1{+}1)^{\text{th}}\text{SO})$-source appearing and switching bifurcations.

(iv_{1a}) The $((2m_1+1)^{\text{th}}\text{SO}, (2n_1+1)^{\text{th}}\text{SO})$-source appearing bifurcation are from a (SO,SO)-source to a $(2m_1+1) \times (2n_1+1)$-equilibrium network as

$$\underbrace{(a_{j_1 j_1 1}, a_{j_2 j_2 1})}_{\text{(SO,SO)-source}} \rightleftharpoons \underbrace{((2m_1+1)^{\text{th}}\text{SO},(2n_1+1)^{\text{th}}\text{SO})}_{((2m_1+1),(2n_1+1))\text{-source}} \rightleftharpoons \bigcup_{r=1}^{2m_1+1} \bigcup_{s=1}^{2n_1+1} \underbrace{(a_{j_1 j_1 r}, a_{j_2 j_2 s})}_{\text{XX}} \tag{3.47}$$

where

$$\bigcup_{r=1}^{2m_1+1} \bigcup_{s=1}^{2n_1+1} \underbrace{(a_{j_1 j_1 r}, a_{j_2 j_2 s})}_{\text{XX}} = \left\{ \begin{array}{cccc} \underbrace{\text{(SO,SO)}}_{\text{source}} & \underbrace{\text{(SI,SO)}}_{\text{saddle}} & \cdots & \underbrace{\text{(SO,SO)}}_{\text{source}} \\ \underbrace{\text{(SO,SI)}}_{\text{saddle}} & \underbrace{\text{(SI,SI)}}_{\text{sink}} & \cdots & \underbrace{\text{(SO,SI)}}_{\text{saddle}} \\ \vdots & \vdots & \vdots & \vdots \\ \underbrace{\text{(SO,SO)}}_{\text{source}} & \underbrace{\text{(SI,SO)}}_{\text{saddle}} & \cdots & \underbrace{\text{(SO,SO)}}_{\text{source}} \end{array} \right\}_{(2n_1+1)\times(2m_1+1)}. \tag{3.48}$$

(iv_{1b}) The $((2m_1+1)^{\text{th}}\text{SO}, (2n_1+1)^{\text{th}}\text{SO})$-source appearing bifurcation is from a $q_1 \times p_1$ to $q_2 \times p_2$ equilibrium network as

$$\begin{aligned} &\bigcup_{r_1=1}^{p_1} \bigcup_{s_1=1}^{q_1} \underbrace{(a_{j_1 j_1 r_1}, a_{j_2 j_2 s_1})}_{(m_{r_1}^{(1)}, n_{s_1}^{(1)})\text{- XX}} \rightleftharpoons \underbrace{((2m_1+1)^{\text{th}}\text{SO},(2n_1+1)^{\text{th}}\text{SO})}_{((2m_1+1),(2n_1+1))\text{-source}} \\ &\rightleftharpoons \bigcup_{r_2=1}^{p_2} \bigcup_{s_2=1}^{q_2} \underbrace{(a_{j_1 j_1 r_2}, a_{j_2 j_2 s_2})}_{(m_{r_2}^{(2)}, n_{s_2}^{(2)})\text{- XX}} \end{aligned} \tag{3.49}$$

where

$$2m_1 + 1 - \sum_{r_1=1}^{p_1} m_{r_1}^{(1)} = 2\sum_{s=1}^{l_1} m_s$$

with l_1-quadratic polynomials without real roots
and s^{th}-quadratic polynomial with power m_s,

$$2n_1 + 1 - \sum_{s_1=1}^{q_1} n_{s_1}^{(1)} = 2\sum_{l=1}^{l_2} n_l$$

with l_2-quadraitc polynomials without real roots
and l^{th}-quadratic polynomial with power n_l;

$$\sum_{r_2=1}^{p_2} m_{r_2}^{(2)} = 2m_1 + 1,\ \sum_{s_2=1}^{q_2} n_{s_2}^{(2)} = 2n_1 + 1;\ \text{for } i = 1, 2,$$

$$\underbrace{((m_{p_i}^{(i)})^{\text{th}}\text{XX},(n_{q_i}^{(i)})^{\text{th}}\text{XX})}_{(m_{p_i}^{(i)},n_{q_i}^{(i)})\text{-XX}} \in \left\{ \begin{array}{l} \underbrace{((2m_{p_i1}^{(i)}+1)^{\text{th}}\text{SO},(2n_{q_i1}^{(i)}+1)^{\text{th}}\text{SO})}_{((2m_{p_i1}^{(i)}+1),(2n_{q_i1}^{(i)}+1))\text{-source}},\ \underbrace{((2m_{p_i1}^{(i)})^{\text{th}}\text{US},(2n_{q_i1}^{(i)})^{\text{th}}\text{US})}_{((2m_{p_i1}^{(i)}),(2n_{q_i1}^{(i)}))\text{-double-saddle}}, \\ \underbrace{((2m_{p_i1}^{(i)})^{\text{th}}\text{US},(2n_{q_i1}^{(i)}+1)^{\text{th}}\text{SO})}_{((2m_{p_i1}^{(i)}),(2n_{q_i1}^{(i)}+1))\text{-upper-saddle-source}},\ \underbrace{((2m_{p_i1}^{(i)}+1)^{\text{th}}\text{SO},(2n_{q_i1}^{(i)})^{\text{th}}\text{US})}_{((2m_{p_i1}^{(i)}+1),(2n_{q_i1}^{(i)}))\text{-upper-saddle-source}} \end{array} \right\};$$

$$\underbrace{((m_{p_i}^{(i)})^{\text{th}}\text{XX},(n_1^{(i)})^{\text{th}}\text{XX})}_{(m_{p_i}^{(i)},n_1^{(i)})\text{-XX}} \in \left\{ \begin{array}{l} \underbrace{((2m_{p_i1}^{(i)}+1)^{\text{th}}\text{SO},(2n_{11}^{(i)}+1)^{\text{th}}\text{SO})}_{((2m_{p_i1}^{(i)}+1),(2n_{11}^{(i)}+1))\text{-source}},\ \underbrace{((2m_{p_i1}^{(i)})^{\text{th}}\text{US},(2n_{11}^{(i)})^{\text{th}}\text{LS})}_{((2m_{p_i1}^{(i)}),(2n_{11}^{(i)}))\text{-double-saddle}}, \\ \underbrace{((2m_{p_i1}^{(i)})^{\text{th}}\text{US},(2n_{11}^{(i)}+1)^{\text{th}}\text{SO})}_{((2m_{p_i1}^{(i)}),(2n_{11}^{(i)}+1))\text{-upper-saddle-source}},\ \underbrace{((2m_{p_i1}^{(i)}+1)^{\text{th}}\text{SO},(2n_{11}^{(i)})^{\text{th}}\text{LS})}_{((2m_{p_i1}^{(i)}+1),(2n_{11}^{(i)}))\text{-lower-saddle-source}} \end{array} \right\};$$

$$\underbrace{((m_1^{(i)})^{\text{th}}\text{XX},(n_{q_i}^{(i)})^{\text{th}}\text{XX})}_{(m_1^{(i)},n_{q_i}^{(i)})\text{-XX}} \in \left\{ \begin{array}{l} \underbrace{((2m_{11}^{(i)}+1)^{\text{th}}\text{SO},(2n_{q_i1}^{(i)}+1)^{\text{th}}\text{SO})}_{((2m_{11}^{(i)}+1),(2n_{q_i1}^{(i)}+1))\text{-source}},\ \underbrace{((2m_{11}^{(i)})^{\text{th}}\text{LS},(2n_{q_i1}^{(i)})^{\text{th}}\text{US})}_{((2m_{11}^{(i)}),(2n_{q_i1}^{(i)}))\text{-double-saddle}}, \\ \underbrace{((2m_{11}^{(i)})^{\text{th}}\text{LS},(2n_{q_i1}^{(i)}+1)^{\text{th}}\text{SO})}_{((2m_{11}^{(i)}),(2n_{q_i1}^{(i)}+1))\text{-lower-saddle-source}},\ \underbrace{((2m_{11}^{(i)}+1)^{\text{th}}\text{SO},(2n_{q_i1}^{(i)})^{\text{th}}\text{US})}_{((2m_{11}^{(i)}+1),(2n_{q_i1}^{(i)}))\text{-upper-saddle-source}} \end{array} \right\};$$

$$\underbrace{((m_1^{(i)})^{\text{th}}\text{XX},(n_1^{(i)})^{\text{th}}\text{XX})}_{(m_1^{(i)},n_1^{(i)})\text{-XX}} \in \left\{ \begin{array}{l} \underbrace{((2m_{11}^{(i)}+1)^{\text{th}}\text{SO},(2n_{11}^{(i)}+1)^{\text{th}}\text{SO})}_{((2m_{11}^{(i)}+1),(2n_{11}^{(i)}+1))\text{-source}},\ \underbrace{((2m_{11}^{(i)})^{\text{th}}\text{LS},(2n_{11}^{(i)})^{\text{th}}\text{LS})}_{((2m_{11}^{(i)}),(2n_{11}^{(i)}))\text{-double-saddle}}, \\ \underbrace{((2m_{11}^{(i)})^{\text{th}}\text{LS},(2n_{11}^{(i)}+1)^{\text{th}}\text{SO})}_{((2m_{11}^{(i)}),(2n_{11}^{(i)}+1))\text{-lower-saddle-source}},\ \underbrace{((2m_{11}^{(i)}+1)^{\text{th}}\text{SO},(2n_{11}^{(i)})^{\text{th}}\text{LS})}_{((2m_{11}^{(i)}+1),(2n_{11}^{(i)}))\text{-lower-saddle-source}} \end{array} \right\}. \tag{3.50}$$

($\text{iv}_{1\text{c}}$) The $((2m_1+1)^{\text{th}}\text{SO}, (2n_1+1)^{\text{th}}\text{SO})$-source switching bifurcation is for the switching of two $q \times p$ equilibrium networks as

$$\bigcup_{r_1=1}^{p}\bigcup_{s_1=1}^{q}\underbrace{(a_{j_1j_1r_1},a_{j_2j_2s_1})}_{(m_{r_1}^{(1)},n_{s_1}^{(1)})\text{-XX}} \rightleftharpoons \underbrace{((2m_1+1)^{\text{th}}\text{SO},(2n_1+1)^{\text{th}}\text{SO})}_{((2m_1+1),(2n_1+1))\text{-source}}$$
$$\rightleftharpoons \bigcup_{r_2=1}^{p}\bigcup_{s_2=1}^{q}\underbrace{(a_{j_1j_1r_2},a_{j_2j_2s_2})}_{(m_{r_2}^{(2)},n_{s_2}^{(2)})\text{-XX}} \tag{3.51}$$

with $\Sigma_{r_i=1}^{p} m_{r_i} = 2m_1 + 1$ and $\Sigma_{s_i=1}^{q} n_{s_i} = 2n_1 + 1$ for $i = 1, 2$.

(iv$_2$) For $a_{j_1j_10} < 0$ and $a_{j_2j_20} > 0$, there is a $((2m_1+1)^{\text{th}}\text{SI}, (2n_1+1)^{\text{th}}\text{SO})$-saddle equilibrium as

$$(a_{j_1j_11},a_{j_2j_21}) = \underbrace{((2m_1+1)^{\text{th}}\text{SI},(2n_1+1)^{\text{th}}\text{SO})}_{((2m_1+1),(2n_1+1))\text{-saddle}}. \tag{3.52}$$

There are the following three $((2m_1+1)^{\text{th}}\text{SI},(2n_1+1)^{\text{th}}\text{SO})$-saddle appearing and switching bifurcations.

(iv$_{2a}$) The $((2m_1+1)^{\text{th}}\text{SI}, (2n_1+1)^{\text{th}}\text{SO})$-saddle appearing bifurcation is from a (SI,SO)-saddle to a $(2m_1 + 1) \times (2n_1 + 1)$ network as

$$\underbrace{(a_{j_1j_11},a_{j_2j_21})}_{\text{(SI,SO)-saddle}} \rightleftharpoons \underbrace{((2m_1+1)^{\text{th}}\text{SI},(2n_1+1)^{\text{th}}\text{SO})}_{((2m_1+1),(2n_1+1))\text{-saddle}} \rightleftharpoons \bigcup_{r=1}^{2m_1+1}\bigcup_{s=1}^{2n_1+1}\underbrace{(a_{j_1j_1r},a_{j_2j_2s})}_{\text{XX}} \tag{3.53}$$

where

$$\bigcup_{r=1}^{2m_1+1}\bigcup_{s=1}^{2n_1+1}\underbrace{(a_{j_1j_1r},a_{j_2j_2s})}_{XX} = \left\{\begin{matrix} \underbrace{\text{(SI,SO)}}_{\text{saddle}} & \underbrace{\text{(SO,SO)}}_{\text{source}} & \cdots & \underbrace{\text{(SI,SO)}}_{\text{saddle}} \\ \underbrace{\text{(SI,SI)}}_{\text{sink}} & \underbrace{\text{(SO,SI)}}_{\text{saddle}} & \cdots & \underbrace{\text{(SI,SI)}}_{\text{sink}} \\ \vdots & \vdots & \vdots & \vdots \\ \underbrace{\text{(SI,SO)}}_{\text{saddle}} & \underbrace{\text{(SO,SO)}}_{\text{source}} & \cdots & \underbrace{\text{(SI,SO)}}_{\text{saddle}} \end{matrix}\right\}_{(2n_1+1)\times(2m_1+1)}. \tag{3.54}$$

(iv$_{2b}$) The $((2m_1+1)^{\text{th}}\text{SI}, (2n_1+1)^{\text{th}}\text{SO})$-saddle appearing bifurcation is from $q_1 \times p_1$ to $q_2 \times p_2$ equilibrium networks as

$$\bigcup_{r_1=1}^{p_1}\bigcup_{s_1=1}^{q_1}\underbrace{(a_{j_1j_1r_1},a_{j_2j_2s_1})}_{(m_{r_1}^{(1)},n_{s_1}^{(1)})\text{-XX}} \rightleftharpoons \underbrace{((2m_1+1)^{\text{th}}\text{SI},(2n_1+1)^{\text{th}}\text{SO})}_{((2m_1+1),(2n_1+1))\text{-saddle}}$$

$$\rightleftharpoons \bigcup_{r_2=1}^{p_2} \bigcup_{s_2=1}^{q_2} \underbrace{(a_{j_1 j_1 r_2}, a_{j_2 j_2 s_2})}_{(m_{r_2}^{(2)}, n_{s_2}^{(2)})\text{-XX}} \tag{3.55}$$

where

$$2m_1 + 1 - \sum_{r_1=1}^{p_1} m_{r_1}^{(1)} = 2\sum_{s=1}^{l_1} m_s$$

with l_1-quadratic polynomials without real roots

and s^{th}-quadratic polynomial with power m_s,

$$2n_1 + 1 - \sum_{s_1=1}^{q_1} n_{s_1}^{(1)} = 2\sum_{l=1}^{l_2} n_l$$

with l_2-quadraitc polynomials without real roots

and l^{th}-quadratic polynomial with power n_l;

$$\sum_{r_2=1}^{p_2} m_{r_2}^{(2)} = 2m_1 + 1,\ \sum_{s_2=1}^{q_2} n_{s_2}^{(2)} = 2n_1 + 1;\ \text{for } i = 1, 2$$

$$\underbrace{((m_{p_i}^{(i)})^{\text{th}}\text{XX},(n_{q_i}^{(i)})^{\text{th}}\text{XX})}_{(m_{p_i}^{(i)}, n_{q_i}^{(i)})\text{-XX}}$$

$$\in \left\{ \begin{array}{l} \underbrace{((2m_{p_i 1}^{(i)}+1)^{\text{th}}\text{SI},(2n_{q_i 1}^{(i)}+1)^{\text{th}}\text{SO})}_{((2m_{p_i 1}^{(i)}+1),(2n_{q_i 1}^{(i)}+1))\text{-saddle}}, \underbrace{((2m_{p_i 1}^{(i)})^{\text{th}}\text{LS},(2n_{q_i 1}^{(i)})^{\text{th}}\text{US})}_{((2m_{p_i 1}^{(i)}),(2n_{q_i 1}^{(i)}))\text{-double-saddle}}, \\ \underbrace{((2m_{p_i 1}^{(i)})^{\text{th}}\text{LS},(2n_{q_i 1}^{(i)}+1)^{\text{th}}\text{SO})}_{((2m_{p_i 1}^{(i)}),(2n_{q_i 1}^{(i)}+1))\text{-lower-saddle-source}}, \underbrace{((2m_{p_i 1}^{(i)}+1)^{\text{th}}\text{SI},(2n_{q_i 1}^{(i)})^{\text{th}}\text{US})}_{((2m_{p_i 1}^{(i)}+1),(2n_{q_i 1}^{(i)}))\text{-upper-saddle-sink}} \end{array} \right\};$$

$$\underbrace{((m_{p_i}^{(i)})^{\text{th}}\text{XX},(n_{1}^{(i)})^{\text{th}}\text{XX})}_{(m_{p_i}^{(i)}, n_{1}^{(i)})\text{-XX}}$$

$$\in \left\{ \begin{array}{l} \underbrace{((2m_{p_i 1}^{(i)}+1)^{\text{th}}\text{SI},(2n_{11}^{(i)}+1)^{\text{th}}\text{SO})}_{((2m_{p_i 1}^{(i)}+1),(2n_{11}^{(i)}+1))\text{-saddle}}, \underbrace{((2m_{p_i 1}^{(i)})^{\text{th}}\text{LS},(2n_{11}^{(i)})^{\text{th}}\text{LS})}_{((2m_{p_i 1}^{(i)}),(2n_{11}^{(i)}))\text{-double-saddle}}, \\ \underbrace{((2m_{p_i 1}^{(i)})^{\text{th}}\text{LS},(2n_{11}^{(i)}+1)^{\text{th}}\text{SO})}_{((2m_{p_i 1}^{(i)}),(2n_{11}^{(i)}+1))\text{-lower-saddle-source}}, \underbrace{((2m_{p_i 1}^{(i)}+1)^{\text{th}}\text{SI},(2n_{11}^{(i)})^{\text{th}}\text{LS})}_{((2m_{p_i 1}^{(i)}+1),(2n_{11}^{(i)}))\text{-lower-saddle-sink}} \end{array} \right\};$$

$$\underbrace{((m_1^{(i)})^{\text{th}}\text{XX},(n_{q_i}^{(i)})^{\text{th}}\text{XX})}_{(m_1^{(i)},n_{q_i}^{(i)})\text{-XX}}$$
$$\in \left\{ \begin{array}{l} \underbrace{((2m_{11}^{(i)}+1)^{\text{th}}\text{SI},(2n_{q_i1}^{(i)}+1)^{\text{th}}\text{SO})}_{((2m_{11}^{(i)}+1),(2n_{q_i1}^{(i)}+1))\text{-saddle}}, \underbrace{((2m_{11}^{(i)})^{\text{th}}\text{US},(2n_{q_i1}^{(i)})^{\text{th}}\text{US})}_{((2m_{11}^{(i)}),(2n_{q_i1}^{(i)}))\text{-double-saddle}}, \\ \underbrace{((2m_{11}^{(i)})^{\text{th}}\text{US},(2n_{q_i1}^{(i)}+1)^{\text{th}}\text{SO})}_{((2m_{11}^{(i)}),(2n_{q_i1}^{(i)}+1))\text{-upper-saddle-source}}, \underbrace{((2m_{11}^{(i)}+1)^{\text{th}}\text{SI},(2n_{q_i1}^{(i)})^{\text{th}}\text{US})}_{((2m_{11}^{(i)}+1),(2n_{q_i1}^{(i)}))\text{-upper-saddle-sink}} \end{array} \right\};$$
$$\underbrace{((m_1^{(i)})^{\text{th}}\text{XX},(n_1^{(i)})^{\text{th}}XX)}_{(m_1^{(i)},n_1^{(i)})\text{-XX}}$$
$$\in \left\{ \begin{array}{l} \underbrace{((2m_{11}^{(i)}+1)^{\text{th}}\text{SI},(2n_{11}^{(i)}+1)^{\text{th}}\text{SO})}_{((2m_{11}^{(i)}+1),(2n_{11}^{(i)}+1))\text{-saddle}}, \underbrace{((2m_{11}^{(i)})^{\text{th}}\text{US},(2n_{11}^{(i)})^{\text{th}}\text{LS})}_{((2m_{11}^{(i)}),(2n_{11}^{(i)}))\text{-double-saddle}}, \\ \underbrace{((2m_{11}^{(i)})^{\text{th}}\text{US},(2n_{11}^{(i)}+1)^{\text{th}}\text{SO})}_{((2m_{11}^{(i)}),(2n_{11}^{(i)}+1))\text{-upper-saddle-source}}, \underbrace{((2m_{11}^{(i)}+1)^{\text{th}}\text{SI},(2n_{11}^{(i)})^{\text{th}}\text{LS})}_{((2m_{11}^{(i)}+1),(2n_{11}^{(i)}))\text{-lower-saddle-sink}} \end{array} \right\}. \tag{3.56}$$

(iv_{2c}) The $((2m_1+1)^{\text{th}}\text{SI}, (2n_1+1)^{\text{th}}\text{SO})$-saddle switching bifurcations is for two $q\times p$ equilibrium networks as

$$\begin{array}{l} \bigcup_{r_1=1}^{p}\bigcup_{s_1=1}^{q} \underbrace{(a_{j_1j_1r_1}, a_{j_2j_2s_1})}_{(m_{r_1}^{(1)},\, n_{s_1}^{(1)})\text{-XX}} \rightleftharpoons \underbrace{((2m_1+1)^{\text{th}}\text{SI},(2n_1+1)^{\text{th}}\text{SO})}_{((2m_1+1),(2n_1+1))\text{-saddle}} \\ \rightleftharpoons \bigcup_{r_2=1}^{p}\bigcup_{s_2=1}^{q} \underbrace{(a_{j_1j_1r_2}, a_{j_2j_2s_2})}_{(m_{r_2}^{(2)},\, n_{s_2}^{(2)})\text{-XX}} \end{array} \tag{3.57}$$

with $\Sigma_{r_i=1}^{p} m_{r_i} = 2m_1 + 1$ and $\Sigma_{s_i=1}^{q} n_{s_i} = 2n_1 + 1$ for $i = 1, 2$.
(iv_3) For $a_{j_1j_10} > 0$ and $a_{j_2j_20} < 0$, there is a $((2m_1+1)^{\text{th}}\text{SO}, (2n_1+1)^{\text{th}}\text{SI})$-saddle equilibrium as

$$(a_{j_1j_11}, a_{j_2j_21}) = \underbrace{((2m_1+1)^{\text{th}}\text{SO},(2n_1+1)^{\text{th}}\text{SI})}_{((2m_1+1),(2n_1+1))\text{-saddle}}. \tag{3.58}$$

There are the following three $((2m_1+1)^{\text{th}}\text{SO}, (2n_1+1)^{\text{th}}\text{SI})$-saddle appearing and switching bifurcations.

(iv_{3a}) The $((2m_1+1)^{\text{th}}\text{SO}, (2n_1+1)^{\text{th}}\text{SI})$-saddle appearing bifurcation are from a (SO,SI)-saddle to a $(2m_1 + 1) \times (2n_1 + 1)$-equilibrium network as

$$\underbrace{(a_{j_1j_1 1}, a_{j_2j_2 1})}_{\text{(SO,SI)-saddle}} \rightleftharpoons \underbrace{((2m_1+1)^{\text{th}}\text{SO},(2n_1+1)^{\text{th}}\text{SI})}_{((2m_1+1),(2n_1+1))\text{-saddle}} \rightleftharpoons \bigcup_{r=1}^{2m_1+1}\bigcup_{s=1}^{2n_1+1} \underbrace{(a_{j_1j_1 r}, a_{j_2j_2 s})}_{\text{XX}} \quad (3.59)$$

where

$$\bigcup_{r=1}^{2m_1+1}\bigcup_{s=1}^{2n_1+1} \underbrace{(a_{j_1j_1 r}, a_{j_2j_2 s})}_{\text{XX}} = \left\{\begin{matrix} \underbrace{\text{(SO,SI)}}_{\text{saddle}} & \underbrace{\text{(SI,SI)}}_{\text{sink}} & \cdots & \underbrace{\text{(SO,SI)}}_{\text{saddle}} \\ \underbrace{\text{(SO,SO)}}_{\text{source}} & \underbrace{\text{(SI,SO)}}_{\text{saddle}} & \cdots & \underbrace{\text{(SO,SO)}}_{\text{source}} \\ \vdots & \vdots & \vdots & \vdots \\ \underbrace{\text{(SO,SI)}}_{\text{saddle}} & \underbrace{\text{(SI,SI)}}_{\text{sink}} & \cdots & \underbrace{\text{(SO,SI)}}_{\text{saddle}} \end{matrix}\right\}_{(2n_1+1)\times(2m_1+1)}. \quad (3.60)$$

(iv_{3b}) The $((2m_1+1)^{\text{th}}\text{SO}, (2n_1+1)^{\text{th}}\text{SI})$-saddle appearing bifurcation is from a $q_1 \times p_1$ to $q_2 \times p_2$ equilibrium network as

$$\begin{aligned} &\bigcup_{r_1=1}^{p_1}\bigcup_{s_1=1}^{q_1} \underbrace{(a_{j_1j_1 r_1}, a_{j_2j_2 s_1})}_{(m_{r_1}^{(1)}, n_{s_1}^{(1)})\text{-XX}} \rightleftharpoons \underbrace{((2m_1+1)^{\text{th}}\text{SO},(2n_1+1)^{\text{th}}\text{SI})}_{((2m_1+1),(2n_1+1))\text{-saddle}} \\ &\rightleftharpoons \bigcup_{r_2=1}^{p_2}\bigcup_{s_2=1}^{q_2} \underbrace{(a_{j_1j_1 r_2}, a_{j_2j_2 s_2})}_{(m_{r_2}^{(2)}, n_{s_2}^{(2)})\text{-XX}} \end{aligned} \quad (3.61)$$

where

$$2m_1 + 1 - \sum_{r_1=1}^{p_1} m_{r_1}^{(1)} = 2\sum_{s=1}^{l_1} m_s$$

with l_1-quadratic polynomials without real roots
and s^{th}-quadratic polynomial with power m_s,

$$2n_1 + 1 - \sum_{r_1=1}^{q_1} n_{s_1}^{(1)} = 2\sum_{l=1}^{l_2} n_l$$

with l_2-quadraitc polynomials without real roots
and l^{th}-quadratic polynomial with power n_l;

$$\sum_{r_2=1}^{p_2} m_{r_2}^{(2)} = 2m_1 + 1, \ \sum_{s_2=1}^{q_2} n_{s_2}^{(2)} = 2n_1 + 1; \text{ for } i = 1, 2,$$

$$
\begin{aligned}
&\underbrace{((m_{p_i}^{(i)})^{\text{th}}\text{XX},(n_{q_i}^{(i)})^{\text{th}}\text{XX})}_{(m_{p_i}^{(i)},n_{q_i}^{(i)})\text{-XX}}\\
&\in \left\{\begin{array}{l}
\underbrace{((2m_{p_i1}^{(i)}+1)^{\text{th}}\text{SO},(2n_{q_i1}^{(i)}+1)^{\text{th}}\text{SI})}_{((2m_{p_i1}^{(i)}+1),(2n_{q_i1}^{(i)}+1))\text{-saddle}},\ \underbrace{((2m_{p_i1}^{(i)})^{\text{th}}\text{US},(2n_{q_i1}^{(i)})^{\text{th}}\text{LS})}_{((2m_{p_i1}^{(i)}),(2n_{q_i1}^{(i)}))\text{-double-saddle}},\\
\underbrace{((2m_{p_i1}^{(i)})^{\text{th}}\text{US},(2n_{q_i1}^{(i)}+1)^{\text{th}}\text{SI})}_{((2m_{p_i1}^{(i)}),(2n_{q_i1}^{(i)}+1))\text{-upper-saddle-sink}},\ \underbrace{((2m_{p_i1}^{(i)}+1)^{\text{th}}\text{SO},(2n_{q_i1}^{(i)})^{\text{th}}\text{LS})}_{((2m_{p_i1}^{(i)}+1),(2n_{q_i1}^{(i)}))\text{-lower-saddle-source}}
\end{array}\right\};\\
&\underbrace{((m_{p_i}^{(i)})^{\text{th}}\text{XX},(n_1^{(i)})^{\text{th}}\text{XX})}_{(m_{p_i}^{(i)},n_1^{(i)})\text{-XX}}\\
&\in \left\{\begin{array}{l}
\underbrace{((2m_{p_i1}^{(i)}+1)^{\text{th}}\text{SO},(2n_{11}^{(i)}+1)^{\text{th}}\text{SI})}_{((2m_{p_i1}^{(i)}+1),(2n_{11}^{(i)}+1))\text{-saddle}},\ \underbrace{((2m_{p_i1}^{(i)})^{\text{th}}\text{US},(2n_{11}^{(i)})^{\text{th}}\text{US})}_{((2m_{p_i1}^{(i)}),(2n_{11}^{(i)}))\text{-double-saddle}},\\
\underbrace{((2m_{p_i1}^{(i)})^{\text{th}}\text{US},(2n_{11}^{(i)}+1)^{\text{th}}\text{SI})}_{((2m_{p_i1}^{(i)}),(2n_{11}^{(i)}+1))\text{-upper-saddle-sink}},\ \underbrace{((2m_{p_i1}^{(i)}+1)^{\text{th}}\text{SO},(2n_{11}^{(i)})^{\text{th}}\text{US})}_{((2m_{p_i1}^{(i)}+1),(2n_{11}^{(i)}))\text{-upper-saddle-source}}
\end{array}\right\};\\
&\underbrace{((m_1^{(i)})^{\text{th}}\text{XX},(n_{q_i}^{(i)})^{\text{th}}\text{XX})}_{(m_1^{(i)},n_{q_i}^{(i)})\text{-XX}}\\
&\in \left\{\begin{array}{l}
\underbrace{((2m_{11}^{(i)}+1)^{\text{th}}\text{SO},(2n_{q_i1}^{(i)}+1)^{\text{th}}\text{SI})}_{((2m_{11}^{(i)}+1),(2n_{q_i1}^{(i)}+1))\text{-saddle}},\ \underbrace{((2m_{11}^{(i)})^{\text{th}}\text{LS},(2n_{q_i1}^{(i)})^{\text{th}}\text{LS})}_{((2m_{11}^{(i)}),(2n_{q_i1}^{(i)}))\text{-double-saddle}},\\
\underbrace{((2m_{11}^{(i)})^{\text{th}}\text{LS},(2n_{q_i1}^{(i)}+1)^{\text{th}}\text{SI})}_{((2m_{11}^{(i)}),(2n_{q_i1}^{(i)}+1))\text{-lower-saddle-sink}},\ \underbrace{((2m_{11}^{(i)}+1)^{\text{th}}\text{SO},(2n_{q_i1}^{(i)})^{\text{th}}\text{LS})}_{((2m_{11}^{(i)}+1),(2n_{q_i1}^{(i)}))\text{-lower-saddle-source}}
\end{array}\right\};\\
&\underbrace{((m_1^{(i)})^{\text{th}}\text{XX},(n_1^{(i)})^{\text{th}}\text{XX})}_{(m_1^{(i)},n_1^{(i)})\text{-XX}}\\
&\in \left\{\begin{array}{l}
\underbrace{((2m_{11}^{(i)}+1)^{\text{th}}\text{SO},(2n_{q_i1}^{(i)}+1)^{\text{th}}\text{SI})}_{((2m_{11}^{(i)}+1),(2n_{q_i1}^{(i)}+1))\text{-saddle}},\ \underbrace{((2m_{p_i1}^{(i)})^{\text{th}}\text{LS},(2n_{q_i1}^{(i)})^{\text{th}}\text{US})}_{((2m_{p_i1}^{(i)}),(2n_{q_i1}^{(i)}))\text{-double-saddle}},\\
\underbrace{((2m_{11}^{(i)})^{\text{th}}\text{LS},(2n_{11}^{(i)}+1)^{\text{th}}\text{SI})}_{((2m_{11}^{(i)}),(2n_{11}^{(i)}+1))\text{-lower-saddle-sink}},\ \underbrace{((2m_{p_i1}^{(i)}+1)^{\text{th}}\text{SI},(2n_{11}^{(i)})^{\text{th}}\text{US})}_{((2m_{p_i1}^{(i)}+1),(2n_{11}^{(i)}))\text{-upper-saddle-sink}}
\end{array}\right\}.
\end{aligned}
\tag{3.62}
$$

(iv_{3c}) The $((2m_1+1)^{\text{th}}\text{SO}, (2n_1+1)^{\text{th}}\text{SI})$-saddle switching bifurcation is for the switching of two $q \times p$ equilibrium networks as

$$\bigcup_{r_1=1}^{p}\bigcup_{r_1=1}^{p}\underbrace{(a_{j_1j_1r_1},a_{j_2j_2s_1})}_{(m_{r_1}^{(1)},n_{s_1}^{(1)})\text{-XX}}\rightleftharpoons\underbrace{((2m_1+1)^{\text{th}}\text{SO},(2n_1+1)^{\text{th}}\text{SI})}_{((2m_1+1),(2n_1+1))\text{-saddle}}$$
$$\rightleftharpoons\bigcup_{r_2=1}^{p}\bigcup_{s_2=1}^{q}\underbrace{(a_{j_1j_1r_2},a_{j_2j_2s_2})}_{(m_{r_2}^{(2)},n_{s_2}^{(2)})\text{-XX}}\tag{3.63}$$

with $\Sigma_{r_i=1}^{p}m_{r_i}=2m_1+1$ and $\Sigma_{s_i=1}^{q}n_{s_i}=2n_1+1$ for $i=1,2$.

(iv_4) For $a_{j_1j_10}<0$ and $a_{j_2j_20}<0$, there is a $((2m_1+1)^{\text{th}}\text{SI},(2n_1+1)^{\text{th}}\text{SI})$-sink equilibrium as

$$(a_{j_1j_11},a_{j_2j_21})=\underbrace{((2m_1+1)^{\text{th}}\text{SI},(2n_1+1)^{\text{th}}\text{SI})}_{((2m_1+1),(2n_1+1))\text{-sink}}.\tag{3.64}$$

There are the following three $((2m_1+1)^{\text{th}}\text{SI},(2n_1+1)^{\text{th}}\text{SI})$-sink appearing and switching bifurcations.

(iv_{4a}) The $((2m_1+1)^{\text{th}}\text{SI},(2n_1+1)^{\text{th}}\text{SI})$-sink appearing bifurcation is from a (SI,SI)-sink to a $(2m_1+1)\times(2n_1+1)$ network as

$$\underbrace{(a_{j_1j_11},a_{j_2j_21})}_{(\text{SI,SI})\text{-sink}}\rightleftharpoons\underbrace{((2m_1+1)^{\text{th}}\text{SI},(2n_1+1)^{\text{th}}\text{SI})}_{((2m_1+1),(2n_1+1))\text{-sink}}\rightleftharpoons\bigcup_{r=1}^{2m_1+1}\bigcup_{s=1}^{2n_1+1}\underbrace{(a_{j_1j_1r},a_{j_2j_2s})}_{\text{XX}}\tag{3.65}$$

where

$$\bigcup_{r=1}^{2m_1+1}\bigcup_{s=1}^{2n_1+1}\underbrace{(a_{j_1j_1r},a_{j_2j_2s})}_{XX}=\left\{\begin{matrix}\underbrace{(\text{SI,SI})}_{\text{sink}} & \underbrace{(\text{SO,SI})}_{\text{saddle}} & \cdots & \underbrace{(\text{SI,SI})}_{\text{sink}}\\ \underbrace{(\text{SI,SO})}_{\text{saddle}} & \underbrace{(\text{SO,SO})}_{\text{source}} & \cdots & \underbrace{(\text{SI,SO})}_{\text{saddle}}\\ \vdots & \vdots & \vdots & \vdots\\ \underbrace{(\text{SI,SI})}_{\text{sink}} & \underbrace{(\text{SO,SI})}_{\text{saddle}} & \cdots & \underbrace{(\text{SI,SI})}_{\text{sink}}\end{matrix}\right\}_{(2n_1+1)\times(2m_1+1)}.\tag{3.66}$$

(iv_{4b}) The $((2m_1+1)^{\text{th}}\text{SI},(2n_1+1)^{\text{th}}\text{SI})$-sink appearing bifurcation is from $q_1\times p_1$ to $q_2\times p_2$ equilibrium networks as

$$\bigcup_{r_1=1}^{p_1}\bigcup_{s_1=1}^{q_1}\underbrace{(a_{j_1j_1r_1},a_{j_2j_2s_1})}_{(m_{r_1}^{(1)},n_{s_1}^{(1)})\text{-XX}}\rightleftharpoons\underbrace{((2m_1+1)^{\text{th}}\text{SI},(2n_1+1)^{\text{th}}\text{SI})}_{((2m_1+1),(2n_1+1))\text{-sink}}$$

$$\rightleftharpoons \bigcup_{r_2=1}^{p_2} \bigcup_{s_2=1}^{q_2} \underbrace{(a_{j_1 j_1 r_2}, a_{j_2 j_2 s_2})}_{(m_{r_2}^{(2)}, n_{s_2}^{(2)})\text{-XX}} \tag{3.67}$$

where

$$2m_1 + 1 - \sum_{r_1=1}^{p_1} m_{r_1}^{(1)} = 2\sum_{s=1}^{l_1} m_s$$

with l_1-quadratic polynomials without real roots
and s^{th}-quadratic polynomial with power m_s,

$$2n_1 + 1 - \sum_{r_1=1}^{q_1} n_{s_1}^{(1)} = 2\sum_{l=1}^{l_2} n_l$$

with l_2-quadraitc polynomials without real roots
and l^{th}-quadratic polynomial with power n_l;

$$\sum_{r_2=1}^{p_2} m_{r_2}^{(2)} = 2m_1 + 1, \ \sum_{s_2=1}^{q_2} n_{s_2}^{(2)} = 2n_1 + 1; \text{ for } i = 1, 2,$$

$$\underbrace{((m_{p_i}^{(i)})^{\text{th}}\text{XX},(n_{q_i}^{(i)})^{\text{th}}\text{XX})}_{(m_{p_i}^{(i)},n_{q_i}^{(i)})\text{-XX}}$$

$$\in \left\{ \begin{array}{l} \underbrace{((2m_{p_i1}^{(i)}+1)^{\text{th}}\text{SI},(2n_{q_i1}^{(i)}+1)^{\text{th}}\text{SI})}_{((2m_{p_i1}^{(i)}+1),(2n_{q_i1}^{(i)}+1))\text{-sink}}, \underbrace{((2m_{p_i1}^{(i)})^{\text{th}}\text{LS},(2n_{q_i1}^{(i)})^{\text{th}}\text{LS})}_{((2m_{p_i1}^{(i)}),(2n_{q_i1}^{(i)}))\text{-double-saddle}}, \\ \underbrace{((2m_{p_i1}^{(i)})^{\text{th}}\text{LS},(2n_{q_i1}^{(i)}+1)^{\text{th}}\text{SI})}_{((2m_{p_i1}^{(i)}),(2n_{q_i1}^{(i)}+1))\text{-lower-saddle-sink}}, \underbrace{((2m_{p_i1}^{(i)}+1)^{\text{th}}\text{SI},(2n_{q_i1}^{(i)})^{\text{th}}\text{LS})}_{((2m_{p_i1}^{(i)}+1),(2n_{q_i1}^{(i)}))\text{-lower-saddle-sink}} \end{array} \right\};$$

$$\underbrace{((m_{p_i}^{(i)})^{\text{th}}\text{XX},(n_1^{(i)})^{\text{th}}\text{XX})}_{(m_{p_i}^{(i)},n_1^{(i)})\text{-XX}}$$

$$\in \left\{ \begin{array}{l} \underbrace{((2m_{p_i1}^{(i)}+1)^{\text{th}}\text{SI},(2n_{11}^{(i)}+1)^{\text{th}}\text{SI})}_{((2m_{p_i1}^{(i)}+1),(2n_{11}^{(i)}+1))\text{-sink}}, \underbrace{((2m_{p_i1}^{(i)})^{\text{th}}\text{LS},(2n_{11}^{(i)})^{\text{th}}\text{US})}_{((2m_{p_i1}^{(i)}),(2n_{11}^{(i)}))\text{-double-saddle}}, \\ \underbrace{((2m_{p_i1}^{(i)})^{\text{th}}\text{LS},(2n_{11}^{(i)}+1)^{\text{th}}\text{SI})}_{((2m_{p_i1}^{(i)}),(2n_{11}^{(i)}+1))\text{-lower-saddle-sink}}, \underbrace{((2m_{p_i1}^{(i)}+1)^{\text{th}}\text{SI},(2n_{11}^{(i)})^{\text{th}}\text{US})}_{((2m_{p_i1}^{(i)}+1),(2n_{11}^{(i)}))\text{-upper-saddle-sink}} \end{array} \right\};$$

$$
\underbrace{((m_1^{(i)})^{\text{th}}\text{XX},(n_{q_i}^{(i)})^{\text{th}}\text{XX})}_{(m_1^{(i)},n_{q_i}^{(i)})\text{-XX}}
\in \left\{ \begin{array}{l}
\underbrace{((2m_{11}^{(i)}+1)^{\text{th}}\text{SI},(2n_{q_i1}^{(i)}+1)^{\text{th}}\text{SI})}_{((2m_{11}^{(i)}+1),(2n_{q_i1}^{(i)}+1))\text{-sink}},
\underbrace{((2m_{11}^{(i)})^{\text{th}}\text{US},(2n_{q_i1}^{(i)})^{\text{th}}\text{LS})}_{((2m_{11}^{(i)}),(2n_{q_i1}^{(i)}))\text{-double-saddle}}, \\
\underbrace{((2m_{11}^{(i)})^{\text{th}}\text{US},(2n_{q_i1}^{(i)}+1)^{\text{th}}\text{SI})}_{((2m_{11}^{(i)}),(2n_{q_i1}^{(i)}+1))\text{-upper-saddle-sink}},
\underbrace{((2m_{11}^{(i)}+1)^{\text{th}}\text{SI},(2n_{q_i1}^{(i)})^{\text{th}}\text{LS})}_{((2m_{11}^{(i)}+1),(2n_{q_i1}^{(i)}))\text{-lower-saddle-sink}}
\end{array} \right\};
$$

$$
\underbrace{((m_1^{(i)})^{\text{th}}\text{XX},(n_1^{(i)})^{\text{th}}\text{XX})}_{(m_1^{(i)},n_1^{(i)})\text{-XX}}
\in \left\{ \begin{array}{l}
\underbrace{((2m_{11}^{(i)}+1)^{\text{th}}\text{SI},(2n_{11}^{(i)}+1)^{\text{th}}\text{SI})}_{((2m_{11}^{(i)}+1),(2n_{11}^{(i)}+1))\text{-sink}},
\underbrace{((2m_{11}^{(i)})^{\text{th}}\text{US},(2n_{11}^{(i)})^{\text{th}}\text{US})}_{((2m_{11}^{(i)}),(2n_{11}^{(i)}))\text{-double-saddle}}, \\
\underbrace{((2m_{11}^{(i)})^{\text{th}}\text{US},(2n_{11}^{(i)}+1)^{\text{th}}\text{SI})}_{((2m_{11}^{(i)}),(2n_{11}^{(i)}+1))\text{-upper-saddle-sink}},
\underbrace{((2m_{11}^{(i)}+1)^{\text{th}}\text{SI},(2n_{11}^{(i)})^{\text{th}}\text{US})}_{((2m_{11}^{(i)}+1),(2n_{11}^{(i)}))\text{-upper-saddle-sink}}
\end{array} \right\}. \tag{3.68}
$$

(iv_{4c}) The $((2m_1+1)^{\text{th}}\text{SI}, (2n_1+1)^{\text{th}}\text{SI})$-sink switching bifurcation is for two $q \times p$ equilibrium networks as

$$
\begin{aligned}
&\bigcup_{r_1=1}^{p}\bigcup_{s_1=1}^{q}\underbrace{(a_{j_1j_1r_1}, a_{j_2j_2s_1})}_{(m_{r_1}^{(1)},\, n_{s_1}^{(1)})\text{-XX}} \rightleftharpoons \underbrace{((2m_1+1)^{\text{th}}\text{SI},(2n_1+1)^{\text{th}}\text{SI})}_{((2m_1+1),(2n_1+1))\text{-sink}} \\
&\rightleftharpoons \bigcup_{r_2=1}^{p}\bigcup_{s_2=1}^{q}\underbrace{(a_{j_1j_1r_2}, a_{j_2j_2s_2})}_{(m_{r_2}^{(2)},\, n_{s_2}^{(2)})\text{-XX}}
\end{aligned} \tag{3.69}
$$

with $\Sigma_{r_i=1}^{p} m_{r_i} = 2m_1 + 1$ and $\Sigma_{s_i=1}^{q} n_{s_i} = 2n_1 + 1$ for $i = 1, 2$.

(v) For $m = 2m_1$ and $n = 2n_1 + 1$, the equilibrium of $(x_{j_1}^*, x_{j_2}^*) = (a_{j_1j_11}, a_{j_2j_21})$ has the following bifurcation properties.

(v_1) For $a_{j_1j_10} > 0$ and $a_{j_2j_20} > 0$, there is a $((2m_1)^{\text{th}}\text{US},(2n_1+1)^{\text{th}}\text{SO})$-upper-saddle-source equilibrium as

$$
(a_{j_1j_11}, a_{j_2j_21}) = \underbrace{((2m_1)^{\text{th}}\text{US},(2n_1+1)^{\text{th}}\text{SO})}_{((2m_1),(2n_1+1))\text{-upper-saddle-source}}. \tag{3.70}
$$

There are three following $((2m_1)^{\text{th}}\text{US},(2n_1+1)^{\text{th}}\text{SO})$-upper-saddle-source appearing and switching bifurcations.

(v_{1a}) The $((2m_1)^{\text{th}}\text{US}, (2n_1+1)^{\text{th}}\text{SO})$-upper-saddle-source appearing bifurcation are from a (pF,SO)-positive source flow to a $(2m_1) \times (2n_1+1)$-equilibrium network as

$$\underbrace{(\dot{x}_{j_1}, a_{j_2j_21})}_{\text{(pF,SO)-psotive source flow}} \rightleftharpoons \underbrace{((2m_1)^{\text{th}}\text{US},(2n_1+1)^{\text{th}}\text{SO})}_{((2m_1),(2n_1+1))\text{-upper-saddle-source}} \rightleftharpoons \bigcup_{r=1}^{2m_1}\bigcup_{s=1}^{2n_1+1} \underbrace{(a_{j_1j_1r}, a_{j_2j_2s})}_{\text{XX}} \tag{3.71}$$

where

$$\bigcup_{r=1}^{2m_1}\bigcup_{s=1}^{2n_1+1} \underbrace{(a_{j_1j_1r}, a_{j_2j_2s})}_{\text{XX}} = \begin{Bmatrix} \underbrace{\text{(SI,SO)}}_{\text{saddle}} & \underbrace{\text{(SO,SO)}}_{\text{source}} & \cdots & \underbrace{\text{(SO,SO)}}_{\text{source}} \\ \underbrace{\text{(SI,SI)}}_{\text{sink}} & \underbrace{\text{(SO,SI)}}_{\text{saddle}} & \cdots & \underbrace{\text{(SO,SI)}}_{\text{saddle}} \\ \vdots & \vdots & \vdots & \vdots \\ \underbrace{\text{(SI,SO)}}_{\text{saddle}} & \underbrace{\text{(SO,SO)}}_{\text{source}} & \cdots & \underbrace{\text{(SO,SO)}}_{\text{source}} \end{Bmatrix}_{(2n_1+1)\times(2m_1)}. \tag{3.72}$$

(v_{1b}) The $((2m_1)^{\text{th}}\text{US}, (2n_1+1)^{\text{th}}\text{SO})$-upper-saddle-source appearing bifurcation is from a $q_1 \times p_1$ to $q_2 \times p_2$ equilibrium network as

$$\begin{aligned} \bigcup_{r_1=1}^{p_1}\bigcup_{s_1=1}^{q_1} \underbrace{(a_{j_1j_1r_1}, a_{j_2j_2s_1})}_{(m_{r_1}^{(1)}, n_{s_1}^{(1)})\text{-XX}} &\rightleftharpoons \underbrace{((2m_1)^{\text{th}}\text{US},(2n_1+1)^{\text{th}}\text{SO})}_{((2m_1),(2n_1+1))\text{-upper-saddle-source}} \\ &\rightleftharpoons \bigcup_{r_2=1}^{p_2}\bigcup_{s_2=1}^{q_2} \underbrace{(a_{j_1j_1r_2}, a_{j_2j_2s_2})}_{(m_{r_2}^{(2)}, n_{s_2}^{(2)})\text{-XX}} \end{aligned} \tag{3.73}$$

where

$$2m_1 - \sum_{r_1=1}^{p_1} m_{r_1}^{(1)} = 2\sum_{s=1}^{l_1} m_s$$

with l_1-quadratic polynomials without real roots and s^{th}-quadratic polynomial with power m_s,

$$2n_1 + 1 - \sum_{s_1=1}^{q_1} n_{s_1}^{(1)} = 2\sum_{l=1}^{l_2} n_l$$

with l_2-quadraitc polynomials without real roots and l^{th}-quadratic polynomial with power n_l;

$$
\begin{array}{l}
\sum_{r_2=1}^{p_2} m_{r_2}^{(2)} = 2m_1+1,\ \sum_{s_2=1}^{q_2} n_{s_2}^{(2)} = 2n_1+1;\ \text{for}\, i=1,2,\\
\underbrace{((m_{p_i}^{(i)})^{\text{th}}\text{XX},(n_{q_i}^{(i)})^{\text{th}}\text{XX})}_{(m_{p_i}^{(i)},n_{q_i}^{(i)})\text{-XX}}\\
\in \left\{ \begin{array}{l}
\underbrace{((2m_{p_i1}^{(i)}+1)^{\text{th}}\text{SO},(2n_{q_i1}^{(i)}+1)^{\text{th}}\text{SO})}_{((2m_{p_i1}^{(i)}+1),(2n_{q_i1}^{(i)}+1))\text{-source}},
\underbrace{((2m_{p_i1}^{(i)})^{\text{th}}\text{US},(2n_{q_i1}^{(i)})^{\text{th}}\text{US})}_{((2m_{p_i1}^{(i)}),(2n_{q_i1}^{(i)}))\text{-double-saddle}},\\
\underbrace{((2m_{p_i1}^{(i)})^{\text{th}}\text{US},(2n_{q_i1}^{(i)}+1)^{\text{th}}\text{SO})}_{((2m_{p_i1}^{(i)}),(2n_{q_i1}^{(i)}+1))\text{-uuper-saddle-source}},
\underbrace{((2m_{p_i1}^{(i)}+1)^{\text{th}}\text{SO},(2n_{q_i1}^{(i)})^{\text{th}}\text{US})}_{((2m_{p_i1}^{(i)}+1),(2n_{q_i1}^{(i)}))\text{-double-saddle}}
\end{array}\right\};\\
\underbrace{((m_{p_i}^{(i)})^{\text{th}}\text{XX},(n_{1}^{(i)})^{\text{th}}\text{XX})}_{(m_{p_i}^{(i)},n_{1}^{(i)})\text{-XX}}\\
\in \left\{ \begin{array}{l}
\underbrace{((2m_{p_i1}^{(i)}+1)^{\text{th}}\text{SO},(2n_{11}^{(i)}+1)^{\text{th}}\text{SO})}_{((2m_{p_i1}^{(i)}+1),(2n_{11}^{(i)}+1))\text{-source}},
\underbrace{((2m_{p_i1}^{(i)})^{\text{th}}\text{US},(2n_{11}^{(i)})^{\text{th}}\text{LS})}_{((2m_{p_i1}^{(i)}),(2n_{11}^{(i)}))\text{-double-saddle}},\\
\underbrace{((2m_{p_i1}^{(i)})^{\text{th}}\text{US},(2n_{11}^{(i)}+1)^{\text{th}}\text{SO})}_{((2m_{p_i1}^{(i)}),(2n_{11}^{(i)}+1))\text{-upper-saddle-source}},
\underbrace{((2m_{p_i1}^{(i)}+1)^{\text{th}}\text{SO},(2n_{11}^{(i)})^{\text{th}}\text{LS})}_{((2m_{p_i1}^{(i)}+1),(2n_{11}^{(i)}))\text{-lower-saddle-source}}
\end{array}\right\};\\
\underbrace{((m_{1}^{(i)})^{\text{th}}\text{XX},(n_{q_i}^{(i)})^{\text{th}}\text{XX})}_{(m_{1}^{(i)},n_{q_i}^{(i)})\text{-XX}}\\
\in \left\{ \begin{array}{l}
\underbrace{((2m_{11}^{(i)}+1)^{\text{th}}\text{SI},(2n_{q_i1}^{(i)}+1)^{\text{th}}\text{SO})}_{((2m_{11}^{(i)}+1),(2n_{q_i1}^{(i)}+1))\text{-saddle}},
\underbrace{((2m_{11}^{(i)})^{\text{th}}\text{US},(2n_{q_i1}^{(i)})^{\text{th}}\text{US})}_{((2m_{11}^{(i)}),(2n_{q_i1}^{(i)}))\text{-double-saddle}},\\
\underbrace{((2m_{11}^{(i)})^{\text{th}}\text{US},(2n_{q_i1}^{(i)}+1)^{\text{th}}\text{SO})}_{((2m_{11}^{(i)}),(2n_{q_i1}^{(i)}+1))\text{-upper-saddle-source}},
\underbrace{((2m_{11}^{(i)}+1)^{\text{th}}\text{SI},(2n_{q_i1}^{(i)})^{\text{th}}\text{US})}_{((2m_{11}^{(i)}+1),(2n_{q_i1}^{(i)}))\text{-upper-saddle-sink}}
\end{array}\right\};\\
\underbrace{((m_{1}^{(i)})^{\text{th}}\text{XX},(n_{1}^{(i)})^{\text{th}}\text{XX})}_{(m_{1}^{(i)},n_{1}^{(i)})\text{-XX}}\\
\in \left\{ \begin{array}{l}
\underbrace{((2m_{11}^{(i)}+1)^{\text{th}}\text{SI},(2n_{11}^{(i)}+1)^{\text{th}}\text{SO})}_{((2m_{11}^{(i)}+1),(2n_{11}^{(i)}+1))\text{-saddle}},
\underbrace{((2m_{11}^{(i)})^{\text{th}}\text{US},(2n_{11}^{(i)})^{\text{th}}\text{LS})}_{((2m_{11}^{(i)}),(2n_{11}^{(i)}))\text{-double-saddle}},\\
\underbrace{((2m_{11}^{(i)})^{\text{th}}\text{US},(2n_{11}^{(i)}+1)^{\text{th}}\text{SO})}_{((2m_{11}^{(i)}),(2n_{11}^{(i)}+1))\text{-upper-saddle-source}},
\underbrace{((2m_{11}^{(i)}+1)^{\text{th}}\text{SI},(2n_{11}^{(i)})^{\text{th}}\text{LS})}_{((2m_{11}^{(i)}+1),(2n_{11}^{(i)}))\text{-lower-saddle-sink}}
\end{array}\right\}.
\end{array}
\tag{3.74}
$$

($\text{v}_{1\text{c}}$) The $((2m_1)^{\text{th}}\text{US},\ (2n_1+1)^{\text{th}}\text{SO})$-upper-saddle-source switching bifurcation is for the switching of two $q \times p$ equilibrium networks as

$$\bigcup_{r_1=1}^{p}\bigcup_{s_1=1}^{q}\underbrace{(a_{j_1j_1r_1},a_{j_2j_2s_1})}_{(m_{r_1}^{(1)},n_{s_1}^{(1)})\text{-XX}} \rightleftharpoons \underbrace{((2m_1)^{\text{th}}\text{US},(2n_1+1)^{\text{th}}\text{SO})}_{((2m_1),(2n_1+1))\text{-upper-saddle-source}}$$
$$\rightleftharpoons \bigcup_{r_2=1}^{p}\bigcup_{s_2=1}^{q}\underbrace{(a_{j_1j_1r_2},a_{j_2j_2s_2})}_{(m_{r_2}^{(2)},n_{s_2}^{(2)})\text{-XX}} \tag{3.75}$$

with $\Sigma_{r_i=1}^{p} m_{r_i} = 2m_1$ and $\Sigma_{s_i=1}^{q} n_{s_i} = 2n_1 + 1$ for $i = 1, 2$.

(v_2) For $a_{j_1j_10} < 0$ and $a_{j_2j_20} > 0$, there is a $((2m_1)^{\text{th}}\text{LS}, (2n_1+1)^{\text{th}}\text{SO})$-lower-saddle-source equilibrium as

$$(a_{j_1j_11},a_{j_2j_21}) = \underbrace{((2m_1)^{\text{th}}\text{LS},(2n_1+1)^{\text{th}}\text{SO})}_{((2m_1),(2n_1+1))\text{-lower-saddle-source}}. \tag{3.76}$$

There are the following three $((2m_1)^{\text{th}}\text{LS}, (2n_1+1)^{\text{th}}\text{SO})$-lower-saddle-source appearing and switching bifurcations.

(v_{2a}) The $((2m_1)^{\text{th}}\text{LS}, (2n_1+1)^{\text{th}}\text{SO})$-lower-saddle-source appearing bifurcation is from an (nF,SO)-negative source flow to a $(2m_1) \times (2n_1 + 1)$ network as

$$\underbrace{(\dot{x}_{j_1},a_{j_2j_21})}_{\text{(nF,SO)-negative source flow}} \rightleftharpoons \underbrace{((2m_1)^{\text{th}}\text{LS},(2n_1+1)^{\text{th}}\text{SO})}_{((2m_1),(2n_1+1))\text{-lower-saddle-source}} \rightleftharpoons \bigcup_{r=1}^{2m_1}\bigcup_{s=1}^{2n_1+1}\underbrace{(a_{j_1j_1r},a_{j_2j_2s})}_{XX} \tag{3.77}$$

where

$$\bigcup_{r=1}^{2m_1}\bigcup_{s=1}^{2n_1+1}\underbrace{(a_{j_1j_1r},a_{j_2j_2s})}_{XX} = \left\{\begin{array}{cccc} \underbrace{(\text{SO,SO})}_{\text{source}} & \underbrace{(\text{SI,SO})}_{\text{saddle}} & \cdots & \underbrace{(\text{SI,SO})}_{\text{saddle}} \\ \underbrace{(\text{SO,SI})}_{\text{saddle}} & \underbrace{(\text{SI,SI})}_{\text{sink}} & \cdots & \underbrace{(\text{SI,SI})}_{\text{sink}} \\ \vdots & \vdots & \vdots & \vdots \\ \underbrace{(\text{SO,SO})}_{\text{source}} & \underbrace{(\text{SI,SO})}_{\text{saddle}} & \cdots & \underbrace{(\text{SI,SO})}_{\text{saddle}} \end{array}\right\}_{(2n_1+1)\times(2m_1)}. \tag{3.78}$$

(v_{2b}) The $((2m_1)^{\text{th}}\text{LS}, (2n_1+1)^{\text{th}}\text{SO})$-lower-saddle-source appearing bifurcation is from $q_1 \times p_1$ to $q_2 \times p_2$ equilibrium networks as

$$\bigcup_{r_1=1}^{p_1}\bigcup_{s_1=1}^{q_1}\underbrace{(a_{j_1j_1r_1},a_{j_2j_2s_1})}_{(m_{r_1}^{(1)},n_{s_1}^{(1)})\text{-XX}} \rightleftharpoons \underbrace{((2m_1)^{\text{th}}\text{LS},(2n_1+1)^{\text{th}}\text{SO})}_{((2m_1),(2n_1+1))\text{-lower-saddle-source}}$$

$$\rightleftharpoons \bigcup_{r_2=1}^{p_2}\bigcup_{s_2=1}^{q_2} \underbrace{(a_{j_1j_1r_2}, a_{j_2j_2s_2})}_{(m_{r_2}^{(2)}, n_{s_2}^{(2)})\text{-XX}} \tag{3.79}$$

where

$$2m_1 - \sum_{r_1=1}^{p_1} m_{r_1}^{(1)} = 2\sum_{s=1}^{l_1} m_s$$

with l_1-quadratic polynomials without real roots

and s^{th}-quadratic polynomial with power m_s,

$$2n_1 + 1 - \sum_{s_1=1}^{q_1} n_{s_1}^{(1)} = 2\sum_{l=1}^{l_2} n_l$$

with l_2-quadraitc polynomials without real roots

and l^{th}-quadratic polynomial with power n_l;

$$\sum_{r_2=1}^{p_2} m_{r_2}^{(2)} = 2m_1, \ \sum_{s_2=1}^{q_2} n_{s_2}^{(2)} = 2n_1 + 1; \text{ for } i = 1, 2,$$

$$\underbrace{((m_{p_i}^{(i)})^{\text{th}}\text{XX},(n_{q_i}^{(i)})^{\text{th}}\text{XX})}_{(m_{p_i}^{(i)}, n_{q_i}^{(i)})\text{-XX}}$$

$$\in \left\{ \begin{array}{l} \underbrace{((2m_{p_i1}^{(i)}+1)^{\text{th}}\text{SI},(2n_{q_i1}^{(i)}+1)^{\text{th}}\text{SO})}_{((2m_{p_i1}^{(i)}+1),(2n_{q_i1}^{(i)}+1))\text{-saddle}}, \underbrace{((2m_{p_i1}^{(i)})^{\text{th}}\text{LS},(2n_{q_i1}^{(i)})^{\text{th}}\text{US})}_{((2m_{p_i1}^{(i)}),(2n_{q_i1}^{(i)}))\text{-double-saddle}}, \\ \underbrace{((2m_{p_i1}^{(i)})^{\text{th}}\text{LS},(2n_{q_i1}^{(i)}+1)^{\text{th}}\text{SO})}_{((2m_{p_i1}^{(i)}),(2n_{q_i1}^{(i)}+1))\text{-lower-saddle-source}}, \underbrace{((2m_{p_i1}^{(i)}+1)^{\text{th}}\text{SI},(2n_{q_i1}^{(i)})^{\text{th}}\text{US})}_{((2m_{p_i1}^{(i)}+1),(2n_{q_i1}^{(i)}))\text{-upper-saddle-sink}} \end{array} \right\};$$

$$\underbrace{((m_{p_i}^{(i)})^{\text{th}}\text{XX},(n_1^{(i)})^{\text{th}}\text{XX})}_{(m_{p_i}^{(i)}, n_1^{(i)})\text{-XX}}$$

$$\in \left\{ \begin{array}{l} \underbrace{((2m_{p_i1}^{(i)}+1)^{\text{th}}\text{SI},(2n_{11}^{(i)}+1)^{\text{th}}\text{SO})}_{((2m_{p_i1}^{(i)}+1),(2n_{11}^{(i)}+1))\text{-saddle}}, \underbrace{((2m_{p_i1}^{(i)})^{\text{th}}\text{LS},(2n_{11}^{(i)})^{\text{th}}\text{LS})}_{((2m_{p_i1}^{(i)}),(2n_{11}^{(i)}))\text{-double-saddle}}, \\ \underbrace{((2m_{p_i1}^{(i)})^{\text{th}}\text{LS},(2n_{11}^{(i)}+1)^{\text{th}}\text{SO})}_{((2m_{p_i1}^{(i)}),(2n_{11}^{(i)}+1))\text{-lower-saddle-source}}, \underbrace{((2m_{p_i1}^{(i)}+1)^{\text{th}}\text{SI},(2n_{11}^{(i)})^{\text{th}}\text{LS})}_{((2m_{p_i1}^{(i)}+1),(2n_{11}^{(i)}))\text{-lower-saddle-sink}} \end{array} \right\};$$

$$\underbrace{((m_1^{(i)})^{\text{th}}\text{XX},(n_{q_i}^{(i)})^{\text{th}}\text{XX})}_{(m_1^{(i)},n_{q_i}^{(i)})\text{-XX}}
\in \left\{ \begin{array}{l} \underbrace{((2m_{11}^{(i)}+1)^{\text{th}}\text{SO},(2n_{q_i 1}^{(i)}+1)^{\text{th}}\text{SO})}_{((2m_{11}^{(i)}+1),(2n_{q_i 1}^{(i)}+1))\text{-source}}, \underbrace{((2m_{11}^{(i)})^{\text{th}}\text{LS},(2n_{q_i 1}^{(i)})^{\text{th}}\text{US})}_{((2m_{11}^{(i)}),(2n_{q_i 1}^{(i)}))\text{-double-saddle}}, \\ \underbrace{((2m_{11}^{(i)})^{\text{th}}\text{LS},(2n_{q_i 1}^{(i)}+1)^{\text{th}}\text{SO})}_{((2m_{11}^{(i)}),(2n_{q_i 1}^{(i)}+1))\text{-lower-saddle-source}}, \underbrace{((2m_{11}^{(i)}+1)^{\text{th}}\text{SO},(2n_{q_i 1}^{(i)})^{\text{th}}\text{US})}_{((2m_{11}^{(i)}+1),(2n_{q_i 1}^{(i)}))\text{-upper-saddle-source}} \end{array} \right\};$$

$$\underbrace{((m_1^{(i)})^{\text{th}}\text{XX},(n_1^{(i)})^{\text{th}}\text{XX})}_{(m_1^{(i)},n_1^{(i)})\text{-XX}}
\in \left\{ \begin{array}{l} \underbrace{((2m_{11}^{(i)}+1)^{\text{th}}\text{SO},(2n_{11}^{(i)}+1)^{\text{th}}\text{SO})}_{((2m_{11}^{(i)}+1),(2n_{11}^{(i)}+1))\text{-source}}, \underbrace{((2m_{11}^{(i)})^{\text{th}}\text{LS},(2n_{11}^{(i)})^{\text{th}}\text{LS})}_{((2m_{11}^{(i)}),(2n_{11}^{(i)}))\text{-double-saddle}}, \\ \underbrace{((2m_{11}^{(i)})^{\text{th}}\text{LS},(2n_{11}^{(i)}+1)^{\text{th}}\text{SO})}_{((2m_{11}^{(i)}),(2n_{11}^{(i)}+1))\text{-lower-saddle-source}}, \underbrace{((2m_{11}^{(i)}+1)^{\text{th}}\text{SO},(2n_{11}^{(i)})^{\text{th}}\text{LS})}_{((2m_{11}^{(i)}+1),(2n_{11}^{(i)}))\text{-lower-saddle-source}} \end{array} \right\}. \tag{3.80}$$

(v_{2c}) The $((2m_1)^{\text{th}}\text{LS}, (2n_1+1)^{\text{th}}\text{SO})$-lower-saddle-source switching bifurcation is for two $q \times p$ equilibrium networks as

$$\begin{aligned} \bigcup_{r_1=1}^{p}\bigcup_{s_1=1}^{q} \underbrace{(a_{j_1 j_1 r_1}, a_{j_2 j_2 s_1})}_{(m_{r_1}^{(1)},\, n_{s_1}^{(1)})\text{-XX}} &\rightleftharpoons \underbrace{((2m_1)^{\text{th}}\text{LS},(2n_1+1)^{\text{th}}\text{SO})}_{((2m_1),(2n_1+1))\text{-lower-saddle-source}} \\ &\rightleftharpoons \bigcup_{r_2=1}^{p}\bigcup_{s_2=1}^{q} \underbrace{(a_{j_1 j_1 r_2}, a_{j_2 j_2 s_2})}_{(m_{r_2}^{(2)},\, n_{s_2}^{(2)})\text{-XX}} \end{aligned} \tag{3.81}$$

with $\Sigma_{r_i=1}^{p} m_{r_i} = 2m_1$ and $\Sigma_{s_i=1}^{q} n_{s_i} = 2n_1 + 1$ for $i = 1, 2$.

(v_3) For $a_{j_1 j_1 0} > 0$ and $a_{j_2 j_2 0} < 0$, there is a $((2m_1)^{\text{th}}\text{US}, (2n_1+1)^{\text{th}}\text{SI})$-upper-saddle-sink equilibrium as

$$(a_{j_1 j_1 1}, a_{j_2 j_2 1}) = \underbrace{((2m_1)^{\text{th}}\text{US},(2n_1+1)^{\text{th}}\text{SI})}_{((2m_1),(2n_1+1))\text{-upper-saddle-sink}}. \tag{3.82}$$

There are the following three $((2m_1)^{\text{th}}\text{US}, (2n_1+1)^{\text{th}}\text{SI})$-upper-saddle-sink appearing and switching bifurcations.

(v_{3a}) The $((2m_1)^{\text{th}}\text{US}, (2n_1+1)^{\text{th}}\text{SI})$-upper-saddle-sink appearing bifurcation is from a (pF,SI)-positive sink flow to a $(2m_1) \times (2n_1 + 1)$-equilibrium network as

$$\underbrace{(\dot{x}_{j_1}, a_{j_2 j_2 1})}_{\text{(pF,SI)-positive sink flow}} \rightleftharpoons \underbrace{((2m_1)^{\text{th}}\text{US},(2n_1+1)^{\text{th}}\text{SI})}_{((2m_1),(2n_1+1))\text{-upper-saddle-sink}} \rightleftharpoons \bigcup_{r=1}^{2m_1} \bigcup_{s=1}^{2n_1+1} \underbrace{(a_{j_1 j_1 r}, a_{j_2 j_2 s})}_{\text{XX}} \quad (3.83)$$

where

$$\bigcup_{r=1}^{2m_1} \bigcup_{s=1}^{2n_1+1} \underbrace{(a_{j_1 j_1 r}, a_{j_2 j_2 s})}_{\text{XX}} = \left\{ \begin{array}{cccc} \underbrace{\text{(SI,SI)}}_{\text{sink}} & \underbrace{\text{(SO,SI)}}_{\text{saddle}} & \cdots & \underbrace{\text{(SO,SI)}}_{\text{saddle}} \\ \underbrace{\text{(SI,SO)}}_{\text{saddle}} & \underbrace{\text{(SO,SO)}}_{\text{source}} & \cdots & \underbrace{\text{(SO,SO)}}_{\text{source}} \\ \vdots & \vdots & \vdots & \vdots \\ \underbrace{\text{(SI,SI)}}_{\text{sink}} & \underbrace{\text{(SO,SI)}}_{\text{saddle}} & \cdots & \underbrace{\text{(SO,SI)}}_{\text{saddle}} \end{array} \right\}_{(2n_1+1)\times(2m_1)} . \quad (3.84)$$

(v_{3b}) The $((2m_1)^{\text{th}}\text{US}, (2n_1+1)^{\text{th}}\text{SI})$-upper-saddle-sink appearing bifurcation is from a $q_1 \times p_1$ to $q_2 \times p_2$ equilibrium network as

$$\begin{aligned} &\bigcup_{r_1=1}^{p_1} \bigcup_{s_1=1}^{q_1} \underbrace{(a_{j_1 j_1 r_1}, a_{j_2 j_2 s_1})}_{(m_{r_1}^{(1)}, n_{s_1}^{(1)})\text{-XX}} \rightleftharpoons \underbrace{((2m_1)^{\text{th}}\text{US},(2n_1+1)^{\text{th}}\text{SI})}_{((2m_1),(2n_1+1))\text{-upper-saddle-sink}} \\ &\rightleftharpoons \bigcup_{r_2=1}^{p_2} \bigcup_{s_2=1}^{q_2} \underbrace{(a_{j_1 j_1 r_2}, a_{j_2 j_2 s_2})}_{(m_{r_2}^{(2)}, n_{s_2}^{(2)})\text{-XX}} \end{aligned} \quad (3.85)$$

where

$$2m_1 - \sum_{r_1=1}^{p_1} m_{r_1}^{(1)} = 2\sum_{s=1}^{l_1} m_s$$

with l_1-quadratic polynomials without real roots

and s^{th}-quadratic polynomial with power m_s,

$$2n_1 + 1 - \sum_{s_1=1}^{q_1} n_{s_1}^{(1)} = 2\sum_{l=1}^{l_2} n_l$$

with l_2-quadraitc polynomials without real roots

and l^{th}-quadratic polynomial with power n_l;

$$\sum_{r_2=1}^{p_2} m_{r_2}^{(2)} = 2m_1, \ \sum_{r_2=1}^{q_2} n_{s_2}^{(2)} = 2n_1 + 1; \ \text{for } i = 1, 2,$$

$$\underbrace{((m_{p_i}^{(i)})^{\text{th}}\text{XX},(n_{q_i}^{(i)})^{\text{th}}\text{XX})}_{(m_{p_i}^{(i)},n_{q_i}^{(i)})\text{-XX}} \in \left\{ \begin{array}{l} \underbrace{((2m_{p_i1}^{(i)}+1)^{\text{th}}\text{SO},(2n_{q_i1}^{(i)}+1)^{\text{th}}\text{SI})}_{((2m_{p_i1}^{(i)}+1),(2n_{q_i1}^{(i)}+1))\text{-saddle}}, \underbrace{((2m_{p_i1}^{(i)})^{\text{th}}\text{US},(2n_{q_i1}^{(i)})^{\text{th}}\text{LS})}_{((2m_{p_i1}^{(i)}),(2n_{q_i1}^{(i)}))\text{-double-saddle}}, \\ \underbrace{((2m_{p_i1}^{(i)})^{\text{th}}\text{US},(2n_{q_i1}^{(i)}+1)^{\text{th}}\text{SI})}_{((2m_{p_i1}^{(i)}),(2n_{q_i1}^{(i)}+1))\text{-upper-saddle-sink}}, \underbrace{((2m_{p_i1}^{(i)}+1)^{\text{th}}\text{SO},(2n_{q_i1}^{(i)})^{\text{th}}\text{LS})}_{((2m_{p_i1}^{(i)}+1),(2n_{q_i1}^{(i)}))\text{-lower-saddle-source}} \end{array} \right\};$$

$$\underbrace{((m_{p_i}^{(i)})^{\text{th}}\text{XX},(n_{1}^{(i)})^{\text{th}}\text{XX})}_{(m_{p_i}^{(i)},n_{1}^{(i)})\text{-XX}} \in \left\{ \begin{array}{l} \underbrace{((2m_{p_i1}^{(i)}+1)^{\text{th}}\text{SO},(2n_{11}^{(i)}+1)^{\text{th}}\text{SI})}_{((2m_{p_i1}^{(i)}+1),(2n_{11}^{(i)}+1))\text{-saddle}}, \underbrace{((2m_{p_i1}^{(i)})^{\text{th}}\text{US},(2n_{11}^{(i)})^{\text{th}}\text{US})}_{((2m_{p_i1}^{(i)}),(2n_{11}^{(i)}))\text{-double-saddle}}, \\ \underbrace{((2m_{p_i1}^{(i)})^{\text{th}}\text{US},(2n_{11}^{(i)}+1)^{\text{th}}\text{SI})}_{((2m_{p_i1}^{(i)}),(2n_{11}^{(i)}+1))\text{-upper-saddle-sink}}, \underbrace{((2m_{p_i1}^{(i)}+1)^{\text{th}}\text{SO},(2n_{11}^{(i)})^{\text{th}}\text{US})}_{((2m_{p_i1}^{(i)}+1),(2n_{11}^{(i)}))\text{-upper-saddle-source}} \end{array} \right\};$$

$$\underbrace{((m_{1}^{(i)})^{\text{th}}\text{XX},(n_{q_i}^{(i)})^{\text{th}}\text{XX})}_{(m_{1}^{(i)},n_{q_i}^{(i)})\text{-XX}} \in \left\{ \begin{array}{l} \underbrace{((2m_{11}^{(i)}+1)^{\text{th}}\text{SI},(2n_{q_i1}^{(i)}+1)^{\text{th}}\text{SI})}_{((2m_{11}^{(i)}+1),(2n_{q_i1}^{(i)}+1))\text{-sink}}, \underbrace{((2m_{11}^{(i)})^{\text{th}}\text{US},(2n_{q_i1}^{(i)})^{\text{th}}\text{LS})}_{((2m_{11}^{(i)}),(2n_{q_i1}^{(i)}))\text{-double-saddle}}, \\ \underbrace{((2m_{11}^{(i)})^{\text{th}}\text{US},(2n_{q_i1}^{(i)}+1)^{\text{th}}\text{SI})}_{((2m_{11}^{(i)}),(2n_{q_i1}^{(i)}+1))\text{-upper-saddle-sink}}, \underbrace{((2m_{11}^{(i)}+1)^{\text{th}}\text{SI},(2n_{q_i1}^{(i)})^{\text{th}}\text{LS})}_{((2m_{11}^{(i)}+1),(2n_{q_i1}^{(i)}))\text{-lower-saddle-sink}} \end{array} \right\};$$

$$\underbrace{((m_{1}^{(i)})^{\text{th}}\text{XX},(n_{1}^{(i)})^{\text{th}}\text{XX})}_{(m_{1}^{(i)},n_{1}^{(i)})\text{-XX}} \in \left\{ \begin{array}{l} \underbrace{((2m_{11}^{(i)}+1)^{\text{th}}\text{SI},(2n_{11}^{(i)}+1)^{\text{th}}\text{SI})}_{((2m_{11}^{(i)}+1),(2n_{11}^{(i)}+1))\text{-sink}}, \underbrace{((2m_{11}^{(i)})^{\text{th}}\text{US},(2n_{11}^{(i)})^{\text{th}}\text{US})}_{((2m_{11}^{(i)}),(2n_{11}^{(i)}))\text{-double-saddle}}, \\ \underbrace{((2m_{11}^{(i)})^{\text{th}}\text{US},(2n_{11}^{(i)}+1)^{\text{th}}\text{SI})}_{((2m_{11}^{(i)}),(2n_{11}^{(i)}+1))\text{-upper-saddle-sink}}, \underbrace{((2m_{11}^{(i)}+1)^{\text{th}}\text{SI},(2n_{11}^{(i)})^{\text{th}}\text{US})}_{((2m_{11}^{(i)}+1),(2n_{11}^{(i)}))\text{-upper-saddle-sink}} \end{array} \right\}. \tag{3.86}$$

(v_{3c}) The $((2m_1)^{\text{th}}\text{US}, (2n_1+1)^{\text{th}}\text{SI})$-upper-saddle-sink switching bifurcation is for the switching of two $q \times p$ equilibrium networks as

$$\bigcup_{r_1=1}^{p}\bigcup_{s_1=1}^{q}\underbrace{(a_{j_1j_1r_1}, a_{j_2j_2s_1})}_{(m_{r_1}^{(1)},\, n_{s_1}^{(1)})\text{-XX}} \rightleftharpoons \underbrace{((2m_1)^{\text{th}}\text{US},(2n_1+1)^{\text{th}}\text{SI})}_{((2m_1),(2n_1+1))\text{-upper-saddle-sink}}$$

$$\rightleftharpoons \bigcup_{r_2=1}^{p}\bigcup_{s_2=1}^{q}\underbrace{(a_{j_1j_1r_2}, a_{j_2j_2s_2})}_{(m_{r_2}^{(2)}, n_{s_2}^{(2)})\text{-XX}} \tag{3.87}$$

where $\Sigma_{r_i=1}^{p} m_{r_i} = 2m_1$, $\Sigma_{s_i=1}^{q} n_{s_i} = 2n_1 + 1$ for $i = 1, 2$.

(v_4) For $a_{j_1j_10} < 0$ and $a_{j_2j_20} < 0$, there is a $((2m_1)^{\text{th}}\text{LS}, (2n_1{+}1)^{\text{th}}\text{SI})$-lower-saddle-sink equilibrium as

$$(a_{j_1j_11}, a_{j_2j_21}) = \underbrace{((2m_1)^{\text{th}}\text{LS}, (2n_1{+}1)^{\text{th}}\text{SI})}_{((2m_1),(2n_1+1))\text{-lower-saddle-sink}} \tag{3.88}$$

There are the following three $((2m_1)^{\text{th}}\text{LS},(2n_1{+}1)^{\text{th}}\text{SI})$-lower-saddle-sink appearing and switching bifurcations.

(v_{4a}) The $((2m_1)^{\text{th}}\text{LS},(2n_1{+}1)^{\text{th}}\text{SI})$-lower-saddle-sink appearing bifurcation is from an (nF,SI)-negative sink flow to a $(2m_1) \times (2n_1 + 1)$ network as

$$\underbrace{(\dot{x}_{j_1}, a_{j_2j_21})}_{\text{(nF,SI)-negative sink flow}} \rightleftharpoons \underbrace{((2m_1)^{\text{th}}\text{LS},(2n_1{+}1)^{\text{th}}\text{SI})}_{((2m_1),(2n_1+1))\text{-lower-saddle-sink}} \rightleftharpoons \bigcup_{r=1}^{2m_1}\bigcup_{s=1}^{2n_1+1}\underbrace{(a_{j_1j_1r}, a_{j_2j_2s})}_{XX} \tag{3.89}$$

where

$$\bigcup_{r=1}^{2m_1}\bigcup_{s=1}^{2n_1+1}\underbrace{(a_{j_1j_1r}, a_{j_2j_2s})}_{XX} = \left\{\begin{matrix} \underbrace{\text{(SO,SI)}}_{\text{saddle}} & \underbrace{\text{(SI,SI)}}_{\text{sink}} & \cdots & \underbrace{\text{(SI,SI)}}_{\text{sink}} \\ \underbrace{\text{(SO,SO)}}_{\text{source}} & \underbrace{\text{(SI,SO)}}_{\text{saddle}} & \cdots & \underbrace{\text{(SI,SO)}}_{\text{saddle}} \\ \vdots & \vdots & \cdots & \vdots \\ \underbrace{\text{(SO,SI)}}_{\text{saddle}} & \underbrace{\text{(SI,SI)}}_{\text{sink}} & \cdots & \underbrace{\text{(SI,SI)}}_{\text{sink}} \end{matrix}\right\}_{(2n_1+1)\times(2m_1)}. \tag{3.90}$$

(v_{4b}) The $((2m_1)^{\text{th}}\text{LS}, (2n_1{+}1)^{\text{th}}\text{SI})$-lower-saddle-sink appearing bifurcation is from $q_1 \times p_1$ to $q_2 \times p_2$ equilibrium networks as

$$\begin{aligned} &\bigcup_{r_1=1}^{p_1}\bigcup_{s_1=1}^{q_1}\underbrace{(a_{j_1j_1r_1}, a_{j_2j_2s_1})}_{(m_{r_1}^{(1)}, n_{s_1}^{(1)})\text{-XX}} \rightleftharpoons \underbrace{((2m_1)^{\text{th}}\text{LS},(2n_1{+}1)^{\text{th}}\text{SI})}_{((2m_1),(2n_1+1))\text{-lower-saddle-sink}} \\ &\rightleftharpoons \bigcup_{r_2=1}^{p_2}\bigcup_{s_2=1}^{q_2}\underbrace{(a_{j_1j_1r_2}, a_{j_2j_2s_2})}_{(m_{r_2}^{(2)}, n_{s_2}^{(2)})\text{-XX}} \end{aligned} \tag{3.91}$$

where

$$2m_1 - \sum_{r_1=1}^{p_1} m_{r_1}^{(1)} = 2\sum_{s=1}^{l_1} m_s$$

with l_1-quadratic polynomials without real roots

and s^{th}-quadratic polynomial with power m_s,

$$2n_1 + 1 - \sum_{s_1=1}^{q_1} n_{s_1}^{(1)} = 2\sum_{l=1}^{l_2} n_l$$

with l_2-quadraitc polynomials without real roots

and l^{th}-quadratic polynomial with power n_l;

$$\sum_{r_2=1}^{p_2} m_{r_2}^{(2)} = 2m_1, \ \sum_{s_2=1}^{q_2} n_{s_2}^{(2)} = 2n_1 + 1; \text{ for } i = 1, 2,$$

$$\underbrace{((m_{p_i}^{(i)})^{\text{th}}\text{XX},(n_{q_i}^{(i)})^{\text{th}}\text{XX})}_{(m_{p_i}^{(i)},n_{q_i}^{(i)})\text{-XX}}$$
$$\in \left\{ \begin{array}{l} \underbrace{((2m_{p_i1}^{(i)}+1)^{\text{th}}\text{SI},(2n_{q_i1}^{(i)}+1)^{\text{th}}\text{SI})}_{((2m_{p_i1}^{(i)}+1),(2n_{q_i1}^{(i)}+1))\text{-sink}}, \underbrace{((2m_{p_i1}^{(i)})^{\text{th}}\text{LS},(2n_{q_i1}^{(i)})^{\text{th}}\text{LS})}_{((2m_{p_i1}^{(i)}),(2n_{q_i1}^{(i)}))\text{-double-saddle}}, \\ \underbrace{((2m_{p_i1}^{(i)})^{\text{th}}\text{LS},(2n_{q_i1}^{(i)}+1)^{\text{th}}\text{SI})}_{((2m_{p_i1}^{(i)}),(2n_{q_i1}^{(i)}+1))\text{-lower-saddle-sink}}, \underbrace{((2m_{p_i1}^{(i)}+1)^{\text{th}}\text{SI},(2n_{q_i1}^{(i)})^{\text{th}}\text{LS})}_{((2m_{p_i1}^{(i)}+1),(2n_{q_i1}^{(i)}))\text{-lower-saddle-sink}} \end{array} \right\},$$

$$\underbrace{((m_{p_i}^{(i)})^{\text{th}}\text{XX},(n_1^{(i)})^{\text{th}}\text{XX})}_{(m_{p_i}^{(i)},n_1^{(i)})\text{-XX}}$$
$$\in \left\{ \begin{array}{l} \underbrace{((2m_{p_i1}^{(i)}+1)^{\text{th}}\text{SI},(2n_{11}^{(i)}+1)^{\text{th}}\text{SI})}_{((2m_{p_i1}^{(i)}+1),(2n_{11}^{(i)}+1))\text{-sink}}, \underbrace{((2m_{p_i1}^{(i)})^{\text{th}}\text{LS},(2n_{11}^{(i)})^{\text{th}}\text{US})}_{((2m_{p_i1}^{(i)}),(2n_{11}^{(i)}))\text{-double-saddle}}, \\ \underbrace{((2m_{p_i1}^{(i)})^{\text{th}}\text{LS},(2n_{11}^{(i)}+1)^{\text{th}}\text{SI})}_{((2m_{p_i1}^{(i)}),(2n_{11}^{(i)}+1))\text{-lower-saddle-sink}}, \underbrace{((2m_{p_i1}^{(i)}+1)^{\text{th}}\text{SI},(2n_{11}^{(i)})^{\text{th}}\text{US})}_{((2m_{p_i1}^{(i)}+1),(2n_{11}^{(i)}))\text{-upper-saddle-sink}} \end{array} \right\};$$

$$\underbrace{((m_1^{(i)})^{\text{th}}\text{XX},(n_{q_i}^{(i)})^{\text{th}}\text{XX})}_{(m_1^{(i)},n_{q_i}^{(i)})\text{-XX}}$$
$$\in \left\{ \begin{array}{l} \underbrace{((2m_{11}^{(i)}+1)^{\text{th}}\text{SO},(2n_{q_i1}^{(i)}+1)^{\text{th}}\text{SI})}_{((2m_{11}^{(i)}+1),(2n_{q_i1}^{(i)}+1))\text{-saddle}}, \underbrace{((2m_{11}^{(i)})^{\text{th}}\text{LS},(2n_{q_i1}^{(i)})^{\text{th}}\text{LS})}_{((2m_{11}^{(i)}),(2n_{q_i1}^{(i)}))\text{-double-saddle}}, \\ \underbrace{((2m_{11}^{(i)})^{\text{th}}\text{LS},(2n_{q_i1}^{(i)}+1)^{\text{th}}\text{SI})}_{((2m_{11}^{(i)}),(2n_{q_i1}^{(i)}+1))\text{-lower-saddle-sink}}, \underbrace{((2m_{11}^{(i)}+1)^{\text{th}}\text{SO},(2n_{q_i1}^{(i)})^{\text{th}}\text{LS})}_{((2m_{11}^{(i)}+1),(2n_{q_i1}^{(i)}))\text{-lower-saddle-source}} \end{array} \right\};$$

$$\underbrace{((m_1^{(i)})^{\text{th}}\text{XX},(n_1^{(i)})^{\text{th}}\text{XX})}_{(m_1^{(i)},n_1^{(i)})\text{-XX}}$$
$$\in \left\{ \begin{array}{l} \underbrace{((2m_{11}^{(i)}+1)^{\text{th}}\text{SO},(2n_{11}^{(i)}+1)^{\text{th}}\text{SI})}_{((2m_{11}^{(i)}+1),(2n_{11}^{(i)}+1))\text{-saddle}}, \underbrace{((2m_{11}^{(i)})^{\text{th}}\text{LS},(2n_{11}^{(i)})^{\text{th}}\text{US})}_{((2m_{11}^{(i)}),(2n_{11}^{(i)}))\text{-double-saddle}}, \\ \underbrace{((2m_{11}^{(i)})^{\text{th}}\text{LS},(2n_{11}^{(i)}+1)^{\text{th}}\text{SI})}_{((2m_{11}^{(i)}),(2n_{11}^{(i)}+1))\text{-lower-saddle-sink}}, \underbrace{((2m_{11}^{(i)}+1)^{\text{th}}\text{SO},(2n_{11}^{(i)})^{\text{th}}\text{US})}_{((2m_{11}^{(i)}+1),(2n_{11}^{(i)}))\text{-upper-saddle-source}} \end{array} \right\}. \tag{3.92}$$

(v_{4c}) The $((2m_1)^{\text{th}}\text{LS}, (2n_1+1)^{\text{th}}\text{SI})$-lower-saddle-sink switching bifurcation is for two $q \times p$ equilibrium networks as

$$\begin{aligned} \bigcup_{r_1=1}^{p}\bigcup_{s_1=1}^{q}\underbrace{(a_{j_1j_1r_1}, a_{j_2j_2s_1})}_{(m_{r_1}^{(1)},n_{s_1}^{(1)})\text{-XX}} &\rightleftharpoons \underbrace{((2m_1)^{\text{th}}\text{LS},(2n_1+1)^{\text{th}}\text{SI})}_{((2m_1),(2n_1+1))\text{-lower-saddle-sink}} \\ &\rightleftharpoons \bigcup_{r_2=1}^{p}\bigcup_{s_2=1}^{q}\underbrace{(a_{j_1j_1r_2}, a_{j_2j_2s_2})}_{(m_{r_2}^{(2)},n_{s_2}^{(2)})\text{-XX}} \end{aligned} \tag{3.93}$$

where $\Sigma_{r_i=1}^{p} m_{r_i} = 2m_1$, $\Sigma_{s_i=1}^{q} n_{s_i} = 2n_1 + 1$ for $i = 1, 2$.

(vi) For $m = 2m_1 + 1$ and $n = 2n_1$, the equilibrium of $(x_{j_1}^*, x_{j_2}^*) = (a_{j_1j_1 1}, a_{j_2j_2 1})$ has the following properties.

(vi_1) For $a_{j_1j_10} > 0$ and $a_{j_2j_20} > 0$, there is a $((2m_1+1)^{\text{th}}\text{SO},(2n_1)^{\text{th}}\text{US})$-upper-saddle-source equilibrium as

$$(a_{j_1j_1 1}, a_{j_2j_2 1}) = \underbrace{((2m_1+1)^{\text{th}}\text{SO},(2n_1)^{\text{th}}\text{US})}_{((2m_1+1),(2n_1))\text{-upper-saddle-source}}. \tag{3.94}$$

There are three following $((2m_1+1)^{\text{th}}\text{SO},(2n_1)^{\text{th}}\text{US})$-upper-saddle-source appearing and switching bifurcations.

(vi_{1a}) The $((2m_1+1)^{\text{th}}\text{SO},(2n_1)^{\text{th}}\text{US})$-upper-saddle-source appearing bifurcation is from an (SO,pF)-positive source flow to a $(2m_1 + 1) \times (2n_1)$-equilibrium network as

$$\begin{aligned} \underbrace{(a_{j_1j_1 1}, \dot{x}_{j_2})}_{\text{(SO,pF)-positive source flow}} &\Rightarrow \underbrace{((2m_1+1)^{\text{th}}\text{SO},(2n_1)^{\text{th}}\text{US})}_{((2m_1+1),(2n_1))\text{-upper-saddle-source}} \\ &\Rightarrow \bigcup_{r=1}^{2m_1+1}\bigcup_{s=1}^{2n_1}\underbrace{(a_{j_1j_1r}, a_{j_2j_2s})}_{\text{XX}} \end{aligned} \tag{3.95}$$

where

$$\bigcup_{r=1}^{2m_1+1}\bigcup_{s=1}^{2n_1}\underbrace{(a_{j_1j_1r},a_{j_2j_2s})}_{\text{XX}} = \left\{\begin{matrix} \underbrace{\text{(SO,SO)}}_{\text{source}} & \underbrace{\text{(SI,SO)}}_{\text{saddle}} & \cdots & \underbrace{\text{(SO,SO)}}_{\text{source}} \\ \underbrace{\text{(SO,SI)}}_{\text{saddle}} & \underbrace{\text{(SI,SI)}}_{\text{sink}} & \cdots & \underbrace{\text{(SO,SI)}}_{\text{saddle}} \\ \vdots & \vdots & \vdots & \vdots \\ \underbrace{\text{(SO,SI)}}_{\text{saddle}} & \underbrace{\text{(SI,SI)}}_{\text{sink}} & \cdots & \underbrace{\text{(SO,SI)}}_{\text{saddle}} \end{matrix}\right\}_{(2n_1)\times(2m_1+1)}. \tag{3.96}$$

(vi_{1b}) The $((2m_1+1)^{\text{th}}\text{SO}, (2n_1)^{\text{th}}\text{US})$-upper-sadddle-source appearing bifurcation is from a $q_1 \times p_1$ to $q_2 \times p_2$ equilibrium network as

$$\begin{aligned} &\bigcup_{r_1=1}^{p_1}\bigcup_{s_1=1}^{q_1}\underbrace{(a_{j_1j_1r_1},a_{j_2j_2s_1})}_{(m_{r_1}^{(1)},n_{s_1}^{(1)})\text{-XX}} \rightleftharpoons \underbrace{((2m_1+1)^{\text{th}}\text{SO},(2n_1)^{\text{th}}\text{US})}_{((2m_1+1),(2n_1))\text{-upper-saddle-source}} \\ &\rightleftharpoons \bigcup_{r_2=1}^{p_2}\bigcup_{s_2=1}^{q_2}\underbrace{(a_{j_1j_1r_2},a_{j_2j_2s_2})}_{(m_{r_2}^{(2)},n_{s_2}^{(2)})\text{-XX}} \end{aligned} \tag{3.97}$$

where

$$2m_1+1-\sum_{r_1=1}^{p_1} m_{r_1}^{(1)} = 2\sum_{s=1}^{l_1} m_s$$

with l_1-quadratic polynomials without real roots

and s^{th}-quadratic polynomial with power m_s,

$$2n_1-\sum_{s_1=1}^{q_1} n_{s_1}^{(1)} = 2\sum_{l=1}^{l_2} n_l$$

with l_2-quadraitc polynomials without real roots

and l^{th}-quadratic polynomial with power n_l;

$$\sum_{r_2=1}^{p_2} m_{r_2}^{(2)} = 2m_1+1,\ \sum_{s_2=1}^{q_2} n_{s_2}^{(2)} = 2n_1;\ \text{for } i=1,2,$$

$$
\begin{aligned}
&\underbrace{((m_{p_i}^{(i)})^{\text{th}}\text{XX},(n_{q_i}^{(i)})^{\text{th}}\text{XX})}_{(m_{p_i}^{(i)},n_{q_i}^{(i)})\text{-XX}} \\
&\in \left\{ \begin{array}{l}
\underbrace{((2m_{p_i1}^{(i)}+1)^{\text{th}}\text{SO},(2n_{q_i1}^{(i)}+1)^{\text{th}}\text{SO})}_{((2m_{p_i1}^{(i)}+1),(2n_{q_i1}^{(i)}+1))\text{-source}},\ \underbrace{((2m_{p_i1}^{(i)})^{\text{th}}\text{US},(2n_{q_i1}^{(i)})^{\text{th}}\text{US})}_{((2m_{p_i1}^{(i)}),(2n_{q_i1}^{(i)}))\text{-double-saddle}}, \\
\underbrace{((2m_{p_i1}^{(i)})^{\text{th}}\text{US},(2n_{q_i1}^{(i)}+1)^{\text{th}}\text{SO})}_{((2m_{p_i1}^{(i)}),(2n_{q_i1}^{(i)}+1))\text{-upper-saddle-source}},\ \underbrace{((2m_{p_i1}^{(i)}+1)^{\text{th}}\text{SO},(2n_{q_i1}^{(i)})^{\text{th}}\text{US})}_{((2m_{p_i1}^{(i)}+1),(2n_{q_i1}^{(i)}))\text{-upper-saddle-source}}
\end{array} \right\}; \\
&\underbrace{((m_{p_i}^{(i)})^{\text{th}}\text{XX},(n_{1}^{(i)})^{\text{th}}\text{XX})}_{(m_{p_i}^{(i)},n_{1}^{(i)})\text{-XX}} \\
&\in \left\{ \begin{array}{l}
\underbrace{((2m_{p_i1}^{(i)}+1)^{\text{th}}\text{SO},(2n_{11}^{(i)}+1)^{\text{th}}\text{SI})}_{((2m_{p_i1}^{(i)}+1),(2n_{11}^{(i)}+1))\text{-saddle}},\ \underbrace{((2m_{p_i1}^{(i)})^{\text{th}}\text{US},(2n_{11}^{(i)})^{\text{th}}\text{US})}_{((2m_{p_i1}^{(i)}),(2n_{11}^{(i)}))\text{-double-saddle}}, \\
\underbrace{((2m_{p_i1}^{(i)})^{\text{th}}\text{US},(2n_{11}^{(i)}+1)^{\text{th}}\text{SI})}_{((2m_{p_i1}^{(i)}),(2n_{11}^{(i)}+1))\text{-upper-saddle-sink}},\ \underbrace{((2m_{p_i1}^{(i)}+1)^{\text{th}}\text{SO},(2n_{11}^{(i)})^{\text{th}}\text{US})}_{((2m_{p_i1}^{(i)}+1),(2n_{11}^{(i)}))\text{-upper-saddle-source}}
\end{array} \right\}; \\
&\underbrace{((m_{1}^{(i)})^{\text{th}}\text{XX},(n_{q_i}^{(i)})^{\text{th}}\text{XX})}_{(m_{1}^{(i)},n_{q_i}^{(i)})\text{-XX}} \\
&\in \left\{ \begin{array}{l}
\underbrace{((2m_{11}^{(i)}+1)^{\text{th}}\text{SO},(2n_{q_i1}^{(i)}+1)^{\text{th}}\text{SO})}_{((2m_{11}^{(i)}+1),(2n_{q_i1}^{(i)}+1))\text{-source}},\ \underbrace{((2m_{11}^{(i)})^{\text{th}}\text{LS},(2n_{q_i1}^{(i)})^{\text{th}}\text{US})}_{((2m_{11}^{(i)}),(2n_{q_i1}^{(i)}))\text{-double-saddle}}, \\
\underbrace{((2m_{11}^{(i)})^{\text{th}}\text{LS},(2n_{q_i1}^{(i)}+1)^{\text{th}}\text{SO})}_{((2m_{11}^{(i)}),(2n_{q_i1}^{(i)}+1))\text{-lower-saddle-source}},\ \underbrace{((2m_{11}^{(i)}+1)^{\text{th}}\text{SO},(2n_{q_i1}^{(i)})^{\text{th}}\text{US})}_{((2m_{11}^{(i)}+1),(2n_{q_i1}^{(i)}))\text{-upper-saddle-source}}
\end{array} \right\}; \\
&\underbrace{((m_{1}^{(i)})^{\text{th}}\text{XX},(n_{1}^{(i)})^{\text{th}}\text{XX})}_{(m_{1}^{(i)},n_{1}^{(i)})\text{-XX}} \\
&\in \left\{ \begin{array}{l}
\underbrace{((2m_{11}^{(i)}+1)^{\text{th}}\text{SO},(2n_{11}^{(i)}+1)^{\text{th}}\text{SI})}_{((2m_{11}^{(i)}+1),(2n_{11}^{(i)}+1))\text{-saddle}},\ \underbrace{((2m_{11}^{(i)})^{\text{th}}\text{LS},(2n_{11}^{(i)})^{\text{th}}\text{US})}_{((2m_{11}^{(i)}),(2n_{11}^{(i)}))\text{-double-saddle}}, \\
\underbrace{((2m_{11}^{(i)})^{\text{th}}\text{LS},(2n_{11}^{(i)}+1)^{\text{th}}\text{SI})}_{((2m_{11}^{(i)}),(2n_{11}^{(i)}+1))\text{-lower-saddle-sink}},\ \underbrace{((2m_{11}^{(i)}+1)^{\text{th}}\text{SO},(2n_{11}^{(i)})^{\text{th}}\text{US})}_{((2m_{11}^{(i)}+1),(2n_{11}^{(i)}))\text{-upper-saddle-source}}
\end{array} \right\}.
\end{aligned}
\tag{3.98}
$$

(vi_{1c}) The $((2m_1+1)^{\text{th}}\text{SO},\ (2n_1)^{\text{th}}\text{US})$-upper-saddle-source switching bifurcation is for the switching of two $q \times p$ equilibrium networks as

$$
\bigcup_{r_1=1}^{p}\bigcup_{s_1=1}^{q}\underbrace{(a_{j_1j_1r_1},\,a_{j_2j_2s_1})}_{(m_{r_1}^{(1)},\,n_{s_1}^{(1)})\text{-XX}} \rightleftharpoons \underbrace{((2m_1+1)^{\text{th}}\text{SO},(2n_1)^{\text{th}}\text{US})}_{((2m_1+1),(2n_1))\text{-upper-saddle-source}}
$$

$$\rightleftharpoons \bigcup_{r_2=1}^{p}\bigcup_{s_2=1}^{q}\underbrace{(a_{j_1j_1r_2}, a_{j_2j_2s_2})}_{(m_{r_2}^{(2)}, n_{s_2}^{(2)})\text{-XX}} \tag{3.99}$$

where $\Sigma_{r_i=1}^{p} m_{r_i} = 2m_1 + 1$, $\Sigma_{s_i=1}^{q} n_{s_i} = 2n_1$ for $i = 1, 2$.

(vi_2) For $a_{j_1j_10} < 0$ and $a_{j_2j_20} > 0$, there is a $((2m_1+1)^{\text{th}}\text{SI}, (2n_1)^{\text{th}}\text{US})$-upper-saddle-sink equilibrium as

$$(a_{j_1j_11}, a_{j_2j_21}) = \underbrace{((2m_1+1)^{\text{th}}\text{SI},(2n_1)^{\text{th}}\text{US})}_{((2m_1+1),(2n_1))\text{-upper-saddle-sink}}. \tag{3.100}$$

There are the following three $((2m_1+1)^{\text{th}}\text{SI}, (2n_1)^{\text{th}}\text{US})$-upper-saddle-sink appearing and switching bifurcations.

($\text{vi}_{2\text{a}}$) The $((2m_1+1)^{\text{th}}\text{SI}, (2n_1)^{\text{th}}\text{US})$-upper-saddle-sink appearing bifurcation is from an (SI,pF)-positive sink flow to a $(2m_1 + 1) \times (2n_1)$ network as

$$\underbrace{(a_{j_1j_11}, \dot{x}_{j_2})}_{\text{(SI,pF)-positive sink flow}} \rightleftharpoons \underbrace{((2m_1+1)^{\text{th}}\text{SI},(2n_1)^{\text{th}}\text{US})}_{((2m_1+1),(2n_1))\text{-upper-saddle-sink}} \rightleftharpoons \bigcup_{r=1}^{2m_1+1}\bigcup_{s=1}^{2n_1}\underbrace{(a_{j_1j_1r}, a_{j_2j_2s})}_{\text{XX}} \tag{3.101}$$

where

$$\bigcup_{r=1}^{2m_1+1}\bigcup_{s=1}^{2n_1}\underbrace{(a_{j_1j_1r}, a_{j_2j_2s})}_{\text{XX}} = \left\{\begin{matrix} \underbrace{\text{(SI,SO)}}_{\text{saddle}} & \underbrace{\text{(SO,SO)}}_{\text{source}} & \cdots & \underbrace{\text{(SI,SO)}}_{\text{saddle}} \\ \underbrace{\text{(SI,SI)}}_{\text{sink}} & \underbrace{\text{(SO,SI)}}_{\text{saddle}} & \cdots & \underbrace{\text{(SI,SI)}}_{\text{sink}} \\ \vdots & \vdots & \vdots & \vdots \\ \underbrace{\text{(SI,SI)}}_{\text{sink}} & \underbrace{\text{(SO,SI)}}_{\text{saddle}} & \cdots & \underbrace{\text{(SI,SI)}}_{\text{sink}} \end{matrix}\right\}_{(2n_1)\times(2m_1+1)}. \tag{3.102}$$

($\text{vi}_{2\text{b}}$) The $((2m_1+1)^{\text{th}}\text{SI}, (2n_1)^{\text{th}}\text{US})$-upper-saddle-sink appearing bifurcation is from $q_1 \times p_1$ to $q_2 \times p_2$ equilibrium networks as

$$\begin{aligned} &\bigcup_{r_1=1}^{p_1}\bigcup_{s_1=1}^{q_1}\underbrace{(a_{j_1j_1r_1}, a_{j_2j_2s_1})}_{(m_{r_1}^{(1)}, n_{s_1}^{(1)})-XX} \rightleftharpoons \underbrace{((2m_1+1)^{\text{th}}\text{SI},(2n_1)^{\text{th}}\text{US})}_{((2m_1+1),(2n_1))\text{-upper-saddle-sink}} \\ &\rightleftharpoons \bigcup_{r_2=1}^{p_2}\bigcup_{s_2=1}^{q_2}\underbrace{(a_{j_1j_1r_2}, a_{j_2j_2s_2})}_{(m_{r_2}^{(2)}, n_{s_2}^{(2)})-XX} \end{aligned} \tag{3.103}$$

where

$$2m_1+1-\sum_{r_1=1}^{p_1} m_{r_1}^{(1)}=2\sum_{s=1}^{l_1} m_s$$

with l_1-quadratic polynomials without real roots

and s^{th}-quadratic polynomial with power m_s,

$$2n_1-\sum_{s_1=1}^{q_1} n_{s_1}^{(1)}=2\sum_{l=1}^{l_2} n_l$$

with l_2-quadraitc polynomials without real roots

and l^{th}-quadratic polynomial with power n_l;

$$\sum_{r_2=1}^{p_2} m_{r_2}^{(2)}=2m_1+1,\ \sum_{s_2=1}^{q_2} n_{s_2}^{(2)}=2n_1;\ \text{for } i=1,2,$$

$$\underbrace{((m_{p_i}^{(i)})^{\text{th}}\text{XX},(n_{q_i}^{(i)})^{\text{th}}\text{XX})}_{(m_{p_i}^{(i)},n_{q_i}^{(i)})\text{-XX}}$$

$$\in\left\{\begin{array}{l}\underbrace{((2m_{p_i1}^{(i)}+1)^{\text{th}}\text{SI},(2n_{q_i1}^{(i)}+1)^{\text{th}}\text{SO})}_{((2m_{p_i1}^{(i)}+1),(2n_{q_i1}^{(i)}+1))\text{-saddle}},\ \underbrace{((2m_{p_i1}^{(i)})^{\text{th}}\text{LS},(2n_{q_i1}^{(i)})^{\text{th}}\text{US})}_{((2m_{p_i1}^{(i)}),(2n_{q_i1}^{(i)}))\text{-double-saddle}},\\ \underbrace{((2m_{p_i1}^{(i)})^{\text{th}}\text{LS},(2n_{q_i1}^{(i)}+1)^{\text{th}}\text{SO})}_{((2m_{p_i1}^{(i)}),(2n_{q_i1}^{(i)}+1))\text{-lower-saddle-source}},\ \underbrace{((2m_{p_i1}^{(i)}+1)^{\text{th}}\text{SI},(2n_{q_i1}^{(i)})^{\text{th}}\text{US})}_{((2m_{p_i1}^{(i)}+1),(2n_{q_i1}^{(i)}))\text{-upper-saddle-sink}}\end{array}\right\};$$

$$\underbrace{((m_{p_i}^{(i)})^{\text{th}}\text{XX},(n_1^{(i)})^{\text{th}}\text{XX})}_{(m_{p_i}^{(i)},n_1^{(i)})\text{-XX}}$$

$$\in\left\{\begin{array}{l}\underbrace{((2m_{p_i1}^{(i)}+1)^{\text{th}}\text{SI},(2n_{11}^{(i)}+1)^{\text{th}}\text{SI})}_{((2m_{p_i1}^{(i)}+1),(2n_{11}^{(i)}+1))\text{-sink}},\ \underbrace{((2m_{p_i1}^{(i)})^{\text{th}}\text{LS},(2n_{11}^{(i)})^{\text{th}}\text{US})}_{((2m_{p_i1}^{(i)}),(2n_{11}^{(i)}))\text{-double-saddle}},\\ \underbrace{((2m_{p_i1}^{(i)})^{\text{th}}\text{LS},(2n_{11}^{(i)}+1)^{\text{th}}\text{SI})}_{((2m_{p_i1}^{(i)}),(2n_{11}^{(i)}+1))\text{-lower-saddle-sink}},\ \underbrace{((2m_{p_i1}^{(i)}+1)^{\text{th}}\text{SI},(2n_{11}^{(i)})^{\text{th}}\text{US})}_{((2m_{p_i1}^{(i)}+1),(2n_{11}^{(i)}))\text{-upper-saddle-sink}}\end{array}\right\};$$

$$\underbrace{((m_1^{(i)})^{\text{th}}\text{XX},(n_{q_i}^{(i)})^{\text{th}}\text{XX})}_{(m_1^{(i)},n_{q_i}^{(i)})\text{-XX}}$$

$$\in\left\{\begin{array}{l}\underbrace{((2m_{11}^{(i)}+1)^{\text{th}}\text{SI},(2n_{q_i1}^{(i)}+1)^{\text{th}}\text{SO})}_{((2m_{11}^{(i)}+1),(2n_{q_i1}^{(i)}+1))\text{-saddle}},\ \underbrace{((2m_{11}^{(i)})^{\text{th}}\text{US},(2n_{q_i1}^{(i)})^{\text{th}}\text{US})}_{((2m_{11}^{(i)}),(2n_{q_i1}^{(i)}))\text{-double-saddle}},\\ \underbrace{((2m_{11}^{(i)})^{\text{th}}\text{US},(2n_{q_i1}^{(i)}+1)^{\text{th}}\text{SO})}_{((2m_{11}^{(i)}),(2n_{q_i1}^{(i)}+1))\text{-upper-saddle-source}},\ \underbrace{((2m_{11}^{(i)}+1)^{\text{th}}\text{SI},(2n_{q_i1}^{(i)})^{\text{th}}\text{US})}_{((2m_{11}^{(i)}+1),(2n_{q_i1}^{(i)}))\text{-upper-saddle-sink}}\end{array}\right\};$$

$$\underbrace{((m_1^{(i)})^{\text{th}}\text{XX},(n_1^{(i)})^{\text{th}}\text{XX})}_{(m_1^{(i)},n_1^{(i)})\text{-XX}}$$

$$\in \left\{ \begin{array}{l} \underbrace{((2m_{11}^{(i)}+1)^{\text{th}}\text{SI},(2n_{11}^{(i)}+1)^{\text{th}}\text{SI})}_{((2m_{11}^{(i)}+1),(2n_{11}^{(i)}+1))\text{-sink}}, \underbrace{((2m_{11}^{(i)})^{\text{th}}\text{US},(2n_{11}^{(i)})^{\text{th}}\text{US})}_{((2m_{11}^{(i)}),(2n_{11}^{(i)}))\text{-double-saddle}}, \\ \underbrace{((2m_{11}^{(i)})^{\text{th}}\text{US},(2n_{11}^{(i)}+1)^{\text{th}}\text{SI})}_{((2m_{11}^{(i)}),(2n_{11}^{(i)}+1))\text{-upper-saddle-sink}}, \underbrace{((2m_{11}^{(i)}+1)^{\text{th}}\text{SI},(2n_{11}^{(i)})^{\text{th}}\text{US})}_{((2m_{11}^{(i)}+1),(2n_{11}^{(i)}))\text{-upper-saddle-sink}} \end{array} \right\}. \tag{3.104}$$

(vi$_{2c}$) The $((2m_1+1)^{\text{th}}\text{SI}, (2n_1)^{\text{th}}\text{US})$-upper-saddle-sink switching bifurcation is for two $q \times p$ equilibrium networks as

$$\begin{aligned} \bigcup_{r_1=1}^{p}\bigcup_{s_1=1}^{q} \underbrace{(a_{j_1j_1r_1}, a_{j_2j_2s_1})}_{(m_{r_1}^{(1)},n_{s_1}^{(1)})\text{-XX}} &\rightleftharpoons \underbrace{((2m_1+1)^{\text{th}}\text{SI},(2n_1)^{\text{th}}\text{US})}_{((2m_1+1),(2n_1))\text{-upper-saddle-sink}} \\ &\rightleftharpoons \bigcup_{r_2=1}^{p}\bigcup_{s_2=1}^{q} \underbrace{(a_{j_1j_1r_2}, a_{j_2j_2s_2})}_{(m_{r_2}^{(2)},n_{s_2}^{(2)})\text{-XX}} \end{aligned} \tag{3.105}$$

where $\Sigma_{r_i=1}^{p} m_{r_i} = 2m_1 + 1$, $\Sigma_{s_i=1}^{q} n_{s_i} = 2n_1$ for $i = 1, 2$.

(vi$_3$) For $a_{j_1j_10} > 0$ and $a_{j_2j_20} < 0$, there is a $((2m_1+1)^{\text{th}}\text{SO}, (2n_1)^{\text{th}}\text{LS})$-lower-saddle-source equilibrium as

$$(a_{j_1j_11}, a_{j_2j_21}) = \underbrace{((2m_1+1)^{\text{th}}\text{SO},(2n_1)^{\text{th}}\text{LS})}_{((2m_1+1),(2n_1))\text{-lower-saddle-source}}. \tag{3.106}$$

There are the following three $((2m_1+1)^{\text{th}}\text{SO}, (2n_1)^{\text{th}}\text{LS})$-lower-saddle-source appearing and switching bifurcations.

(vi$_{3a}$) The $((2m_1+1)^{\text{th}}\text{SO}, (2n_1)^{\text{th}}\text{LS})$-lower-saddle-source appearing bifurcation is from an (SO,nF)-negative source flow to a $(2m_1 + 1) \times (2n_1)$-equilibrium network as

$$\underbrace{(a_{j_1j_11}, \dot{x}_{j_2})}_{\text{(SO,nF)-negative source flow}} \rightleftharpoons \underbrace{((2m_1+1)^{\text{th}}\text{SO},(2n_1)^{\text{th}}\text{LS})}_{((2m_1+1),(2n_1))\text{-lower-saddle-source}} \rightleftharpoons \bigcup_{r=1}^{2m_1+1}\bigcup_{s=1}^{2n_1} \underbrace{(a_{j_1j_1r}, a_{j_2j_2s})}_{\text{XX}} \tag{3.107}$$

where

$$\bigcup_{r=1}^{2m_1+1}\bigcup_{r=1}^{2n_1}\underbrace{(a_{j_1j_1r},a_{j_2j_2s})}_{\text{XX}}=\left\{\begin{matrix}\underbrace{(\text{SO,SI})}_{\text{saddle}} & \underbrace{(\text{SI,SI})}_{\text{sink}} & \cdots & \underbrace{(\text{SO,SI})}_{\text{saddle}}\\ \underbrace{(\text{SO,SO})}_{\text{source}} & \underbrace{(\text{SI,SO})}_{\text{saddle}} & \cdots & \underbrace{(\text{SO,SO})}_{\text{source}}\\ \vdots & \vdots & \vdots & \vdots\\ \underbrace{(\text{SO,SO})}_{\text{source}} & \underbrace{(\text{SI,SO})}_{\text{saddle}} & \cdots & \underbrace{(\text{SO,SO})}_{\text{source}}\end{matrix}\right\}_{(2n_1)\times(2m_1+1)}. \tag{3.108}$$

($\text{vi}_{3\text{b}}$) The $((2m_1+1)^{\text{th}}\text{SO}, (2n_1)^{\text{th}}\text{LS})$-lower-saddle-source appearing bifurcation is from a $q_1 \times p_1$ to $q_2 \times p_2$ equilibrium network as

$$\begin{aligned}&\bigcup_{r_1=1}^{p_1}\bigcup_{s_1=1}^{q_1}\underbrace{(a_{j_1j_1r_1},a_{j_2j_2s_1})}_{(m_{r_1}^{(1)},n_{s_1}^{(1)})\text{-XX}}\rightleftharpoons\underbrace{((2m_1+1)^{\text{th}}\text{SO},(2n_1)^{\text{th}}\text{LS})}_{((2m_1+1),(2n_1))\text{-lower-saddle-source}}\\&\rightleftharpoons\bigcup_{r_2=1}^{p_2}\bigcup_{s_2=1}^{q_2}\underbrace{(a_{j_1j_1r_2},a_{j_2j_2s_2})}_{(m_{r_2}^{(2)},n_{s_2}^{(2)})\text{-XX}}\end{aligned} \tag{3.109}$$

where

$$2m_1+1-\sum_{r_1=1}^{p_1}m_{r_1}^{(1)}=2\sum_{s=1}^{l_1}m_s$$

with l_1-quadratic polynomials without real roots

and s^{th}-quadratic polynomial with power m_s,

$$2n_1-\sum_{s_1=1}^{q_1}n_{s_1}^{(1)}=2\sum_{l=1}^{l_2}n_l$$

with l_2-quadraitc polynomials without real roots

and l^{th}-quadratic polynomial with power n_l;

$$\sum_{r_2=1}^{p_2}m_{r_2}^{(2)}=2m_1+1,\ \sum_{s_2=1}^{q_2}n_{s_2}^{(2)}=2n_1;\ \text{for } i=1,2,$$

$$
\begin{aligned}
&\underbrace{((m_{p_i}^{(i)})^{\text{th}}\text{XX},(n_{q_i}^{(i)})^{\text{th}}\text{XX})}_{(m_{p_i}^{(i)},n_{q_i}^{(i)})\text{-XX}} \\
&\in \left\{ \begin{array}{l}
\underbrace{((2m_{p_i1}^{(i)}+1)^{\text{th}}\text{SO},(2n_{q_i1}^{(i)}+1)^{\text{th}}\text{SI})}_{((2m_{p_i1}^{(i)}+1),(2n_{q_i1}^{(i)}+1))\text{-saddle}},\ \underbrace{((2m_{p_i1}^{(i)})^{\text{th}}\text{US},(2n_{q_i1}^{(i)})^{\text{th}}\text{LS})}_{((2m_{p_i1}^{(i)}),(2n_{q_i1}^{(i)}))\text{-double-saddle}}, \\
\underbrace{((2m_{p_i1}^{(i)})^{\text{th}}\text{US},(2n_{q_i1}^{(i)}+1)^{\text{th}}\text{SI})}_{((2m_{p_i1}^{(i)}),(2n_{q_i1}^{(i)}+1))\text{-upper-saddle-sink}},\ \underbrace{((2m_{p_i1}^{(i)}+1)^{\text{th}}\text{SO},(2n_{q_i1}^{(i)})^{\text{th}}\text{LS})}_{((2m_{p_i1}^{(i)}+1),(2n_{q_i1}^{(i)}))\text{-lower-saddle-source}}
\end{array} \right\}; \\
&\underbrace{((m_{p_i}^{(i)})^{\text{th}}\text{XX},(n_{1}^{(i)})^{\text{th}}\text{XX})}_{(m_{p_i}^{(i)},n_{1}^{(i)})\text{-XX}} \\
&\in \left\{ \begin{array}{l}
\underbrace{((2m_{p_i1}^{(i)}+1)^{\text{th}}\text{SO},(2n_{11}^{(i)}+1)^{\text{th}}\text{SO})}_{((2m_{p_i1}^{(i)}+1),(2n_{11}^{(i)}+1))\text{-source}},\ \underbrace{((2m_{p_i1}^{(i)})^{\text{th}}\text{US},(2n_{11}^{(i)})^{\text{th}}\text{LS})}_{((2m_{p_i1}^{(i)}),(2n_{11}^{(i)}))\text{-double-saddle}}, \\
\underbrace{((2m_{p_i1}^{(i)})^{\text{th}}\text{US},(2n_{11}^{(i)}+1)^{\text{th}}\text{SO})}_{((2m_{p_i1}^{(i)}),(2n_{11}^{(i)}+1))\text{-upper-saddle-source}},\ \underbrace{((2m_{p_i1}^{(i)}+1)^{\text{th}}\text{SO},(2n_{11}^{(i)})^{\text{th}}\text{LS})}_{((2m_{p_i1}^{(i)}+1),(2n_{11}^{(i)}))\text{-lower-saddle-source}}
\end{array} \right\}; \\
&\underbrace{((m_{1}^{(i)})^{\text{th}}\text{XX},(n_{q_i}^{(i)})^{\text{th}}\text{XX})}_{(m_{1}^{(i)},n_{q_i}^{(i)})\text{-XX}} \\
&\in \left\{ \begin{array}{l}
\underbrace{((2m_{11}^{(i)}+1)^{\text{th}}\text{SO},(2n_{q_i1}^{(i)}+1)^{\text{th}}\text{SI})}_{((2m_{11}^{(i)}+1),(2n_{q_i1}^{(i)}+1))\text{-saddle}},\ \underbrace{((2m_{11}^{(i)})^{\text{th}}\text{LS},(2n_{q_i1}^{(i)})^{\text{th}}\text{LS})}_{((2m_{11}^{(i)}),(2n_{q_i1}^{(i)}))\text{-double-saddle}}, \\
\underbrace{((2m_{11}^{(i)})^{\text{th}}\text{LS},(2n_{q_i1}^{(i)}+1)^{\text{th}}\text{SI})}_{((2m_{11}^{(i)}),(2n_{q_i1}^{(i)}+1))\text{-lower-saddle-sink}},\ \underbrace{((2m_{11}^{(i)}+1)^{\text{th}}\text{SO},(2n_{q_i1}^{(i)})^{\text{th}}\text{LS})}_{((2m_{11}^{(i)}+1),(2n_{q_i1}^{(i)}))\text{-lower-saddle-source}}
\end{array} \right\}; \\
&\underbrace{((m_{1}^{(i)})^{\text{th}}\text{XX},(n_{1}^{(i)})^{\text{th}}\text{XX})}_{(m_{1}^{(i)},n_{1}^{(i)})\text{-XX}} \\
&\in \left\{ \begin{array}{l}
\underbrace{((2m_{11}^{(i)}+1)^{\text{th}}\text{SO},(2n_{11}^{(i)}+1)^{\text{th}}\text{SO})}_{((2m_{11}^{(i)}+1),(2n_{11}^{(i)}+1))\text{-source}},\ \underbrace{((2m_{11}^{(i)})^{\text{th}}\text{LS},(2n_{11}^{(i)})^{\text{th}}\text{LS})}_{((2m_{11}^{(i)}),(2n_{11}^{(i)}))\text{-double-saddle}}, \\
\underbrace{((2m_{11}^{(i)})^{\text{th}}\text{LS},(2n_{11}^{(i)}+1)^{\text{th}}\text{SO})}_{((2m_{11}^{(i)}),(2n_{11}^{(i)}+1))\text{-lower-saddle-source}},\ \underbrace{((2m_{11}^{(i)}+1)^{\text{th}}\text{SO},(2n_{11}^{(i)})^{\text{th}}\text{LS})}_{((2m_{11}^{(i)}+1),(2n_{11}^{(i)}))\text{-lower-saddle-source}}
\end{array} \right\}.
\end{aligned}
\tag{3.110}
$$

(vi_{3c}) The $((2m_1+1)^{\text{th}}\text{SO}, (2n_1)^{\text{th}}\text{LS})$-lower-saddle-source switching bifurcation is for the switching of two $q \times p$ equilibrium networks as

$$
\bigcup_{r_1=1}^{p_1} \bigcup_{s_1=1}^{q_1} \underbrace{(a_{j_1 j_1 r_1}, a_{j_2 j_2 s_1})}_{(m_{r_1}^{(1)},\, n_{s_1}^{(1)})\text{-XX}} \rightleftharpoons \underbrace{((2m_1+1)^{\text{th}}\text{SO},(2n_1)^{\text{th}}\text{LS})}_{((2m_1+1),(2n_1))\text{-lower-saddle-sink}}
$$

$$\rightleftharpoons \bigcup_{r_2=1}^{p_2} \bigcup_{s_2=1}^{q_2} \underbrace{(a_{j_1 j_1 r_2}, a_{j_2 j_2 s_2})}_{(m_{r_2}^{(2)}, n_{s_2}^{(2)})\text{-XX}} \tag{3.111}$$

where $\Sigma_{r_i=1}^{p} m_{r_i} = 2m_1 + 1$, $\Sigma_{s_i=1}^{q} n_{s_i} = 2n_1$ for $i = 1, 2$.

(vi$_4$) For $a_{j_1 j_1 0} < 0$ and $a_{j_2 j_2 0} < 0$, there is a $((2m_1+1)^{\text{th}}\text{SI},(2n_1)^{\text{th}}\text{LS})$-lower-saddle-sink equilibrium as

$$(a_{j_1 j_1 1}, a_{j_2 j_2 1}) = \underbrace{((2m_1+1)^{\text{th}}\text{SI}, (2n_1)^{\text{th}}\text{LS})}_{((2m_1+1),(2n_1))\text{-lower-saddle-sink}}. \tag{3.112}$$

There are the following three $((2m_1+1)^{\text{th}}\text{SI},(2n_1)^{\text{th}}\text{LS})$-lower-saddle-sink appearing and switching bifurcations.

(vi$_{4a}$) The $((2m_1+1)^{\text{th}}\text{SI},(2n_1)^{\text{th}}\text{LS})$-lower-saddle-sink appearing bifurcation is from an (SI,nF)-negative sink flow to a $(2m_1 + 1) \times (2n_1)$ network as

$$\underbrace{(a_{j_1 j_1 1}, \dot{x}_{j_2})}_{\text{(SI,nF)-negative sink flow}} \rightleftharpoons \underbrace{((2m_1+1)^{\text{th}}\text{SI},(2n_1)^{\text{th}}\text{LS})}_{((2m_1+1),(2n_1))\text{-lower-saddle-sink}} \rightleftharpoons \bigcup_{r=1}^{2m_1+1} \bigcup_{s=1}^{2n_1} \underbrace{(a_{j_1 j_1 r}, a_{j_2 j_2 s})}_{\text{XX}} \tag{3.113}$$

where

$$\bigcup_{r=1}^{2m_1+1} \bigcup_{s=1}^{2n_1} \underbrace{(a_{j_1 j_1 r}, a_{j_2 j_2 s})}_{XX} = \left\{ \begin{matrix} \underbrace{\text{(SI,SI)}}_{\text{sink}} & \underbrace{\text{(SO,SI)}}_{\text{saddle}} & \cdots & \underbrace{\text{(SI,SI)}}_{\text{sink}} \\ \underbrace{\text{(SI,SO)}}_{\text{saddle}} & \underbrace{\text{(SO,SO)}}_{\text{source}} & \cdots & \underbrace{\text{(SI,SO)}}_{\text{saddle}} \\ \vdots & \vdots & \vdots & \vdots \\ \underbrace{\text{(SI,SO)}}_{\text{saddle}} & \underbrace{\text{(SO,SO)}}_{\text{source}} & \cdots & \underbrace{\text{(SI,SO)}}_{\text{saddle}} \end{matrix} \right\}_{(2n_1)\times(2m_1+1)}. \tag{3.114}$$

(vi$_{4b}$) The $((2m_1+1)^{\text{th}}\text{SI},(2n_1)^{\text{th}}\text{LS})$-lower-saddle-sink appearing bifurcation is from $q_1 \times p_1$ to $q_2 \times p_2$ equilibrium networks as

$$\begin{aligned} &\bigcup_{r_1=1}^{p_1} \bigcup_{s_1=1}^{q_1} \underbrace{(a_{j_1 j_1 r_1}, a_{j_2 j_2 s_1})}_{(m_{r_1}^{(1)}, n_{s_1}^{(1)})\text{-XX}} \rightleftharpoons \underbrace{((2m_1+1)^{\text{th}}\text{SI},(2n_1)^{\text{th}}\text{LS})}_{((2m_1+1),(2n_1))\text{-lower-saddle-sink}} \\ &\rightleftharpoons \bigcup_{r_2=1}^{p_2} \bigcup_{s_2=1}^{q_2} \underbrace{(a_{j_1 j_1 r_2}, a_{j_2 j_2 s_2})}_{(m_{r_2}^{(2)}, n_{s_2}^{(2)})\text{-XX}} \end{aligned} \tag{3.115}$$

where

$$2m_1 + 1 - \sum_{r_1=1}^{p_1} m_{r_1}^{(1)} = 2\sum_{s=1}^{l_1} m_s$$

with l_1-quadratic polynomials without real roots

and s^{th}-quadratic polynomial with power m_s,

$$2n_1 - \sum_{s_1=1}^{q_1} n_{s_1}^{(1)} = 2\sum_{l=1}^{l_2} n_l$$

with l_2-quadraitc polynomials without real roots

and l^{th}-quadratic polynomial with power n_l;

$$\sum_{r_2=1}^{p_2} m_{r_2}^{(2)} = 2m_1 + 1,\ \sum_{s_2=1}^{q_2} n_{s_2}^{(2)} = 2n_1;\ \text{for } i = 1, 2,$$

$$\underbrace{((m_{p_i}^{(i)})^{\text{th}}\text{XX},(n_{q_i}^{(i)})^{\text{th}}\text{XX})}_{(m_{p_i}^{(i)},\, n_{q_i}^{(i)})\text{-XX}}$$

$$\in \left\{ \begin{array}{l} \underbrace{((2m_{p_i1}^{(i)}+1)^{\text{th}}\text{SI},(2n_{q_i1}^{(i)}+1)^{\text{th}}\text{SI})}_{((2m_{p_i1}^{(i)}+1),\,(2n_{q_i1}^{(i)}+1))\text{-sink}},\ \underbrace{((2m_{p_i1}^{(i)})^{\text{th}}\text{LS},(2n_{q_i1}^{(i)})^{\text{th}}\text{LS})}_{((2m_{p_i1}^{(i)}),\,(2n_{q_i1}^{(i)}))\text{-double-saddle}}, \\ \underbrace{((2m_{p_i1}^{(i)})^{\text{th}}\text{LS},(2n_{q_i1}^{(i)}+1)^{\text{th}}\text{SI})}_{((2m_{p_i1}^{(i)}),\,(2n_{q_i1}^{(i)}+1))\text{-lower-saddle-sink}},\ \underbrace{((2m_{p_i1}^{(i)}+1)^{\text{th}}\text{SI},(2n_{q_i1}^{(i)})^{\text{th}}\text{LS})}_{((2m_{p_i1}^{(i)}+1),\,(2n_{q_i1}^{(i)}))\text{-lower-saddle-sink}} \end{array} \right\};$$

$$\underbrace{((m_{p_i}^{(i)})^{\text{th}}\text{XX},(n_{1}^{(i)})^{\text{th}}\text{XX})}_{(m_{p_i}^{(i)},\, n_{1}^{(i)})\text{-XX}}$$

$$\in \left\{ \begin{array}{l} \underbrace{((2m_{p_i1}^{(i)}+1)^{\text{th}}\text{SI},(2n_{11}^{(i)}+1)^{\text{th}}\text{SO})}_{((2m_{p_i1}^{(i)}+1),\,(2n_{11}^{(i)}+1))\text{-saddle}},\ \underbrace{((2m_{p_i1}^{(i)})^{\text{th}}\text{LS},(2n_{11}^{(i)})^{\text{th}}\text{LS})}_{((2m_{p_i1}^{(i)}),\,(2n_{11}^{(i)}))\text{-double-saddle}}, \\ \underbrace{((2m_{p_i1}^{(i)})^{\text{th}}\text{LS},(2n_{11}^{(i)}+1)^{\text{th}}\text{SO})}_{((2m_{p_i1}^{(i)}),\,(2n_{11}^{(i)}+1))\text{-lower-saddle-source}},\ \underbrace{((2m_{p_i1}^{(i)}+1)^{\text{th}}\text{SI},(2n_{11}^{(i)})^{\text{th}}\text{LS})}_{((2m_{p_i1}^{(i)}+1),\,(2n_{11}^{(i)}))\text{-lower-saddle-sink}} \end{array} \right\};$$

$$\underbrace{((m_{1}^{(i)})^{\text{th}}\text{XX},(n_{q_i}^{(i)})^{\text{th}}\text{XX})}_{(m_{1}^{(i)},\, n_{q_i}^{(i)})\text{-XX}}$$

$$\in \left\{ \begin{array}{l} \underbrace{((2m_{11}^{(i)}+1)^{\text{th}}\text{SI},(2n_{q_i1}^{(i)}+1)^{\text{th}}\text{SI})}_{((2m_{11}^{(i)}+1),\,(2n_{q_i1}^{(i)}+1))\text{-sink}},\ \underbrace{((2m_{11}^{(i)})^{\text{th}}\text{US},(2n_{q_i1}^{(i)})^{\text{th}}\text{LS})}_{((2m_{11}^{(i)}),\,(2n_{q_i1}^{(i)}))\text{-double-saddle}}, \\ \underbrace{((2m_{11}^{(i)})^{\text{th}}\text{US},(2n_{q_i1}^{(i)}+1)^{\text{th}}\text{SI})}_{((2m_{11}^{(i)}),\,(2n_{q_i1}^{(i)}+1))\text{-upper-saddle-sink}},\ \underbrace{((2m_{11}^{(i)}+1)^{\text{th}}\text{SI},(2n_{q_i1}^{(i)})^{\text{th}}\text{LS})}_{((2m_{11}^{(i)}+1),\,(2n_{q_i1}^{(i)}))\text{-lower-saddle-sink}} \end{array} \right\};$$

$$\underbrace{((m_1^{(i)})^{\text{th}}\text{XX},(n_1^{(i)})^{\text{th}}\text{XX})}_{(m_1^{(i)},\,n_1^{(i)})\text{-XX}}$$
$$\in \left\{ \begin{array}{l} \underbrace{((2m_{11}^{(i)}+1)^{\text{th}}\text{SI},(2n_{11}^{(i)}+1)^{\text{th}}\text{SO})}_{((2m_{11}^{(i)}+1),\,(2n_{11}^{(i)}+1))\text{-saddle}}, \underbrace{((2m_{11}^{(i)})^{\text{th}}\text{US},(2n_{11}^{(i)})^{\text{th}}\text{LS})}_{((2m_{11}^{(i)}),\,(2n_{11}^{(i)}))\text{-double-saddle}}, \\ \underbrace{((2m_{11}^{(i)})^{\text{th}}\text{US},(2n_{11}^{(i)}+1)^{\text{th}}\text{SO})}_{((2m_{11}^{(i)}),\,(2n_{11}^{(i)}+1))\text{-upper-saddle-source}}, \underbrace{((2m_{11}^{(i)}+1)^{\text{th}}\text{SI},(2n_{11}^{(i)})^{\text{th}}\text{LS})}_{((2m_{11}^{(i)}+1),\,(2n_{11}^{(i)}))\text{-lower-saddle-sink}} \end{array} \right\}. \tag{3.116}$$

(vi_{4c}) The $((2m_1+1)^{\text{th}}\text{SI},(2n_1)^{\text{th}}\text{LS})$-lower-saddle-sink switching bifurcation is for two $q \times p$ equilibrium networks as

$$\bigcup_{r_1=1}^{p}\bigcup_{s_1=1}^{q}\underbrace{(a_{j_1j_1r_1},a_{j_2j_2s_1})}_{(m_{r_1}^{(1)},\,n_{s_1}^{(1)})\text{-XX}} \rightleftharpoons \underbrace{((2m_1+1)^{\text{th}}\text{SI},(2n_1)^{\text{th}}\text{LS})}_{((2m_1+1),(2n_1))\text{-lower-saddle-sink}}$$
$$\rightleftharpoons \bigcup_{r_2=1}^{p}\bigcup_{s_2=1}^{q}\underbrace{(a_{j_1j_1r_2},a_{j_2j_2s_2})}_{(m_{r_2}^{(2)},\,n_{s_2}^{(2)})\text{-XX}} \tag{3.117}$$

where $\Sigma_{r_i=1}^{p} m_{r_i} = 2m_1 + 1,\ \Sigma_{s_i=1}^{q} n_{s_i} = 2n_1$ for $i = 1, 2$.

(vii) For $m = 2m_1$ and $n = 2n_1$, the equilibrium of $(x_{j_1}^*, x_{j_2}^*) = (a_{j_1j_1 1}, a_{j_2j_2 1})$ has the following properties.

(vii_1) For $a_{j_1j_1 0} > 0$ and $a_{j_2j_2 0} > 0$, there is a $((2m_1)^{\text{th}}\text{US},(2n_1)^{\text{th}}\text{US})$-double-saddle equilibrium as

$$(a_{j_1j_1 1}, a_{j_2j_2 1}) = \underbrace{((2m_1)^{\text{th}}\text{US},(2n_1)^{\text{th}}\text{US})}_{((2m_1),(2n_1))\text{-double-saddle}}. \tag{3.118}$$

There are three following $((2m_1)^{\text{th}}\text{US},(2n_1)^{\text{th}}\text{US})$-double-saddle appearing and switching bifurcations.

(vii_{1a}) The $((2m_1)^{\text{th}}\text{US}, (2n_1)^{\text{th}}\text{US})$-double-saddle appearing bifurcation is from a (pF,pF)-positive flow to a $(2m_1) \times (2n_1)$-equilibrium network as

$$\underbrace{(\dot{x}_{j_1},\dot{x}_{j_2})}_{(\text{pF,pF})\text{-flow}} \rightleftharpoons \underbrace{((2m_1)^{\text{th}}\text{US},(2n_1)^{\text{th}}\text{US})}_{((2m_1),(2n_1))\text{-double-saddle}} \rightleftharpoons \bigcup_{r=1}^{2m_1}\bigcup_{s=1}^{2n_1}\underbrace{(a_{j_1j_1r},a_{j_2j_2s})}_{\text{XX}} \tag{3.119}$$

where

$$\bigcup_{r=1}^{2m_1}\bigcup_{s=1}^{2n_1}\underbrace{(a_{j_1j_1r},a_{j_2j_2s})}_{XX}=\left\{\begin{matrix}\underbrace{(\text{SI,SO})}_{\text{saddle}} & \underbrace{(\text{SO,SO})}_{\text{source}} & \cdots & \underbrace{(\text{SO,SO})}_{\text{source}}\\ \underbrace{(\text{SI,SI})}_{\text{sink}} & \underbrace{(\text{SO,SI})}_{\text{saddle}} & \cdots & \underbrace{(\text{SO,SI})}_{\text{saddle}}\\ \vdots & \vdots & \vdots & \vdots\\ \underbrace{(\text{SI,SI})}_{\text{sink}} & \underbrace{(\text{SO,SI})}_{\text{saddle}} & \cdots & \underbrace{(\text{SO,SI})}_{\text{saddle}}\end{matrix}\right\}_{(2n_1)\times(2m_1)}. \tag{3.120}$$

(vii$_{1b}$) The $((2m_1)^{\text{th}}\text{US},(2n_1)^{\text{th}}\text{US})$-double-saddle appearing bifurcation is from a $q_1\times p_1$ to $q_2\times p_2$ equilibrium network as

$$\begin{aligned}&\bigcup_{r_1=1}^{p_1}\bigcup_{s_1=1}^{q_1}\underbrace{(a_{j_1j_1r_1},a_{j_2j_2s_1})}_{(m_{r_1}^{(1)},n_{s_1}^{(1)})\text{-XX}}\rightleftharpoons\underbrace{((2m_1)^{\text{th}}\text{US},(2n_1)^{\text{th}}\text{US})}_{((2m_1),(2n_1))\text{-double-saddle}}\\&\rightleftharpoons\bigcup_{r_2=1}^{p_2}\bigcup_{s_2=1}^{q_2}\underbrace{(a_{j_1j_1r_2},a_{j_2j_2s_2})}_{(m_{r_2}^{(2)},n_{s_2}^{(2)})\text{-XX}}\end{aligned} \tag{3.121}$$

where

$$2m_1-\sum_{r_1=1}^{p_1}m_{r_1}^{(1)}=2\sum_{s=1}^{l_1}m_s$$

with l_1-quadratic polynomials without real roots

and s^{th}-quadratic polynomial with power m_s,

$$2n_1-\sum_{s_1=1}^{q_1}n_{s_1}^{(1)}=2\sum_{l=1}^{l_2}n_l$$

with l_2-quadraitc polynomials without real roots

and l^{th}-quadratic polynomial with power n_l;

$$\sum_{r_2=1}^{p_2}m_{r_2}^{(2)}=2m_1,\ \sum_{s_2=1}^{q_2}n_{s_2}^{(2)}=2n_1;\ \text{for } i=1,2,$$

$$
\begin{array}{l}
\underbrace{((m_{p_i}^{(i)})^{\text{th}}\text{XX},(n_{q_i}^{(i)})^{\text{th}}\text{XX})}_{(m_{p_i}^{(i)},n_{q_i}^{(i)})\text{-XX}} \\
\in \left\{ \begin{array}{l}
\underbrace{((2m_{p_i1}^{(i)}+1)^{\text{th}}\text{SO},(2n_{q_i1}^{(i)}+1)^{\text{th}}\text{SO})}_{((2m_{p_i1}^{(i)}+1),(2n_{q_i1}^{(i)}+1))\text{-source}}, \underbrace{((2m_{p_i1}^{(i)})^{\text{th}}\text{US},(2n_{q_i1}^{(i)})^{\text{th}}\text{US})}_{((2m_{p_i1}^{(i)}),(2n_{q_i1}^{(i)}))\text{-double-saddle}}, \\
\underbrace{((2m_{p_i1}^{(i)})^{\text{th}}\text{US},(2n_{q_i1}^{(i)}+1)^{\text{th}}\text{SO})}_{((2m_{p_i1}^{(i)}),(2n_{q_i1}^{(i)}+1))\text{-upper-saddle-source}}, \underbrace{((2m_{p_i1}^{(i)}+1)^{\text{th}}\text{SO},(2n_{q_i1}^{(i)})^{\text{th}}\text{US})}_{((2m_{p_i1}^{(i)}+1),(2n_{q_i1}^{(i)}))\text{-upper-saddle-source}}
\end{array} \right\}; \\
\underbrace{((m_{p_i}^{(i)})^{\text{th}}\text{XX},(n_{1}^{(i)})^{\text{th}}\text{XX})}_{(m_{p_i}^{(i)},n_{1}^{(i)})\text{-XX}} \\
\in \left\{ \begin{array}{l}
\underbrace{((2m_{p_i1}^{(i)}+1)^{\text{th}}\text{SO},(2n_{11}^{(i)}+1)^{\text{th}}\text{SI})}_{((2m_{p_i1}^{(i)}+1),(2n_{11}^{(i)}+1))\text{-saddle}}, \underbrace{((2m_{p_i1}^{(i)})^{\text{th}}\text{US},(2n_{11}^{(i)})^{\text{th}}\text{US})}_{((2m_{p_i1}^{(i)}),(2n_{11}^{(i)}))\text{-double-saddle}}, \\
\underbrace{((2m_{p_i1}^{(i)})^{\text{th}}\text{US},(2n_{11}^{(i)}+1)^{\text{th}}\text{SI})}_{((2m_{p_i1}^{(i)}),(2n_{11}^{(i)}+1))\text{-upper-saddle-sink}}, \underbrace{((2m_{p_i1}^{(i)}+1)^{\text{th}}\text{SO},(2n_{11}^{(i)})^{\text{th}}\text{US})}_{((m_{p_i1}^{(i)}+1),(2n_{11}^{(i)}))\text{-upper-saddle-source}}
\end{array} \right\}; \\
\underbrace{((m_{1}^{(i)})^{\text{th}}\text{XX},(n_{q_i}^{(i)})^{\text{th}}\text{XX})}_{(m_{1}^{(i)},n_{q_i}^{(i)})\text{-XX}} \\
\in \left\{ \begin{array}{l}
\underbrace{((2m_{11}^{(i)}+1)^{\text{th}}\text{SI},(2n_{q_i1}^{(i)}+1)^{\text{th}}\text{SO})}_{((2m_{11}^{(i)}+1),(2n_{q_i1}^{(i)}+1))\text{-saddle}}, \underbrace{((2m_{11}^{(i)})^{\text{th}}\text{US},(2n_{q_i1}^{(i)})^{\text{th}}\text{US})}_{((2m_{11}^{(i)}),(2n_{q_i1}^{(i)}))\text{-double-saddle}}, \\
\underbrace{((2m_{11}^{(i)})^{\text{th}}\text{US},(2n_{q_i1}^{(i)}+1)^{\text{th}}\text{SO})}_{((2m_{11}^{(i)}),(2n_{q_i1}^{(i)}+1))\text{-upper-saddle-source}}, \underbrace{((2m_{11}^{(i)}+1)^{\text{th}}\text{SI},(2n_{q_i1}^{(i)})^{\text{th}}\text{US})}_{((2m_{11}^{(i)}+1),(2n_{q_i1}^{(i)}))\text{-upper-saddle-sink}}
\end{array} \right\}; \\
\underbrace{((m_{1}^{(i)})^{\text{th}}\text{XX},(n_{1}^{(i)})^{\text{th}}\text{XX})}_{(m_{1}^{(i)},n_{1}^{(i)})\text{-XX}} \\
\in \left\{ \begin{array}{l}
\underbrace{((2m_{11}^{(i)}+1)^{\text{th}}\text{SI},(2n_{11}^{(i)}+1)^{\text{th}}\text{SI})}_{((2m_{11}^{(i)}+1),(2n_{11}^{(i)}+1))\text{-sink}}, \underbrace{((2m_{11}^{(i)})^{\text{th}}\text{US},(2n_{11}^{(i)})^{\text{th}}\text{US})}_{((2m_{11}^{(i)}),(2n_{11}^{(i)}))\text{-double-saddle}}, \\
\underbrace{((2m_{11}^{(i)})^{\text{th}}\text{US},(2n_{11}^{(i)}+1)^{\text{th}}\text{SI})}_{((2m_{11}^{(i)}),(2n_{11}^{(i)}+1))\text{-upper-saddle-sink}}, \underbrace{((2m_{11}^{(i)}+1)^{\text{th}}\text{SI},(2n_{11}^{(i)})^{\text{th}}\text{US})}_{((2m_{11}^{(i)}+1),(2n_{11}^{(i)}))\text{-upper-saddle-sink}}
\end{array} \right\}.
\end{array}
\tag{3.122}
$$

(vii$_{1c}$) The $((2m_1)^{\text{th}}\text{US}, (2n_1)^{\text{th}}\text{US})$-double-saddle switching bifurcation is for the switching of two $q \times p$ equilibrium networks as

$$
\bigcup_{r_1=1}^{p}\bigcup_{s_1=1}^{q}\underbrace{(a_{j_1j_1r_1}, a_{j_2j_2s_1})}_{(m_{r_1}^{(1)}, n_{s_1}^{(1)})-XX} \rightleftharpoons \underbrace{((2m_1)^{\text{th}}\text{US},(2n_1)^{\text{th}}\text{US})}_{((2m_1),(2n_1))\text{-double-saddle}}
$$

$$\rightleftharpoons \bigcup_{r_2=1}^{p}\bigcup_{s_2=1}^{q}\underbrace{(a_{j_1j_1r_2}, a_{j_2j_2s_2})}_{(m_{r_2}^{(2)}, n_{s_2}^{(2)})-XX} \tag{3.123}$$

where $\Sigma_{r_i=1}^{p} m_{r_i} = 2m_1$, $\Sigma_{s_i=1}^{q} n_{s_i} = 2n_1$ for $i = 1, 2$.

(vii$_2$) For $a_{j_1j_10} < 0$ and $a_{j_2j_20} > 0$, there is a $((2m_1)^{\text{th}}$LS, $(2n_1)^{\text{th}}$US)-double-saddle equilibrium as

$$(a_{j_1j_11}, a_{j_2j_21}) = \underbrace{((2m_1)^{\text{th}}\text{LS},(2n_1)^{\text{th}}\text{US})}_{((2m_1),(2n_1))\text{-double-saddle}}. \tag{3.124}$$

There are the following three $((2m_1)^{\text{th}}$LS, $(2n_1)^{\text{th}}$US)-double-saddle appearing and switching bifurcations.

(vii$_{2a}$) The $((2m_1)^{\text{th}}$LS, $(2n_1)^{\text{th}}$US)-double-saddle appearing bifurcation is from an (nF,pF)-flow to a $(2m_1) \times (2n_1)$ network as

$$\underbrace{(\dot{x}_{j_1}, \dot{x}_{j_2})}_{(\text{nF,pF})\text{-flow}} \rightleftharpoons \underbrace{((2m_1)^{\text{th}}\text{LS},(2n_1)^{\text{th}}\text{US})}_{((2m_1),(2n_1))\text{-double-saddle}} \rightleftharpoons \bigcup_{r=1}^{2m_1}\bigcup_{s=1}^{2n_1}\underbrace{(a_{j_1j_1r}, a_{j_2j_2s})}_{XX} \tag{3.125}$$

where

$$\bigcup_{r=1}^{2m_1}\bigcup_{s=1}^{2n_1}\underbrace{(a_{j_1j_1r}, a_{j_2j_2s})}_{XX} = \left\{\begin{matrix} \underbrace{(\text{SO,SO})}_{\text{source}} & \underbrace{(\text{SI,SO})}_{\text{saddle}} & \cdots & \underbrace{(\text{SI,SO})}_{\text{saddle}} \\ \underbrace{(\text{SO,SI})}_{\text{saddle}} & \underbrace{(\text{SI,SI})}_{\text{sink}} & \cdots & \underbrace{(\text{SI,SI})}_{\text{sink}} \\ \vdots & \vdots & \vdots & \vdots \\ \underbrace{(\text{SO,SI})}_{\text{saddle}} & \underbrace{(\text{SI,SI})}_{\text{sink}} & \cdots & \underbrace{(\text{SI,SI})}_{\text{sink}} \end{matrix}\right\}_{(2n_1)\times(2m_1)}. \tag{3.126}$$

(vii$_{2b}$) The $((2m_1)^{\text{th}}$LS, $(2n_1)^{\text{th}}$US)-double-saddle appearing bifurcation is from $q_1 \times p_1$ to $q_2 \times p_2$ equilibrium networks as

$$\begin{aligned} &\bigcup_{r_1=1}^{p_1}\bigcup_{s_1=1}^{q_1}\underbrace{(a_{j_1j_1r_1}, a_{j_2j_2s_1})}_{(m_{r_1}^{(1)}, n_{s_1}^{(1)})\text{-XX}} \rightleftharpoons \underbrace{((2m_1)^{\text{th}}\text{LS},(2n_1)^{\text{th}}\text{US})}_{((2m_1),(2n_1))\text{-double-saddle}} \\ &\rightleftharpoons \bigcup_{r_2=1}^{p_2}\bigcup_{s_2=1}^{q_2}\underbrace{(a_{j_1j_1r_2}, a_{j_2j_2s_2})}_{(m_{r_2}^{(2)}, n_{s_2}^{(2)})\text{-XX}} \end{aligned} \tag{3.127}$$

where

$$2m_1 - \sum_{r_1=1}^{p_1} m_{r_1}^{(1)} = 2\sum_{s=1}^{l_1} m_s$$

with l_1-quadratic polynomials without real roots

and s^{th}-quadratic polynomial with power m_s,

$$2n_1 - \sum_{s_1=1}^{q_1} n_{s_1}^{(1)} = 2\sum_{l=1}^{l_2} n_l$$

with l_2-quadraitc polynomials without real roots

and l^{th}-quadratic polynomial with power n_l;

$$\sum_{r_2=1}^{p_2} m_{r_2}^{(2)} = 2m_1, \quad \sum_{s_2=1}^{q_2} n_{s_2}^{(2)} = 2n_1; \text{ for } i = 1, 2,$$

$$\underbrace{((m_{p_i}^{(i)})^{\text{th}}\text{XX},(n_{q_i}^{(i)})^{\text{th}}\text{XX})}_{(m_{p_i}^{(i)},\, n_{q_i}^{(i)})\text{-XX}} \in \left\{ \begin{array}{l} \underbrace{((2m_{p_i1}^{(i)}+1)^{\text{th}}\text{SI},(2n_{q_i1}^{(i)}+1)^{\text{th}}\text{SO})}_{((2m_{p_i1}^{(i)}+1),\,(2n_{q_i1}^{(i)}+1))\text{-saddle}}, \underbrace{((2m_{p_i1}^{(i)})^{\text{th}}\text{LS},(2n_{q_i1}^{(i)})^{\text{th}}\text{US})}_{((2m_{p_i1}^{(i)}),\,(2n_{q_i1}^{(i)}))\text{-double-saddle}}, \\ \underbrace{((2m_{p_i1}^{(i)})^{\text{th}}\text{LS},(2n_{q_i1}^{(i)}+1)^{\text{th}}\text{SO})}_{((2m_{p_i1}^{(i)}),\,(2n_{q_i1}^{(i)}+1))\text{-lower-saddle-source}}, \underbrace{((2m_{p_i1}^{(i)}+1)^{\text{th}}\text{SI},(2n_{q_i1}^{(i)})^{\text{th}}\text{US})}_{((2m_{p_i1}^{(i)}+1),\,(2n_{q_i1}^{(i)}))\text{-upper-saddle-sink}} \end{array} \right\};$$

$$\underbrace{((m_{p_i}^{(i)})^{\text{th}}\text{XX},(n_{1}^{(i)})^{\text{th}}\text{XX})}_{(m_{p_i}^{(i)},\, n_{1}^{(i)})\text{-XX}} \in \left\{ \begin{array}{l} \underbrace{((2m_{p_i1}^{(i)}+1)^{\text{th}}\text{SI},(2n_{11}^{(i)}+1)^{\text{th}}\text{SI})}_{((2m_{p_i1}^{(i)}+1),\,(2n_{11}^{(i)}+1))\text{-sink}}, \underbrace{((2m_{p_i1}^{(i)})^{\text{th}}\text{LS},(2n_{11}^{(i)})^{\text{th}}\text{US})}_{((2m_{p_i1}^{(i)}),\,(2n_{11}^{(i)}))\text{-double-saddle}}, \\ \underbrace{((2m_{p_i1}^{(i)})^{\text{th}}\text{LS},(2n_{11}^{(i)}+1)^{\text{th}}\text{SI})}_{((2m_{p_i1}^{(i)}),\,(2n_{11}^{(i)}+1))\text{-lower-saddle-sink}}, \underbrace{((2m_{p_i1}^{(i)}+1)^{\text{th}}\text{SI},(2n_{11}^{(i)})^{\text{th}}\text{US})}_{((2m_{p_i1}^{(i)}+1),\,(2n_{11}^{(i)}))\text{-upper-saddle-sink}} \end{array} \right\};$$

$$\underbrace{((m_{1}^{(i)})^{\text{th}}\text{XX},(n_{q_i}^{(i)})^{\text{th}}\text{XX})}_{(m_{1}^{(i)},\, n_{q_i}^{(i)})\text{-XX}} \in \left\{ \begin{array}{l} \underbrace{((2m_{11}^{(i)}+1)^{\text{th}}\text{SO},(2n_{q_i1}^{(i)}+1)^{\text{th}}\text{SO})}_{((2m_{11}^{(i)}+1),\,(2n_{q_i1}^{(i)}+1))\text{-source}}, \underbrace{((2m_{11}^{(i)})^{\text{th}}\text{LS},(2n_{q_i1}^{(i)})^{\text{th}}\text{US})}_{((2m_{11}^{(i)}),\,(2n_{q_i1}^{(i)}))\text{-double-saddle}}, \\ \underbrace{((2m_{11}^{(i)})^{\text{th}}\text{LS},(2n_{q_i1}^{(i)}+1)^{\text{th}}\text{SO})}_{((2m_{11}^{(i)}),\,(2n_{q_i1}^{(i)}+1))\text{-lower-saddle-source}}, \underbrace{((2m_{11}^{(i)}+1)^{\text{th}}\text{SO},(2n_{q_i1}^{(i)})^{\text{th}}\text{US})}_{((2m_{11}^{(i)}+1),\,(2n_{q_i1}^{(i)}))\text{-upper-saddle-source}} \end{array} \right\};$$

$$\underbrace{((m_1^{(i)})^{\text{th}}\text{XX},(n_1^{(i)})^{\text{th}}\text{XX})}_{(m_1^{(i)},\,n_1^{(i)})\text{-XX}}$$

$$\in \left\{ \begin{array}{l} \underbrace{((2m_{11}^{(i)}+1)^{\text{th}}\text{SO},(2n_{11}^{(i)}+1)^{\text{th}}\text{SI})}_{((2m_{11}^{(i)}+1),\,(2n_{11}^{(i)}+1))\text{-saddle}},\ \underbrace{((2m_{11}^{(i)})^{\text{th}}\text{LS},(2n_{11}^{(i)})^{\text{th}}\text{US})}_{((2m_{11}^{(i)}),\,(2n_{11}^{(i)}))\text{-double-saddle}}, \\ \underbrace{((2m_{11}^{(i)})^{\text{th}}\text{LS},(2n_{11}^{(i)}+1)^{\text{th}}\text{SI})}_{((2m_{11}^{(i)}),\,(2n_{11}^{(i)}+1))\text{-lower-saddle-sink}},\ \underbrace{((2m_{11}^{(i)}+1)^{\text{th}}\text{SO},(2n_{11}^{(i)})^{\text{th}}\text{US})}_{((2m_{11}^{(i)}+1),\,(2n_{11}^{(i)}))\text{-upper-saddle-source}} \end{array} \right\}. \tag{3.128}$$

(vii_{2c}) The $((2m_1)^{\text{th}}\text{LS}, (2n_1)^{\text{th}}\text{US})$-double-saddle switching bifurcation is for two $q \times p$ equilibrium networks as

$$\begin{aligned} \bigcup_{r_1=1}^{p}\bigcup_{s_1=1}^{q} \underbrace{(a_{j_1j_1r_1}, a_{j_2j_2s_1})}_{(m_{r_1}^{(1)},\,n_{s_1}^{(1)})\text{-XX}} &\rightleftharpoons \underbrace{((2m_1)^{\text{th}}\text{LS},(2n_1)^{\text{th}}\text{US})}_{((2m_1),(2n_1))\text{-double-saddle}} \\ &\rightleftharpoons \bigcup_{r_2=1}^{p}\bigcup_{s_2=1}^{q} \underbrace{(a_{j_1j_1r_2}, a_{j_2j_2s_2})}_{(m_{r_2}^{(2)},\,n_{s_2}^{(2)})\text{-XX}} \end{aligned} \tag{3.129}$$

where $\Sigma_{r_i=1}^{p} m_{r_i} = 2m_1$, $\Sigma_{s_i=1}^{q} n_{s_i} = 2n_1$ for $i = 1, 2$.

(vii_3) For $a_{j_1j_10} > 0$ and $a_{j_2j_20} < 0$, there is a $((2m_1)^{\text{th}}\text{US}, (2n_1)^{\text{th}}\text{LS})$ -double-saddle equilibrium as

$$(a_{j_1j_11}, a_{j_2j_21}) = \underbrace{((2m_1)^{\text{th}}\text{US},(2n_1)^{\text{th}}\text{LS})}_{((2m_1),(2n_1))\text{-double-saddle}}. \tag{3.130}$$

There are the following three $((2m_1)^{\text{th}}\text{US}, (2n_1)^{\text{th}}\text{LS})$-double-saddle appearing and switching bifurcations.

(vii_{3a}) The $((2m_1)^{\text{th}}\text{US}, (2n_1)^{\text{th}}\text{LS})$-double-saddle appearing bifurcation is from an (pF,nF)-flow to a $(2m_1) \times (2n_1)$-equilibrium network as

$$\underbrace{(\dot{x}_{j_1}, \dot{x}_{j_2})}_{(\text{pF,nF})\text{-flow}} \rightleftharpoons \underbrace{((2m_1)^{\text{th}}\text{US},(2n_1)^{\text{th}}\text{LS})}_{((2m_1),(2n_1))\text{-double-saddle}} \rightleftharpoons \bigcup_{r=1}^{2m_1}\bigcup_{s=1}^{2n_1} \underbrace{(a_{j_1j_1r}, a_{j_2j_2s})}_{\text{XX}} \tag{3.131}$$

where

$$\bigcup_{r=1}^{2m_1}\bigcup_{s=1}^{2n_1}\underbrace{(a_{j_1j_1r},a_{j_2j_2s})}_{XX}=\left\{\begin{array}{cccc}\underbrace{(\mathrm{SI,SI})}_{\text{sink}} & \underbrace{(\mathrm{SO,SI})}_{\text{saddle}} & \cdots & \underbrace{(\mathrm{SO,SI})}_{\text{saddle}}\\ \underbrace{(\mathrm{SI,SO})}_{\text{saddle}} & \underbrace{(\mathrm{SO,SO})}_{\text{source}} & \cdots & \underbrace{(\mathrm{SO,SO})}_{\text{source}}\\ \vdots & \vdots & \vdots & \vdots\\ \underbrace{(\mathrm{SI,SO})}_{\text{saddle}} & \underbrace{(\mathrm{SO,SO})}_{\text{source}} & \cdots & \underbrace{(\mathrm{SO,SO})}_{\text{source}}\end{array}\right\}_{(2n_1)\times(2m_1)}. \tag{3.132}$$

(vii$_{3b}$) The $((2m_1)^{\text{th}}\text{US}, (2n_1)^{\text{th}}\text{LS})$-double-saddle appearing bifurcation is from a $q_1 \times p_1$ to $q_2 \times p_2$ equilibrium network as

$$\begin{aligned}&\bigcup_{r_1=1}^{p_1}\bigcup_{s_1=1}^{q_1}\underbrace{(a_{j_1j_1r_1},a_{j_2j_2s_1})}_{(m_{r_1}^{(1)},n_{s_1}^{(1)})\text{-XX}} \rightleftharpoons \underbrace{((2m_1)^{\text{th}}\text{US},(2n_1)^{\text{th}}\text{LS})}_{((2m_1),(2n_1))\text{-double-saddle}}\\ &\rightleftharpoons \bigcup_{r_2=1}^{p_2}\bigcup_{s_2=1}^{q_2}\underbrace{(a_{j_1j_1r_2},a_{j_2j_2s_2})}_{(m_{r_2}^{(2)},n_{s_2}^{(2)})\text{-XX}}\end{aligned} \tag{3.133}$$

where

$$2m_1-\sum_{r_1=1}^{p_1}m_{r_1}^{(1)}=2\sum_{s=1}^{l_1}m_s$$

with l_1-quadratic polynomials without real roots

and s^{th}-quadratic polynomial with power m_s,

$$2n_1-\sum_{s_1=1}^{q_1}n_{s_1}^{(1)}=2\sum_{l=1}^{l_2}n_l$$

with l_2-quadraitc polynomials without real roots

and l^{th}-quadratic polynomial with power n_l;

$$\sum_{r_2=1}^{p_2}m_{r_2}^{(2)}=2m_1,\ \sum_{s_2=1}^{q_2}n_{s_2}^{(2)}=2n_1;\ \text{for } i=1,2,$$

$$
\begin{aligned}
&\underbrace{((m_{p_i}^{(i)})^{\text{th}}\text{XX},(n_{q_i}^{(i)})^{\text{th}}\text{XX})}_{(m_{p_i}^{(i)},n_{q_i}^{(i)})\text{-XX}} \\
&\in \left\{\begin{array}{l}
\underbrace{((2m_{p_i1}^{(i)}+1)^{\text{th}}\text{SO},(2n_{q_i1}^{(i)}+1)^{\text{th}}\text{SI})}_{((2m_{p_i1}^{(i)}+1),(2n_{q_i1}^{(i)}+1))\text{-saddle}},\ \underbrace{((2m_{p_i1}^{(i)})^{\text{th}}\text{US},(2n_{q_i1}^{(i)})^{\text{th}}\text{LS})}_{((2m_{p_i1}^{(i)}),(2n_{q_i1}^{(i)}))\text{-double-saddle}}, \\
\underbrace{((2m_{p_i1}^{(i)})^{\text{th}}\text{US},(2n_{q_i1}^{(i)}+1)^{\text{th}}\text{SI})}_{((2m_{p_i1}^{(i)}),(2n_{q_i1}^{(i)}+1))\text{-upper-saddle-sink}},\ \underbrace{((2m_{p_i1}^{(i)}+1)^{\text{th}}\text{SO},(2n_{q_i1}^{(i)})^{\text{th}}\text{LS})}_{((2m_{p_i1}^{(i)}+1),(2n_{q_i1}^{(i)}))\text{-lower-saddle-source}}
\end{array}\right\}; \\
&\underbrace{((m_{p_i}^{(i)})^{\text{th}}\text{XX},(n_{1}^{(i)})^{\text{th}}\text{XX})}_{(m_{p_i}^{(i)},n_{1}^{(i)})\text{-XX}} \\
&\in \left\{\begin{array}{l}
\underbrace{((2m_{p_i1}^{(i)}+1)^{\text{th}}\text{SO},(2n_{11}^{(i)}+1)^{\text{th}}\text{SO})}_{((2m_{p_i1}^{(i)}+1),(2n_{11}^{(i)}+1))\text{-source}},\ \underbrace{((2m_{p_i1}^{(i)})^{\text{th}}\text{US},(2n_{11}^{(i)})^{\text{th}}\text{LS})}_{((2m_{p_i1}^{(i)}),(2n_{11}^{(i)}))\text{-double-saddle}}, \\
\underbrace{((2m_{p_i1}^{(i)})^{\text{th}}\text{US},(2n_{11}^{(i)}+1)^{\text{th}}\text{SO})}_{((2m_{p_i1}^{(i)}),(2n_{11}^{(i)}+1))\text{-upper-saddle-source}},\ \underbrace{((2m_{p_i1}^{(i)}+1)^{\text{th}}\text{SO},(2n_{11}^{(i)})^{\text{th}}\text{LS})}_{((2m_{p_i1}^{(i)}+1),(2n_{11}^{(i)}))\text{-lower-saddle-source}}
\end{array}\right\}; \\
&\underbrace{((m_{1}^{(i)})^{\text{th}}\text{XX},(n_{q_i}^{(i)})^{\text{th}}\text{XX})}_{(m_{1}^{(i)},n_{q_i}^{(i)})\text{-XX}} \\
&\in \left\{\begin{array}{l}
\underbrace{((2m_{11}^{(i)}+1)^{\text{th}}\text{SI},(2n_{q_i1}^{(i)}+1)^{\text{th}}\text{SI})}_{((2m_{11}^{(i)}+1),(2n_{q_i1}^{(i)}+1))\text{-sink}},\ \underbrace{((2m_{11}^{(i)})^{\text{th}}\text{US},(2n_{q_i1}^{(i)})^{\text{th}}\text{LS})}_{((2m_{11}^{(i)}),(2n_{q_i1}^{(i)}))\text{-double-saddle}}, \\
\underbrace{((2m_{11}^{(i)})^{\text{th}}\text{US},(2n_{q_i1}^{(i)}+1)^{\text{th}}\text{SI})}_{((2m_{11}^{(i)}),(2n_{q_i1}^{(i)}+1))\text{-upper-saddle-sink}},\ \underbrace{((2m_{11}^{(i)}+1)^{\text{th}}\text{SI},(2n_{q_i1}^{(i)})^{\text{th}}\text{LS})}_{((2m_{11}^{(i)}+1),(2n_{q_i1}^{(i)}))\text{-lower-saddle-sink}}
\end{array}\right\}; \\
&\underbrace{((m_{1}^{(i)})^{\text{th}}\text{XX},(n_{1}^{(i)})^{\text{th}}\text{XX})}_{(m_{1}^{(i)},n_{1}^{(i)})\text{-XX}} \\
&\in \left\{\begin{array}{l}
\underbrace{((2m_{11}^{(i)}+1)^{\text{th}}\text{SI},(2n_{11}^{(i)}+1)^{\text{th}}\text{SO})}_{((2m_{11}^{(i)}+1),(2n_{11}^{(i)}+1))\text{-saddle}},\ \underbrace{((2m_{11}^{(i)})^{\text{th}}\text{US},(2n_{11}^{(i)})^{\text{th}}\text{LS})}_{((2m_{11}^{(i)}),(2n_{11}^{(i)}))\text{-double-saddle}}, \\
\underbrace{((2m_{11}^{(i)})^{\text{th}}\text{US},(2n_{11}^{(i)}+1)^{\text{th}}\text{SO})}_{((2m_{11}^{(i)}),(2n_{11}^{(i)}+1))\text{-upper-saddle-source}},\ \underbrace{((2m_{11}^{(i)}+1)^{\text{th}}\text{SI},(2n_{11}^{(i)})^{\text{th}}\text{LS})}_{((2m_{11}^{(i)}+1),(2n_{11}^{(i)}))\text{-lower-saddle-sink}}
\end{array}\right\}.
\end{aligned}
\tag{3.134}
$$

(vii_{3c}) The $((2m_1)^{\text{th}}\text{US},(2n_1)^{\text{th}}\text{LS})$-double-saddle switching bifurcation is for the switching of two $q \times p$ equilibrium networks as

$$
\bigcup_{r_1=1}^{p_1}\bigcup_{s_1=1}^{q_1}\underbrace{(a_{j_1j_1r_1},a_{j_2j_2s_1})}_{(m_{r_1}^{(1)},n_{s_1}^{(1)})\text{-XX}} \rightleftharpoons \underbrace{((2m_1)^{\text{th}}\text{US},(2n_1)^{\text{th}}\text{LS})}_{((2m_1),(2n_1))\text{-double-saddle}}
$$

$$\rightleftharpoons \bigcup_{r_2=1}^{p_2} \bigcup_{s_2=1}^{q_2} \underbrace{(a_{j_1 j_1 r_2}, a_{j_2 j_2 s_2})}_{(m_{r_2}^{(2)}, n_{s_2}^{(2)})\text{-XX}} \tag{3.135}$$

where $\Sigma_{r_i=1}^{p} m_{r_i} = 2m_1$, $\Sigma_{s_i=1}^{q} n_{s_i} = 2n_1$ for $i = 1, 2$.

(vii$_4$) For $a_{j_1 j_1 0} < 0$and $a_{j_2 j_2 0} < 0$, there is a ((2m_1)thLS,(2n_1)thLS)-double-saddle equilibrium as

$$(a_{j_1 j_1 1}, a_{j_2 j_2 1}) = \underbrace{((2m_1)^{\text{th}}\text{LS}, (2n_1)^{\text{th}}\text{LS})}_{((2m_1),(2n_1))\text{-double-saddle}} \tag{3.136}$$

There are the following three ((2m_1)thLS,(2n_1)thLS)-double-saddle appearing and switching bifurcations.

(vi$_{4a}$) The ((2m_1)thLS,(2n_1)thLS)-double-saddle appearing bifurcation is from a (nF,nF)-flow to a $(2m_1) \times (2n_1)$ network as

$$\underbrace{(\dot{x}_{j_1}, \dot{x}_{j_2})}_{\text{(nF,nF)-flow}} \rightleftharpoons \underbrace{((2m_1)^{\text{th}}\text{LS},(2n_1)^{\text{th}}\text{LS})}_{((2m_1),(2n_1))\text{-double-saddle}} \rightleftharpoons \bigcup_{r=1}^{2m_1} \bigcup_{s=1}^{2n_1} \underbrace{(a_{j_1 j_1 r}, a_{j_2 j_2 s})}_{\text{XX}} \tag{3.137}$$

where

$$\bigcup_{r=1}^{2m_1} \bigcup_{s=1}^{2n_1} \underbrace{(a_{j_1 j_1 r}, a_{j_2 j_2 s})}_{XX} = \left\{ \begin{matrix} \underbrace{(\text{SO,SI})}_{\text{saddle}} & \underbrace{(\text{SI,SI})}_{\text{sink}} & \cdots & \underbrace{(\text{SI,SI})}_{\text{sink}} \\ \underbrace{(\text{SO,SO})}_{\text{source}} & \underbrace{(\text{SI,SO})}_{\text{saddle}} & \cdots & \underbrace{(\text{SI,SO})}_{\text{saddle}} \\ \vdots & \vdots & \vdots & \vdots \\ \underbrace{(\text{SO,SO})}_{\text{source}} & \underbrace{(\text{SI,SO})}_{\text{saddle}} & \cdots & \underbrace{(\text{SI,SO})}_{\text{saddle}} \end{matrix} \right\}_{(2n_1)\times(2m_1)} . \tag{3.138}$$

(vii$_{4b}$) The ((2m_1)thLS,(2n_1)thLS)-double-saddle appearing bifurcation is from $q_1 \times p_1$ to $q_2 \times p_2$ equilibrium networks as

$$\begin{aligned} &\bigcup_{r_1=1}^{p_1} \bigcup_{s_1=1}^{q_1} \underbrace{(a_{j_1 j_1 r_1}, a_{j_2 j_2 s_1})}_{(m_{r_1}^{(1)}, n_{s_1}^{(1)})\text{-XX}} \rightleftharpoons \underbrace{((2m_1)^{\text{th}}\text{SI},(2n_1)^{\text{th}}\text{LS})}_{((2m_1),(2n_1))\text{-double-saddle}} \\ &\rightleftharpoons \bigcup_{r_2=1}^{p_2} \bigcup_{s_2=1}^{q_2} \underbrace{(a_{j_1 j_1 r_2}, a_{j_2 j_2 s_2})}_{(m_{r_2}^{(2)}, n_{s_2}^{(2)})\text{-XX}} \end{aligned} \tag{3.139}$$

where

$$2m_1 - \sum_{r_1=1}^{p_1} m_{r_1}^{(1)} = 2\sum_{s=1}^{l_1} m_s$$

with l_1-quadratic polynomials without real roots

and s^{th}-quadratic polynomial with power m_s,

$$2n_1 - \sum_{s_1=1}^{q_1} n_{s_1}^{(1)} = 2\sum_{l=1}^{l_2} n_l$$

with l_2-quadraitc polynomials without real roots

and l^{th}-quadratic polynomial with power n_l;

$$\sum_{r_2=1}^{p_2} m_{r_2}^{(2)} = 2m_1, \sum_{s_2=1}^{q_2} n_{s_2}^{(2)} = 2n_1; \text{ for } i = 1, 2,$$

$$\underbrace{((m_{p_i}^{(i)})^{\text{th}}\text{XX},(n_{q_i}^{(i)})^{\text{th}}\text{XX})}_{(m_{p_i}^{(i)},n_{q_i}^{(i)})\text{-XX}}$$

$$\in \left\{ \begin{array}{l} \underbrace{((2m_{p_i1}^{(i)}+1)^{\text{th}}\text{SI},(2n_{q_i1}^{(i)}+1)^{\text{th}}\text{SI})}_{((2m_{p_i1}^{(i)}+1),(2n_{q_i1}^{(i)}+1))\text{-sink}}, \underbrace{((2m_{p_i1}^{(i)})^{\text{th}}\text{LS},(2n_{q_i1}^{(i)})^{\text{th}}\text{LS})}_{((2m_{p_i1}^{(i)}),(2n_{q_i1}^{(i)}))\text{-double-saddle}}, \\ \underbrace{((2m_{p_i1}^{(i)})^{\text{th}}\text{LS},(2n_{q_i1}^{(i)}+1)^{\text{th}}\text{SI})}_{((2m_{p_i1}^{(i)}),(2n_{q_i1}^{(i)}+1))\text{-lower-saddle-sink}}, \underbrace{((2m_{p_i1}^{(i)}+1)^{\text{th}}\text{SI},(2n_{q_i1}^{(i)})^{\text{th}}\text{LS})}_{((2m_{p_i1}^{(i)}+1),(2n_{q_i1}^{(i)}))\text{-lower-saddle-sink}} \end{array} \right\};$$

$$\underbrace{((m_{p_i}^{(i)})^{\text{th}}\text{XX},(n_1^{(i)})^{\text{th}}\text{XX})}_{(m_{p_i}^{(i)},n_1^{(i)})\text{-XX}}$$

$$\in \left\{ \begin{array}{l} \underbrace{((2m_{p_i1}^{(i)}+1)^{\text{th}}\text{SI},(2n_{11}^{(i)}+1)^{\text{th}}\text{SO})}_{((2m_{p_i1}^{(i)}+1),(2n_{11}^{(i)}+1))\text{-saddle}}, \underbrace{((2m_{p_i1}^{(i)})^{\text{th}}\text{LS},(2n_{11}^{(i)})^{\text{th}}\text{LS})}_{((2m_{p_i1}^{(i)}),(2n_{11}^{(i)}))\text{-double-saddle}}, \\ \underbrace{((2m_{p_i1}^{(i)})^{\text{th}}\text{LS},(2n_{11}^{(i)}+1)^{\text{th}}\text{SO})}_{((2m_{p_i1}^{(i)}),(2n_{11}^{(i)}+1))\text{-lower-saddle-source}}, \underbrace{((2m_{p_i1}^{(i)}+1)^{\text{th}}\text{SI},(2n_{11}^{(i)})^{\text{th}}\text{LS})}_{((2m_{p_i1}^{(i)}+1),(2n_{11}^{(i)}))\text{-lower-saddle-sink}} \end{array} \right\};$$

$$\underbrace{((m_1^{(i)})^{\text{th}}\text{XX},(n_{q_i}^{(i)})^{\text{th}}\text{XX})}_{(m_1^{(i)},n_{q_i}^{(i)})\text{-XX}}$$

$$\in \left\{ \begin{array}{l} \underbrace{((2m_{11}^{(i)}+1)^{\text{th}}\text{SO},(2n_{q_i1}^{(i)}+1)^{\text{th}}\text{SI})}_{((2m_{11}^{(i)}+1),(2n_{q_i1}^{(i)}+1))\text{-saddle}}, \underbrace{((2m_{11}^{(i)})^{\text{th}}\text{LS},(2n_{q_i1}^{(i)})^{\text{th}}\text{LS})}_{((2m_{11}^{(i)}),(2n_{q_i1}^{(i)}))\text{-double-saddle}}, \\ \underbrace{((2m_{11}^{(i)})^{\text{th}}\text{LS},(2n_{q_i1}^{(i)}+1)^{\text{th}}\text{SI})}_{((2m_{11}^{(i)}),(2n_{q_i1}^{(i)}+1))\text{-lower-saddle-sink}}, \underbrace{((2m_{11}^{(i)}+1)^{\text{th}}\text{SO},(2n_{q_i1}^{(i)})^{\text{th}}\text{LS})}_{((2m_{11}^{(i)}+1),(2n_{q_i1}^{(i)}))\text{-lower-saddle-sourc}} \end{array} \right\};$$

$$\underbrace{((m_1^{(i)})^{\text{th}}\text{XX},(n_1^{(i)})^{\text{th}}\text{XX})}_{(m_1^{(i)},n_1^{(i)})\text{-XX}}$$
$$\in \left\{ \begin{array}{l} \underbrace{((2m_{11}^{(i)}+1)^{\text{th}}\text{SO},(2n_{11}^{(i)}+1)^{\text{th}}\text{SO})}_{((2m_{11}^{(i)}+1),(2n_{11}^{(i)}+1))\text{-source}}, \underbrace{((2m_{11}^{(i)})^{\text{th}}\text{LS},(2n_{11}^{(i)})^{\text{th}}\text{LS})}_{((2m_{11}^{(i)}),(2n_{11}^{(i)}))\text{-double-saddle}}, \\ \underbrace{((2m_{11}^{(i)})^{\text{th}}\text{LS},(2n_{11}^{(i)}+1)^{\text{th}}\text{SO})}_{((2m_{11}^{(i)}),(2n_{11}^{(i)}+1))\text{-lower-saddle-source}}, \underbrace{((2m_{11}^{(i)}+1)^{\text{th}}\text{SO},(2n_{11}^{(i)})^{\text{th}}\text{LS})}_{((2m_{11}^{(i)}+1),(2n_{11}^{(i)}))\text{-lower-saddle-source}} \end{array} \right\}. \tag{3.140}$$

(vii_{4c}) The $((2m_1)^{\text{th}}\text{LS},(2n_1)^{\text{th}}\text{LS})$-double-saddle switching bifurcation is for two $q \times p$ equilibrium networks as

$$\begin{aligned} &\bigcup_{r_1=1}^{p}\bigcup_{s_1=1}^{q}\underbrace{(a_{j_1j_1r_1}, a_{j_2j_2s_1})}_{(m_{r_1}^{(1)}, n_{s_1}^{(1)})-XX} \rightleftharpoons \underbrace{((2m_1)^{\text{th}}\text{LS},(2n_1)^{\text{th}}\text{LS})}_{((2m_1),(2n_1))\text{-double-saddle}} \\ &\rightleftharpoons \bigcup_{r_2=1}^{p}\bigcup_{s_2=1}^{q}\underbrace{(a_{j_1j_1r_2}, a_{j_2j_2s_2})}_{(m_{r_2}^{(2)}, n_{s_2}^{(2)})-XX} \end{aligned} \tag{3.141}$$

where $\Sigma_{r_i=1}^{p} m_{r_i} = 2m_1$, $\Sigma_{s_i=1}^{q} n_{s_i} = 2n_1$ for $i = 1, 2$.

3.2 Proof of Theorem 3.1

Consider a self-univariate polynomial dynamical system as

$$\begin{aligned} &\dot{x}_{j_1} = a_{j_1j_10}(x_{j_1} - a_{j_1j_11})^m,\ \dot{x}_{j_2} = a_{j_2j_20}(x_{j_2} - a_{j_2j_21})^n, \\ &j_1, j_2 \in \{1, 2\}; j_1 \neq j_2. \end{aligned}$$

In phase space,

$$\frac{dx_{j_1}}{dx_{j_2}} = \frac{a_{j_1j_10}}{a_{j_2j_20}}\frac{(x_{j_1} - a_{j_1j_11})^m}{(x_{j_2} - a_{j_2j_21})^n},$$

and

$$\frac{dx_{j_2}}{(x_{j_2} - a_{j_2j_21})^n} = \frac{a_{j_2j_20}}{a_{j_1j_10}}\frac{dx_{j_1}}{(x_{j_1} - a_{j_1j_11})^m},$$

With the initial condition (x_{j_10}, x_{j_20}), the integration of the foregoing equation gives the first integral manifold for for $m, n \neq 1$ as

$$\frac{1}{n-1}\left[\frac{1}{(x_{j_2} - a_{j_2j_21})^{n-1}} - \frac{1}{(x_{j_20} - a_{j_2j_21})^{n-1}}\right]$$

$$= \frac{a_{j_2j_20}}{a_{j_1j_10}} \frac{1}{m-1} [\frac{1}{(x_{j_1} - a_{j_1j_11})^{m-1}} - \frac{1}{(x_{j_10} - a_{j_1j_11})^{m-1}}].$$

For $n = 1$ and $m \neq 1$, the first integral manifold becomes

$$-\ln \frac{|x_{j_2} - a_{j_2j_21}|}{|x_{j_20} - a_{j_2j_21}|} = \frac{a_{j_2j_20}}{a_{j_1j_10}} \frac{1}{m-1} [\frac{1}{(x_{j_1} - a_{j_1j_11})^{m-1}} - \frac{1}{(x_{j_10} - a_{j_1j_11})^{m-1}}].$$

For $m = 1$ and $n \neq 1$, the first integral manifold becomes

$$\frac{1}{n-1} [\frac{1}{(x_{j_2} - a_{j_2j_21})^{n-1}} - \frac{1}{(x_{j_20} - a_{j_2j_21})^{n-1}}] = -\frac{a_{j_2j_20}}{a_{j_1j_10}} \ln \frac{|x_{j_1} - a_{j_1j_11}|}{|x_{j_10} - a_{j_1j_11}|}.$$

For $m, n = 1$, the first integral manifold becomes

$$\ln \frac{|x_{j_2} - a_{j_2j_21}|}{|x_{j_20} - a_{j_2j_21}|} = \frac{a_{j_2j_20}}{a_{j_1j_10}} \ln \frac{|x_{j_1} - a_{j_1j_11}|}{|x_{j_10} - a_{j_1j_11}|}.$$

The corresponding variational equation at $(x_{j_1}^*, x_{j_2}^*) = (a_{j_1j_11}, a_{j_2j_21})$ is

$$\Delta\dot{x}_{j_1} = a_{j_1j_10}(\Delta x_{j_1})^m, \ \Delta\dot{x}_{j_2} = a_{j_2j_20}(\Delta x_{j_2})^n.$$

(i_1) For $m = 2m_1 + 1$ and $n = 2n_1 + 1$, the flows at $x_{j_1}^* = a_{j_1j_11}$ are the $(2m_1 + 1)^{\text{th}}$-order source and sink for $a_{j_1j_10} > 0$ and $a_{j_1j_10} < 0$, respectively; and the flows at $x_{j_2}^* = a_{j_2j_21}$ are the $(2n_1 + 1)^{\text{th}}$-order source and sink for $a_{j_2j_20} > 0$ and $a_{j_2j_20} < 0$, respectively.

Therefore, for $m = 2m_1 + 1$ and $n = 2n_1 + 1$, the equilibrium of $(x_{j_1}^*, x_{j_2}^*) = (a_{j_1j_11}, a_{j_2j_21})$ has the following properties as in Eqs. (3.3)–(3.6).

- For $a_{j_1j_10} > 0$ and $a_{j_2j_20} > 0$,

$$(a_{j_1j_11}, a_{j_2j_21}) = \underbrace{((2m_1{+}1)^{\text{th}}\text{SO},(2n_1{+}1)^{\text{th}}\text{SO})}_{((2m_1+1),(2n_1+1))\text{-source}}.$$

- For $a_{j_1j_10} < 0$ and $a_{j_2j_20} > 0$,

$$(a_{j_1j_11}, a_{j_2j_21}) = \underbrace{((2m_1{+}1)^{\text{th}}\text{SI},(2n_1{+}1)^{\text{th}}\text{SO})}_{((2m_1+1),(2n_1+1))\text{-saddle}}.$$

- For $a_{j_1j_10} > 0$ and $a_{j_2j_20} < 0$,

$$(a_{j_1j_11}, a_{j_2j_21}) = \underbrace{((2m_1{+}1)^{\text{th}}\text{SO},(2n_1{+}1)^{\text{th}}\text{SI})}_{((2m_1+1),(2n_1+1))\text{-saddle}}.$$

- For $a_{j_1j_10} < 0$ and $a_{j_2j_20} < 0$,

$$(a_{j_1j_11}, a_{j_2j_21}) = \underbrace{((2m_1+1)^{\text{th}}\text{SI},(2n_1+1)^{\text{th}}\text{SI})}_{((2m_1+1),(2n_1+1))\text{-sink}}.$$

(i_2) For $m = 2m_1$ and $n = 2n_1 + 1$, the flows at $x_{j_1}^* = a_{j_1j_11}$ are the $(2m_1)^{\text{th}}$-order upper-saddle and lower-saddle for $a_{j_1j_10} > 0$ and $a_{j_1j_10} < 0$, respectively; and the flows at $x_{j_2}^* = a_{j_2j_21}$ are the $(2n_1 + 1)^{\text{th}}$-order source and sink for $a_{j_2j_20} > 0$ and $a_{j_2j_20} < 0$, respectively.

Therefore, for $m = 2m_1$ and $n = 2n_1 + 1$, the equilibrium of $(x_{j_1}^*, x_{j_2}^*) = (a_{j_1j_11}, a_{j_2j_21})$ has the following properties as in Eqs. (3.17)–(3.20).

- For $a_{j_1j_10} > 0$ and $a_{j_2j_20} > 0$,

$$(a_{j_1j_11}, a_{j_2j_21}) = \underbrace{((2m_1)^{\text{th}}\text{US},(2n_1+1)^{\text{th}}\text{SO})}_{((2m_1),(2n_1+1))\text{-upper-saddle source}}.$$

- For $a_{j_1j_10} < 0$ and $a_{j_2j_20} > 0$,

$$(a_{j_1j_11}, a_{j_2j_21}) = \underbrace{((2m_1)^{\text{th}}\text{LS},(2n_1+1)^{\text{th}}\text{SO})}_{((2m_1),(2n_1+1))\text{-lower-saddle source}}.$$

- For $a_{j_1j_10} > 0$ and $a_{j_2j_20} < 0$,

$$(a_{j_1j_11}, a_{j_2j_21}) = \underbrace{((2m_1)^{\text{th}}\text{US},(2\text{n}_1+1)^{\text{th}}\text{SI})}_{((2m_1),(2n_1+1))\text{-upper-saddle sink}}.$$

- For $a_{j_1j_10} < 0$ and $a_{j_2j_20} < 0$,

$$(a_{j_1j_11}, a_{j_2j_21}) = \underbrace{((2m_1)^{\text{th}}\text{LS},(2n_1+1)^{\text{th}}\text{SI})}_{((2m_1),(2n_1+1))\text{-lower-saddle sink}}.$$

(i_3) For $m = 2m_1 + 1$ and $n = 2n_1$, the flows at $x_{j_1}^* = a_{j_1j_11}$ are the $(2m_1 + 1)^{\text{th}}$-order source and sink for $a_{j_1j_10} > 0$ and $a_{j_1j_10} < 0$, respectively; and the flows at $x_{j_2}^* = a_{j_2j_21}$ are the $(2n_1)^{\text{th}}$-order upper-saddle and lower-saddle for $a_{j_2j_20} > 0$ and $a_{j_2j_20} < 0$, respectively.

Therefore, for $m = 2m_1 + 1$ and $n = 2n_1$, the equilibrium of $(x_{j_1}^*, x_{j_2}^*) = (a_{j_1j_11}, a_{j_2j_21})$ has the following properties as in Eqs. (3.21)–(3.24).

- For $a_{j_1j_10} > 0$ and $a_{j_2j_20} > 0$,

$$(a_{j_1 j_1 1}, a_{j_2 j_2 1}) = \underbrace{((2m_1+1)^{\text{th}}\text{SO},(2n_1)^{\text{th}}\text{US})}_{((2m_1+1),(2n_1)\text{-upper-saddle source}}.$$

- For $a_{j_1 j_1 0} < 0$ and $a_{j_2 j_2 0} > 0$,

$$(a_{j_1 j_1 1}, a_{j_2 j_2 1}) = \underbrace{((2m_1+1)^{\text{th}}\text{SI},(2n_1)^{\text{th}}\text{US})}_{((2m_1+1),(2n_1))\text{-upper-saddle sink}}.$$

- For $a_{j_1 j_1 0} > 0$ and $a_{j_2 j_2 0} < 0$,

$$(a_{j_1 j_1 1}, a_{j_2 j_2 1}) = \underbrace{((2m_1+1)^{\text{th}}\text{SO},(2n_1)^{\text{th}}\text{LS})}_{((2m_1+1),(2n_1))\text{-lower-saddle source}}.$$

- For $a_{j_1 j_1 0} < 0$ and $a_{j_2 j_2 0} > 0$,

$$(a_{j_1 j_1 1}, a_{j_2 j_2 1}) = \underbrace{((2m_1+1)^{\text{th}}\text{SI},(2n_1)^{\text{th}}\text{LS})}_{((2m_1+1),(2n_1))\text{-lower-saddle sink}}.$$

(i_4) For $m = 2m_1$ and $n = 2n_1$, the flows at $x_{j_1}^* = a_{j_1 j_1 1}$ are the $(2m_1)^{\text{th}}$-order upper-saddle and lower-saddle for $a_{j_1 j_1 0} > 0$ and $a_{j_1 j_1 0} < 0$, respectively; and the flows at $x_{j_2}^* = a_{j_2 j_2 1}$ are the $(2n_1)^{\text{th}}$-order upper-saddle and lower-saddle for $a_{j_2 j_2 0} > 0$ and $a_{j_2 j_2 0} < 0$, respectively.

Therefore, for $m = 2m_1$ and $n = 2n_1$, the equilibrium of $(x_{j_1}^*, x_{j_2}^*) = (a_{j_1 j_1 1}, a_{j_2 j_2 1})$ has the following properties as in Eqs. (3.25)–(3.28).

- For $a_{j_1 j_1 0} > 0$ and $a_{j_2 j_2 0} > 0$,

$$(a_{j_1 j_1 1}, a_{j_2 j_2 1}) = \underbrace{((2m_1)^{\text{th}}\text{US},(2n_1)^{\text{th}}\text{US})}_{((2m_1),(2n_1))\text{-double saddle}}.$$

- For $a_{j_1 j_1 0} < 0$ and $a_{j_2 j_2 0} > 0$,

$$(a_{j_1 j_1 1}, a_{j_2 j_2 1}) = \underbrace{((2m_1)^{\text{th}}\text{LS},(2n_1)^{\text{th}}\text{US})}_{((2m_1),(2n_1))\text{-double saddle}}.$$

- For $a_{j_1 j_1 0} > 0$ and $a_{j_2 j_2 0} < 0$,

$$(a_{j_1 j_1 1}, a_{j_2 j_2 1}) = \underbrace{((2m_1)^{\text{th}}\text{US},(2\text{n}_1)^{\text{th}}\text{LS})}_{((2m_1),(2n_1))\text{-upper-saddle sink}}.$$

- For $a_{j_1 j_1 0} < 0$ and $a_{j_2 j_2 0} < 0$,

$$(a_{j_1j_11}, a_{j_2j_21}) = \underbrace{((2m_1)^{\text{th}}\text{LS},(2n_1)^{\text{th}}\text{LS})}_{((2m_1),(2n_1))\text{-double-saddle}}.$$

(ii) Consider a self-univariate polynomial dynamical system $\Sigma_{r=1}^{p} m_r = m$ and $\Sigma_{s=1}^{q} n_s = n$ as

$$\dot{x}_{j_1} = a_{j_1j_10}\prod_{r=1}^{p}(x_{j_1} - a_{j_1j_1r})^{m_r},\ \dot{x}_{j_2} = a_{j_2j_20}\prod_{s=1}^{q}(x_{j_2} - a_{j_2j_2s})^{n_s}.$$

In phase space, $\sum_{r=1}^{p} m_r = m$ and $\Sigma_{s=1}^{q} n_s = n$

$$\frac{dx_{j_2}}{dx_{j_1}} = \frac{a_{j_2j_20}\ (x_{j_2} - a_{j_2j_2s_1})^{n_{s_1}} \prod_{s_2=1,s_2\neq s_1}^{q}(x_{j_2} - a_{j_2j_2s_2})^{n_{s_2}}}{a_{j_1j_10}\ (x_{j_1} - a_{j_1j_1r_1})^{m_{r_1}} \prod_{r_2=1,r_2\neq r_1}^{p}(x_{j_1} - a_{j_1j_1r_2})^{m_{r_2}}},$$

and

$$\sum_{s_1=1}^{q}\sum_{l_2=1}^{n_{s_1}} \frac{B_{l_2s_1}dx_{j_2}}{(x_{j_2} - a_{j_2j_2s_1})^{l_2}} = \frac{a_{j_2j_20}}{a_{j_1j_10}}\sum_{r_1=1}^{p}\sum_{l_1=1}^{m_{r_1}} \frac{A_{l_1r_1}dx_{j_1}}{(x_{j_1} - a_{j_1j_1r_1})^{l_1}},$$

where

$$\begin{aligned}
A_{l_1m_{r_1}} &= \frac{1}{(m_{r_1}-l_1)!}\frac{d^{m_{r_1}-l_1}}{dx_{j_1}^{m_{r_1}-l_1}}\frac{1}{\prod_{r_2=1,r_2\neq r_1}^{p}(x_{j_1} - a_{j_1j_1r_2})^{m_{r_2}}}\Big|_{x_{j_1}^*=a_{j_1j_1r_1}},\\
A_{m_{r_1}m_{r_1}} &= \frac{1}{\prod_{r_2=1,r_2\neq r_1,}^{p}(a_{j_1j_1r_1} - a_{j_1j_1r_2})^{m_{r_2}}};\\
B_{l_2n_{s_1}} &= \frac{1}{(n_{s_1}-l_2)!}\frac{d^{n_{s_1}-l_2}}{dx_{j_2}^{n_{s_1}-l_2}}\frac{1}{\prod_{s_2=1,s_2\neq s_1}^{q}(x_{j_2} - a_{j_2j_2s_2})^{n_{s_2}}}\Big|_{x_{j_2}^*=a_{j_2j_2s_1}},\\
B_{n_{s_1}n_{s_1}} &= \frac{1}{\prod_{s_2=1,s_2\neq s_1}^{q}(a_{j_2j_2s_1} - a_{j_2j_2s_2})^{n_{s_2}}}.
\end{aligned}$$

With the initial condition (x_{j_10}, x_{j_20}), the integration of the foregoing equation gives the first integral manifold as

$$\begin{aligned}
&\sum_{s_1=1}^{q}\sum_{l_2=1}^{n_{s_1}} \frac{B_{l_2n_{s_1}}}{l_2-1}\Big[\frac{1}{(x_{j_2} - a_{j_2j_2s_1})^{l_2-1}} - \frac{1}{(x_{j_20} - a_{j_2j_2s_1})^{l_2-1}}\Big]\\
&= \frac{a_{j_2j_20}}{a_{j_1j_10}}\sum_{r_1=1}^{p}\sum_{l_1=1}^{m_{r_1}} \frac{A_{l_1m_{r_1}}}{l_1-1}\Big[\frac{1}{(x_{j_1} - a_{j_1j_1r_1})^{l_1-1}} - \frac{1}{(x_{j_10} - a_{j_1j_1r_1})^{l_1-1}}\Big],
\end{aligned}$$

for $l_1, l_2 \neq 1$;

$$\lim_{l_2\to 1}\frac{1}{(l_2-1)}\Big[\frac{1}{(x_{j_2}-a_{j_2j_2s_1})^{l_2-1}}-\frac{1}{(x_{j_20}-a_{j_2j_2s_1})^{l_2-1}}\Big]=-\ln\frac{|x_{j_2}-a_{j_2j_2s_1}|}{|x_{j_20}-a_{j_2j_2s_1}|},$$

for $l_2 = 1$;

$$\lim_{l_1\to 1}\frac{1}{l_1-1}\Big[\frac{1}{(x_{j_1}-a_{j_1j_1r_1})^{l_1-1}}-\frac{1}{(x_{j_10}-a_{j_1j_1r_1})^{l_1-1}}\Big]=-\ln\frac{|x_{j_1}-a_{j_1j_1r_1}|}{|x_{j_10}-a_{j_1j_1r_1}|},$$

for $l_1 = 1$.

The singular equilibrium network with $\Sigma_{r_i=1}^{p} m_{r_i} = m$ and $\Sigma_{r_i=1}^{q} n_{s_i} = n$ is defined as

$$\bigcup_{r=1}^{p}\bigcup_{s=1}^{q}\underbrace{(a_{j_1j_1r},a_{j_2j_2s})}_{(m_r,n_s)\text{-XX}}\equiv\left\{\begin{array}{ccc}(a_{j_1j_11},a_{j_2j_2q}) & \cdots & (a_{j_1j_1p},a_{j_2j_2q})\\ \vdots & \cdots & \vdots\\ (a_{j_1j_11},a_{j_2j_21}) & \cdots & (a_{j_1j_1p},a_{j_2j_21})\end{array}\right\}_{q\times p}$$

$$=\left\{\begin{array}{ccc}\underbrace{((m_1)^{\text{th}}\text{XX},(n_q)^{\text{th}}\text{XX})}_{(m_1,n_q)\text{-XX}} & \cdots & \underbrace{((m_p)^{\text{th}}\text{XX},(n_q)^{\text{th}}\text{XX})}_{(m_{p_i}^{(i)},n_{q_i}^{(i)})\text{-XX}}\\ \vdots & \vdots & \vdots\\ \underbrace{((m_1)^{\text{th}}\text{XX},(n_1)^{\text{th}}\text{XX})}_{(m_1,n_1)\text{-XX}} & \cdots & \underbrace{((m_p)^{\text{th}}\text{XX},(n_1)^{\text{th}}\text{XX})}_{(m_p,n_1)\text{-XX}}\end{array}\right\}_{q\times p}.$$

The corresponding variational equations at $(x_{j_1}^*, x_{j_2}^*) = (a_{j_1j_1s_1}, a_{j_2j_2s_2})$ are

$$\Delta\dot{x}_{j_1}=a_{j_1j_10}\prod_{r_2=1,r_2\neq r_1}^{p}(a_{j_1j_1r_1}-a_{j_1j_1r_2})^{m_{r_2}}(\Delta x_{j_1})^{m_{r_1}},$$

$$\Delta\dot{x}_{j_2}=a_{j_2j_20}\prod_{s_2=1,s_2\neq s_1}^{q}(a_{j_2j_2s_1}-a_{j_2j_2s_2})^{n_{s_2}}(\Delta x_{j_2})^{n_{s_1}}.$$

(ii$_1$) For $m_{r_1} = 2m_{r_11}+1$ and $n_{s_1} = 2n_{s_11}+1$, the flows at $x_{j_1}^* = a_{j_1j_1s_1}$ are $(2m_{r_11}+1)^{\text{th}}$-source and sink in the x_{j_1}-direction for $a_{j_1j_10}\prod_{r_2=1,r_2\neq r}^{p}(a_{j_1j_1r_1}-a_{j_1j_1r_2})^{m_{r_2}} > 0$ and $a_{j_1j_10}\prod_{r_2=1,r_2\neq r_1}^{p}(a_{j_1j_1r_1}-a_{j_1j_1r_2})^{m_{r_2}} < 0$, respectively; and the flows at $x_{j_2}^* = a_{j_2j_2s_2}$ are $(2n_{s_11}+1)^{\text{th}}$-source and sink in the x_{j_2}-direction for $a_{j_2j_20}\prod_{s_2=1,s_2\neq s_1}^{q}(a_{j_2j_2s_1}-a_{j_2j_2s_2})^{n_{s_2}} > 0$ and $a_{j_2j_20}\prod_{s_2=1,s_2\neq s_1}^{q}(a_{j_2j_2s_1}-a_{j_2j_2s_2})^{n_{s_2}} < 0$, respectively.

Therefore, for $m_{r_1} = 2m_{r_11}+1$ and $n_{s_1} = 2n_{s_11}+1$, the equilibrium of $(x_{j_1}^*, x_{j_2}^*) = (a_{j_1j_1r_1}, a_{j_2j_2s_1})$ has the following properties as in Eqs. (3.23)–(3.26).

- For $a_{j_1j_10}\prod_{r_2=1,r_2\neq s_1}^{p}(a_{j_1j_1r_1}-a_{j_1j_1r_2})^{m_{r_2}} > 0$ and $a_{j_2j_20}\prod_{s_2=1,s_2\neq s_1}^{q}(a_{j_2j_2s_1}-a_{j_2j_2s_2})^{n_{s_2}} > 0$,

$$(a_{j_1j_1r_1}, a_{j_2j_2s_1}) = \underbrace{((2m_{r_11}+1)^{\text{th}}\text{SO},(2n_{s_11}+1)^{\text{th}}\text{SO})}_{((2m_{r_11}+1),(2n_{s_11}+1))\text{-source}}.$$

- For $a_{j_1j_10}\prod_{r_2=1,r_2\neq s_1}^{p}(a_{j_1j_1r_1}-a_{j_1j_1r_2})^{m_{r_2}}<0$ and $a_{j_2j_20}\prod_{s_2=1,s_2\neq s_1}^{q}(a_{j_2j_2s_1}-a_{j_2j_2s_2})^{n_{s_2}}>0$,

$$(a_{j_1j_1r_1}, a_{j_2j_2s_1}) = \underbrace{((2m_{r_11}+1)^{\text{th}}\text{SI},(2n_{s_11}+1)^{\text{th}}\text{SO})}_{((2m_{r_11}+1),(2n_{s_11}+1))\text{-saddle}}.$$

- For $a_{j_1j_10}\prod_{r_2=1,r_2\neq s_1}^{p}(a_{j_1j_1r_1}-a_{j_1j_1r_2})^{m_{r_2}}>0$ and $a_{j_2j_20}\prod_{s_2=1,s_2\neq s_1}^{q}(a_{j_2j_2s_1}-a_{j_2j_2s_2})^{n_{s_2}}<0$,

$$(a_{j_1j_1r_1}, a_{j_2j_2s_1}) = \underbrace{((2m_{r_11}+1)^{\text{th}}\text{SO},(2n_{s_11}+1)^{\text{th}}\text{SI})}_{((2m_{r_11}+1),(2n_{s_11}+1))\text{-saddle}}.$$

- For $a_{j_1j_10}\prod_{r_2=1,r_2\neq s_1}^{p}(a_{j_1j_1r_1}-a_{j_1j_1r_2})^{m_{r_2}}<0$ and $a_{j_2j_20}\prod_{s_2=1,s_2\neq s_1}^{q}(a_{j_2j_2s_1}-a_{j_2j_2s_2})^{n_{s_2}}<0$,

$$(a_{j_1j_1r_1}, a_{j_2j_2s_1}) = \underbrace{((2m_{r_11}+1)^{\text{th}}\text{SI},(2n_{s_11}+1)^{\text{th}}\text{SI})}_{((2m_{r_11}+1),(2n_{s_11}+1))\text{-sink}}.$$

(ii$_2$) For $m_{r_1}=2m_{r_11}$ and $n_{s_1}=2n_{s_11}+1$, the flows at $x_{j_1}^*=a_{j_1j_1s_1}$ are $(2m_{r_11})^{\text{th}}$-upper-saddle and lower-saddle in the x_{j_1}-direction for $a_{j_1j_10}\prod_{r_2=1,r_2\neq r}^{p}(a_{j_1j_1r_1}-a_{j_1j_1r_2})^{m_{r_2}}>0$ and $a_{j_1j_10}\prod_{r_2=1,r_2\neq r}^{p}(a_{j_1j_1r_1}-a_{j_1j_1r_2})^{m_{r_2}}<0$, respectively; and the flows at $x_{j_2}^*=a_{j_2j_2s_2}$ are $(2n_{s_11}+1)^{\text{th}}$-source and sink for $a_{j_2j_20}\prod_{s_2=1,s_2\neq s_1}^{q}(a_{j_2j_2s_1}-a_{j_2j_2s_2})^{n_{s_2}}>0$ and $a_{j_2j_20}\prod_{s_2=1,s_2\neq s_1}^{q}(a_{j_2j_2s_1}-a_{j_2j_2s_2})^{n_{s_2}}<0$ in the x_{j_2}-direction, respectively.

Therefore, for $m_{r_1}=2m_{r_11}$ and $n_{s_1}=2n_{s_11}+1$, the equilibrium of $(x_{j_1}^*, x_{j_2}^*)=(a_{j_1j_1r_1}, a_{j_2j_2s_1})$ has the following properties, as in Eqs.(3.27)–(3.30).

- For $a_{j_1j_10}\prod_{r_2=1,r_2\neq s_1}^{p}(a_{j_1j_1r_1}-a_{j_1j_1r_2})^{m_{r_2}}>0$ and $a_{j_2j_20}\prod_{s_2=1,s_2\neq s_1}^{q}(a_{j_2j_2s_1}-a_{j_2j_2s_2})^{n_{s_2}}>0$,

$$(a_{j_1j_1r_1}, a_{j_2j_2s_1}) = \underbrace{((2m_{r_11})^{\text{th}}\text{US},(2n_{s_11}+1)^{\text{th}}\text{SO})}_{((2m_{r_11}),(2n_{s_11}+1))\text{-upper-saddle source}}.$$

- For $a_{j_1j_10}\prod_{r_2=1,r_2\neq s_1}^{p}(a_{j_1j_1r_1}-a_{j_1j_1r_2})^{m_{r_2}}<0$ and $a_{j_2j_20}\prod_{s_2=1,s_2\neq s_1}^{q}(a_{j_2j_2s_1}-a_{j_2j_2s_2})^{n_{s_2}}>0$,

$$(a_{j_1j_1r_1}, a_{j_2j_2s_1}) = \underbrace{((2m_{r_11})^{\text{th}}\text{LS},(2n_{s_11}+1)^{\text{th}}\text{SO})}_{((2m_{r_11}),(2n_{s_11}+1))\text{-lower-saddle source}}.$$

- For $a_{j_1j_10}\prod_{r_2=1,r_2\neq s_1}^{p}(a_{j_1j_1r_1}-a_{j_1j_1r_2})^{m_{r_2}}>0$ and $a_{j_2j_20}\prod_{s_2=1,s_2\neq s_1}^{q}(a_{j_2j_2s_1}-a_{j_2j_2s_2})^{n_{s_2}}<0$,

$$(a_{j_1j_1r_1},a_{j_2j_2s_1})=\underbrace{((2m_{r_11})^{\text{th}}\text{US},(2n_{s_11}+1)^{\text{th}}\text{SI})}_{((2m_{r_11}),(2n_{s_11}+1))\text{-upper-saddle sink}}.$$

- For $a_{j_1j_10}\prod_{r_2=1,r_2\neq s_1}^{p}(a_{j_1j_1r_1}-a_{j_1j_1r_2})^{m_{r_2}}<0$ and $a_{j_2j_20}\prod_{s_2=1,s_2\neq s_1}^{q}(a_{j_2j_2s_1}-a_{j_2j_2s_2})^{n_{s_2}}<0$,

$$(a_{j_1j_1r_1},a_{j_2j_2s_1})=\underbrace{((2m_{r_11})^{\text{th}}\text{LS},(2n_{s_11}+1)^{\text{th}}\text{SI})}_{((2m_{r_11}),(2n_{s_11}+1))\text{-lower-saddle-sink}}.$$

(ii$_3$) For $m_{r_1}=2m_{r_11}+1$ and $n_{s_1}=2n_{s_11}$, the flows at $x_{j_1}^*=a_{j_1j_1s_1}$ are $(2m_{r_11}+1)^{\text{th}}$ source and sink in the x_{j_1}-direction for $a_{j_1j_10}\prod_{r_2=1,r_2\neq r}^{p}(a_{j_1j_1r_1}-a_{j_1j_1r_2})^{m_{r_2}}>0$ and $a_{j_1j_10}\prod_{r_2=1,r_2\neq r}^{p}(a_{j_1j_1r_1}-a_{j_1j_1r_2})^{m_{r_2}}<0$, respectively; and the flows at $x_{j_2}^*=a_{j_2j_2s_2}$ are $(2n_{s_11})^{\text{th}}$ upper-saddle and lower-saddle for $a_{j_2j_20}\prod_{s_2=1,s_2\neq s_1}^{q}(a_{j_2j_2s_1}-a_{j_2j_2s_2})^{n_{s_2}}>0$ and $a_{j_2j_20}\prod_{s_2=1,s_2\neq s_1}^{q}(a_{j_2j_2s_1}-a_{j_2j_2s_2})^{n_{s_2}}<0$ in the x_{j_2}-direction, respectively.

Therefore, for $m_{r_1}=2m_{r_11}+1$ and $n_{s_1}=2n_{s_11}$, the equilibrium of $(x_{j_1}^*,x_{j_2}^*)=(a_{j_1j_1r_1},a_{j_2j_2s_1})$ has the following properties as in Eqs. (3.31)–(3.34).

- For $a_{j_1j_10}\prod_{r_2=1,r_2\neq s_1}^{p}(a_{j_1j_1r_1}-a_{j_1j_1r_2})^{m_{r_2}}>0$ and $a_{j_2j_20}\prod_{s_2=1,s_2\neq s_1}^{q}(a_{j_2j_2s_1}-a_{j_2j_2s_2})^{n_{s_2}}>0$,

$$(a_{j_1j_1r_1},a_{j_2j_2s_1})=\underbrace{((2m_{r_11}+1)^{\text{th}}\text{SO},(2n_{s_11})^{\text{th}}\text{US})}_{((2m_{r_11}+1),(2n_{s_11}))\text{-upper-saddle source}}.$$

- For $a_{j_1j_10}\prod_{r_2=1,r_2\neq s_1}^{p}(a_{j_1j_1r_1}-a_{j_1j_1r_2})^{m_{r_2}}<0$ and $a_{j_2j_20}\prod_{s_2=1,s_2\neq s_1}^{q}(a_{j_2j_2s_1}-a_{j_2j_2s_2})^{n_{s_2}}>0$,

$$(a_{j_1j_1r_1},a_{j_2j_2s_1})=\underbrace{((2m_{r_11}+1)^{\text{th}}\text{SI},(2n_{s_11})^{\text{th}}\text{US})}_{((2m_{r_11}+1),(2n_{s_11}))\text{-upper-saddle sink}}.$$

- For $a_{j_1j_10}\prod_{r_2=1,r_2\neq s_1}^{p}(a_{j_1j_1r_1}-a_{j_1j_1r_2})^{m_{r_2}}>0$ and $a_{j_2j_20}\prod_{s_2=1,s_2\neq s_1}^{q}(a_{j_2j_2s_1}-a_{j_2j_2s_2})^{n_{s_2}}<0$,

$$(a_{j_1j_1r_1},a_{j_2j_2s_1})=\underbrace{((2m_{r_11}+1)^{\text{th}}\text{SO},(2n_{s_11})^{\text{th}}\text{LS})}_{((2m_{r_11}+1),(2n_{s_11}))\text{-lower-saddle source}}.$$

- For $a_{j_1j_10}\prod_{r_2=1,r_2\neq s_1}^{p}(a_{j_1j_1r_1}-a_{j_1j_1r_2})^{m_{r_2}}<0$ and $a_{j_2j_20}\prod_{s_2=1,s_2\neq s_1}^{q}(a_{j_2j_2s_1}-a_{j_2j_2s_2})^{n_{s_2}}<0$,

$$(a_{j_1 j_1 r_1}, a_{j_2 j_2 s_1}) = \underbrace{((2m_{r_1 1}+1)^{\text{th}}\text{SI},(2n_{s_1 1})^{\text{th}}\text{LS})}_{((2m_{r_1 1}+1),(2n_{s_1 1}))\text{-lower-saddle sink}}.$$

(ii$_4$) For $m_{r_1} = 2m_{r_1 1}$ and $n_{s_1} = 2n_{s_1 1}$, the flows at $x_{j_1}^* = a_{j_1 j_1 s_1}$ are $(2m_{r_1 1})^{\text{th}}$-upper-saddle and lower-saddle in the x_{j_1}-direction for $a_{j_1 j_1 0} \prod_{r_2=1, r_2 \neq r_1}^{p} (a_{j_1 j_1 r_1} - a_{j_1 j_1 r_2})^{m_{r_2}} > 0$ and $a_{j_1 j_1 0} \prod_{r_2=1, r_2 \neq r_1}^{p} (a_{j_1 j_1 r_1} - a_{j_1 j_1 r_2})^{m_{r_2}} < 0$, respectively; and the flows at $x_{j_2}^* = a_{j_2 j_2 s_2}$ are $(2n_{s_1 1})^{\text{th}}$ upper-saddle and lower-saddle for $a_{j_2 j_2 0} \prod_{s_2=1, s_2 \neq s_1}^{q} (a_{j_2 j_2 s_1} - a_{j_2 j_2 s_2})^{n_{s_2}} > 0$ and $a_{j_2 j_2 0} \prod_{s_2=1, s_2 \neq s_1}^{q} (a_{j_2 j_2 s_1} - a_{j_2 j_2 s_2})^{n_{s_2}} < 0$ in the x_{j_2}-direction, respectively.

Therefore, for $m_{r_1} = 2m_{r_1 1}$ and $n_{s_1} = 2n_{s_1 1}$, the equilibrium of $(x_{j_1}^*, x_{j_2}^*) = (a_{j_1 j_1 r_1}, a_{j_2 j_2 s_1})$ has the following properties as in Eqs. (3.35)–(3.38).

- For $a_{j_1 j_1 0} \prod_{r_2=1, r_2 \neq s_1}^{p} (a_{j_1 j_1 r_1} - a_{j_1 j_1 r_2})^{m_{r_2}} > 0$ and $a_{j_2 j_2 0} \prod_{s_2=1, s_2 \neq s_1}^{q} (a_{j_2 j_2 s_1} - a_{j_2 j_2 s_2})^{n_{s_2}} > 0$,

$$\left(a_{j_1 j_1 l_1}, a_{j_2 j_2 s_1}\right) = \underbrace{\left((2m_{r_1 1})^{\text{th}}\ \text{US},\ (2n_{s_1 1})^{\text{th}}\ \text{US}\right)}_{((2m_{2_1}+1),(2n_{s_1}))\text{-double-saddle}}.$$

- For $a_{j_1 j_1 0} \prod_{r_2=1, r_2 \neq s_1}^{p} (a_{j_1 j_1 r_1} - a_{j_1 j_1 r_2})^{m_{r_2}} < 0$ and $a_{j_2 j_2 0} \prod_{s_2=1, s_2 \neq s_1}^{q} (a_{j_2 j_2 s_1} - a_{j_2 j_2 s_2})^{n_{s_2}} > 0$,

$$(a_{j_1 j_1 r_1}, a_{j_2 j_2 s_1}) = \underbrace{((2m_{r_1 1})^{\text{th}}\text{LS},(2n_{s_1 1})^{\text{th}}\text{US})}_{((2m_{r_1 1}),(2n_{s_1 1}))\text{-double-saddle}}.$$

- For $a_{j_1 j_1 0} \prod_{r_2=1, r_2 \neq s_1}^{p} (a_{j_1 j_1 r_1} - a_{j_1 j_1 r_2})^{m_{r_2}} > 0$ and $a_{j_2 j_2 0} \prod_{s_2=1, s_2 \neq s_1}^{q} (a_{j_2 j_2 s_1} - a_{j_2 j_2 s_2})^{n_{s_2}} < 0$,

$$(a_{j_1 j_1 r_1}, a_{j_2 j_2 s_1}) = \underbrace{((2m_{r_1 1})^{\text{th}}\text{US},(2n_{s_1 1})^{\text{th}}\text{LS})}_{((2m_{r_1 1}),(2n_{s_1 1}))\text{-doubel-saddle}}.$$

- For $a_{j_1 j_1 0} \prod_{r_2=1, r_2 \neq s_1}^{p} (a_{j_1 j_1 r_1} - a_{j_1 j_1 r_2})^{m_{r_2}} < 0$ and $a_{j_2 j_2 0} \prod_{s_2=1, s_2 \neq s_1}^{q} (a_{j_2 j_2 s_1} - a_{j_2 j_2 s_2})^{n_{s_2}} < 0$,

$$(a_{j_1 j_1 r_1}, a_{j_2 j_2 s_1}) = \underbrace{((2m_{r_1 1})^{\text{th}}\text{LS},(2n_{s_1 1})^{\text{th}}\text{LS})}_{((2m_{r_1 1}),(2n_{s_1 1}))\text{-double-saddle}}.$$

(iii) Consider a self-univariate polynomial dynamical system as

$$\dot{x}_{j_1} = a_{j_1j_10}\prod_{s_1=1}^{m}(x_{j_1} - a_{j_1j_1s_1}), \dot{x}_{j_2} = a_{j_2j_20}\prod_{l_1=1}^{n}(x_{j_2} - a_{j_2j_2l_1}),$$
$$j_1, j_2 \in \{1, 2\}; j_1 \neq j_2.$$

In phase space,

$$\frac{dx_{j_2}}{dx_{j_1}} = \frac{a_{j_2j_20}\prod_{l_1=1}^{n}(x_{j_2} - a_{j_2j_21})}{a_{j_1j_10}\prod_{s_1=1}^{m}(x_{j_1} - a_{j_1j_1s_1})},$$

and

$$\frac{dx_{j_2}}{\prod_{l_1=1}^{n}(x_{j_2} - a_{j_2j_21})} = \frac{a_{j_2j_20}}{a_{j_1j_10}}\frac{dx_{j_1}}{\prod_{s_1=1}^{m}(x_{j_1} - a_{j_1j_1s_1})}.$$

With the initial condition (x_{j_10}, x_{j_20}), the integration of the foregoing equation gives the first integral manifold as

$$\sum_{l_1=1}^{n}\frac{1}{\prod_{l_2=1,l_2\neq l_1}^{n}(a_{j_2j_2l_1} - a_{j_2j_2l_2})}\ln\frac{|x_{j_2} - a_{j_2j_2l_1}|}{|x_{j_20} - a_{j_2j_2l_1}|}$$
$$= \frac{a_{j_2j_20}}{a_{j_1j_10}}\sum_{s_1=1}^{m}\frac{1}{\prod_{s_2=1,s_2\neq s_1}^{m}(a_{j_1j_1s_1} - a_{j_1j_1s_2})}\ln\frac{|x_{j_1} - a_{j_1j_1s_1}|}{|x_{j_10} - a_{j_1j_1s_1}|}.$$

The simple-equilibrium network is defined as

$$\bigcup_{s=1}^{m}\bigcup_{l=1}^{n}\underbrace{(a_{j_1j_1s}, a_{j_2j_2l})}_{XX}$$
$$\equiv \left\{\begin{matrix} (a_{j_1j_11}, a_{j_2j_2n}) & (a_{j_1j_12}, a_{j_2j_2n}) & \cdots & (a_{j_1j_1m}, a_{j_2j_2n}) \\ (a_{j_1j_11}, a_{j_2j_2(n-1)}) & (a_{j_1j_12}, a_{j_2j_2(n-1)}) & \cdots & (a_{j_1j_1m}, a_{j_2j_2(n-1)}) \\ \vdots & \vdots & \cdots & \vdots \\ (a_{j_1j_11}, a_{j_2j_21}) & (a_{j_1j_12}, a_{j_2j_21}) & \cdots & (a_{j_1j_1m}, a_{j_2j_21}) \end{matrix}\right\}_{n\times m}.$$

The variational equation at $(x_{j_1}^*, x_{j_2}^*) = (a_{j_1j_1s_1}, a_{j_2j_2l_1})$ is given by

$$\Delta\dot{x}_{j_1} = a_{j_1j_10}\prod_{s_2=1,s_2\neq s_1}^{m}(a_{j_1j_1s_1} - a_{j_1j_1s_2})\Delta x_{j_1},$$
$$\Delta\dot{x}_{j_2} = a_{j_2j_20}\prod_{l_2=1,l_2\neq l_1}^{n}(a_{j_2j_2l_1} - a_{j_2j_2l_2})\Delta x_{j_2}.$$

The flows at $x_{j_1}^* = a_{j_1j_1s_1}$ are source and sink for $a_{j_1j_10}\prod_{s_2=1,s_2\neq s_1}^{m}(a_{j_1j_1s_1}x_{j_1} - a_{j_1j_1s_2}) > 0$ and $a_{j_1j_10}\prod_{s_2=1,s_2\neq s_1}^{m}(a_{j_1j_1s_1}x_{j_1} - a_{j_1j_1s_2}) < 0$, in the x_{j_1}-direction respectively; and the flows at $x_{j_2}^* = a_{j_2j_2l_1}$ are source and sink for $a_{j_2j_20}\prod_{l_2=1,l_2\neq l_1}^{n}(a_{j_2j_2l_1} - a_{j_2j_2l_2}) > 0$ and $a_{j_2j_20}\prod_{l_2=1,l_2\neq l_1}^{n}(a_{j_2j_2l_1} - a_{j_2j_2l_2}) < 0$ in the x_{j_2}-direction, respectively.

Therefore, the equilibrium of $(x_{j_1}^*, x_{j_2}^*) = (a_{j_1j_1s_1}, a_{j_2j_2l_1})$ $(s_1, s_2 \in \{1, 2, \cdots, m\}, s_1 \neq s_2; l_1, l_2 \in \{1, 2, \cdots, n\}, l_1 \neq l_2;)$ possesses the following properties as in Eqs.(3.42)-(3.45).

- For $a_{j_1j_10}\prod_{s_2=1,s_2\neq s_1}^{m}(a_{j_1j_1s_1} - a_{j_1j_1s_2}) > 0$ and $a_{j_2j_20}\prod_{l_2=1,l_2\neq l_1}^{n}(a_{j_2j_2l_1} - a_{j_2j_2l_2}) > 0$,

$$(a_{j_1j_1s_1}, a_{j_2j_2l_1}) = \underbrace{(\text{SO,SO})}_{\text{source}}.$$

The equilibrium of $(x_{j_1}^*, x_{j_2}^*) = (a_{j_1j_1s_1}, a_{j_2j_2l_1})$ is an (SO,SO)-source.

- For $a_{j_1j_10}\prod_{s_2=1,s_2\neq s_1}^{m}(a_{j_1j_1s_1} - a_{j_1j_1s_2}) < 0$ and $a_{j_2j_20}\prod_{l_2=1,l_2\neq l_1}^{n}(a_{j_2j_2l_1} - a_{j_2j_2l_2}) > 0$,

$$(a_{j_1j_1s_1}, a_{j_2j_2l_1}) = \underbrace{(\text{SI,SO})}_{\text{saddle}}.$$

The equilibrium of $(x_{j_1}^*, x_{j_2}^*) = (a_{j_1j_1s_1}, a_{j_2j_2l_1})$ is an (SI,SO)-saddle.

- For $a_{j_1j_10}\prod_{s_2=1,s_2\neq s_1}^{m}(a_{j_1j_1s_1} - a_{j_1j_1s_2}) > 0$ and $a_{j_2j_20}\prod_{l_2=1,l_2\neq l_1}^{n}(a_{j_2j_2l_1} - a_{j_2j_2l_2}) < 0$,

$$(a_{j_1j_1s_1}, a_{j_2j_2l_1}) = \underbrace{(\text{SO,SI})}_{\text{saddle}}.$$

The equilibrium of $(x_{j_1}^*, x_{j_2}^*) = (a_{j_1j_1s_1}, a_{j_2j_2l_1})$ is called an (SO,SI)-saddle.

- For $a_{j_1j_10}\prod_{s_2=1,s_2\neq s_1}^{m}(a_{j_1j_1s_1} - a_{j_1j_1s_2}) < 0$ and $a_{j_2j_20}\prod_{l_2=1,l_2\neq l_1}^{n}(a_{j_2j_2l_1} - a_{j_2j_2l_2}) < 0$,

$$(a_{j_1j_1s_1}, a_{j_2j_2l_1}) = \underbrace{(\text{SI,SI})}_{\text{sink}}.$$

The equilibrium of $(x_{j_1}^*, x_{j_2}^*) = (a_{j_1j_1s_1}, a_{j_2j_2l_1})$ is an (SI,SI)-sink.

(iv) For $m = 2m_1 + 1$ and $n = 2n_1 + 1$, the bifurcation process is discussed through differential equations as follows. There are three cases (I)-(III).

Case I: Consider a dynamical system having a single equilibrium (A_1) as

$$\dot{x}_{j_1} = a_{j_1j_10}(x_{j_1} - b_{j_1j_11})\prod_{s=1}^{m_1}[(x_{j_1} - a_{j_1j_1}^{(s)})^2 + \Delta_{j_1j_1}^{(s)}],$$
$$\dot{x}_{j_2} = a_{j_2j_20}(x_{j_2} - b_{j_2j_21})\prod_{l=1}^{n_1}[(x_{j_2} - a_{j_2j_2}^{(l)})^2 + \Delta_{j_2j_2}^{(l)}];$$

where $a_{j_1j_11} = b_{j_1j_11}$ and $a_{j_2j_21} = b_{j_2j_21}$. Once

$$\Delta_{j_1j_1}^{(s)} = \Delta_{j_1j_1} = 0\,(s = 1, 2, \cdots, m_1),$$
$$\Delta_{j_2j_2}^{(l)} = \Delta_{j_2j_2} = 0\,(l = 1, 2, \cdots, n_1),$$

if

$$a_{j_1j_11} = b_{j_1j_11} = a_{j_1j_1}^{(s)}\,(s = 1, 2, \cdots, m_1),$$
$$a_{j_2j_21} = b_{j_2j_21} = a_{j_2j_2}^{(l)}\,(l = 1, 2, \cdots, n_1),$$

the foregoing differential equations becomes a dynamical system with a singular equilibrium (B_1) as

$$\dot{x}_{j_1} = a_{j_1j_10}(x_{j_1} - a_{j_1j_11})^{2m_1+1},\ \dot{x}_{j_2} = a_{j_2j_20}(x_{j_2} - a_{j_2j_21})^{2n_1+1}.$$

If

$$\Delta_{j_1j_1}^{(s)} = \Delta_{j_1j_1} + \delta_s\,(s = 1, 2, \cdots, m_1),$$
$$\Delta_{j_2j_2}^{(l)} = \Delta_{j_2j_2} + \delta_l\,(l = 1, 2, \cdots, n_1),$$
$$\delta_s > 0\,and\,\delta_l > 0,$$

then

$$a_{j_1j_11}^{(s)}, a_{j_1j_12}^{(s)} = a_{j_1j_11} \pm \varepsilon_s,\ (s = 1, 2, \cdots, m_1),$$
$$a_{j_2j_21}^{(l)}, a_{j_2j_22}^{(l)} = a_{j_2j_21} \pm \varepsilon_l,\ (l = 1, 2, \cdots, n_1);$$
$$\left\{a_{j_1j_11}, a_{j_1j_12}, \cdots, a_{j_1j_1(2m_1+1)}\right\} = \text{sort}\left\{b_{j_1j_11}, a_{j_1j_11}^{(s)}, a_{j_1j_12}^{(s)}\middle| s = 1, 2, \cdots, m_1\right\},$$
$$\left\{a_{j_2j_21}, a_{j_2j_21}, \cdots, a_{j_2j_2(2n_1+1)}\right\} = \text{sort}\left\{b_{j_2j_21}, a_{j_2j_21}^{(l)}, a_{j_2j_22}^{(l)}\middle| l = 1, 2, \cdots, n_1\right\}.$$

Thus, the differential equation becomes a dynamical system with a non-singular equilibrium network (C_1) as

$$\dot{x}_{j_1} = a_{j_1j_10}\prod_{s=1}^{2m_1+1}(x_{j_1} - a_{j_1j_1s}),\ \dot{x}_{j_2} = a_{j_2j_20}\prod_{l=1}^{2n_1+1}(x_{j_2} - a_{j_2j_2l}).$$

From the above bifurcation process analysis, at least, (m_1+1)-parameter variations in the x_{j_1}-direction and (n_1+1)-parameter variations in the x_{j_2}-direction are engaged in such a bifurcation from a single equilibrium to the equilibrium network of $(2m_1+1)\times(2n_1+1)$ through the $(2m_1+1)^{\text{th}}$-source and sink bifurcations in the x_{j_1}-direction and the $(2n_1+1)^{\text{th}}$-source and sink bifurcations in the x_{j_2}-direction. *With both of them, the higher-order source, sink, and saddle bifurcations are developed.*

Thus, the appearing or vanishing bifurcation route is as follows.

$$\underbrace{(a_{j_1j_1 1},a_{j_2j_2 1})}_{\text{XX}} \rightleftharpoons \underbrace{(a_{j_1j_1 1},a_{j_2j_2 1})}_{((2m_1+1),(2n_1+1))\text{-XX}} \rightleftharpoons \bigcup_{s=1}^{2m_1+1}\bigcup_{l=1}^{2n_1+1}\underbrace{(a_{j_1j_1 s},a_{j_2j_2 l})}_{\text{XX}}.$$

Case II: From Case (I), consider a dynamical system having singular equilibriums (A_2) as

$$\begin{aligned}
\dot{x}_{j_1} &= a_{j_1j_10}\prod_{r_1=1}^{p_1}(x_{j_1}-b_{j_1j_1r_1})^{m_{r_1}^{(1)}}\prod_{s=1}^{l_1}[(x_{j_1}-a_{j_1j_1}^{(s)})^2+\Delta_{j_1j_1}^{(s)}]^{m_s},\\
\dot{x}_{j_2} &= a_{j_2j_20}\prod_{s_1=1}^{q_1}(x_{j_2}-b_{j_2j_2s_1})^{n_{s_1}^{(1)}}\prod_{l=1}^{l_2}[(x_{j_2}-a_{j_2j_2}^{(l)})^2+\Delta_{j_2j_2}^{(l)}]^{n_l};
\end{aligned}$$

where

$$\begin{aligned}
&\{a_{j_1j_1r_1}|r_1=1,2,\cdots,p_1\}=\text{sort}\{b_{j_1j_1r_1}|r_1=1,2,\cdots,p_1\},\\
&\{a_{j_2j_2s_1}|s_1=1,2,\cdots,q_1\}=\text{sort}\{b_{j_2j_2s_1}|s_1=1,2,\cdots,q_1\};\\
&2m_1+1-\sum_{r_1=1}^{p_1}m_{r_1}^{(1)}=2\sum_{s=1}^{l_1}m_s\\
&\text{with } l_1\text{-quadratic polynomials without real roots}\\
&\text{and } s^{\text{th}}\text{-quadratic polynomial with power } m_s,\\
&2n_1+1-\sum_{s_1=1}^{q_1}n_{s_1}^{(1)}=2\sum_{l=1}^{l_2}n_l\\
&\text{with } l_2\text{-quadraitc polynomials without real roots}\\
&\text{and } l^{\text{th}}\text{-quadratic polynomial with power } n_l.
\end{aligned}$$

Once

$$\Delta_{j_1j_1}^{(s)}=0\,(s=1,2,\cdots,l_1)\text{ and }\Delta_{j_2j_2}^{(l)}=0\,(l=1,2,\cdots,l_2),$$

if

$$a_{j_1j_11}\equiv a_{j_1j_1r_1}=a_{j_1j_1}^{(s)}\,(r_1=1,2,\cdots,p_1;s=1,2,\cdots,l_1),$$

$$a_{j_2 j_2 1} \equiv a_{j_2 j_2 s_1} = a_{j_2 j_2}^{(l)} (s_1 = 1, 2, \cdots, q_1; l = 1, 2, \cdots, l_2).$$

the foregoing differential equations becomes a dynamical system with a singular equilibrium (B_2) as

$$\dot{x}_{j_1} = a_{j_1 j_1 0}(x_{j_1} - a_{j_1 j_1 1})^{2m_1+1}, \; \dot{x}_{j_2} = a_{j_2 j_2 0}(x_{j_2} - a_{j_2 j_2 1})^{2n_1+1}.$$

If

$$\begin{aligned} &\Delta_{j_1 j_1}^{(s)} = \Delta_{j_1 j_1} + \delta_s \; (s = 1, 2, \cdots, l_1), \\ &\Delta_{j_2 j_2}^{(l)} = \Delta_{j_2 j_2} + \delta_l \; (l = 1, 2, \cdots, l_2), \\ &\delta_s > 0 \, and \, \delta_l > 0, \end{aligned}$$

then

$$\begin{aligned} &a_{j_1 j_1 1}^{(s)}, a_{j_1 j_1 2}^{(s)} = a_{j_1 j_1 1} \pm \varepsilon_s, \; (s = 1, 2, \cdots, l_1), \\ &a_{j_2 j_2 1}^{(l)}, a_{j_2 j_2 2}^{(l)} = a_{j_2 j_2 1} \pm \varepsilon_l, \; (l = 1, 2, \cdots, l_2); \\ &\left\{ a_{j_1 j_1 1}, a_{j_1 j_1 2}, \cdots, a_{j_1 j_1 p_2} \right\} \\ &= \text{sort} \left\{ a_{j_1 j_1 1}, a_{j_1 j_1 2}, \cdots, a_{j_1 j_1 p_1}; a_{j_1 j_1 1}^{(s)}, a_{j_1 j_1 2}^{(s)} \middle| s = 1, 2, \cdots, l_1 \right\}, \\ &\left\{ a_{j_2 j_2 1}, a_{j_2 j_2 1}, \cdots, a_{j_2 j_2 q_2} \right\} \\ &= \text{sort} \left\{ a_{j_2 j_2 1}, a_{j_2 j_2 1}, \cdots, a_{j_2 j_2 q_1}; a_{j_2 j_2 1}^{(l)}, a_{j_2 j_2 2}^{(l)} \middle| l = 1, 2, \cdots, l_2 \right\}. \end{aligned}$$

Thus, the differential equation becomes a dynamical system with singular equilibriums (C_2) as

$$\dot{x}_{j_1} = a_{j_1 j_1 0} \prod_{r_2=1}^{p_2} (x_{j_1} - a_{j_1 j_1 r_2})^{m_{r_2}^{(2)}}, \; \dot{x}_{j_2} = a_{j_2 j_2 0} \prod_{s_2=1}^{q_2} (x_{j_2} - a_{j_2 j_2 s_2})^{n_{s_2}^{(2)}};$$

where

$$\sum_{r_2=1}^{p_2} m_{r_2}^{(2)} = 2m_1 + 1, \; \sum_{s_2=1}^{q_2} n_{s_2}^{(2)} = 2n_1 + 1.$$

From the above bifurcation process analysis, at least, p_2-parameter variations in the x_{j_1}-direction and q_2-parameter variations in the x_{j_2}-direction are engaged in such a bifurcation from a $p_1 \times q_1$ equilibrium network to a $p_2 \times q_2$ equilibrium network through the $(2m_1 + 1)^{\text{th}}$-source and sink bifurcations in the x_{j_1}-direction and the $(2n_1 + 1)^{\text{th}}$-source and sink bifurcations in the x_{j_2}-direction.

Thus, the appearing or vanishing bifurcation route from a $p_1 \times q_1$ to a $p_2 \times q_2$ equilibrium network is expressed as

$$\bigcup_{r_1=1}^{p_1}\bigcup_{s_1=1}^{q_1}\underbrace{(a_{j_1j_1r_1}, a_{j_2j_2s_1})}_{(m_{r_1}^{(1)}, n_{s_1}^{(1)})\text{-XX}} \rightleftharpoons \underbrace{(a_{j_1j_1}1, a_{j_2j_2}1)}_{((2m_1+1),(2n_1+1))\text{-XX}} \rightleftharpoons \bigcup_{r_2=1}^{p_2}\bigcup_{s_2=1}^{q_2}\underbrace{(a_{j_1j_1r_2}, a_{j_2j_2s_2})}_{(m_{r_2}^{(2)}, n_{s_2}^{(2)})\text{-XX}}.$$

From case (I), the singular equilibriums can be formed through nonsingular equilibrium networks. Thus,

$$\underbrace{(a_{j_1j_1r_1}, a_{j_2j_2s_1})}_{(m_{r_1}^{(1)}, n_{s_1}^{(1)})\text{-XX}} \rightleftharpoons \bigcup_{s=1}^{m_{r_1}^{(1)}}\bigcup_{l=1}^{n_{s_1}^{(1)}}\underbrace{(a_{j_1j_1s}, a_{j_2j_2l})}_{\text{XX}}$$

$$\equiv \begin{Bmatrix} (a_{j_1j_1}1, a_{j_2j_2n_{s_1}^{(1)}}) & (a_{j_1j_1}2, a_{j_2j_2n_{s_1}^{(1)}}) & \cdots & (a_{j_1j_1m_{r_1}^{(1)}}, a_{j_2j_2n_{s_1}^{(1)}}) \\ (a_{j_1j_1}1, a_{j_2j_2(n_{s_1}^{(1)}-1)}) & (a_{j_1j_1}2, a_{j_2j_2(n_{s_1}^{(1)}-1)}) & \cdots & (a_{j_1j_1m_{r_1}^{(1)}}, a_{j_2j_2(n_{s_1}^{(1)}-1)}) \\ \vdots & \vdots & \cdots & \vdots \\ (a_{j_1j_1}1, a_{j_2j_2}1) & (a_{j_1j_1}2, a_{j_2j_2}1) & \cdots & (a_{j_1j_1m_{r_1}^{(1)}}, a_{j_2j_2}1) \end{Bmatrix}_{n_{s_1}^{(1)}\times m_{r_1}^{(1)}},$$

and

$$\underbrace{(a_{j_1j_1r_2}, a_{j_2j_2s_2})}_{(m_{r_2}^{(2)}, n_{s_2}^{(2)})\text{-XX}} \rightleftharpoons \bigcup_{s=1}^{m_{r_2}^{(2)}}\bigcup_{l=1}^{n_{s_2}^{(2)}}\underbrace{(a_{j_1j_1s}, a_{j_2j_2l})}_{\text{XX}}$$

$$\equiv \begin{Bmatrix} (a_{j_1j_1}1, a_{j_2j_2n_{s_2}^{(2)}}) & (a_{j_1j_1}2, a_{j_2j_2n_{s_2}^{(2)}}) & \cdots & (a_{j_1j_1m_{r_2}^{(2)}}, a_{j_2j_2n_{s_2}^{(2)}}) \\ (a_{j_1j_1}1, a_{j_2j_2(n_{s_2}^{(2)}-1)}) & (a_{j_1j_1}2, a_{j_2j_2(n_{s_2}^{(2)}-1)}) & \cdots & (a_{j_1j_1m_{r_2}^{(2)}}, a_{j_2j_2(n_{s_2}^{(2)}-1)}) \\ \vdots & \vdots & \cdots & \vdots \\ (a_{j_1j_1}1, a_{j_2j_2}1) & (a_{j_1j_1}2, a_{j_2j_2}1) & \cdots & (a_{j_1j_1m_{r_2}^{(2)}}, a_{j_2j_2}1) \end{Bmatrix}_{n_{s_2}^{(2)}\times m_{r_2}^{(2)}}.$$

From the above definition, the corner singular equilibriums for $i = 1, 2$ are determined by

$$\underbrace{((m_{p_i}^{(i)})^{\text{th}}\text{XX},(n_{q_i}^{(i)})^{\text{th}}\text{XX})}_{(m_{p_i}^{(i)},n_{q_i}^{(i)})\text{-XX}} \rightleftharpoons \bigcup_{s=1}^{m_{p_i}^{(i)}}\bigcup_{l=1}^{n_{q_i}^{(i)}}\underbrace{(a_{j_1j_1s}, a_{j_2j_2l})}_{\text{XX}},$$

$$\underbrace{((m_{p_i}^{(i)})^{\text{th}}\text{XX},(n_1^{(i)})^{\text{th}}\text{XX})}_{(m_{p_i}^{(i)},n_1^{(i)})\text{-XX}} \rightleftharpoons \bigcup_{s=1}^{m_{p_i}^{(i)}}\bigcup_{l=1}^{n_1^{(i)}}\underbrace{(a_{j_1j_1s}, a_{j_2j_2l})}_{\text{XX}},$$

$$\underbrace{((m_1^{(i)})^{\text{th}}\text{XX},(n_{q_i 1}^{(i)})^{\text{th}}\text{XX})}_{(m_1^{(i)},n_{q_i}^{(i)})\text{-XX}} \rightleftharpoons \bigcup_{s=1}^{m_1^{(i)}}\bigcup_{l=1}^{n_{q_i}^{(i)}} \underbrace{(a_{j_1j_1s}, a_{j_2j_2l})}_{\text{XX}},$$

$$\underbrace{((m_1^{(i)})^{\text{th}}\text{XX},(n_1^{(i)})^{\text{th}}\text{XX})}_{(m_1^{(i)},n_1^{(i)})\text{-XX}} \rightleftharpoons \bigcup_{s=1}^{m_1^{(i)}}\bigcup_{l=1}^{n_1^{(i)}} \underbrace{(a_{j_1j_1s}, a_{j_2j_2l})}_{\text{XX}}.$$

Case III: Consider two dynamical systems with the same equilibriums with locations switched (A_3,C_3) for $i = 1, 2$ as

$$\dot{x}_{j_1} = a_{j_1j_10}\prod_{r_i=1}^{p}(x_{j_1} - a_{j_1j_1r_i})^{m_{r_i}^{(i)}},\ \dot{x}_{j_2} = a_{j_2j_20}\prod_{s_i=1}^{q}(x_{j_2} - a_{j_2j_2s_i})^{n_{s_i}^{(i)}};$$

where

$$\sum_{r_i=1}^{p} m_{r_i}^{(i)} = 2m_1 + 1,\ \sum_{s_i=1}^{q} n_{s_i}^{(i)} = 2n_1 + 1.$$

Consider a dynamical system as a singular equilibrium (B_3) as

$$\dot{x}_{j_1} = a_{j_1j_10}(x_{j_1} - a_{j_1j_11})^{2m_1+1},\ \dot{x}_{j_2} = a_{j_2j_20}(x_{j_2} - a_{j_2j_21})^{2n_1+1}.$$

For $r_i = 1, 2, \cdots, p$ and $s_i = 1, 2, \cdots, q\,(i = 1, 2)$, define two functions as

$$\Delta_{j_1j_1}^{(r_1r_2)} = (a_{j_1j_1r_1} - a_{j_1j_1r_2})^2, (r_1, r_2 \in \{1, 2, \cdots, p\}, r_1 \neq r_2),$$
$$\Delta_{j_2j_2}^{(s_1s_2)} = (a_{j_2j_2s_1} - a_{j_2j_2s_2})^2, (s_1, s_2 \in \{1, 2, \cdots, q\}, s_1 \neq s_2).$$

For $r_i = 1, 2, \cdots, p$ and $s_i = 1, 2, \cdots, q\,(i = 1, 2)$, if

$$\Delta_{j_1j_1}^{(r_1r_2)} = (a_{j_1j_1r_1} - a_{j_1j_1r_2})^2 = 0,$$
$$\Delta_{j_2j_2}^{(s_1s_2)} = (a_{j_2j_2s_1} - a_{j_2j_2s_2})^2 = 0,$$

two equilibriums of $(a_{j_1j_1r_1}, a_{j_2j_2s_1})$ and $(a_{j_1j_1r_2}, a_{j_2j_2s_2})$ switching at point $(a_{j_1j_11}, a_{j_2j_21})$ with the same order singularity are given through

$$\underbrace{(a_{j_1j_1r_1}, a_{j_2j_2s_1})}_{(m_{r_1}^{(1)},n_{s_1}^{(1)})\text{-XX}} \rightleftharpoons \underbrace{(a_{j_1j_11}, a_{j_2j_21})}_{((2m_1+1),(2n_1+1))\text{-XX}} \rightleftharpoons \underbrace{(a_{j_1j_1r_2}, a_{j_2j_2s_2})}_{(m_{r_2}^{(2)},n_{s_2}^{(2)})\text{-XX}}$$
$$\left\{a_{j_1j_1r_1} \rightleftharpoons a_{j_1j_1r_2} \middle| m_{r_1}^{(1)} = m_{r_2}^{(2)};\ r_1, r_2 \in \{1, 2, \cdots, p\}, r_1 \neq r_2\right\},$$
$$\left\{a_{j_2j_2s_1} \rightleftharpoons a_{j_2j_2s_2} \middle| n_{s_1}^{(1)} = n_{s_2}^{(2)};\ s_1, s_2 \in \{1, 2, \cdots, q\}, s_1 \neq s_2\right\}.$$

and

$$
\begin{aligned}
& a_{j_1 j_1 r_1} = a_{j_1 j_1 1} - \varepsilon_{r_1}, a_{j_1 j_1 r_2} = a_{j_1 j_1 1} + \varepsilon_{r_2}, \\
& (a_{j_1 j_1 r_1} = a_{j_1 j_1 1} + \varepsilon_{r_1}, a_{j_1 j_1 r_2} = a_{j_1 j_1 1} - \varepsilon_{r_2}); \\
& a_{j_2 j_2 s_1} = a_{j_2 j_2 1} - \varepsilon_{s_1}, a_{j_2 j_2 s_2} = a_{j_2 j_2 1} + \varepsilon_{s_2}, \\
& (a_{j_2 j_2 s_1} = a_{j_2 j_2 1} + \varepsilon_{s_1}, a_{j_2 j_2 s_2} = a_{j_2 j_2 1} - \varepsilon_{s_2}) \\
& \varepsilon_{r_1}, \varepsilon_{r_2} > 0 \text{ and } \varepsilon_{s_1}, \varepsilon_{s_2} > 0; \\
& r_i = 1, 2, \cdots, p; s_i = 1, 2, \cdots, q \, (i = 1, 2).
\end{aligned}
$$

From case (I), the singular equilibriums can be formed through nonsingular equilibrium networks as presented in Case (II), and the corner singular equilibriums for $i = 1, 2$ are determined similarly.

In summary, from the cases (I)–(III), the equilibrium of $(x^*_{j_1}, x^*_{j_2}) = (a_{j_1 j_1 1}, a_{j_2 j_2 1})$ for $m = 2m_1 + 1$ and $n = 2n_1 + 1$ has the bifurcation properties as stated in (iv_1)-(iv_4) through Eqs. (3.46)–(3.69).

(v) For $m = 2m_1$ and $n = 2n_1 + 1$, the bifurcation process is discussed through differential equations as similar as in (iv). There are three cases (I)–(III).

Case I: Consider a dynamical system having a 1-dimesnional flow (A_1) as

$$
\begin{aligned}
\dot{x}_{j_1} &= a_{j_1 j_1 0} \prod_{s=1}^{m_1} [(x_{j_1} - a^{(s)}_{j_1 j_1})^2 + \Delta^{(s)}_{j_1 j_1}], \\
\dot{x}_{j_2} &= a_{j_2 j_2 0} (x_{j_2} - b_{j_2 j_2 1}) \prod_{l=1}^{n_1} [(x_{j_2} - a^{(l)}_{j_2 j_2})^2 + \Delta^{(l)}_{j_2 j_2}];
\end{aligned}
$$

where $a_{j_2 j_2 1} = b_{j_2 j_2 1}$. Once

$$
\Delta^{(s)}_{j_1 j_1} = 0 \, (s = 1, 2, \cdots, m_1) \text{ and } \Delta^{(l)}_{j_2 j_2} = 0 \, (l = 1, 2, \cdots, n_1),
$$

if

$$
a_{j_1 j_1 1} = a^{(s)}_{j_1 j_1} \, (s = 1, 2, \cdots, m_1) \text{ and } a_{j_2 j_2 1} = b_{j_2 j_2 1} = a^{(l)}_{j_2 j_2} \, (l = 1, 2, \cdots, n_1),
$$

the foregoing differential equation becomes a dynamical system with a singular equilibrium (B_1) as

$$
\dot{x}_{j_1} = a_{j_1 j_1 0} (x_{j_1} - a_{j_1 j_1 1})^{2m_1}, \; \dot{x}_{j_2} = a_{j_2 j_2 0} (x_{j_2} - a_{j_2 j_2 1})^{2n_1 + 1}.
$$

If

$$
\Delta^{(s)}_{j_1 j_1} = \Delta_{j_1 j_1} + \delta_s \, (s = 1, 2, \cdots, m_1),
$$

$$\Delta_{j_2j_2}^{(l)} = \Delta_{j_2j_2} + \delta_l \; (l = 1, 2, \cdots, n_1),$$
$$\delta_s > 0 \text{ and } \delta_l > 0,$$

then

$$a_{j_1j_11}^{(s)}, a_{j_1j_12}^{(s)} = a_{j_1j_11} \pm \varepsilon_s, \; (s = 1, 2, \cdots, m_1),$$
$$a_{j_2j_21}^{(l)}, a_{j_2j_22}^{(l)} = a_{j_2j_21} \pm \varepsilon_l, \; (l = 1, 2, \cdots, n_1);$$
$$\left\{ a_{j_1j_11}, a_{j_1j_12}, \cdots, a_{j_1j_1(2m_1)} \right\} = \text{sort}\left\{ a_{j_1j_11}^{(s)}, a_{j_1j_12}^{(s)} \middle| s = 1, 2, \cdots, m_1 \right\},$$
$$\left\{ a_{j_2j_21}, a_{j_2j_21}, \cdots, a_{j_2j_2(2n_1+1)} \right\} = \text{sort}\left\{ b_{j_2j_21}, a_{j_2j_21}^{(l)}, a_{j_2j_22}^{(l)} \middle| l = 1, 2, \cdots, n_1 \right\}.$$

Thus, the differential equation becomes a dynamical system with a non-singular equilibrium network (C_1) as

$$\dot{x}_{j_1} = a_{j_1j_10} \prod_{s=1}^{2m_1} (x_{j_1} - a_{j_1j_1s}), \; \dot{x}_{j_2} = a_{j_2j_20} \prod_{l=1}^{2n_1+1} (x_{j_2} - a_{j_2j_2l}).$$

From the above bifurcation process analysis, at least, (m_1)-parameter variations in the x_{j_1}-direction and $(n_1 + 1)$-parameter variations in the x_{j_2}-direction are engaged in such a bifurcation from a source or sink flow to the equilibrium network of $(2m_1) \times (2n_1 + 1)$ through the $(2m_1)^{\text{th}}$-upper-saddle and lower-saddle bifurcations in the x_{j_1}-direction and the $(2n_1 + 1)^{\text{th}}$-source and sink bifurcations in the x_{j_2}-direction. *With both of them, the higher-order saddle-source and saddle-sink bifurcations are developed.*

Thus, the appearing or vanishing bifurcation route is as follows.

$$\underbrace{(\dot{x}_{j_1}, a_{j_2j_21})}_{\text{XX}} \rightleftharpoons \underbrace{(a_{j_1j_11}, a_{j_2j_21})}_{((2m_1),(2n_1+1))\text{-XX}} \rightleftharpoons \bigcup_{s=1}^{2m_1} \bigcup_{l=1}^{2n_1+1} \underbrace{(a_{j_1j_1s}, a_{j_2j_2l})}_{\text{XX}}.$$

Case II: From Case (I), as in Case (II) for (iv), consider a dynamical system having singular equilibriums (A_2) as

$$\dot{x}_{j_1} = a_{j_1j_10} \prod_{r_1=1}^{p_1} (x_{j_1} - b_{j_1j_1r_1})^{m_{r_1}^{(1)}} \prod_{s=1}^{l_1} [(x_{j_1} - a_{j_1j_1}^{(s)})^2 + \Delta_{j_1j_1}^{(s)}]^{m_s},$$

$$\dot{x}_{j_2} = a_{j_2j_20} \prod_{s_1=1}^{q_1} (x_{j_2} - b_{j_2j_2s_1})^{n_{s_1}^{(1)}} \prod_{l=1}^{l_2} [(x_{j_2} - a_{j_2j_2}^{(l)})^2 + \Delta_{j_2j_2}^{(l)}]^{n_l};$$

where

$$\begin{aligned}
&\{a_{j_1j_1r_1}|r_1=1,2,\cdots,p_1\}=\mathrm{sort}\{b_{j_1j_1r_1}|r_1=1,2,\cdots,p_1\},\\
&\{a_{j_2j_2s_1}|s_1=1,2,\cdots,q_1\}=\mathrm{sort}\{b_{j_2j_2s_1}|s_1=1,2,\cdots,q_1\};\\
&2m_1-\sum_{r_1=1}^{p_1}m_{r_1}^{(1)}=2\sum_{s=1}^{l_1}m_s\\
&\text{with } l_1\text{-quadratic polynomials without real roots}\\
&\text{and } s^{\text{th}}\text{-quadratic polynomial with power } m_s,\\
&2n_1+1-\sum_{s_1=1}^{q_1}n_{s_1}^{(1)}=2\sum_{l=1}^{l_2}n_l\\
&\text{with } l_2\text{-quadraitc polynomials without real roots}\\
&\text{and } l^{\text{th}}\text{-quadratic polynomial with power } n_l.
\end{aligned}$$

Once

$$\Delta_{j_1j_1}^{(s)}=0\,(s=1,2,\cdots,l_1)\text{ and }\Delta_{j_2j_2}^{(l)}=0\,(l=1,2,\cdots,l_2),$$

if

$$\begin{aligned}
a_{j_1j_1 1}&\equiv a_{j_1j_1r_1}=a_{j_1j_1}^{(s)}\,(r_1=1,2,\cdots,p_1;s=1,2,\cdots,l_1),\\
a_{j_2j_2 1}&\equiv a_{j_2j_2s_1}=a_{j_2j_2}^{(l)}(s_1=1,2,\cdots,q_1;l=1,2,\cdots,l_2),
\end{aligned}$$

the foregoing differential equations becomes a dynamical system with a singular equilibrium (B_2) as

$$\dot{x}_{j_1}=a_{j_1j_1 0}(x_{j_1}-a_{j_1j_1 1})^{2m_1},\ \dot{x}_{j_2}=a_{j_2j_2 0}(x_{j_2}-a_{j_2j_2 1})^{2n_1+1}.$$

If

$$\begin{aligned}
&\Delta_{j_1j_1}^{(s)}=\Delta_{j_1j_1}+\delta_s\,(s=1,2,\cdots,l_1);\\
&\Delta_{j_2j_2}^{(l)}=\Delta_{j_2j_2}+\delta_l\,(l=1,2,\cdots,l_2)\\
&\delta_s>0\,and\,\delta_l>0
\end{aligned}$$

then

$$\begin{aligned}
&a_{j_1j_1 1}^{(s)},a_{j_1j_1 2}^{(s)}=a_{j_1j_1 1}\pm\varepsilon_s,\ (s=1,2,\cdots,l_1),\\
&a_{j_2j_2 1}^{(l)},a_{j_2j_2 2}^{(l)}=a_{j_2j_2 1}\pm\varepsilon_l,\ (l=1,2,\cdots,l_2);\\
&\{a_{j_1j_1 1},a_{j_1j_1 2},\cdots,a_{j_1j_1p_2}\}\\
&=\mathrm{sort}\{a_{j_1j_1 1},a_{j_1j_1 2},\cdots,a_{j_1j_1p_1};a_{j_1j_1 1}^{(s)},a_{j_1j_1 1}^{(s)}|s=1,2,\cdots,l_1\},\\
&\{a_{j_2j_2 1},a_{j_2j_2 1},\cdots,a_{j_2j_2q_2}\}\\
&=\mathrm{sort}\{a_{j_2j_2 1},a_{j_2j_2 1},\cdots,a_{j_2j_2q_1};a_{j_2j_2 1}^{(l)},a_{j_2j_2 2}^{(l)}|l=1,2,\cdots,l_2\}.
\end{aligned}$$

Thus, the differential equation becomes a dynamical system with singular equilibriums (C_2) as

$$\dot{x}_{j_1} = a_{j_1j_10}\prod_{r_2=1}^{p_2}(x_{j_1} - a_{j_1j_1r_2})^{m_{r_2}^{(2)}},\ \dot{x}_{j_2} = a_{j_2j_20}\prod_{s_2=1}^{q_2}(x_{j_2} - a_{j_2j_2s_2})^{n_{s_2}^{(2)}};$$

where

$$\sum_{r_2=1}^{p_2} m_{r_2}^{(2)} = 2m_1,\ \sum_{s_2=1}^{q_2} n_{s_2}^{(2)} = 2n_1 + 1.$$

From the above bifurcation process analysis, at least, p_2-parameter variations in the x_{j_1}-direction and q_2-parameter variations in the x_{j_2}-direction are engaged in such a bifurcation from a $p_1 \times q_1$ equilibrium network to a $p_2 \times q_2$ equilibrium network through the $(2m_1)^{\text{th}}$-upper-saddle and lower-saddle bifurcations in the x_{j_1}-direction and the $(2n_1+1)^{\text{th}}$-source and sink bifurcations in the x_{j_2}-direction.

Therefore, the appearing or vanishing bifurcation route from a $p_1 \times q_1$ to a $p_2 \times q_2$ equilibrium network is expressed as

$$\bigcup_{r_1=1}^{p_1}\bigcup_{s_1=1}^{q_1}\underbrace{(a_{j_1j_1r_1}, a_{j_2j_2s_1})}_{(m_{r_1}^{(1)}, n_{s_1}^{(1)})\text{-XX}} \rightleftharpoons \underbrace{(a_{j_1j_11}, a_{j_2j_21})}_{((2m_1),(2n_1+1))\text{-XX}} \rightleftharpoons \bigcup_{r_2=1}^{p_2}\bigcup_{s_2=1}^{q_2}\underbrace{(a_{j_1j_1r_2}, a_{j_2j_2s_2})}_{(m_{r_2}^{(2)}, n_{s_2}^{(2)})\text{-XX}}.$$

The corner singular equilibriums for $i = 1, 2$ can be determined as in Case (II) for (iv).

Case III, consider two dynamical systems with the same equilibriums with locations switched (A_3,C_3) for $i = 1, 2$ as

$$\dot{x}_{j_1} = a_{j_1j_10}\prod_{r_i=1}^{p}(x_{j_1} - a_{j_1j_1r_i})^{m_{r_i}^{(i)}},\ \dot{x}_{j_2} = a_{j_2j_20}\prod_{s_i=1}^{q}(x_{j_2} - a_{j_2j_2s_i})^{n_{s_i}^{(i)}};$$

where

$$\sum_{r_i=1}^{p} m_{r_i}^{(i)} = 2m_1,\ \sum_{s_i=1}^{q} n_{s_i}^{(i)} = 2n_1 + 1.$$

Consider a dynamical system as a singular equilibrium (B_3) as

$$\dot{x}_{j_1} = a_{j_1j_10}(x_{j_1} - a_{j_1j_11})^{2m_1},\ \dot{x}_{j_2} = a_{j_2j_20}(x_{j_2} - a_{j_2j_21})^{2n_1+1}.$$

For $r_i = 1, 2, \cdots, p$ and $s_i = 1, 2, \cdots, q\,(i = 1, 2)$, define two functions as

$$\Delta_{j_1j_1}^{(r_1r_2)} = (a_{j_1j_1r_1} - a_{j_1j_1r_2})^2, (r_1, r_2 \in \{1, 2, \cdots, p\}, r_1 \neq r_2),$$

$$\Delta_{j_2j_2}^{(s_1s_2)} = (a_{j_2j_2s_1} - a_{j_2j_2s_2})^2, (s_1, s_2 \in \{1, 2, \cdots, q\}, s_1 \neq s_2).$$

For $r_i = 1, 2, \cdots, p$ and $s_i = 1, 2, \cdots, q\,(i = 1, 2)$, if

$$\begin{aligned} \Delta_{j_1j_1}^{(r_1r_2)} &= (a_{j_1j_1r_1} - a_{j_1j_1r_2})^2 = 0, \\ \Delta_{j_2j_2}^{(s_1s_2)} &= (a_{j_2j_2s_1} - a_{j_2j_2s_2})^2 = 0, \end{aligned}$$

two equilibriums of $(a_{j_1j_1r_1}, a_{j_2j_2s_1})$ and $(a_{j_1j_1r_2}, a_{j_2j_2s_2})$ switching at point $(a_{j_1j_11}, a_{j_2j_21})$ with the same order singularity are given through

$$\begin{aligned} &\underbrace{(a_{j_1j_1r_1}, a_{j_2j_2s_1})}_{(m_{r_1}^{(1)}, n_{s_1}^{(1)})\text{-XX}} \rightleftharpoons \underbrace{(a_{j_1j_11}, a_{j_2j_21})}_{((2m_1),(2n_1+1))\text{-XX}} \rightleftharpoons \underbrace{(a_{j_1j_1r_2}, a_{j_2j_2s_2})}_{(m_{r_2}^{(2)}, n_{s_2}^{(2)})\text{-XX}} \\ &\left\{a_{j_1j_1r_1} \rightleftharpoons a_{j_1j_1r_2} \middle| m_{r_1}^{(1)} = m_{r_2}^{(2)}; r_1, r_2 \in \{1, 2, \cdots, p\}, r_1 \neq r_2\right\}, \\ &\left\{a_{j_2j_2s_1} \rightleftharpoons a_{j_2j_2s_2} \middle| n_{s_1}^{(1)} = n_{s_2}^{(2)}; s_1, s_2 \in \{1, 2, \cdots, q\}, s_1 \neq s_2\right\}; \end{aligned}$$

and

$$\begin{aligned} &a_{j_1j_1r_1} = a_{j_1j_11} - \varepsilon_{r_1}, a_{j_1j_1r_2} = a_{j_1j_11} + \varepsilon_{r_2}, \\ &(a_{j_1j_1r_1} = a_{j_1j_11} + \varepsilon_{r_1}, a_{j_1j_1r_2} = a_{j_1j_11} - \varepsilon_{r_2}); \\ &a_{j_2j_2s_1} = a_{j_2j_21} - \varepsilon_{s_1}, a_{j_2j_2s_2} = a_{j_2j_21} + \varepsilon_{s_2}, \\ &(a_{j_2j_2s_1} = a_{j_2j_21} + \varepsilon_{s_1}, a_{j_2j_2s_2} = a_{j_2j_21} - \varepsilon_{s_2}) \\ &\varepsilon_{r_1}, \varepsilon_{r_2} > 0 \text{ and } \varepsilon_{s_1}, \varepsilon_{s_2} > 0; \\ &r_i = 1, 2, \cdots, p; s_i = 1, 2, \cdots, q\,(i = 1, 2). \end{aligned}$$

From case (I), the singular equilibriums can be formed through nonsingular equilibrium networks as presented in Case (II), and the corner singular equilibriums for $i = 1, 2$ are determined similarly.

In summary, from the cases (I)–(III), the equilibrium of $(x_{j_1}^*, x_{j_2}^*) = (a_{j_1j_11}, a_{j_2j_21})$ for $m = 2m_1$ and $n = 2n_1 + 1$ has the bifurcation properties as stated in (v_1)-(v_4) through Eqs.(3.70)–(3.93).

(vi) For $m = 2m_1 + 1$ and $n = 2n_1$, this case is quite similar to (v) for $m = 2m_1$ and $n = 2n_1 + 1$, the bifurcation process is discussed through differential equations as follows. There are three cases (I)–(III).

Case I: Consider a dynamical system having a single equilibrium (A_1) as

$$\begin{aligned} \dot{x}_{j_1} &= a_{j_1j_10}(x_{j_1} - b_{j_1j_11}) \prod_{s=1}^{m_1} [(x_{j_1} - a_{j_1j_1}^{(s)})^2 + \Delta_{j_1j_1}^{(s)}], \\ \dot{x}_{j_2} &= a_{j_2j_20} \prod_{l=1}^{n_1} [(x_{j_2} - a_{j_2j_2}^{(l)})^2 + \Delta_{j_2j_2}^{(l)}]; \end{aligned}$$

where $a_{j_1j_11} = b_{j_1j_11}$. Once

$$\Delta_{j_1j_1}^{(s)} = \Delta_{j_1j_1} = 0\,(s = 1, 2, \cdots, m_1) \text{ and } \Delta_{j_2j_2}^{(l)} = \Delta_{j_2j_2} = 0\,(l = 1, 2, \cdots, n_1),$$

if

$$a_{j_1j_11} = b_{j_1j_11} = a_{j_1j_1}^{(s)}\ (s = 1, 2, \cdots, m_1) \text{ and } a_{j_2j_21} = a_{j_2j_2}^{(l)}\,(l = 1, 2, \cdots, n_1),$$

the foregoing differential equations becomes a dynamical system with a singular equilibrium (B_1) as

$$\dot{x}_{j_1} = a_{j_1j_10}(x_{j_1} - a_{j_1j_11})^{2m_1+1},\ \dot{x}_{j_2} = a_{j_2j_20}(x_{j_2} - a_{j_2j_21})^{2n_1}.$$

If

$$\begin{aligned} &\Delta_{j_1j_1}^{(s)} = \Delta_{j_1j_1} + \delta_s\ (s = 1, 2, \cdots, m_1);\\ &\Delta_{j_2j_2}^{(l)} = \Delta_{j_2j_2} + \delta_l\ (l = 1, 2, \cdots, n_1)\\ &\delta_s > 0\, and\ \delta_l > 0. \end{aligned}$$

then

$$\begin{aligned} &a_{j_1j_11}^{(s)}, a_{j_1j_12}^{(s)} = a_{j_1j_11} \pm \varepsilon_s,\ (s = 1, 2, \cdots, m_1),\\ &a_{j_2j_21}^{(l)}, a_{j_2j_22}^{(l)} = a_{j_2j_21} \pm \varepsilon_l,\ (l = 1, 2, \cdots, n_1);\\ &\{a_{j_1j_11}, a_{j_1j_12}, \cdots, a_{j_1j_1(2m_1+1)}\} = \text{sort}\{b_{j_1j_11}, a_{j_1j_11}^{(s)}, a_{j_1j_12}^{(s)} | s = 1, 2, \cdots, m_1\},\\ &\{a_{j_2j_21}, a_{j_2j_21}, \cdots, a_{j_2j_2(2n_1)}\} = \text{sort}\{a_{j_2j_21}^{(l)}, a_{j_2j_22}^{(l)} | l = 1, 2, \cdots, n_1\}. \end{aligned}$$

Thus, the differential equation becomes a dynamical system with a non-singular equilibrium network (C_1) as

$$\dot{x}_{j_1} = a_{j_1j_10} \prod_{s=1}^{2m_1+1} (x_{j_1} - a_{j_1j_1s}),\ \dot{x}_{j_2} = a_{j_2j_20} \prod_{l=1}^{2n_1} (x_{j_2} - a_{j_2j_2l}).$$

From the above bifurcation process analysis, at least, (m_1+1)-parameter variations in the x_{j_1}-direction and (n_1)-parameter variations in the x_{j_2}-direction are engaged in such a bifurcation from a source and sink flow to the equilibrium network of $(2m_1 + 1) \times (2n_1)$ through the $(2m_1 + 1)^{\text{th}}$-source and sink bifurcations in the x_{j_1}-direction and the $(2n_1)^{\text{th}}$-upper-saddle and lower-saddle bifurcations in the x_{j_2}-direction. *With both of them, the higher-order saddle-source and saddle-sink bifurcations are also developed.*

Thus the appearing or vanishing bifurcation route is given as follows.

$$\underbrace{(a_{j_1j_1 1}, \dot{x}_{j_2})}_{\text{XX}} \rightleftharpoons \underbrace{(a_{j_1j_1 1}, a_{j_2j_2 1})}_{((2m_1+1),(2n_1))\text{-XX}} \rightleftharpoons \bigcup_{s=1}^{2m_1+1}\bigcup_{l=1}^{2n_1}\underbrace{(a_{j_1j_1 s}, a_{j_2j_2 l})}_{\text{XX}}.$$

Case II: From Case (I), consider a dynamical system having singular equilibriums (A_2) as

$$\dot{x}_{j_1} = a_{j_1j_1 0}\prod_{r_1=1}^{p_1}(x_{j_1} - b_{j_1j_1 r_1})^{m_{r_1}^{(1)}}\prod_{s=1}^{l_1}[(x_{j_1} - a_{j_1j_1}^{(s)})^2 + \Delta_{j_1j_1}^{(s)}]^{m_s},$$
$$\dot{x}_{j_2} = a_{j_2j_2 0}\prod_{s_1=1}^{q_1}(x_{j_2} - b_{j_2j_2 s_1})^{n_{s_1}^{(1)}}\prod_{l=1}^{l_2}[(x_{j_2} - a_{j_2j_2}^{(l)})^2 + \Delta_{j_2j_2}^{(l)}]^{n_l};$$

where

$$\{a_{j_1j_1 r_1}|r_1 = 1, 2, \cdots, p_1\} = \text{sort}\{b_{j_1j_1 r_1}|r_1 = 1, 2, \cdots, p_1\},$$
$$\{a_{j_2j_2 s_1}|s_1 = 1, 2, \cdots, q_1\} = \text{sort}\{b_{j_2j_2 s_1}|s_1 = 1, 2, \cdots, q_1\};$$
$$2m_1 + 1 - \sum_{r_1=1}^{p_1} m_{r_1}^{(1)} = 2\sum_{s=1}^{l_1} m_s$$

with l_1-quadratic polynomials without real roots
and s^{th}-quadratic polynomial with power m_s,

$$2n_1 - \sum_{s_1=1}^{q_1} n_{s_1}^{(1)} = 2\sum_{l=1}^{l_2} n_l$$

with l_2-quadraitc polynomials without real roots
and l^{th}-quadratic polynomial with power n_l.

Once

$$\Delta_{j_1j_1}^{(s)} = 0\,(s = 1, 2, \cdots, l_1) \text{ and } \Delta_{j_2j_2}^{(l)} = 0\,(l = 1, 2, \cdots, l_2),$$

if

$$a_{j_1j_1 1} \equiv a_{j_1j_1 r_1} = a_{j_1j_1}^{(s)}\,(r_1 = 1, 2, \cdots, p_1; s = 1, 2, \cdots, l_1),$$
$$a_{j_2j_2 1} \equiv a_{j_2j_2 s_1} = a_{j_2j_2}^{(l)}(s_1 = 1, 2, \cdots, q_1; l = 1, 2, \cdots, l_2),$$

the foregoing differential equations becomes a dynamical system with a singular equilibrium (B_2) as

$$\dot{x}_{j_1} = a_{j_1j_1 0}(x_{j_1} - a_{j_1j_1 1})^{2m_1+1},\ \dot{x}_{j_2} = a_{j_2j_2 0}(x_{j_2} - a_{j_2j_2 1})^{2n_1}.$$

If

$$\Delta^{(s)}_{j_1j_1} = \Delta_{j_1j_1} + \delta_s \; (s = 1, 2, \cdots, l_1);$$
$$\Delta^{(l)}_{j_2j_2} = \Delta_{j_2j_2} + \delta_l \; (l = 1, 2, \cdots, l_2)$$
$$\delta_s > 0 \text{ and } \delta_l > 0,$$

then

$$\begin{aligned}
&a^{(s)}_{j_1j_11}, a^{(s)}_{j_1j_12} = a_{j_1j_11} \pm \varepsilon_s, \; (s = 1, 2, \cdots, l_1),\\
&a^{(l)}_{j_2j_21}, a^{(l)}_{j_2j_22} = a_{j_2j_21} \pm \varepsilon_l, \; (l = 1, 2, \cdots, l_2);\\
&\{a_{j_1j_11}, a_{j_1j_12}, \cdots, a_{j_1j_1p_2}\}\\
&= \text{sort}\{\, a_{j_1j_11}, a_{j_1j_12}, \cdots, a_{j_1j_1p_1}; a^{(s)}_{j_1j_11}, a^{(s)}_{j_1j_12} | s = 1, 2, \cdots, l_1\},\\
&\{a_{j_2j_21}, a_{j_2j_21}, \cdots, a_{j_2j_2q_2}\}\\
&= \text{sort}\{a_{j_2j_21}, a_{j_2j_21}, \cdots, a_{j_2j_2q_1}; a^{(l)}_{j_2j_21}, a^{(l)}_{j_2j_22} | l = 1, 2, \cdots, l_2\}.
\end{aligned}$$

Thus, the differential equation becomes a dynamical system with singular equilibriums (C_2) as

$$\dot{x}_{j_1} = a_{j_1j_10} \prod_{r_2=1}^{p_2} (x_{j_1} - a_{j_1j_1r_2})^{m^{(2)}_{r_2}},$$
$$\dot{x}_{j_2} = a_{j_2j_20} \prod_{s_2=1}^{q_2} (x_{j_2} - a_{j_2j_2s_2})^{n^{(2)}_{s_2}};$$

where

$$\sum_{r_2=1}^{p_2} m^{(2)}_{r_2} = 2m_1 + 1, \; \sum_{s_2=1}^{q_2} n^{(2)}_{s_2} = 2n_1.$$

From the above bifurcation process analysis, at least, p_2-parameter variations in the x_{j_1}-direction and q_2-parameter variations in the x_{j_2}-direction are engaged in such a bifurcation from a $p_1 \times q_1$ equilibrium network to a $p_2 \times q_2$ equilibrium network through the $(2m_1 + 1)^{\text{th}}$-source and sink bifurcations in the x_{j_1}-direction and the $(2n_1)^{\text{th}}$-upper-saddle and lower-saddle bifurcations in the x_{j_2}-direction.

Thus, the appearing or vanishing bifurcation route from a $p_1 \times q_1$ to a $p_2 \times q_2$ equilibrium network is expressed as

$$\bigcup_{r_1=1}^{p_1} \bigcup_{s_1=1}^{q_1} \underbrace{(a_{j_1j_1r_1}, a_{j_2j_2s_1})}_{(m^{(1)}_{r_1}, n^{(1)}_{s_1})\text{-XX}} \rightleftharpoons \underbrace{(a_{j_1j_11}, a_{j_2j_21})}_{((2m_1+1),(2n_1))\text{-XX}} \rightleftharpoons \bigcup_{r_2=1}^{p_2} \bigcup_{s_2=1}^{q_2} \underbrace{(a_{j_1j_1r_2}, a_{j_2j_2s_2})}_{(m^{(2)}_{r_2}, n^{(2)}_{s_2})\text{-XX}}.$$

From case (I), the singular equilibriums can be formed through nonsingular equilibrium networks, and the corner singular equilibriums for $i = 1, 2$ are determined.

Case III: Consider two dynamical systems with the same equilibriums with locations switched (A_3,C_3) for $i = 1, 2$ as

$$\dot{x}_{j_1} = a_{j_1 j_1 0} \prod_{r_i=1}^{p} (x_{j_1} - a_{j_1 j_1 r_i})^{m_{r_i}^{(i)}},\ \dot{x}_{j_2} = a_{j_2 j_2 0} \prod_{s_i=1}^{q} (x_{j_2} - a_{j_2 j_2 s_i})^{n_{s_i}^{(i)}};$$

where

$$\sum_{r_i=1}^{p} m_{r_i}^{(i)} = 2m_1 + 1,\ \sum_{s_i=1}^{q} n_{s_i}^{(i)} = 2n_1.$$

Consider a dynamical system as a singular equilibrium (B_3) as

$$\dot{x}_{j_1} = a_{j_1 j_1 0}(x_{j_1} - a_{j_1 j_1 1})^{2m_1+1},\ \dot{x}_{j_2} = a_{j_2 j_2 0}(x_{j_2} - a_{j_2 j_2 1})^{2n_1}.$$

For $r_i = 1, 2, \cdots, p$ and $s_i = 1, 2, \cdots, q\,(i = 1, 2)$, define two functions as

$$\begin{aligned}
\Delta_{j_1 j_1}^{(r_1 r_2)} &= (a_{j_1 j_1 r_1} - a_{j_1 j_1 r_2})^2, (r_1, r_2 \in \{1, 2, \cdots, p\}, r_1 \neq r_2),\\
\Delta_{j_2 j_2}^{(s_1 s_2)} &= (a_{j_2 j_2 s_1} - a_{j_2 j_2 s_2})^2, (s_1, s_2 \in \{1, 2, \cdots, q\}, s_1 \neq s_2).
\end{aligned}$$

For $r_i = 1, 2, \cdots, p$ and $s_i = 1, 2, \cdots, q\,(i = 1, 2)$, if

$$\begin{aligned}
\Delta_{j_1 j_1}^{(r_1 r_2)} &= (a_{j_1 j_1 r_1} - a_{j_1 j_1 r_2})^2 = 0,\\
\Delta_{j_2 j_2}^{(s_1 s_2)} &= (a_{j_2 j_2 s_1} - a_{j_2 j_2 s_2})^2 = 0,
\end{aligned}$$

two equilibriums of $(a_{j_1 j_1 r_1}, a_{j_2 j_2 s_1})$ and $(a_{j_1 j_1 r_2}, a_{j_2 j_2 s_2})$ switching at point $(a_{j_1 j_1 1}, a_{j_2 j_2 1})$ with the same order singularity are given through

$$\begin{aligned}
&\underbrace{(a_{j_1 j_1 r_1}, a_{j_2 j_2 s_1})}_{(m_{r_1}^{(1)}, n_{s_1}^{(1)})-XX} \rightleftharpoons \underbrace{(a_{j_1 j_1 1}, a_{j_2 j_2 1})}_{((2m_1+1),(2n_1))\text{-XX}} \rightleftharpoons \underbrace{(a_{j_1 j_1 r_2}, a_{j_2 j_2 s_2})}_{(m_{r_2}^{(2)}, n_{s_2}^{(2)})-XX},\\
&\left\{a_{j_1 j_1 r_1} \rightleftharpoons a_{j_1 j_1 r_2} \middle| m_{r_1}^{(1)} = m_{r_2}^{(2)}; r_1, r_2 \in \{1, 2, \cdots, p\}, r_1 \neq r_2\right\},\\
&\left\{(a_{j_2 j_2 s_1} \rightleftharpoons a_{j_2 j_2 s_2} \middle| n_{s_1}^{(1)} = n_{s_2}^{(2)}; s_1, s_2 \in \{1, 2, \cdots, q\}, s_1 \neq s_2\right\},
\end{aligned}$$

and

$$\begin{aligned}
&a_{j_1 j_1 r_1} = a_{j_1 j_1 1} - \varepsilon_{r_1}, a_{j_1 j_1 r_2} = a_{j_1 j_1 1} + \varepsilon_{r_2},\\
&(a_{j_1 j_1 r_1} = a_{j_1 j_1 1} + \varepsilon_{r_1}, a_{j_1 j_1 r_2} = a_{j_1 j_1 1} - \varepsilon_{r_2});\\
&a_{j_2 j_2 s_1} = a_{j_2 j_2 1} - \varepsilon_{s_1}, a_{j_2 j_2 s_2} = a_{j_2 j_2 1} + \varepsilon_{s_2},
\end{aligned}$$

$$
\begin{aligned}
&(a_{j_2j_2s_1} = a_{j_2j_21} + \varepsilon_{s_1}, a_{j_2j_2s_2} = a_{j_2j_21} - \varepsilon_{s_2})\\
&\varepsilon_{r_1}, \varepsilon_{r_2} > 0 \text{ and } \varepsilon_{s_1}, \varepsilon_{s_2} > 0;\\
&r_i = 1, 2, \cdots, p; s_i = 1, 2, \cdots, q\,(i = 1, 2).
\end{aligned}
$$

From case (I), the singular equilibriums can be formed through nonsingular equilibrium networks as presented in Case (II), and the corner singular equilibriums for $i = 1, 2$ are determined similarly.

In summary, from the cases (I)–(III), the equilibrium of $(x^*_{j_1}, x^*_{j_2}) = (a_{j_1j_11}, a_{j_2j_21})$ for $m = 2m_1 + 1$ and $n = 2n_1$ has the bifurcation properties as stated in (vi_1)-(vi_4) through Eqs. (3.94)–(3.117).

(vii) For $m = 2m_1$ and $n = 2n_1$, this case is quite similar to (iv) for $m = 2m_1 + 1$ and $n = 2n_1 + 1$, the bifurcation process is discussed through differential equations as follows. There are three cases (I)–(III).

Case I: Consider a dynamical system having a single equilibrium (A_1) as

$$
\dot{x}_{j_1} = a_{j_1j_10} \prod_{s=1}^{m_1} [(x_{j_1} - a^{(s)}_{j_1j_1})^2 + \Delta^{(s)}_{j_1j_1}],
$$

$$
\dot{x}_{j_2} = a_{j_2j_20} \prod_{l=1}^{n_1} [(x_{j_2} - a^{(l)}_{j_2j_2})^2 + \Delta^{(l)}_{j_2j_2}].
$$

Once

$$
\Delta^{(s)}_{j_1j_1} = \Delta_{j_1j_1} = 0\,(s = 1, 2, \cdots, m_1) \text{ and } \Delta^{(l)}_{j_2j_2} = \Delta_{j_2j_2} = 0\,(l = 1, 2, \cdots, n_1),
$$

if

$$
a_{j_1j_11} = a^{(s)}_{j_1j_1}\,(s = 1, 2, \cdots, m_1)\, and\, a_{j_2j_21} = a^{(l)}_{j_2j_2}\,(l = 1, 2, \cdots, n_1),
$$

the foregoing differential equations becomes a dynamical system with a singular equilibrium (B_1) as

$$
\dot{x}_{j_1} = a_{j_1j_10}(x_{j_1} - a_{j_1j_11})^{2m_1},\ \dot{x}_{j_2} = a_{j_2j_20}(x_{j_2} - a_{j_2j_21})^{2n_1}.
$$

If

$$
\begin{aligned}
&\Delta^{(s)}_{j_1j_1} = \Delta_{j_1j_1} + \delta_s\,(s = 1, 2, \cdots, m_1);\\
&\Delta^{(l)}_{j_2j_2} = \Delta_{j_2j_2} + \delta_l\,(l = 1, 2, \cdots, n_1)\\
&\delta_s > 0 \text{ and } \delta_l > 0,
\end{aligned}
$$

then

$$\begin{aligned}
&a^{(s)}_{j_1j_1 1}, a^{(s)}_{j_1j_1 2} = a_{j_1j_1 1} \pm \varepsilon_s, \ (s = 1, 2, \cdots, m_1), \\
&a^{(l)}_{j_2j_2 1}, a^{(l)}_{j_2j_2 2} = a_{j_2j_2 1} \pm \varepsilon_l, \ (l = 1, 2, \cdots, n_1); \\
&\left\{a_{j_1j_1 1}, a_{j_1j_1 2}, \cdots, a_{j_1j_1 (2m_1)}\right\} = \text{sort}\left\{b_{j_1j_1 1}, a^{(s)}_{j_1j_1 1}, a^{(s)}_{j_1j_1 2} \middle| s = 1, 2, \cdots, m_1\right\}, \\
&\left\{a_{j_2j_2 1}, a_{j_2j_2 1}, \cdots, a_{j_2j_2 (2n_1)}\right\} = \text{sort}\left\{a^{(l)}_{j_2j_2 1}, a^{(l)}_{j_2j_2 2} \middle| l = 1, 2, \cdots, n_1\right\}.
\end{aligned}$$

Thus, the differential equation becomes a dynamical system with a non-singular equilibrium network (C_1) as

$$\dot{x}_{j_1} = a_{j_1j_1 0} \prod_{s=1}^{2m_1} (x_{j_1} - a_{j_1j_1 s}), \ \dot{x}_{j_2} = a_{j_2j_2 0} \prod_{l=1}^{2n_1} (x_{j_2} - a_{j_2j_2 l}).$$

From the above bifurcation process analysis, at least, m_1-parameter variations in the x_{j_1}-direction and n_1-parameter variations in the x_{j_2}-direction are engaged in such a bifurcation from a flow to the equilibrium network of $(2m_1) \times (2n_1)$ through the $(2m_1)^{\text{th}}$-upper-saddle and lower-saddle bifurcations in the x_{j_1}-direction and the $(2n_1)^{\text{th}}$-upper-saddle and lower-saddle bifurcations in the x_{j_2}-direction. *With both of them, the higher-order double-saddle bifurcations are developed.*

Thus the appearing or vanishing bifurcation route is as follows.

$$\underbrace{(\dot{x}_{j_1}, \dot{x}_{j_2})}_{\text{XX}} \rightleftharpoons \underbrace{(a_{j_1j_1 1}, a_{j_2j_2 1})}_{((2m_1),(2n_1))\text{-XX}} \rightleftharpoons \bigcup_{s=1}^{2m_1} \bigcup_{l=1}^{2n_1} \underbrace{(a_{j_1j_1 s}, a_{j_2j_2 l})}_{XX}.$$

Case II: From Case (I), consider a dynamical system having singular equilibriums (A_2) as

$$\begin{aligned}
\dot{x}_{j_1} &= a_{j_1j_1 0} \prod_{r_1=1}^{p_1} (x_{j_1} - b_{j_1j_1 r_1})^{m^{(1)}_{r_1}} \prod_{s=1}^{l_1} [(x_{j_1} - a^{(s)}_{j_1j_1})^2 + \Delta^{(s)}_{j_1j_1}]^{m_s}, \\
\dot{x}_{j_2} &= a_{j_2j_2 0} \prod_{s_1=1}^{q_1} (x_{j_2} - b_{j_2j_2 s_1})^{n^{(1)}_{s_1}} \prod_{l=1}^{l_2} [(x_{j_2} - a^{(l)}_{j_2j_2})^2 + \Delta^{(l)}_{j_2j_2}]^{n_l};
\end{aligned}$$

where

$$\begin{aligned}
&\left\{a_{j_1j_1 r_1} \middle| r_1 = 1, 2, \cdots, p_1\right\} = \text{sort}\left\{b_{j_1j_1 r_1} \middle| r_1 = 1, 2, \cdots, p_1\right\}, \\
&\left\{a_{j_2j_2 s_1} \middle| s_1 = 1, 2, \cdots, q_1\right\} = \text{sort}\left\{b_{j_2j_2 s_1} \middle| s_1 = 1, 2, \cdots, q_1\right\}; \\
&2m_1 - \sum_{r_1=1}^{p_1} m^{(1)}_{r_1} = 2\sum_{s=1}^{l_1} m_s
\end{aligned}$$

with l_1-quadratic polynomials without real roots

and s^{th}-quadratic polynomial with power m_s,

$$2n_1 - \sum_{s_1=1}^{q_1} n_{s_1}^{(1)} = 2\sum_{l=1}^{l_2} n_l$$

with l_2-quadraitc polynomials without real roots
and l^{th}-quadratic polynomial with power n_l.

Once

$$\Delta_{j_1j_1}^{(s)} = 0\,(s = 1, 2, \cdots, l_1) \text{ and } \Delta_{j_2j_2}^{(l)} = 0\,(l = 1, 2, \cdots, l_2),$$

if

$$\begin{aligned} a_{j_1j_11} &\equiv a_{j_1j_1r_1} = a_{j_1j_1}^{(s)}\,(r_1 = 1, 2, \cdots, p_1; s = 1, 2, \cdots, l_1), \\ a_{j_2j_21} &\equiv a_{j_2j_2s_1} = a_{j_2j_2}^{(l)}\,(s_1 = 1, 2, \cdots, q_1; l = 1, 2, \cdots, l_2). \end{aligned}$$

the foregoing differential equations becomes a dynamical system with a singular equilibrium (B_2) as

$$\dot{x}_{j_1} = a_{j_1j_10}(x_{j_1} - a_{j_1j_11})^{2m_1},\ \dot{x}_{j_2} = a_{j_2j_20}(x_{j_2} - a_{j_2j_21})^{2n_1}.$$

If

$$\begin{aligned} &\Delta_{j_1j_1}^{(s)} = \Delta_{j_1j_1} + \delta_s\,(s = 1, 2, \cdots, l_1); \\ &\Delta_{j_2j_2}^{(l)} = \Delta_{j_2j_2} + \delta_l\,(l = 1, 2, \cdots, l_2) \\ &\delta_s > 0\,\text{and}\,\delta_l > 0, \end{aligned}$$

then

$$\begin{aligned} &a_{j_1j_11}^{(s)}, a_{j_1j_12}^{(s)} = a_{j_1j_11} \pm \varepsilon_s,\ (s = 1, 2, \cdots, l_1), \\ &a_{j_2j_21}^{(l)}, a_{j_2j_22}^{(l)} = a_{j_2j_21} \pm \varepsilon_l,\ (l = 1, 2, \cdots, l_2); \\ &\left\{a_{j_1j_11}, a_{j_1j_12}, \cdots, a_{j_1j_1p_2}\right\} \\ &= \text{sort}\left\{a_{j_1j_11}, a_{j_1j_12}, \cdots, a_{j_1j_1p_1}; a_{j_1j_11}^{(s)}, a_{j_1j_12}^{(s)} \middle| s = 1, 2, \cdots, l_1\right\}, \\ &\left\{a_{j_2j_21}, a_{j_2j_21}, \cdots, a_{j_2j_2q_2}\right\} \\ &= \text{sort}\left\{a_{j_2j_21}, a_{j_2j_21}, \cdots, a_{j_2j_2q_1}; a_{j_2j_21}^{(l)}, a_{j_2j_22}^{(l)} \middle| l = 1, 2, \cdots, l_2\right\}. \end{aligned}$$

Thus, the differential equation becomes a dynamical system with singular equilibriums (C_2) as

$$\dot{x}_{j_1} = a_{j_1j_10}\prod_{r_2=1}^{p_2}(x_{j_1} - a_{j_1j_1r_2})^{m_{r_2}^{(2)}},\ \dot{x}_{j_2} = a_{j_2j_20}\prod_{s_2=1}^{q_2}(x_{j_2} - a_{j_2j_2s_2})^{n_{s_2}^{(2)}};$$

where

$$\sum_{r_2=1}^{p_2} m_{r_2}^{(2)} = 2m_1, \sum_{s_2=1}^{q_2} n_{s_2}^{(2)} = 2n_1.$$

From the above bifurcation process analysis, at least, p_2-parameter variations in the x_{j_1}-direction and q_2-parameter variations in the x_{j_2}-direction are engaged in such a bifurcation from a $p_1 \times q_1$ equilibrium network to a $p_2 \times q_2$ equilibrium network through the $(2m_1)^{\text{th}}$-upper-saddle and lower-saddle bifurcations in the x_{j_1}-direction and the $(2n_1)^{\text{th}}$-upper-saddle and lower-saddle bifurcations in the x_{j_2}-direction.

Thus the appearing or vanishing bifurcation route from a $p_1 \times q_1$ to a $p_2 \times q_2$ equilibrium network is expressed as

$$\bigcup_{r_1=1}^{p_1} \bigcup_{s_1=1}^{q_1} \underbrace{(a_{j_1 j_1 r_1}, a_{j_2 j_2 s_1})}_{(m_{r_1}^{(1)}, n_{s_1}^{(1)})\text{-XX}} \rightleftharpoons \underbrace{(a_{j_1 j_1 1}, a_{j_2 j_2 1})}_{((2m_1),(2n_1))\text{-XX}} \rightleftharpoons \bigcup_{r_2=1}^{p_2} \bigcup_{s_2=1}^{q_2} \underbrace{(a_{j_1 j_1 r_2}, a_{j_2 j_2 s_2})}_{(m_{r_2}^{(2)}, n_{s_2}^{(2)})\text{-XX}}.$$

From case (I), the singular equilibriums can be formed through nonsingular equilibrium networks, and the corner singular equilibriums for $i = 1, 2$ are determined.

Case III: Consider two dynamical systems with the same equilibriums with locations switched (A_3,C_3) for $i = 1, 2$ as

$$\dot{x}_{j_1} = a_{j_1 j_1 0} \prod_{r_i=1}^{p} (x_{j_1} - a_{j_1 j_1 r_i})^{m_{r_i}^{(i)}}, \ \dot{x}_{j_2} = a_{j_2 j_2 0} \prod_{s_i=1}^{q} (x_{j_2} - a_{j_2 j_2 s_i})^{n_{s_i}^{(i)}};$$

where

$$\sum_{r_i=1}^{p} m_{r_i}^{(i)} = 2m_1, \sum_{s_i=1}^{q} n_{s_i}^{(i)} = 2n_1.$$

Consider a dynamical system as a singular equilibrium (B_3) as

$$\dot{x}_{j_1} = a_{j_1 j_1 0}(x_{j_1} - a_{j_1 j_1 1})^{2m_1}, \ \dot{x}_{j_2} = a_{j_2 j_2 0}(x_{j_2} - a_{j_2 j_2 1})^{2n_1}.$$

For $r_i = 1, 2, \cdots, p$ and $s_i = 1, 2, \cdots, q \, (i = 1, 2)$, define two functions as

$$\begin{aligned}
\Delta_{j_1 j_1}^{(r_1 r_2)} &= (a_{j_1 j_1 r_1} - a_{j_1 j_1 r_2})^2 \ (r_1, r_2 \in \{1, 2, \cdots, p\}, r_1 \neq r_2), \\
\Delta_{j_2 j_2}^{(s_1 s_2)} &= (a_{j_2 j_2 s_1} - a_{j_2 j_2 s_2})^2 \ (s_1, s_2 \in \{1, 2, \cdots, q\}, s_1 \neq s_2).
\end{aligned}$$

For $r_i = 1, 2, \cdots, p$ and $s_i = 1, 2, \cdots, q \, (i = 1, 2)$, if

$$\Delta_{j_1 j_1}^{(r_1 r_2)} = (a_{j_1 j_1 r_1} - a_{j_1 j_1 r_2})^2 = 0,$$

$$\Delta_{j_2j_2}^{(s_1s_2)} = (a_{j_2j_2s_1} - a_{j_2j_2s_2})^2 = 0,$$

two equilibriums of $(a_{j_1j_1r_1}, a_{j_2j_2s_1})$ and $(a_{j_1j_1r_2}, a_{j_2j_2s_2})$ switching at point $(a_{j_1j_1 1}, a_{j_2j_2 1})$ with the same order singularity are given through

$$\underbrace{(a_{j_1j_1r_1}, a_{j_2j_2s_1})}_{(m_{r_1}^{(1)}, n_{s_1}^{(1)})\text{-XX}} \rightleftharpoons \underbrace{(a_{j_1j_1 1}, a_{j_2j_2 1})}_{((2m_1),(2n_1))\text{-XX}} \rightleftharpoons \underbrace{(a_{j_1j_1r_2}, a_{j_2j_2s_2})}_{(m_{r_2}^{(2)}, n_{s_2}^{(2)})\text{-XX}},$$

$$\{a_{j_1j_1r_1} \rightleftharpoons a_{j_1j_1r_2} | m_{r_1}^{(1)} = m_{r_2}^{(2)}; r_1, r_2 \in \{1, 2, \cdots, p\}, r_1 \neq r_2\},$$
$$\{a_{j_2j_2s_1} \rightleftharpoons a_{j_2j_2s_2} | n_{s_1}^{(1)} = n_{s_2}^{(2)}; s_1, s_2 \in \{1, 2, \cdots, q\}, s_1 \neq s_2\},$$

and

$$\begin{aligned}
&a_{j_1j_1r_1} = a_{j_1j_1 1} - \varepsilon_{r_1}, a_{j_1j_1r_2} = a_{j_1j_1 1} + \varepsilon_{r_2},\\
&(a_{j_1j_1r_1} = a_{j_1j_1 1} + \varepsilon_{r_1}, a_{j_1j_1r_2} = a_{j_1j_1 1} - \varepsilon_{r_2});\\
&a_{j_2j_2s_1} = a_{j_2j_2 1} - \varepsilon_{s_1}, a_{j_2j_2s_2} = a_{j_2j_2 1} + \varepsilon_{s_2},\\
&(a_{j_2j_2s_1} = a_{j_2j_2 1} + \varepsilon_{s_1}, a_{j_2j_2s_2} = a_{j_2j_2 1} - \varepsilon_{s_2})\\
&\varepsilon_{r_1}, \varepsilon_{r_2} > 0 \text{ and } \varepsilon_{s_1}, \varepsilon_{s_2} > 0;\\
&r_i = 1, 2, \cdots, p; s_i = 1, 2, \cdots, q\,(i = 1, 2).
\end{aligned}$$

From case (I), the singular equilibriums can be formed through nonsingular equilibrium networks as presented in Case (II), and the corner singular equilibriums for $i = 1, 2$ are determined similarly.

In summary, from the cases (I)-(III), the equilibrium of $(x_{j_1}^*, x_{j_2}^*) = (a_{j_1j_1 1}, a_{j_2j_2 1})$ for $m = 2m_1$ and $n = 2n_1$ has the bifurcation properties as stated in (vii_1)–(vii_4) through Eqs. (3.118)–(3.141).

In the end, the theorem is proved. ■

3.3 Bifurcations to Homoclinic Networks Without Centers

As in Luo [1], in this section, four types of bifurcations are presented for non-singular equilibriums.

3.3.1 $((2m_1 + 1), (2n_1 + 1))$-*Source, Saddles, and Sink Bifurcations*

Consider a 2-dimensional singular system of the $(2m_1 + 1) \times (2n_1 + 1)$-type as

$$\dot{x}_1 = a_{110}(x_1 - a_{111})^{2m_1+1}, \dot{x}_2 = a_{220}(x_2 - a_{221})^{2n_1+1}. \tag{3.142}$$

The first integral manifold is

$$\begin{aligned}
&\frac{1}{2n_1}[\frac{1}{(x_2-a_{221})^{2n_1}}-\frac{1}{(x_{20}-a_{221})^{2n_1}}]\\
&=\frac{a_{220}}{a_{110}}\frac{1}{2m_1}[\frac{1}{(x_1-a_{111})^{2m_1}}-\frac{1}{(x_{10}-a_{111})^{2m_1}}],\\
&\text{for } m_1,n_1\neq 0,\\
&\frac{1}{2n_1}[\frac{1}{(x_2-a_{221})^{2n_1}}-\frac{1}{(x_{20}-a_{221})^{2n_1}}]=-\frac{a_{220}}{a_{110}}\ln\frac{|x_1-a_{111}|}{|x_{10}-a_{111}|},\\
&\text{for } n_1\neq 0 \text{ and } m_1=0,\\
&-\ln\frac{|x_2-a_{221}|}{|x_{20}-a_{221}|}=\frac{a_{220}}{a_{110}}\frac{1}{2m_1}[\frac{1}{(x_1-a_{111})^{2m_1}}-\frac{1}{(x_{10}-a_{111})^{2m_1}}],\\
&\text{for } n_1=0 \text{ and } m_1\neq 0,\\
&\ln\frac{|x_2-a_{221}|}{|x_{20}-a_{221}|}=\frac{a_{220}}{a_{110}}\ln\frac{|x_1-a_{111}|}{|x_{10}-a_{111}|},\text{ for } m_1,n_1=0.
\end{aligned}\tag{3.143}$$

Phase portraits of the singular equilibriums for the 2-dimensional singular system of the $((2m_1+1),(2n_1+1))$-type are presented in Fig. 3.1a–d for $(a_{110}>0, a_{220}>0)$, $(a_{110}<0, a_{220}>0)$, $(a_{110}>0, a_{220}<0)$ and $(a_{110}<0, a_{220}<0)$.

$$\begin{aligned}
(a_{111},a_{221})&=\underbrace{((2m_1+1)^{\text{th}}\text{SO},(2n_1+1)^{\text{th}}\text{SO})}_{((2m_1+1),(2n_1+1))\text{-source}}\text{ for } a_{110}>0 \text{ and } a_{220}>0,\\
(a_{111},a_{221})&=\underbrace{((2m_1+1)^{\text{th}}\text{SI},(2n_1+1)^{\text{th}}\text{SO})}_{((2m_1+1),(2n_1+1))\text{-saddle}}\text{ for } a_{110}<0 \text{ and } a_{220}>0,\\
(a_{111},a_{221})&=\underbrace{((2m_1+1)^{\text{th}}\text{SO},(2n_1+1)^{\text{th}}\text{SI})}_{((2m_1+1),(2n_1+1))\text{-saddle}}\text{ for } a_{110}>0 \text{ and } a_{220}<0,\\
(a_{111},a_{221})&=\underbrace{((2m_1+1)^{\text{th}}\text{SI},(2n_1+1)^{\text{th}}\text{SI})}_{((2m_1+1),(2n_1+1))\text{-sink}}\text{ for } a_{110}<0 \text{ and } a_{220}<0.
\end{aligned}\tag{3.144}$$

From the above bifurcations, there is a non-singular system with $(2m_1+1)\times(2n_1+1)$-equilibriums as

$$\dot{x}_1=a_{110}\prod_{s=1}^{2m_1+1}(x_1-a_{11s})\text{ and }\dot{x}_2=a_{220}\prod_{l=1}^{2n_1+1}(x_2-a_{22l}).\tag{3.145}$$

The first integral manifold is given by

$$\sum_{l_1=1}^{2n_1+1}\frac{1}{\prod_{l_2=1,l_2\neq l_1}^{2n_1+1}(a_{22l_1}-a_{22l_2})}\ln\frac{|x_2-a_{22l_1}|}{|x_{20}-a_{22l_1}|}$$

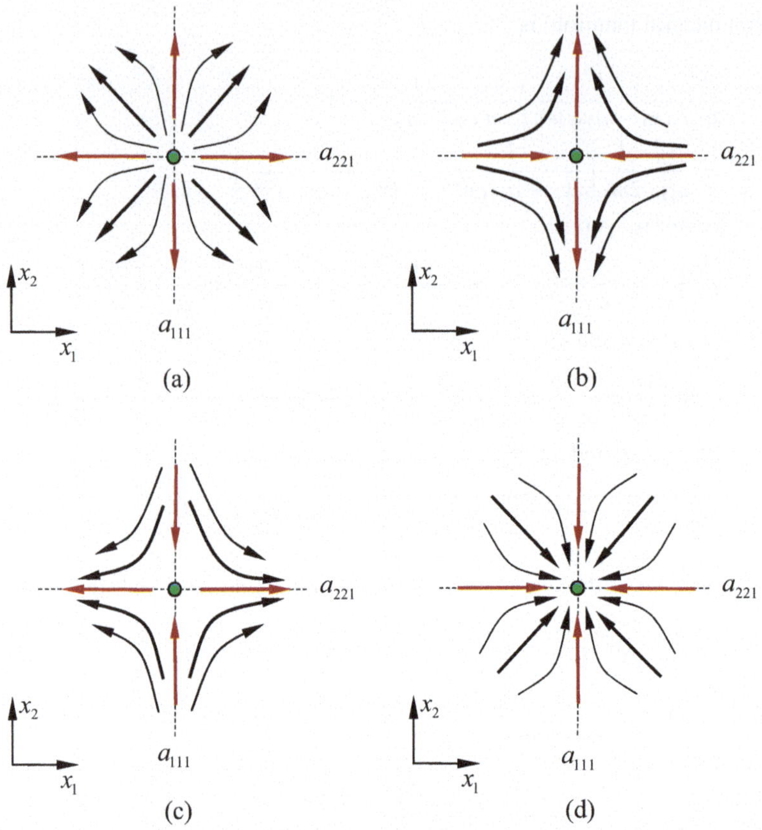

Fig. 3.1 Phase portraits of the $((2m_1+1),(2n_1+1))$-source, saddles, and sink for 2-dimensional systems at $(x_1^*, x_2^*) = (a_{111}, a_{221})$. **a** $(a_{110} > 0, a_{220} > 0)$, **b** $(a_{110} < 0, a_{220} > 0)$, **c** $(a_{110} > 0, a_{220} < 0)$, **d** $(a_{110} < 0, a_{220} < 0)$

$$= \frac{a_{220}}{a_{110}} \sum_{s_1=1}^{2m_1+1} \frac{1}{\prod_{s_2=1, s_2\neq s_1}^{2m_1+1}(a_{11s_1} - a_{11s_2})} \ln \frac{|x_1 - a_{11s_1}|}{|x_{10} - a_{11s_1}|}. \tag{3.146}$$

Phase portraits for the 2-dimensional systems near the simple equilibriums of the saddles, sinks and sources are presented in Fig. 3.2a–d for $(a_{110} > 0, a_{220} > 0)$, $(a_{110} < 0, a_{220} > 0)$, $(a_{110} > 0, a_{220} < 0)$ and $(a_{110} > 0, a_{220} < 0)$. For all cases, the $(2m_1+1)\times(2n_1+1)$-simple equilibriums are based on the $((2m_1+1),(2n_1+1))$-source, saddles and sink bifurcations.

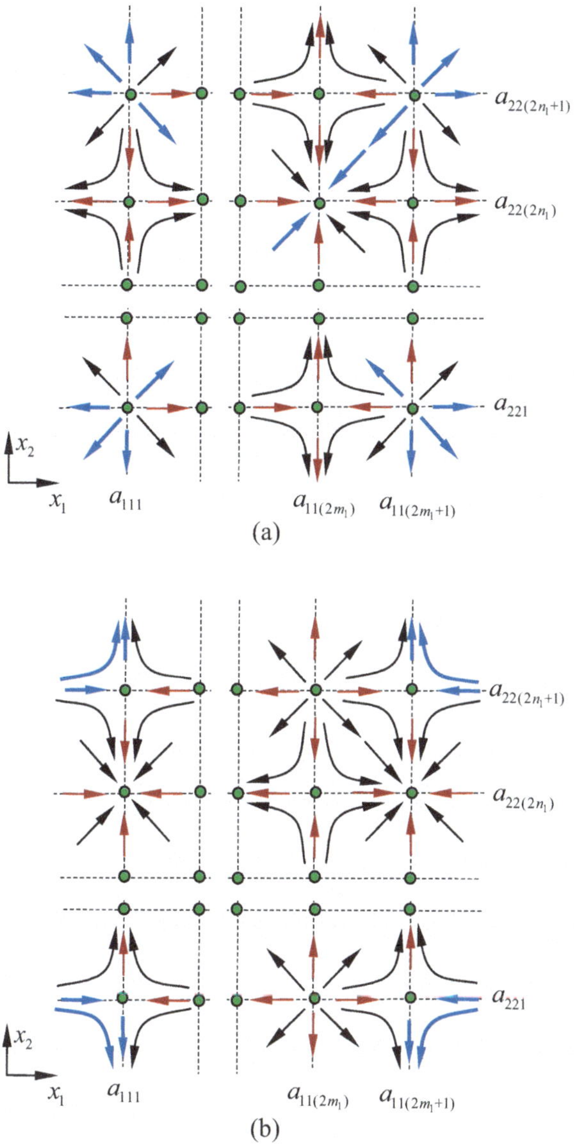

Fig. 3.2 Phase portraits for 2-dimensional systems with $x_1^* = a_{111}, a_{112}, \cdots, a_{11(2m_1+1)}$ and $x_2^* = a_{221}, a_{222}, \cdots, a_{22(2n_1+1)}$. The four networks of sink, source and saddle: **a** ($a_{110} > 0, a_{220} > 0$), **b** ($a_{110} < 0, a_{220} > 0$), **c** ($a_{110} > 0, a_{220} < 0$), **d** ($a_{110} > 0, a_{220} < 0$)

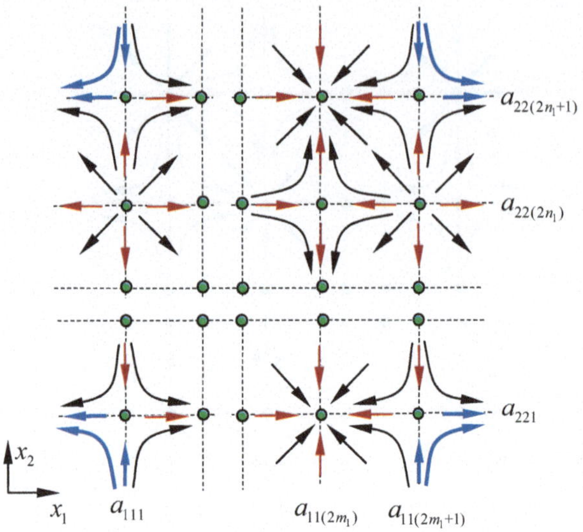

(c)

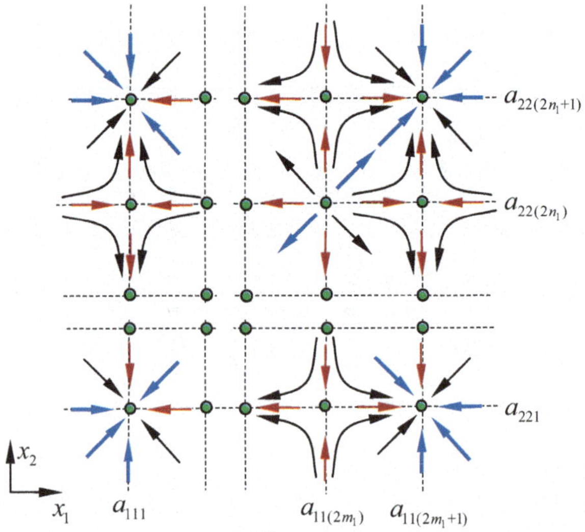

(d)

Fig. 3.2 (continued)

$$\bigcup_{s=1}^{2m_1+1}\bigcup_{l=1}^{2n_1+1}\underbrace{(a_{j_1j_1s},a_{j_2j_2l})}_{\mathrm{XX}}=\left\{\begin{array}{cccc}\underbrace{(\mathrm{SO,SO})}_{\text{source}} & \cdots & \underbrace{(\mathrm{SI,SO})}_{\text{saddle}} & \underbrace{(\mathrm{SO,SO})}_{\text{source}}\\ \underbrace{(\mathrm{SO,SI})}_{\text{saddle}} & \cdots & \underbrace{(\mathrm{SI,SI})}_{\text{sink}} & \underbrace{(\mathrm{SO,SI})}_{\text{saddle}}\\ \vdots & \vdots & \vdots & \vdots\\ \underbrace{(\mathrm{SO,SO})}_{\text{source}} & \cdots & \underbrace{(\mathrm{SI,SO})}_{\text{saddle}} & \underbrace{(\mathrm{SO,SO})}_{\text{source}}\end{array}\right\}_{(2n_1+1)\times(2m_1+1)}$$

for $a_{110}>0$, $a_{220}>0$; (3.147)

$$\bigcup_{s=1}^{2m_1+1}\bigcup_{l=1}^{2n_1+1}\underbrace{(a_{j_1j_1s},a_{j_2j_2l})}_{\mathrm{XX}}=\left\{\begin{array}{cccc}\underbrace{(\mathrm{SI,SO})}_{\text{saddle}} & \cdots & \underbrace{(\mathrm{SO,SO})}_{\text{source}} & \underbrace{(\mathrm{SI,SO})}_{\text{saddle}}\\ \underbrace{(\mathrm{SI,SI})}_{\text{sink}} & \cdots & \underbrace{(\mathrm{SO,SI})}_{\text{saddle}} & \underbrace{(\mathrm{SI,SI})}_{\text{sink}}\\ \vdots & \vdots & \vdots & \vdots\\ \underbrace{(\mathrm{SI,SO})}_{\text{saddle}} & \cdots & \underbrace{(\mathrm{SO,SO})}_{\text{source}} & \underbrace{(\mathrm{SI,SO})}_{\text{saddle}}\end{array}\right\}_{(2n_1+1)\times(2m_1+1)}$$

for $a_{110}<0$, $a_{220}>0$; (3.148)

$$\bigcup_{s=1}^{2m_1+1}\bigcup_{l=1}^{2n_1+1}\underbrace{(a_{j_1j_1s},a_{j_2j_2l})}_{\mathrm{XX}}=\left\{\begin{array}{cccc}\underbrace{(\mathrm{SO,SI})}_{\text{saddle}} & \cdots & \underbrace{(\mathrm{SI,SI})}_{\text{sink}} & \underbrace{(\mathrm{SO,SI})}_{\text{saddle}}\\ \underbrace{(\mathrm{SO,SO})}_{\text{source}} & \cdots & \underbrace{(\mathrm{SI,SO})}_{\text{saddle}} & \underbrace{(\mathrm{SO,SO})}_{\text{source}}\\ \vdots & \vdots & \vdots & \vdots\\ \underbrace{(\mathrm{SO,SI})}_{\text{saddle}} & \cdots & \underbrace{(\mathrm{SI,SI})}_{\text{sink}} & \underbrace{(\mathrm{SO,SI})}_{\text{saddle}}\end{array}\right\}_{(2n_1+1)\times(2m_1+1)}$$

for $a_{110}>0$, $a_{220}<0$; (3.149)

$$\bigcup_{s=1}^{2m_1+1}\bigcup_{l=1}^{2n_1+1}\underbrace{(a_{j_1j_1s},a_{j_2j_2l})}_{\mathrm{XX}}=\left\{\begin{array}{cccc}\underbrace{(\mathrm{SO,SI})}_{\text{saddle}} & \cdots & \underbrace{(\mathrm{SI,SO})}_{\text{saddle}} & \underbrace{(\mathrm{SI,SI})}_{\text{sink}}\\ \underbrace{(\mathrm{SI,SO})}_{\text{saddle}} & \cdots & \underbrace{(\mathrm{SO,SO})}_{\text{source}} & \underbrace{(\mathrm{SI,SO})}_{\text{saddle}}\\ \vdots & \vdots & \vdots & \vdots\\ \underbrace{(\mathrm{SO,SI})}_{\text{saddle}} & \cdots & \underbrace{(\mathrm{SI,SO})}_{\text{saddle}} & \underbrace{(\mathrm{SI,SI})}_{\text{sink}}\end{array}\right\}_{(2n_1+1)\times(2m_1+1)}$$

for $a_{110}<0$, $a_{220}<0$. (3.150)

3.3.2 $((2m_1), (2n_1+1))$-Saddle-Source and Saddle-Sink Bifurcations

Consider a 2-dimensional singular system of the $(2m_1) \times (2n_1+1)$-type as

$$\dot{x}_1 = a_{110}(x_1 - a_{111})^{2m_1}, \dot{x}_2 = a_{220}(x_2 - a_{221})^{2n_1+1}. \tag{3.151}$$

The first integral manifold is

$$\begin{aligned}
&\frac{1}{2n_1}\Big[\frac{1}{(x_2 - a_{221})^{2n_1}} - \frac{1}{(x_{20} - a_{221})^{2n_1}}\Big] \\
&= \frac{a_{220}}{a_{110}}\frac{1}{2m_1 - 1}\Big[\frac{1}{(x_1 - a_{111})^{2m_1-1}} - \frac{1}{(x_{10} - a_{111})^{2m_1-1}}\Big], \\
&\text{for } m_1, n_1 \neq 0, \\
&-\ln\frac{|x_2 - a_{221}|}{|x_{20} - a_{221}|} = \frac{a_{220}}{a_{110}}\frac{1}{2m_1 - 1}\Big[\frac{1}{(x_1 - a_{111})^{2m_1-1}} - \frac{1}{(x_{10} - a_{111})^{2m_1-1}}\Big], \\
&\text{for } n_1 = 0 \text{ and } m_1 \neq 0.
\end{aligned} \tag{3.152}$$

Phase portraits of the singular saddle-sources and saddle-sinks for the 2-dimensional singular system of the $((2m_1), (2n_1+1))$-type are presented in Fig. 3.3a–d for $(a_{110} > 0, a_{220} > 0)$, $(a_{110} < 0, a_{220} > 0)$, $(a_{110} > 0, a_{220} < 0)$ and $(a_{110} < 0, a_{220} < 0)$.

$$\begin{aligned}
(a_{111}, a_{221}) &= \underbrace{((2m_1)^{\text{th}}\text{US},(2n_1+1)^{\text{th}}\text{SO})}_{((2m_1),(2n_1+1))\text{-upper-saddle-source}} \text{ for } a_{110} > 0 \text{ and } a_{220} > 0, \\
(a_{111}, a_{221}) &= \underbrace{((2m_1)^{\text{th}}\text{LS},(2n_1+1)^{\text{th}}\text{SO})}_{((2m_1),(2n_1+1))\text{-lower-saddle-source}} \text{ for } a_{110} < 0 \text{ and } a_{220} > 0, \\
(a_{111}, a_{221}) &= \underbrace{((2m_1)^{\text{th}}\text{US},(2n_1+1)^{\text{th}}\text{SI})}_{((2m_1),(2n_1+1))\text{-upper-saddle-sink}} \text{ for } a_{110} > 0 \text{ and } a_{220} < 0, \\
(a_{111}, a_{221}) &= \underbrace{((2m_1)^{\text{th}}\text{LS},(2n_1+1)^{\text{th}}\text{SI})}_{((2m_1),(2n_1+1))\text{-lower-saddle-sink}} \text{ for } a_{110} < 0 \text{ and } a_{220} < 0.
\end{aligned} \tag{3.153}$$

From the above bifurcations, there is a non-singular system with $(2m_1) \times (2n_1+1)$-equilibriums as

$$\dot{x}_1 = a_{110}\prod_{s=1}^{2m_1}(x_1 - a_{11s}) \text{ and } \dot{x}_2 = a_{220}\prod_{l=1}^{2n_1+1}(x_1 - a_{22l}). \tag{3.154}$$

The first integral manifold is given by

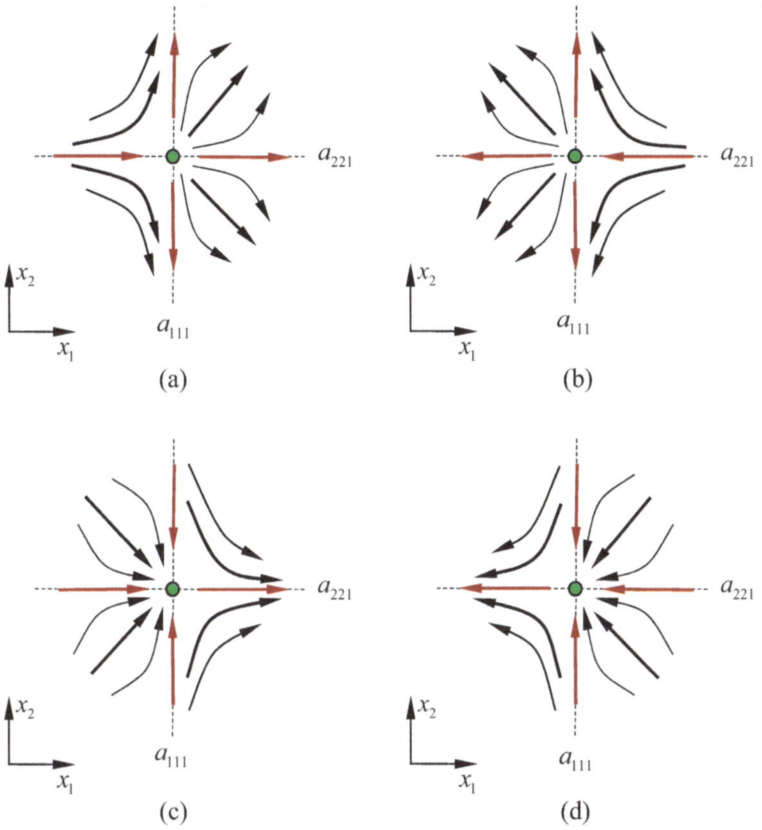

Fig. 3.3 Phase portraits of the $((2m_1), (2n_1+1))$-saddle-source and saddle-sink for 2-dimensional systems at $(x_1^*, x_2^*) = (a_{111}, a_{221})$. **a** $(a_{110} > 0, a_{220} > 0)$, **b** $(a_{110} < 0, a_{220} > 0)$, **c** $(a_{110} > 0, a_{220} < 0)$, **d** $(a_{110} < 0, a_{220} < 0)$

$$\sum_{l_1=1}^{2n_1+1} \frac{1}{\prod_{l_2=1, l_2 \neq l_1}^{2n_1+1} (a_{22l_1} - a_{22l_2})} \ln \frac{|x_2 - a_{22l_1}|}{|x_{20} - a_{22l_1}|}$$
$$= \frac{a_{220}}{a_{110}} \sum_{s_1=1}^{2m_1} \frac{1}{\prod_{s_2=1, s_2 \neq s_1}^{2m_1} (a_{11s_1} - a_{11s_2})} \ln \frac{|x_1 - a_{11s_1}|}{|x_{10} - a_{11s_1}|}. \tag{3.155}$$

Phase portraits for the 2-dimensional systems of $((2m_1), (2n_1+1))$-type with the saddles, sinks and sources are presented in Fig. 3.4a–d for $(a_{110} > 0, a_{220} > 0)$, $(a_{110} < 0, a_{220} > 0)$, $(a_{110} > 0, a_{220} < 0)$ and $(a_{110} < 0, a_{220} < 0)$. For all cases, the $(2m_1) \times (2n_1+1)$-simple equilibriums are based on the $((2m_1), (2n_1+1))$-saddle-source and saddle-sink bifurcations.

$$\bigcup_{s=1}^{2m_1}\bigcup_{l=1}^{2n_1+1}\underbrace{(a_{j_1j_1s},a_{j_2j_2l})}_{\mathrm{XX}}=\left\{\begin{array}{cccc}\underbrace{(\mathrm{SI},\mathrm{SO})}_{\text{saddle}} & \cdots & \underbrace{(\mathrm{SI},\mathrm{SO})}_{\text{saddle}} & \underbrace{(\mathrm{SO},\mathrm{SO})}_{\text{source}}\\ \underbrace{(\mathrm{SI},\mathrm{SI})}_{\text{sink}} & \cdots & \underbrace{(\mathrm{SI},\mathrm{SI})}_{\text{sink}} & \underbrace{(\mathrm{SO},\mathrm{SI})}_{\text{saddle}}\\ \vdots & \vdots & \vdots & \vdots\\ \underbrace{(\mathrm{SI},\mathrm{SO})}_{\text{saddle}} & \cdots & \underbrace{(\mathrm{SI},\mathrm{SO})}_{\text{saddle}} & \underbrace{(\mathrm{SO},\mathrm{SO})}_{\text{source}}\end{array}\right\}_{(2n_1+1)\times(2m_1)}$$

for $a_{110}>0,\ a_{220}>0$; (3.156)

$$\bigcup_{s=1}^{2m_1}\bigcup_{l=1}^{2n_1+1}\underbrace{(a_{j_1j_1s},a_{j_2j_2l})}_{\mathrm{XX}}=\left\{\begin{array}{cccc}\underbrace{(\mathrm{SO},\mathrm{SO})}_{\text{source}} & \cdots & \underbrace{(\mathrm{SO},\mathrm{SO})}_{\text{source}} & \underbrace{(\mathrm{SI},\mathrm{SO})}_{\text{saddle}}\\ \underbrace{(\mathrm{SO},\mathrm{SI})}_{\text{saddle}} & \cdots & \underbrace{(\mathrm{SO},\mathrm{SI})}_{\text{saddle}} & \underbrace{(\mathrm{SI},\mathrm{SI})}_{\text{sink}}\\ \vdots & \vdots & \vdots & \vdots\\ \underbrace{(\mathrm{SO},\mathrm{SO})}_{\text{source}} & \cdots & \underbrace{(\mathrm{SO},\mathrm{SO})}_{\text{source}} & \underbrace{(\mathrm{SI},\mathrm{SO})}_{\text{saddle}}\end{array}\right\}_{(2n_1+1)\times(2m_1)}$$

for $a_{110}<0,\ a_{220}>0$; (3.157)

$$\bigcup_{s=1}^{2m_1}\bigcup_{l=1}^{2n_1+1}\underbrace{(a_{j_1j_1s},a_{j_2j_2l})}_{\mathrm{XX}}=\left\{\begin{array}{cccc}\underbrace{(\mathrm{SI},\mathrm{SI})}_{\text{sink}} & \cdots & \underbrace{(\mathrm{SI},\mathrm{SI})}_{\text{sink}} & \underbrace{(\mathrm{SO},\mathrm{SI})}_{\text{saddle}}\\ \underbrace{(\mathrm{SI},\mathrm{SO})}_{\text{saddle}} & \cdots & \underbrace{(\mathrm{SI},\mathrm{SO})}_{\text{saddle}} & \underbrace{(\mathrm{SO},\mathrm{SO})}_{\text{source}}\\ \vdots & \vdots & \vdots & \vdots\\ \underbrace{(\mathrm{SI},\mathrm{SI})}_{\text{sink}} & \cdots & \underbrace{(\mathrm{SI},\mathrm{SI})}_{\text{sink}} & \underbrace{(\mathrm{SO},\mathrm{SI})}_{\text{saddle}}\end{array}\right\}_{(2n_1+1)\times(2m_1)}$$

for $a_{110}>0,\ a_{220}<0$; (3.158)

$$\bigcup_{s=1}^{2m_1}\bigcup_{l=1}^{2n_1+1}\underbrace{(a_{j_1j_1s},a_{j_2j_2l})}_{\mathrm{XX}}=\left\{\begin{array}{cccc}\underbrace{(\mathrm{SI},\mathrm{SO})}_{\text{saddle}} & \cdots & \underbrace{(\mathrm{SI},\mathrm{SO})}_{\text{saddle}} & \underbrace{(\mathrm{SI},\mathrm{SI})}_{\text{sink}}\\ \underbrace{(\mathrm{SO},\mathrm{SO})}_{\text{source}} & \cdots & \underbrace{(\mathrm{SO},\mathrm{SO})}_{\text{source}} & \underbrace{(\mathrm{SI},\mathrm{SO})}_{\text{saddle}}\\ \vdots & \vdots & \vdots & \vdots\\ \underbrace{(\mathrm{SI},\mathrm{SO})}_{\text{saddle}} & \cdots & \underbrace{(\mathrm{SI},\mathrm{SO})}_{\text{saddle}} & \underbrace{(\mathrm{SI},\mathrm{SI})}_{\text{sink}}\end{array}\right\}_{(2n_1+1)\times(2m_1)}$$

for $a_{110}<0,\ a_{220}<0$. (3.159)

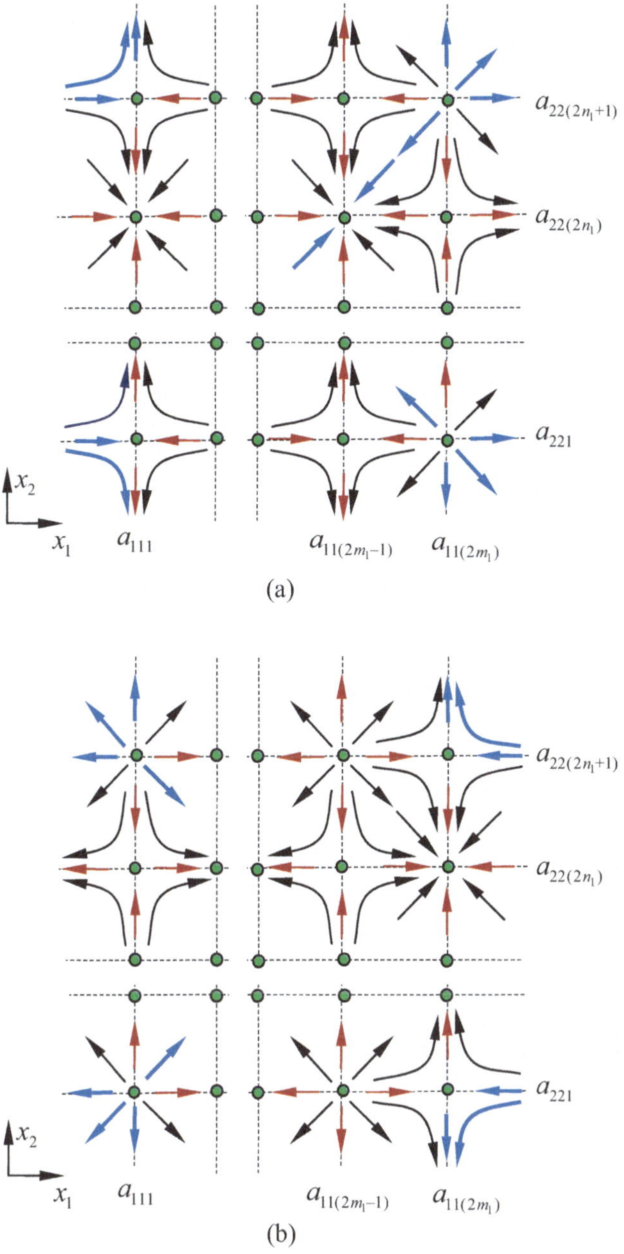

Fig. 3.4 Phase portraits for 2-dimensional systems with $x_1^* = a_{111}, a_{112}, \cdots, a_{11(2m_1)}$ and with $x_2^* = a_{221}, a_{222}, \cdots, a_{22(2n_1+1)}$. The four networks of sinks, sources, and saddles: **a** ($a_{110} > 0, a_{220} > 0$), **b** ($a_{110} < 0, a_{220} > 0$), **c** ($a_{110} > 0, a_{220} < 0$), **d** ($a_{110} < 0, a_{220} < 0$)

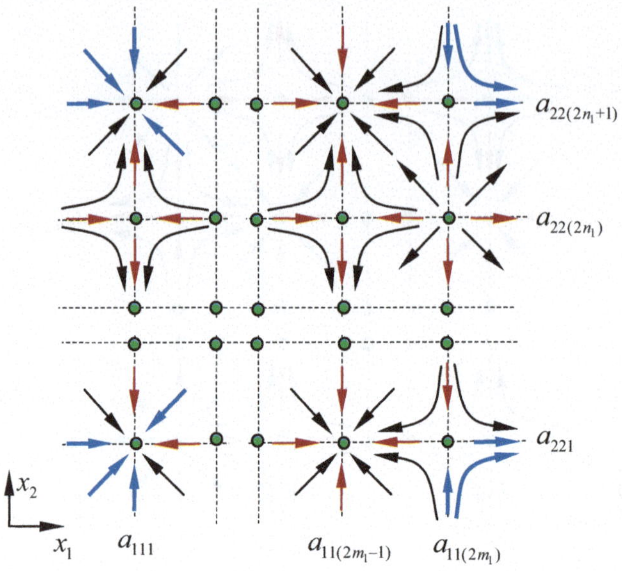

(c)

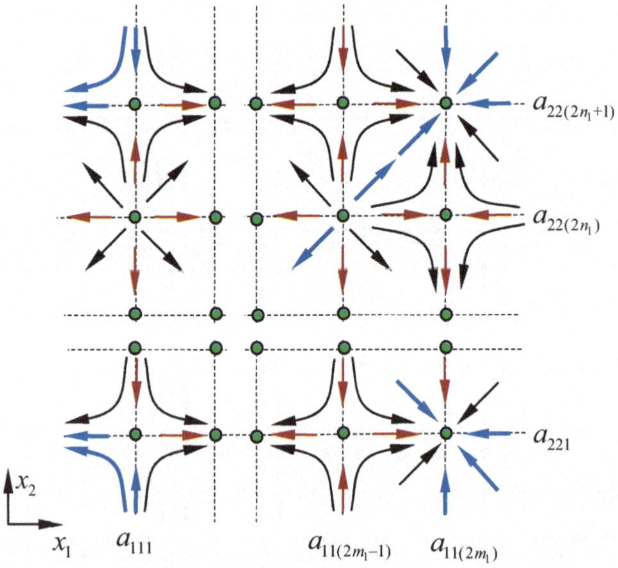

(d)

Fig. 3.4 (continued)

3.3.3 $((2m_1+1),(2n_1))$-Saddle-Source and Saddle-Sink Bifurcations

Consider a 2-dimensional singular system of the $(2m_1+1)\times(2n_1)$-type as

$$\dot{x}_1 = a_{110}(x_1 - a_{111})^{2m_1+1}, \dot{x}_2 = a_{220}(x_2 - a_{221})^{2n_1}. \tag{3.160}$$

The first integral manifold is.

$$\begin{aligned}
&\frac{1}{2n_1-1}\left[\frac{1}{(x_2-a_{221})^{2n_1-1}} - \frac{1}{(x_{20}-a_{221})^{2n_1-1}}\right] \\
&= \frac{a_{220}}{a_{110}}\frac{1}{2m_1}\left[\frac{1}{(x_1-a_{111})^{2m_1}} - \frac{1}{(x_{10}-a_{111})^{2m_1}}\right], \\
&\text{for } m_1, n_1 \neq 0, \\
&\frac{1}{2n_1}\left[\frac{1}{(x_2-a_{221})^{2n_1-1}} - \frac{1}{(x_{20}-a_{221})^{2n_1-1}}\right] = -\frac{a_{220}}{a_{110}}\ln\frac{|x_1-a_{111}|}{|x_{10}-a_{111}|}, \\
&\text{for } n_1 \neq 0 \text{ and } m_1 = 0.
\end{aligned} \tag{3.162}$$

Phase portraits of the singular equilibriums for the 2-dimensional singular system of the $((2m_1+1),(2n_1))$-type are presented in Fig. 3.5a–d for $(a_{110}>0, a_{220}>0)$, $(a_{110}<0, a_{220}>0)$, $(a_{110}>0, a_{220}<0)$ and $(a_{110}<0, a_{220}<0)$.

$$\begin{aligned}
(a_{111}, a_{221}) &= \underbrace{((2m_1{+}1)^{\text{th}}\text{SO},(2n_1)^{\text{th}}\text{US})}_{((2m_1+1),(2n_1))\text{-upper-saddle-source}} \text{ for } a_{110}>0 \text{ and } a_{220}>0, \\
(a_{111}, a_{221}) &= \underbrace{((2m_1{+}1)^{\text{th}}\text{SI},(2n_1)^{\text{th}}\text{US})}_{((2m_1+1),(2n_1))\text{-upper-saddle-sink}} \text{ for } a_{110}<0 \text{ and } a_{220}>0, \\
(a_{111}, a_{221}) &= \underbrace{((2m_1{+}1)^{\text{th}}\text{SO},(2n_1)^{\text{th}}\text{LS})}_{((2m_1+1),(2n_1))\text{-lower-saddle-source}} \text{ for } a_{110}>0 \text{ and } a_{220}<0, \\
(a_{111}, a_{221}) &= \underbrace{((2m_1{+}1)^{\text{th}}\text{SI},(2n_1)^{\text{th}}\text{LS})}_{((2m_1+1),(2n_1))\text{-lower-saddle-sink}} \text{ for } a_{110}<0 \text{ and } a_{220}<0.
\end{aligned} \tag{3.163}$$

From the above bifurcations, there is a non-singular system with $(2m_1+1)\times(2n_1)$-equilibriums as

$$\dot{x}_1 = a_{110}\prod_{s=1}^{2m_1+1}(x_1 - a_{11s}) \text{ and } \dot{x}_2 = a_{220}\prod_{l=1}^{2n_1}(x_1 - a_{22l}). \tag{3.164}$$

The first integral manifold is given by

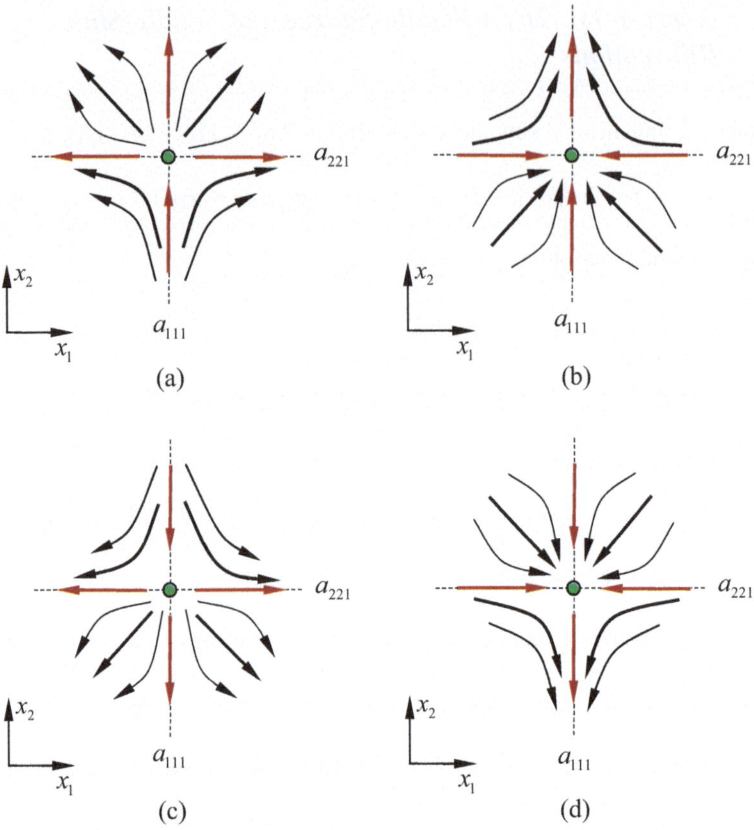

Fig. 3.5 Phase portraits of the $((2m_1+1), (2n_1))$-saddle-source and saddle-sink for 2-dimensional systems at $(x_1^*, x_2^*) = (a_{111}, a_{221})$. **a** $(a_{110} > 0, a_{220} > 0)$, **b** $(a_{110} < 0, a_{220} > 0)$, **c** $(a_{110} > 0, a_{220} < 0)$, **d** $(a_{110} < 0, a_{220} < 0)$

$$\sum_{l_1=1}^{2n_1} \frac{1}{\prod_{l_2=1, l_2 \neq l_1}^{2n_1} (a_{22l_1} - a_{22l_2})} \ln \frac{|x_2 - a_{22l_1}|}{|x_{20} - a_{22l_1}|}$$
$$= \frac{a_{220}}{a_{110}} \sum_{s_1=1}^{2m_1+1} \frac{1}{\prod_{s_2=1, s_2 \neq s_1}^{2m_1+1} (a_{11s_1} - a_{11s_2})} \ln \frac{|x_1 - a_{11s_1}|}{|x_{10} - a_{11s_1}|}. \quad (3.165)$$

Phase portraits for the 2-dimensional systems near the simple equilibriums of the saddles, sinks and sources are presented in Fig. 3.6a–d for $(a_{110} > 0, a_{220} > 0)$, $(a_{110} < 0, a_{220} > 0)$, $(a_{110} > 0, a_{220} < 0)$ and $(a_{110} < 0, a_{220} < 0)$. For all cases, the $(2m_1 + 1) \times (2n_1)$-simple equilibriums are based on the $((2m_1 + 1), (2n_1))$-saddle-source and saddle-sink bifurcations.

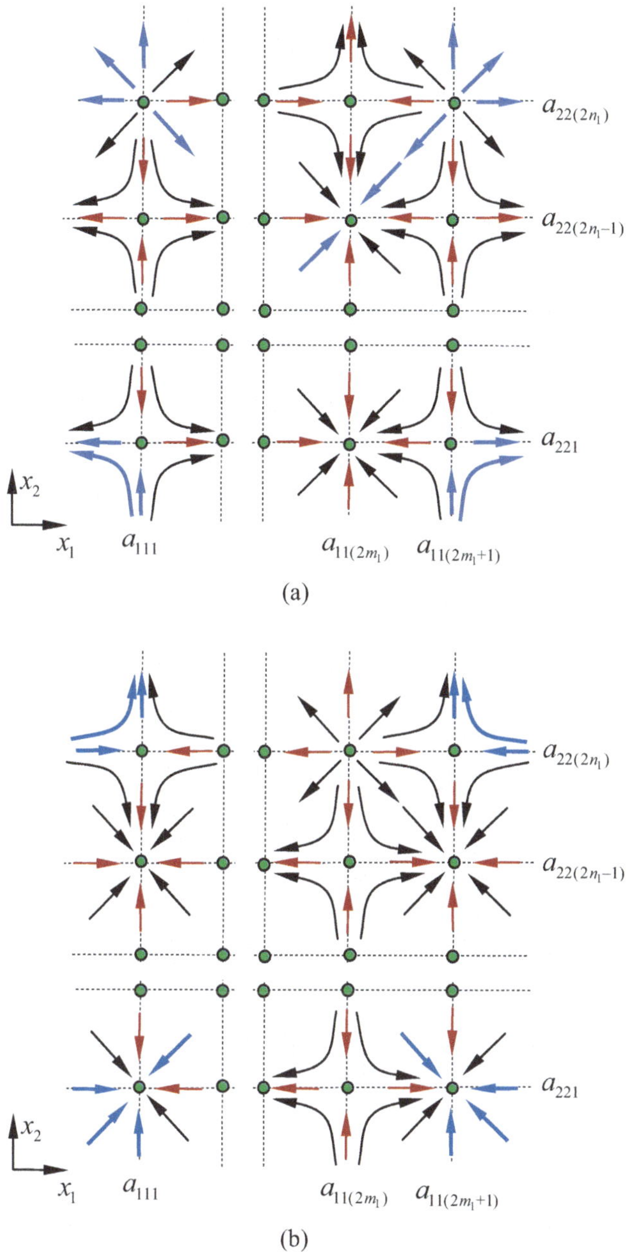

Fig. 3.6 Phase portraits for 2-dimensional systems with $x_1^* = a_{111}, a_{112}, \cdots, a_{11(2m_1+1)}$ and $x_2^* = a_{221}, a_{222}, \cdots, a_{22(2n_1)}$. The four networks of sink, source and saddle: **a** ($a_{110} > 0, a_{220} > 0$), **b** ($a_{110} < 0, a_{220} > 0$), **c** ($a_{110} > 0, a_{220} < 0$), **d** ($a_{110} < 0, a_{220} < 0$)

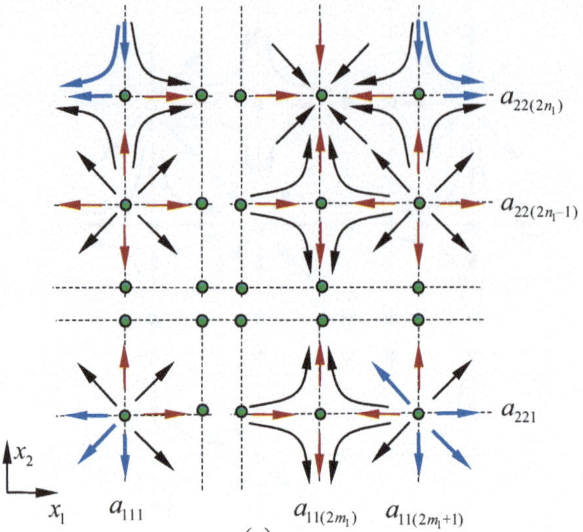

(c)

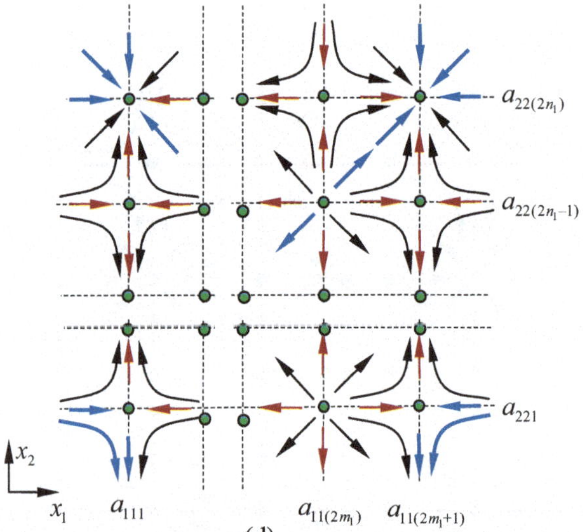

(d)

Fig. 3.6 (continued)

$$\bigcup_{s=1}^{2m_1+1}\bigcup_{l=1}^{2n_1}\underbrace{(a_{j_1j_1s},a_{j_2j_2l})}_{XX}=\left\{\begin{array}{cccc}\underbrace{(SO,SO)}_{source}\cdots & & \underbrace{(SI,SO)}_{saddle} & \underbrace{(SO,SO)}_{source}\\ \underbrace{(SO,SI)}_{saddle}\cdots & & \underbrace{(SI,SI)}_{sink} & \underbrace{(SO,SI)}_{saddle}\\ \vdots & \vdots & \vdots & \vdots\\ \underbrace{(SO,SI)}_{saddle}\cdots & & \underbrace{(SI,SI)}_{sink} & \underbrace{(SO,SI)}_{saddle}\end{array}\right\}_{(2n_1)\times(2m_1+1)}$$

for $a_{110} > 0$, $a_{220} > 0$; (3.166)

$$\bigcup_{s=1}^{2m_1+1}\bigcup_{l=1}^{2n_1}\underbrace{(a_{j_1j_1s},a_{j_2j_2l})}_{XX}=\left\{\begin{array}{cccc}\underbrace{(SI,SO)}_{saddle}\cdots & & \underbrace{(SO,SO)}_{source} & \underbrace{(SI,SO)}_{saddle}\\ \underbrace{(SI,SI)}_{sink}\cdots & & \underbrace{(SO,SI)}_{saddle} & \underbrace{(SI,SI)}_{sink}\\ \vdots & \vdots & \vdots & \vdots\\ \underbrace{(SI,SI)}_{sink}\cdots & & \underbrace{(SO,SI)}_{saddle} & \underbrace{(SI,SI)}_{sink}\end{array}\right\}_{(2n_1)\times(2m_1+1)}$$

for $a_{110} < 0$, $a_{220} > 0$; (3.167)

$$\bigcup_{s=1}^{2m_1+1}\bigcup_{l=1}^{2n_1}\underbrace{(a_{j_1j_1s},a_{j_2j_2l})}_{XX}=\left\{\begin{array}{cccc}\underbrace{(SO,SI)}_{saddle}\cdots & & \underbrace{(SI,SI)}_{sink} & \underbrace{(SO,SI)}_{saddle}\\ \underbrace{(SO,SO)}_{source}\cdots & & \underbrace{(SI,SO)}_{saddle} & \underbrace{(SO,SO)}_{source}\\ \vdots & \vdots & \vdots & \vdots\\ \underbrace{(SO,SO)}_{source}\cdots & & \underbrace{(SI,SO)}_{saddle} & \underbrace{(SO,SO)}_{source}\end{array}\right\}_{(2n_1)\times(2m_1+1)}$$

for $a_{110} > 0$, $a_{220} < 0$; (3.168)

$$\bigcup_{s=1}^{2m_1+1}\bigcup_{l=1}^{2n_1}\underbrace{(a_{j_1j_1s},a_{j_2j_2l})}_{XX}=\left\{\begin{array}{cccc}\underbrace{(SI,SI)}_{sink}\cdots & & \underbrace{(SI,SO)}_{saddle} & \underbrace{(SI,SI)}_{sink}\\ \underbrace{(SI,SO)}_{saddle}\cdots & & \underbrace{(SO,SO)}_{source} & \underbrace{(SI,SO)}_{saddle}\\ \vdots & \vdots & \vdots & \vdots\\ \underbrace{(SI,SO)}_{saddle}\cdots & & \underbrace{(SO,SO)}_{source} & \underbrace{(SI,SO)}_{saddle}\end{array}\right\}_{(2n_1)\times(2m_1+1)}$$

for $a_{110} < 0$, $a_{220} < 0$. (3.169)

3.3.4 $((2m_1), (2n_1))$-Double-Saddle Bifurcations

Consider a 2-dimensional singular system of the $(2m_1) \times (2n_1)$-type as

$$\dot{x}_1 = a_{110}(x_1 - a_{111})^{2m_1}, \dot{x}_2 = a_{220}(x_2 - a_{221})^{2n_1}. \tag{3.170}$$

The first integral manifold is

$$\begin{aligned} &\frac{1}{2n_1 - 1}\Big[\frac{1}{(x_2 - a_{221})^{2n_1-1}} - \frac{1}{(x_{20} - a_{221})^{2n_1-1}}\Big] \\ &= \frac{a_{220}}{a_{110}}\frac{1}{2m_1 - 1}\Big[\frac{1}{(x_1 - a_{111})^{2m_1-1}} - \frac{1}{(x_{10} - a_{111})^{2m_1-1}}\Big], \\ &\text{for } m_1, n_1 \neq 0. \end{aligned} \tag{3.171}$$

Phase portraits of the singular equilibriums for the 2-dimensional singular system of the $((2m_1), (2n_1))$-type are presented in Fig. 3.7a–d for $(a_{110} > 0, a_{220} > 0)$, $(a_{110} < 0, a_{220} > 0)$, $(a_{110} > 0, a_{220} < 0)$ and $(a_{110} < 0, a_{220} < 0)$.

$$\begin{aligned} (a_{111}, a_{221}) &= \underbrace{((2m_1)^{\text{th}}\text{US},(2n_1)^{\text{th}}\text{US})}_{((2m_1),(2n_1))\text{-double-saddle}} \text{ for } a_{110} > 0 \text{ and } a_{220} > 0, \\ (a_{111}, a_{221}) &= \underbrace{((2m_1)^{\text{th}}\text{LS},(2n_1)^{\text{th}}\text{US})}_{((2m_1),(2n_1))\text{-double-saddle}} \text{ for } a_{110} < 0 \text{ and } a_{220} > 0, \\ (a_{111}, a_{221}) &= \underbrace{((2m_1)^{\text{th}}\text{US},(2n_1)^{\text{th}}\text{LS})}_{((2m_1),(2n_1))\text{-double-saddle}} \text{ for } a_{110} > 0 \text{ and } a_{220} < 0, \\ (a_{111}, a_{221}) &= \underbrace{((2m_1)^{\text{th}}\text{LS},(2n_1)^{\text{th}}\text{LS})}_{((2m_1),(2n_1))\text{-double-saddle}} \text{ for } a_{110} < 0 \text{ and } a_{220} < 0. \end{aligned} \tag{3.172}$$

From the above bifurcations, there is a non-singular system with $(2m_1) \times (2n_1)$-equilibriums as

$$\dot{x}_1 = a_{110}\prod_{s=1}^{2m_1}(x_1 - a_{11s}) \text{ and } \dot{x}_2 = a_{220}\prod_{l=1}^{2n_1}(x_1 - a_{22l}). \tag{3.173}$$

The first integral manifold is given by

$$\begin{aligned} &\sum_{l_1=1}^{2n_1}\frac{1}{\prod_{l_2=1, l_2\neq l_1}^{2n_1}(a_{22l_1} - a_{22l_2})}\ln\frac{|x_2 - a_{22l_1}|}{|x_{20} - a_{22l_1}|} \\ &= \frac{a_{220}}{a_{110}}\sum_{s_1=1}^{2m_1}\frac{1}{\prod_{s_2=1, s_2\neq s_1}^{2m_1}(a_{11s_1} - a_{11s_2})}\ln\frac{|x_1 - a_{11s_1}|}{|x_{10} - a_{11s_1}|}. \end{aligned} \tag{3.174}$$

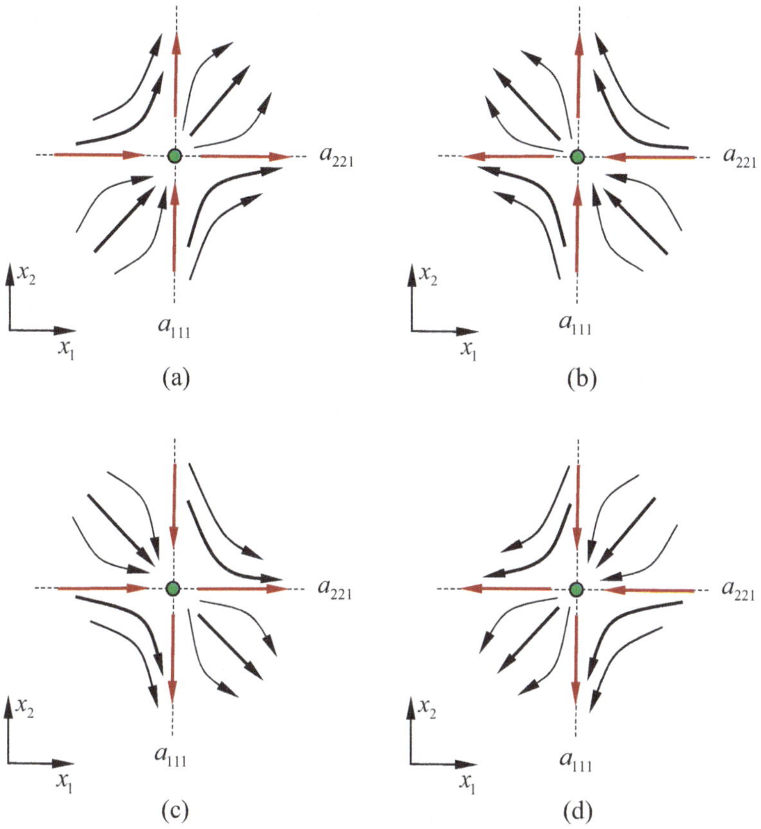

Fig. 3.7 Phase portraits of the $((2m_1),(2n_1))$-double-saddles for 2-dimensional systems at $(x_1^*, x_2^*) = (a_{111}, a_{221})$. **a** $(a_{110} > 0, a_{220} > 0)$, **b** $(a_{110} < 0, a_{220} > 0)$, **c** $(a_{110} > 0, a_{220} < 0)$, **d** $(a_{110} < 0, a_{220} < 0)$

Phase portraits for the 2-dimensional systems near the simple equilibriums of the saddles, sinks and sources are presented in Fig. 3.8a–d for $(a_{110} > 0, a_{220} > 0)$, $(a_{110} < 0, a_{220} > 0)$, $(a_{110} > 0, a_{220} < 0)$ and $(a_{110} < 0, a_{220} < 0)$. For all cases, the $(2m_1) \times (2n_1)$ simple equilibriums are based on the $((2m_1),(2n_1))$-double-saddle bifurcations.

$$\bigcup_{s=1}^{2m_1}\bigcup_{l=1}^{2n_1}\underbrace{(a_{j_1j_1s}, a_{j_2j_2l})}_{\text{XX}} = \left\{\begin{array}{cccc} \underbrace{\text{(SI,SO)}}_{\text{saddle}} & \cdots & \underbrace{\text{(SI,SO)}}_{\text{saddle}} & \underbrace{\text{(SO,SO)}}_{\text{source}} \\ \underbrace{\text{(SI,SI)}}_{\text{sink}} & \cdots & \underbrace{\text{(SI,SI)}}_{\text{sink}} & \underbrace{\text{(SO,SI)}}_{\text{saddle}} \\ \vdots & \vdots & \vdots & \vdots \\ \underbrace{\text{(SI,SI)}}_{\text{sink}} & \cdots & \underbrace{\text{(SI,SI)}}_{\text{sink}} & \underbrace{\text{(SO,SI)}}_{\text{saddle}} \end{array}\right\}_{(2n_1)\times(2m_1)}$$

for $a_{110} > 0, a_{220} > 0$; (3.175)

$$\bigcup_{s=1}^{2m_1}\bigcup_{l=1}^{2n_1}\underbrace{(a_{j_1j_1s}, a_{j_2j_2l})}_{\text{XX}} = \begin{Bmatrix} \underbrace{\text{(SO,SO)}}_{\text{source}} & \cdots & \underbrace{\text{(SO,SO)}}_{\text{source}} & \underbrace{\text{(SI,SO)}}_{\text{saddle}} \\ \underbrace{\text{(SO,SI)}}_{\text{saddle}} & \cdots & \underbrace{\text{(SO,SI)}}_{\text{saddle}} & \underbrace{\text{(SI,SI)}}_{\text{sink}} \\ \vdots & \vdots & \vdots & \vdots \\ \underbrace{\text{(SO,SI)}}_{\text{saddle}} & \cdots & \underbrace{\text{(SO,SI)}}_{\text{saddle}} & \underbrace{\text{(SI,SI)}}_{\text{sink}} \end{Bmatrix}_{(2n_1)\times(2m_1)}$$

for $a_{110} < 0, a_{220} > 0$; (3.176)

$$\bigcup_{s=1}^{2m_1}\bigcup_{l=1}^{2n_1}\underbrace{(a_{j_1j_1s}, a_{j_2j_2l})}_{\text{XX}} = \begin{Bmatrix} \underbrace{\text{(SI,SI)}}_{\text{sink}} & \cdots & \underbrace{\text{(SI,SI)}}_{\text{sink}} & \underbrace{\text{(SO,SI)}}_{\text{saddle}} \\ \underbrace{\text{(SI,SO)}}_{\text{saddle}} & \cdots & \underbrace{\text{(SI,SO)}}_{\text{saddle}} & \underbrace{\text{(SO,SO)}}_{\text{source}} \\ \vdots & \vdots & \vdots & \vdots \\ \underbrace{\text{(SI,SO)}}_{\text{saddle}} & \cdots & \underbrace{\text{(SI,SO)}}_{\text{saddle}} & \underbrace{\text{(SO,SO)}}_{\text{source}} \end{Bmatrix}_{(2n_1)\times(2m_1)}$$

for $a_{110} > 0, a_{220} < 0$; (3.177)

$$\bigcup_{s=1}^{2m_1}\bigcup_{l=1}^{2n_1}\underbrace{(a_{j_1j_1s}, a_{j_2j_2l})}_{\text{XX}} = \begin{Bmatrix} \underbrace{\text{(SI,SO)}}_{\text{saddle}} & \cdots & \underbrace{\text{(SI,SO)}}_{\text{saddle}} & \underbrace{\text{(SI,SI)}}_{\text{sink}} \\ \underbrace{\text{(SO,SO)}}_{\text{source}} & \cdots & \underbrace{\text{(SO,SO)}}_{\text{source}} & \underbrace{\text{(SI,SO)}}_{\text{saddle}} \\ \vdots & \vdots & \vdots & \vdots \\ \underbrace{\text{(SO,SO)}}_{\text{source}} & \cdots & \underbrace{\text{(SO,SO)}}_{\text{source}} & \underbrace{\text{(SI,SO)}}_{\text{saddle}} \end{Bmatrix}_{(2n_1)\times(2m_1)}$$

for $a_{110} < 0, a_{220} < 0$. (3.178)

Similarly, the higher-singular appearing bifurcations are from a singular dynamical system of $p_1 \times q_1$-equilibriums to a singular dynamical system of $p_2 \times q_2$-equilibriums. The higher-singular switching bifurcations are from equilibriums switching of two singular dynamical system of $p \times q$-equilibriums. Such two cases can be done by readers.

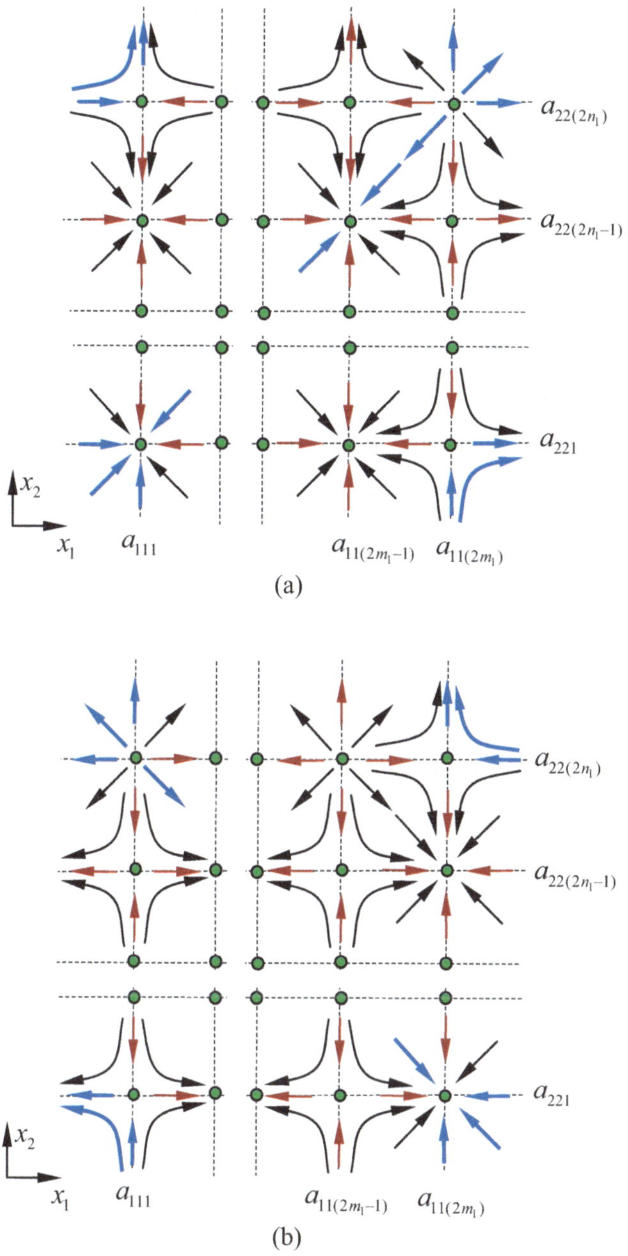

Fig. 3.8 Phase portraits for 2-dimensional systems with $x_1^* = a_{111}, a_{112}, \cdots, a_{11(2m_1)}$ and $x_2^* = a_{221}, a_{222}, \cdots, a_{22(2n_1)}$. The four networks of sink, source and saddle: **a** ($a_{110} > 0, a_{220} > 0$), **b** ($a_{110} < 0, a_{220} > 0$), **c** ($a_{110} > 0, a_{220} < 0$), **d** ($a_{110} < 0, a_{220} < 0$)

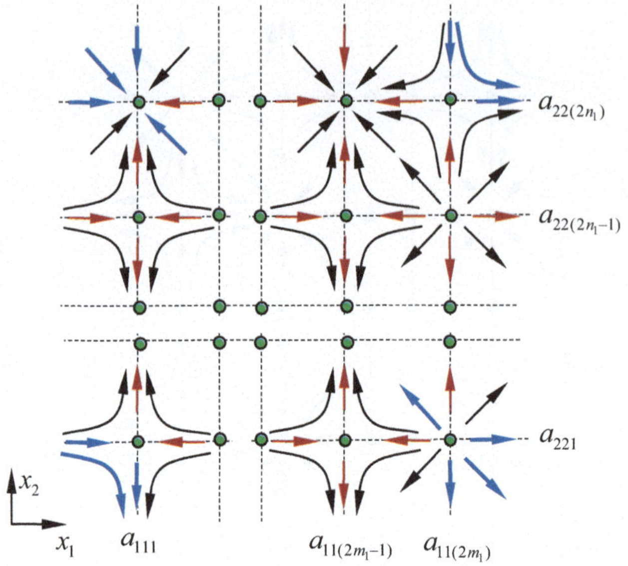

(c)

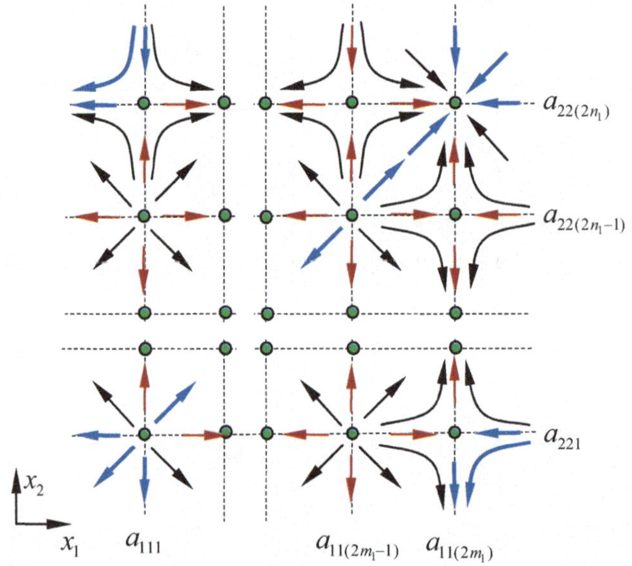

(d)

Fig. 3.8 (continued)

Reference

1. Luo, A. C. J. (2024). Bifurcations of homoclinic networks in polynomial systems. *International Journal of Bifurcation and Chaos*, 34(3), 2430006 (153 pages).

Chapter 4
Homoclinic Networks with Centers

In this chapter, the homoclinic networks of positive and negative saddles with clockwise and counter-clockwise limit cycles in crossing-univariate polynomial systems are studied secondly, and the numbers of saddles and centers are determined through a theorem and the first integral manifolds are determined through polynomial functions. The corresponding proof of the theorem is given, and a few illustrations of networks of saddles and centers are given to show the corresponding geometric structures. Such homoclinic networks of saddles and centers are without any sources and sinks.

4.1 Saddles and Centers

In this section, as in Luo [1], consider homoclinic networks with maximized clockwise and counter-clockwise centers and positive and negative saddles in two-dimensional polynomial nonlinear systems with crossing-variables vector fields. Consider a polynomial system with crossing-univariate polynomial vector fields in two directions. The corresponding dynamical behaviors will be presented through the following theorem.

Theorem 4.1 *Consider a crossing-univariate polynomial system as*

$$\begin{aligned}
\dot{x}_{j_1} &= a_{j_1j_20}\prod_{s_1=1}^{m}\left(x_{j_2}-a_{j_1j_2s_1}\right),\\
\dot{x}_{j_2} &= a_{j_2j_10}\prod_{l_1=1}^{n}\left(x_{j_1}-a_{j_2j_1l_1}\right),\\
&j_1, j_2 \in \{1,2\}; j_1 \neq j_2.
\end{aligned} \tag{4.1}$$

A. C. J. Luo, *Limit Cycles and Homoclinic Networks in Two-Dimensional Polynomial Systems*, https://doi.org/10.1007/978-981-97-2617-2_4

The first integral manifold is

$$
\begin{aligned}
&a_{j_2j_10}\left\{\frac{1}{n+1}\left[(x_{j_1}-a_{j_2j_1l_1})^{n+1}-(x_{j_10}-a_{j_2j_1l_1})^{n+1}\right]\right.\\
&\left.+\sum_{k=1}^{n-1}\frac{1}{n-k+1}b_{j_2j_1k}\left[(x_{j_1}-a_{j_2j_1l_1})^{n-k+1}-(x_{j_10}-a_{j_2j_1l_1})^{n-k+1}\right]\right\}\\
&=a_{j_1j_20}\left\{\frac{1}{m+1}\left[(x_{j_2}-a_{j_1j_2s_1})^{m+1}-(x_{j_20}-a_{j_1j_2s_1})^{m+1}\right]\right.\\
&\left.+\sum_{k=1}^{m-1}\frac{1}{m-k+1}b_{j_1j_2k}\left[(x_{j_2}-a_{j_1j_2s_1})^{m-k+1}-(x_{j_10}-a_{j_1j_2s_1})^{m-k+1}\right]\right\}.
\end{aligned}
\tag{4.2}
$$

where

$$
\begin{aligned}
b_{j_2j_11}&=\sum_{l_2=1,l_2\neq l_1}^{n}\left(a_{j_2j_1l_1}-a_{j_2j_1l_2}\right),\\
b_{j_2j_12}&=\sum_{\substack{l_2,l_3=1;l_2,l_3\neq l_1\\(l_2<l_3)}}^{n}\prod_{r=2}^{3}\left(a_{j_2j_1l_1}-a_{j_2j_1l_r}\right),\cdots,\\
b_{j_2j_1k}&=\sum_{\substack{l_2,l_3,\cdots,l_{k+1}=1;\\l_2,l_3,\cdots,l_{k+1}\neq l_1\\(l_2<l_3<\cdots<l_{k+1})}}^{n}\prod_{r=2}^{k+1}\left(a_{j_2j_1l_1}-a_{j_2j_1l_r}\right),\cdots,\\
b_{j_2j_1(n-1)}&=\prod_{l_2=1,l_2\neq l_1}^{n}\left(a_{j_2j_1l_1}-a_{j_2j_1l_2}\right);\\
b_{j_1j_21}&=\sum_{s_2=1,s_2\neq s_1}^{m}\left(a_{j_1j_2s_1}-a_{j_1j_2s_2}\right),\\
b_{j_1j_22}&=\sum_{\substack{s_2,s_3=1;s_2,s_3\neq s_1\\(s_2<s_3)}}^{m}\prod_{r=2}^{3}\left(a_{j_1j_2s_1}-a_{j_1j_2s_r}\right),\cdots,\\
b_{j_1j_2k}&=\sum_{\substack{s_2,s_3,\cdots,s_{k+1}=1;\\s_2,s_3,\cdots,s_{k+1}\neq s_1\\(s_2<s_3<\cdots<s_{k+1})}}^{m}\prod_{r=2}^{k+1}\left(a_{j_1j_2s_1}-a_{j_1j_2s_r}\right),\cdots,\\
b_{j_1j_2(m-1)}&=\prod_{s_1=1,s_2\neq s_1}^{m}\left(a_{j_1j_2s_1}-a_{j_1j_2s_2}\right).
\end{aligned}
\tag{4.3}
$$

The equilibrium of $(x_{j_2}^*, x_{j_1}^*) = (a_{j_1j_2s_1}, a_{j_2j_1l_1})$ $(s_1, s_2 \in \{1, 2, \cdots, m\}, s_1 \neq s_2; l_1, l_2 \in \{1, 2, \cdots, n\}, l_1 \neq l_2)$ *has the following properties.*

- For $a_{j_1j_20}\prod_{s_2=1,s_2\neq s_1}^{m}\left(a_{j_1j_2s_1}-a_{j_1j_2s_2}\right)>0$ and $a_{j_2j_10}\prod_{l_2=1,l_2\neq l_1}^{n}\left(a_{j_2j_1l_1}-a_{j_2j_1l_2}\right)>0$,

$$\left(a_{j_ij_25_1},a_{j_2j_11}\right)=\underbrace{(\mathrm{UP}_+,\mathrm{UP}_+)}_{\text{positive saddle}}. \tag{4.4}$$

The equilibrium of $(x^*_{j_2},x^*_{j_1})=(a_{j_1j_2s_1},a_{j_2j_1l_1})$ is a (UP$_+$, UP$_+$)-positive saddle.

- For $a_{j_1j_20}\prod_{s_2=1,s_2\neq s_1}^{m}\left(a_{j_1j_2s_1}-a_{j_1j_2s_2}\right)<0$ and $a_{j_2j_10}\prod_{l_2=1,l_2\neq l_1}^{n}\left(a_{j_2j_1l_1}-a_{j_2j_1l_2}\right)>0$,

$$\left(a_{j_ij_2s_1},a_{j_2j_11}\right)=\underbrace{(\mathrm{DP}_+,\mathrm{DP}_-)}_{\text{CCW center}}. \tag{4.5}$$

The equilibrium of $(x^*_{j_2},x^*_{j_1})=(a_{j_1j_2s_1},a_{j_2j_1l_1})$ is a (DP$_+$,DP$_-$)-counter-clockwise center.

- For $a_{j_1j_20}\prod_{s_2=1,s_2\neq s_1}^{m}\left(a_{j_1j_2s_1}-a_{j_1j_2s_2}\right)>0$ and $a_{j_2j_10}\prod_{l_2=1,l_2\neq l_1}^{n}\left(a_{j_2j_1l_1}-a_{j_2j_1l_2}\right)<0$,

$$(a_{j_1j_2s_1},a_{j_2j_1l_1})=\underbrace{(\mathrm{DP}_-,\mathrm{DP}_+)}_{\text{CW center}}. \tag{4.6}$$

The equilibrium of $(x^*_{j_2},x^*_{j_1})=(a_{j_1j_2s_1},a_{j_2j_1l_1})$ is a (DP$_-$, DP$_+$)-clockwise center.

- For $a_{j_1j_20}\prod_{s_2=1,s_2\neq s_1}^{m}\left(a_{j_1j_2s_1}-a_{j_1j_2s_2}\right)<0$ and $a_{j_2j_10}\prod_{l_2=1,l_2\neq l_1}^{n}\left(a_{j_2j_1l_1}-a_{j_2j_1l_2}\right)<0$,

$$(a_{j_1j_2s_1},a_{j_2j_1l_1})=\underbrace{(\mathrm{UP}_-,\mathrm{UP}_-)}_{\text{negative saddle}}. \tag{4.7}$$

The equilibrium of $(x^*_{j_2},x^*_{j_1})=(a_{j_1j_2s_1},a_{j_2j_1l_1})$ is a (UP$_-$,UP$_-$)-negative saddle.

Define a notation as

$$\bigcup_{s=1}^{m}\bigcup_{l=1}^{n}\underbrace{\left(a_{j_1j_2s},a_{j_2j_1l}\right)}_{\mathrm{XX}}
=\begin{Bmatrix}\left(a_{j_1j_2m},a_{j_2j_11}\right) & \left(a_{j_1j_2m},a_{j_2j_12}\right) & \cdots & \left(a_{j_1j_2m},a_{j_2j_1n}\right)\\ \left(a_{j_1j_2(m-1)},a_{j_2j_11}\right) & \left(a_{j_1j_2(m-1)},a_{j_2j_12}\right) & \cdots & \left(a_{j_1j_2(m-1)},a_{j_2j_1n}\right)\\ \vdots & \vdots & \vdots & \vdots\\ \left(a_{j_1j_21},a_{j_2j_21}\right) & \left(a_{j_1j_11},a_{j_2j_12}\right) & \cdots & \left(a_{j_1j_21},a_{j_2j_1n}\right)\end{Bmatrix}_{m\times n} \tag{4.8}$$

(i_1) *For* $m = 2m_1 + 1$ and $n = 2n_1 + 1$, *the equilibrium networks with* $(2m_1 + 1) \times (2n_1 + 1)$ *equilibriums have the following properties.*

- *For* $a_{j_1 j_2 0} > 0$ *and* $a_{j_2 j_1 0} > 0$,

$$\bigcup_{s=1}^{2m_1+1} \bigcup_{l=1}^{2n_1+1} \underbrace{(a_{j_1 j_2 s}, a_{j_2 j_1 l})}_{\mathrm{XX}} = \left\{ \begin{matrix} \underbrace{(\mathrm{UP}_+, \mathrm{UP}_+)}_{\text{saddle }(+)} & \underbrace{(\mathrm{DP}_-, \mathrm{DP}_+)}_{\text{center (CW)}} & \cdots & \underbrace{(\mathrm{UP}_+, \mathrm{UP}_+)}_{\text{saddle }(+)} \\ \underbrace{(\mathrm{DP}_+, \mathrm{DP}_-)}_{\text{center (CCW)}} & \underbrace{(\mathrm{UP}_-, \mathrm{UP}_-)}_{\text{saddle }(-)} & \cdots & \underbrace{(\mathrm{DP}_+, \mathrm{DP}_-)}_{\text{center (CCW)}} \\ \vdots & \vdots & \vdots & \vdots \\ \underbrace{(\mathrm{UP}_+, \mathrm{UP}_+)}_{\text{saddle }(+)} & \underbrace{(\mathrm{DP}_-, \mathrm{DP}_+)}_{\text{center (CW)}} & \cdots & \underbrace{(\mathrm{UP}_+, \mathrm{UP}_+)}_{\text{saddle }(+)} \end{matrix} \right\}_{(2m_1+1)\times(2n_1+1)}. \quad (4.9)$$

  In the network, the number of $(\mathrm{DP}_+, \mathrm{DP}_-)$*-counter-clockwise centers is* $m_1 \times (n_1 + 1)$; *the number of* $(\mathrm{DP}_-, \mathrm{DP}_+)$*-clockwise centers is* $(m_1 + 1) \times n_1$; *the number of* $(\mathrm{UP}_+, \mathrm{UP}_+)$*-positive saddles is* $(m_1 + 1) \times (n_1 + 1)$; *and the number of* $(\mathrm{UP}_-, \mathrm{UP}_-)$*-negative saddles is* $m_1 \times n_1$.
- *For* $a_{j_1 j_2 0} < 0$ *and* $a_{j_2 j_1 0} > 0$,

$$\bigcup_{s=1}^{2m_1+1} \bigcup_{l=1}^{2n_1+1} \underbrace{(a_{j_1 j_2 s}, a_{j_2 j_1 l})}_{\mathrm{XX}} = \left\{ \begin{matrix} \underbrace{(\mathrm{DP}_+, \mathrm{DP}_-)}_{\text{center (CCW)}} & \underbrace{(\mathrm{UP}_-, \mathrm{UP}_-)}_{\text{saddle }(-)} & \cdots & \underbrace{(\mathrm{DP}_+, \mathrm{DP}_-)}_{\text{center (CCW)}} \\ \underbrace{(\mathrm{UP}_+, \mathrm{UP}_+)}_{\text{saddle }(+)} & \underbrace{(\mathrm{DP}_-, \mathrm{DP}_+)}_{\text{center (CW)}} & \cdots & \underbrace{(\mathrm{UP}_+, \mathrm{UP}_+)}_{\text{saddle }(+)} \\ \vdots & \vdots & \vdots & \vdots \\ \underbrace{(\mathrm{DP}_+, \mathrm{DP}_-)}_{\text{center (CCW)}} & \underbrace{(\mathrm{UP}_-, \mathrm{UP}_-)}_{\text{saddle }(-)} & \cdots & \underbrace{(\mathrm{DP}_+, \mathrm{DP}_-)}_{\text{center (CCW)}} \end{matrix} \right\}_{(2m_1+1)\times(2n_1+1)}. \quad (4.10)$$

In the network, the number of $(\mathrm{DP}_+, \mathrm{DP}_-)$*-counter-clockwise centers is* $(m_1+1)\times(n_1+1)$*; the number of* $(\mathrm{DP}_-, \mathrm{DP}_+)$*-clockwise centers is* $m_1 \times n_1$*; the number of* $(\mathrm{UP}_+, \mathrm{UP}_+)$*-positive saddles is* $m_1 \times (n_1+1)$*; and the number of* $(\mathrm{UP}_-, \mathrm{UP}_-)$*-negative saddles is* $(m_1+1)\times n_1$.

- *For* $a_{j_1j_20} > 0$ *and* $a_{j_2j_10} < 0$,

$$\bigcup_{s=1}^{2m_1+1}\bigcup_{l=1}^{2n_1+1}\underbrace{(a_{j_1j_2s}, a_{j_2j_1l})}_{\mathrm{XX}} = \left\{\begin{array}{cccc} \underbrace{(\mathrm{DP}_-, \mathrm{DP}_+)}_{\text{center (CW)}} & \underbrace{(\mathrm{UP}_+, \mathrm{UP}_+)}_{\text{saddle }(+)} & \cdots & \underbrace{(\mathrm{DP}_-, \mathrm{DP}_+)}_{\text{center (CW)}} \\ \underbrace{(\mathrm{UP}_-, \mathrm{UP}_-)}_{\text{saddle }(-)} & \underbrace{(\mathrm{DP}_+, \mathrm{DP}_-)}_{\text{center (CCW)}} & \cdots & \underbrace{(\mathrm{UP}_-, \mathrm{UP}_-)}_{\text{saddle }(-)} \\ \vdots & \vdots & \vdots & \vdots \\ \underbrace{(\mathrm{DP}_-, \mathrm{DP}_+)}_{\text{center (CW)}} & \underbrace{(\mathrm{UP}_+, \mathrm{UP}_+)}_{\text{saddle }(+)} & \cdots & \underbrace{(\mathrm{DP}_-, \mathrm{DP}_+)}_{\text{center (CW)}} \end{array}\right\}_{(2m_1+1)\times(2n_1+1)}. \tag{4.11}$$

In the network, the number of $(\mathrm{DP}_+, \mathrm{DP}_-)$*-counter-clockwise centers is* $m_1 \times n_1$*; the number of* $(\mathrm{DP}_-, \mathrm{DP}_+)$*-clockwise centers is* $(m_1+1)\times(n_1+1)$*; the number of* $(\mathrm{UP}_+, \mathrm{UP}_+)$*-positive saddles is* $(m_1+1)\times n_1$*; and the number of* $(\mathrm{UP}_-, \mathrm{UP}_-)$*-negative saddles is* $m_1 \times (n_1+1)$.

- *For* $a_{j_1j_20} < 0$ *and* $a_{j_2j_10} < 0$,

$$\bigcup_{s=1}^{2m_1+1}\bigcup_{l=1}^{2n_1+1}\underbrace{(a_{j_1j_2s}, a_{j_2j_1l})}_{\mathrm{XX}} = \left\{\begin{array}{cccc} \underbrace{(\mathrm{UP}_-, \mathrm{UP}_-)}_{\text{saddle }(-)} & \underbrace{(\mathrm{DP}_+, \mathrm{DP}_-)}_{\text{center (CCW)}} & \cdots & \underbrace{(\mathrm{UP}_-, \mathrm{UP}_-)}_{\text{saddle }(-)} \\ \underbrace{(\mathrm{DP}_-, \mathrm{DP}_+)}_{\text{center (CW)}} & \underbrace{(\mathrm{UP}_+, \mathrm{UP}_+)}_{\text{saddle }(+)} & \cdots & \underbrace{(\mathrm{DP}_-, \mathrm{DP}_+)}_{\text{center (CW)}} \\ \vdots & \vdots & \vdots & \vdots \\ \underbrace{(\mathrm{UP}_-, \mathrm{UP}_-)}_{\text{saddle }(-)} & \underbrace{(\mathrm{DP}_+, \mathrm{DP}_-)}_{\text{center (CCW)}} & \cdots & \underbrace{(\mathrm{UP}_-, \mathrm{UP}_-)}_{\text{saddle }(-)} \end{array}\right\}_{(2m_1+1)\times(2n_1+1)}. \tag{4.12}$$

In the network, the number of $(\mathrm{DP}_+, \mathrm{DP}_-)$*-counter-clockwise centers is* $(m_1+1)\times n_1$; *the number of* $(\mathrm{DP}_-, \mathrm{DP}_+)$*-clockwise centers is* $m_1 \times (n_1+1)$; *the number of* $(\mathrm{UP}_+, \mathrm{UP}_+)$*-positive saddles is* $m_1 \times n_1$; *and the number of* $(\mathrm{UP}_-, \mathrm{UP}_-)$*-negative saddles is* $(m_1+1)\times(n_1+1)$.

(i_2) The numbers of positive and negative saddles, clockwise and counter-clockwise centers for $(2m_1+1)\times(2n_1+1)$-equilibriums are summarized in Table 4.1.

(i_3) With global centers and saddles, the total numbers of limit cycles and saddles are obtained as follows.

- *For* $a_{j_1j_20} > 0$ *and* $a_{j_2j_10} > 0$,

$$\begin{aligned} N_{LC} &= m_1 \times (n_1+1) + (m_1+1)\times n_1 = 2m_1n_1 + m_1 + n_1;\\ N_{SD} &= (m_1+1)\times(n_1+1) + m_1\times n_1 + 1 = 2m_1n_1 + m_1 + n_1 + 2. \end{aligned} \tag{4.13}$$

- *For* $a_{j_1j_20} < 0$ *and* $a_{j_2j_10} > 0$,

$$\begin{aligned} N_{LC} &= (m_1+1)\times(n_1+1) + m_1\times n_1 + 1 = 2m_1n_1 + m_1 + n_1 + 2;\\ N_{SD} &= m_1 \times (n_1+1) + (m_1+1)\times n_1 = 2m_1n_1 + m_1 + n_1. \end{aligned} \tag{4.14}$$

- *For* $a_{j_1j_20} > 0$ *and* $a_{j_2j_10} < 0$,

$$\begin{aligned} N_{LC} &= (m_1+1)\times(n_1+1) + m_1\times n_1 + 1 = 2m_1n_1 + m_1 + n_1 + 2;\\ N_{SD} &= m_1 \times (n_1+1) + (m_1+1)\times n_1 = 2m_1n_1 + m_1 + n_1. \end{aligned} \tag{4.15}$$

- *For* $a_{j_1j_20} < 0$ *and* $a_{j_2j_10} < 0$,

$$N_{LC} = m_1 \times (n_1+1) + (m_1+1)\times n_1 = 2m_1n_1 + m_1 + n_1;$$

Table 4.1 Numbers of positive and negative saddles, and clockwise and counter-clockwise centers in network for $(2m_1+1)\times(2n_1+1)$-equilibriums

$(a_{j_1j_20}, a_{j_2j_10})$	Numbers			
	$(\mathrm{DP}_+, \mathrm{DP}_-)$-center (CCW)	$(\mathrm{DP}_-, \mathrm{DP}_+)$-center (CW)	$(\mathrm{UP}_+, \mathrm{UP}_+)$-saddle (+)	$(\mathrm{UP}_-, \mathrm{UP}_-)$-saddle (−)
$(+,+)$	$m_1\times(n_1+1)$	$(m_1+1)\times n_1$	$(m_1+1)\times(n_1+1)$	$m_1\times n_1$
$(-,+)$	$(m_1+1)\times(n_1+1)$	$m_1\times n_1$	$m_1\times(n_1+1)$	$(m_1+1)\times n_1$
$(+,-)$	$m_1\times n_1$	$(m_1+1)\times(n_1+1)$	$(m_1+1)\times n_1$	$m_1\times(n_1+1)$
$(-,-)$	$(m_1+1)\times n_1$	$m_1\times(n_1+1)$	$m_1\times n_1$	$(m_1+1)\times(n_1+1)$

$$N_{SD} = (m_1 + 1) \times (n_1 + 1) + m_1 \times n_1 + 1 = 2m_1n_1 + m_1 + n_1 + 2. \tag{4.16}$$

The detailed distributions are listed in Table 4.2.

(ii$_1$) *For* $m = 2m_1$ and $n = 2n_1 + 1$, *the equilibrium networks with* $(2m_1) \times (2n_1 + 1)$ *equilibriums have the following properties.*

- *For* $a_{j_1j_20} > 0$ *and* $a_{j_2j_10} > 0$,

$$\bigcup_{s=1}^{2m_1}\bigcup_{l=1}^{2n_1+1} \underbrace{(a_{j_1j_2s}, a_{j_2j_1l})}_{\text{XX}} = \left\{\begin{matrix} \underbrace{(\text{UP}_+, \text{UP}_+)}_{\text{saddle }(+)} & \underbrace{(\text{DP}_-, \text{DP}_+)}_{\text{center (CW)}} & \cdots & \underbrace{(\text{UP}_+, \text{UP}_+)}_{\text{saddle }(+)} \\ \underbrace{(\text{DP}_+, \text{DP}_-)}_{\text{center (CCW)}} & \underbrace{(\text{UP}_-, \text{UP}_-)}_{\text{saddle }(-)} & \cdots & \underbrace{(\text{DP}_+, \text{DP}_-)}_{\text{center (CCW)}} \\ \vdots & \vdots & \vdots & \vdots \\ \underbrace{(\text{DP}_+, \text{DP}_-)}_{\text{center (CCW)}} & \underbrace{(\text{UP}_-, \text{UP}_-)}_{\text{saddle }(-)} & \cdots & \underbrace{(\text{DP}_+, \text{DP}_-)}_{\text{center (CCW)}} \end{matrix}\right\}_{(2m_1)\times(2n_1+1)}. \tag{4.17}$$

In the network, the number of $(\text{DP}_+, \text{DP}_-)$*-counter-clockwise centers is* $m_1 \times (n_1+1)$; *the number of* $(\text{DP}_-, \text{DP}_+)$*-clockwise centers is* $m_1 \times n_1$; *the number of* $(\text{UP}_+, \text{UP}_+)$*-positive saddles is* $m_1 \times (n_1+1)$; *and the number of* $(\text{UP}_-, \text{UP}_-)$*-negative saddles is* $m_1 \times n_1$.

Table 4.2 Total numbers of limit cycles and saddles in network for $(2m_1 + 1) \times (2n_1 + 1)$-equilibriums

$(a_{j_1j_20}, a_{j_2j_10})$	Total numbers			
	Limit cycles (CCW)	Limit cycles (CW)	Saddles (+)	Saddles (−)
$(+, +)$	$m_1 \times (n_1 + 1)$	$(m_1 + 1) \times n_1$	$(m_1 + 1) \times (n_1 + 1) + 1^*$	$m_1 \times n_1$
$(-, +)$	$(m_1 + 1) \times (n_1 + 1) + 1^*$	$m_1 \times n_1$	$m_1 \times (n_1 + 1)$	$(m_1 + 1) \times n_1$
$(+, -)$	$m_1 \times n_1$	$(m_1 + 1) \times (n_1 + 1) + 1^*$	$(m_1 + 1) \times n_1$	$m_1 \times (n_1 + 1)$
$(-, -)$	$(m_1 + 1) \times n_1$	$m_1 \times (n_1 + 1)$	$m_1 \times n_1$	$(m_1 + 1) \times (n_1 + 1) + 1^*$

* For one global cycle or saddle

- *For* $a_{j_1j_20} < 0$ *and* $a_{j_2j_10} > 0$,

$$\bigcup_{s=1}^{2m_1}\bigcup_{l=1}^{2n_1+1}\underbrace{(a_{j_1j_2s}, a_{j_2j_1l})}_{\text{XX}} = \left\{\begin{array}{cccc} \underbrace{(\mathrm{DP}_+,\mathrm{DP}_-)}_{\text{center (CCW)}} & \underbrace{(\mathrm{UP}_-,\mathrm{UP}_-)}_{\text{saddle }(-)} & \cdots & \underbrace{(\mathrm{DP}_+,\mathrm{DP}_-)}_{\text{center (CCW)}} \\ \underbrace{(\mathrm{UP}_+,\mathrm{UP}_+)}_{\text{saddle }(+)} & \underbrace{(\mathrm{DP}_-,\mathrm{DP}_+)}_{\text{center (CW)}} & \cdots & \underbrace{(\mathrm{UP}_+,\mathrm{UP}_+)}_{\text{saddle }(+)} \\ \vdots & \vdots & \vdots & \vdots \\ \underbrace{(\mathrm{UP}_+,\mathrm{UP}_+)}_{\text{saddle }(+)} & \underbrace{(\mathrm{DP}_-,\mathrm{DP}_+)}_{\text{center (CW)}} & \cdots & \underbrace{(\mathrm{UP}_+,\mathrm{UP}_+)}_{\text{saddle }(+)} \end{array}\right\}_{(2m_1)\times(2n_1+1)}. \tag{4.18}$$

In the network, the number of $(\mathrm{DP}_+, \mathrm{DP}_-)$*-counter-clockwise centers is* $m_1 \times (n_1+1)$*; the number of* $(\mathrm{DP}_-, \mathrm{DP}_+)$*-clockwise centers is* $m_1 \times n_1$*; the number of* $(\mathrm{UP}_+, \mathrm{UP}_+)$*-positive saddles is* $m_1 \times (n_1+1)$*; and the number of* $(\mathrm{UP}_-, \mathrm{UP}_-)$*-negative saddles is* $m_1 \times n_1$.
- *For* $a_{j_1j_20} > 0$ *and* $a_{j_2j_10} < 0$,

$$\bigcup_{s=1}^{2m_1}\bigcup_{l=1}^{2n_1+1}\underbrace{(a_{j_1j_2s}, a_{j_2j_1l})}_{\text{XX}} = \left\{\begin{array}{cccc} \underbrace{(\mathrm{DP}_-,\mathrm{DP}_+)}_{\text{center (CW)}} & \underbrace{(\mathrm{UP}_+,\mathrm{UP}_+)}_{\text{saddle }(+)} & \cdots & \underbrace{(\mathrm{DP}_-,\mathrm{DP}_+)}_{\text{center (CW)}} \\ \underbrace{(\mathrm{UP}_-,\mathrm{UP}_-)}_{\text{saddle }(-)} & \underbrace{(\mathrm{DP}_+,\mathrm{DP}_-)}_{\text{center (CCW)}} & \cdots & \underbrace{(\mathrm{UP}_-,\mathrm{UP}_-)}_{\text{saddle }(-)} \\ \vdots & \vdots & \vdots & \vdots \\ \underbrace{(\mathrm{UP}_-,\mathrm{UP}_-)}_{\text{saddle }(-)} & \underbrace{(\mathrm{DP}_+,\mathrm{DP}_-)}_{\text{center (CCW)}} & \cdots & \underbrace{(\mathrm{UP}_-,\mathrm{UP}_-)}_{\text{saddle }(-)} \end{array}\right\}_{(2m_1)\times(2n_1+1)}. \tag{4.19}$$

In the network, the number of $(\mathrm{DP}_+, \mathrm{DP}_-)$*-counter-clockwise centers is* $m_1 \times n_1$*; the number of* $(\mathrm{DP}_-, \mathrm{DP}_+)$*-clockwise centers is* $m_1 \times (n_1+1)$*; the number of* $(\mathrm{UP}_+, \mathrm{UP}_+)$*-positive saddles is* $m_1 \times n_1$*; and the number of* $(\mathrm{UP}_-, \mathrm{UP}_-)$*-negative saddles is* $m_1 \times (n_1+1)$.
- *For* $a_{j_1j_20} < 0$ *and* $a_{j_2j_10} < 0$,

Table 4.3 Numbers of positive and negative saddles, and clockwise and counter-clockwise centers in network for $(2m_1) \times (2n_1+1)$-equilibriums

$(a_{j_1j_20}, a_{j_2j_10})$	Numbers			
	$(\mathrm{DP}_+, \mathrm{DP}_-)$-center (CCW)	$(\mathrm{DP}_-, \mathrm{DP}_+)$-center (CW)	$(\mathrm{UP}_+, \mathrm{UP}_+)$-saddle (+)	$(\mathrm{UP}_-, \mathrm{UP}_-)$-saddle (−)
$(+,+)$	$m_1 \times (n_1+1)$	$m_1 \times n_1$	$m_1 \times (n_1+1)$	$m_1 \times n_1$
$(-,+)$	$m_1 \times (n_1+1)$	$m_1 \times n_1$	$m_1 \times (n_1+1)$	$m_1 \times n_1$
$(+,-)$	$m_1 \times n_1$	$m_1 \times (n_1+1)$	$m_1 \times n_1$	$m_1 \times (n_1+1)$
$(-,-)$	$m_1 \times n_1$	$m_1 \times (n_1+1)$	$m_1 \times n_1$	$m_1 \times (n_1+1)$

$$\bigcup_{s=1}^{2m_1}\bigcup_{l=1}^{2n_1+1}\underbrace{(a_{j_1j_2s}, a_{j_2j_1l})}_{\mathrm{XX}} = \left\{\begin{matrix} \underbrace{(\mathrm{UP}_-,\mathrm{UP}_-)}_{\text{saddle }(-)} & \underbrace{(\mathrm{DP}_+,\mathrm{DP}_-)}_{\text{center (CCW)}} & \cdots & \underbrace{(\mathrm{UP}_-,\mathrm{UP}_-)}_{\text{saddle }(-)} \\ \underbrace{(\mathrm{DP}_-,\mathrm{DP}_+)}_{\text{center (CW)}} & \underbrace{(\mathrm{UP}_+,\mathrm{UP}_+)}_{\text{saddle }(+)} & \cdots & \underbrace{(\mathrm{DP}_-,\mathrm{DP}_+)}_{\text{center (CW)}} \\ \vdots & \vdots & \vdots & \vdots \\ \underbrace{(\mathrm{DP}_-,\mathrm{DP}_+)}_{\text{center (CW)}} & \underbrace{(\mathrm{UP}_+,\mathrm{UP}_+)}_{\text{saddle }(+)} & \cdots & \underbrace{(\mathrm{DP}_-,\mathrm{DP}_+)}_{\text{center (CW)}} \end{matrix}\right\}_{(2m_1)\times(2n_1+1)}. \quad (4.20)$$

In the network, the number of $(\mathrm{DP}_+, \mathrm{DP}_-)$*-counter-clockwise centers is* $m_1 \times n_1$; *the number of* $(\mathrm{DP}_-, \mathrm{DP}_+)$*-clockwise centers is* $m_1 \times (n_1+1)$; *the number of* $(\mathrm{UP}_+, \mathrm{UP}_+)$*-positive saddles is* $m_1 \times n_1$; *and the number of* $(\mathrm{UP}_-, \mathrm{UP}_-)$*-negative saddles is* $m_1 \times (n_1+1)$.

(ii$_2$) *The numbers of positive and negative saddles, clockwise and counter-clockwise centers for* $(2m_1) \times (2n_1+1)$*-equilibriums are summarized in* Table 4.3.

(ii$_3$) Without global centers and saddles existing, the total numbers of limit cycles and saddles are obtained as follows.

- *For* $a_{j_1j_20} > 0$ *and* $a_{j_2j_10} > 0$,

$$\begin{aligned} N_{LC} &= m_1 \times (n_1+1) + m_1 \times n_1 = m_1(2n_1+1); \\ N_{SD} &= m_1 \times (n_1+1) + m_1 \times n_1 = m_1(2n_1+1). \end{aligned} \quad (4.21)$$

- *For* $a_{j_1j_20} < 0$ *and* $a_{j_2j_10} > 0$,

$$\begin{aligned} N_{LC} &= m_1 \times (n_1+1) + m_1 \times n_1 = m_1(2n_1+1); \\ N_{SD} &= m_1 \times (n_1+1) + m_1 \times n_1 = m_1(2n_1+1). \end{aligned} \quad (4.22)$$

- *For* $a_{j_1j_20} > 0$ *and* $a_{j_2j_10} < 0$,

$$\begin{aligned}N_{LC} &= m_1 \times (n_1+1) + m_1 \times n_1 = m_1(2n_1+1);\\ N_{SD} &= m_1 \times (n_1+1) + m_1 \times n_1 = m_1(2n_1+1).\end{aligned} \tag{4.23}$$

- *For* $a_{j_1j_20} < 0$ *and* $a_{j_2j_10} < 0$,

$$\begin{aligned}N_{LC} &= m_1 \times (n_1+1) + m_1 \times n_1 = m_1(2n_1+1);\\ N_{SD} &= m_1 \times (n_1+1) + m_1 \times n_1 = m_1(2n_1+1).\end{aligned} \tag{4.24}$$

The detailed descriptions are given in Table 4.4.

(iii$_1$) *For* $m = 2m_1+1$ and $n = 2n_1$, *the equilibrium networks with* $(2m_1+1)\times(2n_1)$ *equilibriums have the following properties.*

- *For* $a_{j_1j_20} > 0$ *and* $a_{j_2j_10} > 0$,

$$\bigcup_{s=1}^{2m_1+1}\bigcup_{l=1}^{2n_1}\underbrace{(a_{j_1j_2s}, a_{j_2j_1l})}_{\text{XX}} = \left\{\begin{matrix} \underbrace{(\text{DP}_-,\text{DP}_+)}_{\text{center (CW)}} & \underbrace{(\text{UP}_+,\text{UP}_+)}_{\text{saddle (+)}} & \cdots & \underbrace{(\text{UP}_+,\text{UP}_+)}_{\text{saddle (+)}} \\ \underbrace{(\text{UP}_-,\text{UP}_-)}_{\text{saddle (−)}} & \underbrace{(\text{DP}_+,\text{DP}_-)}_{\text{center (CCW)}} & \cdots & \underbrace{(\text{DP}_+,\text{DP}_-)}_{\text{center (CCW)}} \\ \vdots & \vdots & \vdots & \vdots \\ \underbrace{(\text{DP}_-,\text{DP}_+)}_{\text{center (CW)}} & \underbrace{(\text{UP}_+,\text{UP}_+)}_{\text{saddle (+)}} & \cdots & \underbrace{(\text{UP}_+,\text{UP}_+)}_{\text{saddle (+)}} \end{matrix}\right\}_{(2m_1+1)\times(2n_1)}. \tag{4.25}$$

In the network, the number of $(\text{DP}_+, \text{DP}_-)$*-counter-clockwise centers is* $m_1 \times n_1$; *the number of* $(\text{DP}_-, \text{DP}_+)$*-clockwise centers is* $(m_1+1)\times n_1$; *the number of* $(\text{UP}_+, \text{UP}_+)$*-positive saddles is* $(m_1+1)\times n_1$; *and the number of* $(\text{UP}_-, \text{UP}_-)$*-negative saddles is* $m_1 \times n_1$.

Table 4.4 Numbers of clockwise and counter-clockwise limit cycles in network for $(2m_1)\times(2n_1+1)$-equilibriums

$(a_{j_1j_20}, a_{j_2j_10})$	Total numbers			
	Limit cycles (CCW)	Limit cycles (CW)	Saddles (+)	Saddles (−)
(+, +)	$m_1 \times (n_1+1)$	$m_1 \times n_1$	$m_1 \times (n_1+1)$	$m_1 \times n_1$
(−, +)	$m_1 \times (n_1+1)$	$m_1 \times n_1$	$m_1 \times (n_1+1)$	$m_1 \times n_1$
(+, −)	$m_1 \times n_1$	$m_1 \times (n_1+1)$	$m_1 \times n_1$	$m_1 \times (n_1+1)$
(−, −)	$m_1 \times n_1$	$m_1 \times (n_1+1)$	$m_1 \times n_1$	$m_1 \times (n_1+1)$

Notice: No global cycles or saddles but global parabola-saddle existence

- *For* $a_{j_1j_20} < 0$ *and* $a_{j_2j_10} > 0$,

$$\bigcup_{s=1}^{2m_1+1}\bigcup_{l=1}^{2n_1}\underbrace{(a_{j_1j_2s}, a_{j_2j_1l})}_{\text{XX}}$$
$$= \begin{Bmatrix} \underbrace{(\text{UP}_-,\text{UP}_-)}_{\text{saddle }(-)} & \underbrace{(\text{DP}_+,\text{DP}_-)}_{\text{center (CCW)}} & \cdots & \underbrace{(\text{DP}_+,\text{DP}_-)}_{\text{center (CCW)}} \\ \underbrace{(\text{DP}_-,\text{DP}_+)}_{\text{center (CW)}} & \underbrace{(\text{UP}_+,\text{UP}_+)}_{\text{saddle }(+)} & \cdots & \underbrace{(\text{UP}_+,\text{UP}_+)}_{\text{saddle }(+)} \\ \vdots & \vdots & \vdots & \vdots \\ \underbrace{(\text{UP}_-,\text{UP}_-)}_{\text{saddle }(-)} & \underbrace{(\text{DP}_+,\text{DP}_-)}_{\text{center (CCW)}} & \cdots & \underbrace{(\text{DP}_+,\text{DP}_-)}_{\text{center (CCW)}} \end{Bmatrix}_{(2m_1+1)\times(2n_1)}. \tag{4.26}$$

In the network, the number of $(\text{DP}_+, \text{DP}_-)$*-counter-clockwise centers is* $(m_1 + 1) \times n_1$; *the number of* $(\text{DP}_-, \text{DP}_+)$*-clockwise centers is* $m_1 \times n_1$; *the number of* $(\text{UP}_+, \text{UP}_+)$*-positive saddles is* $m_1 \times n_1$; *and the number of* $(\text{UP}_-, \text{UP}_-)$*-negative saddles is* $(m_1 + 1) \times n_1$.

- *For* $a_{j_1j_20} > 0$ *and* $a_{j_2j_10} < 0$,

$$\bigcup_{s=1}^{2m_1+1}\bigcup_{l=1}^{2n_1}\underbrace{(a_{j_1j_2s}, a_{j_2j_1l})}_{\text{XX}}$$
$$= \begin{Bmatrix} \underbrace{(\text{UP}_+,\text{UP}_+)}_{\text{saddle }(+)} & \underbrace{(\text{DP}_-,\text{DP}_+)}_{\text{center (CW)}} & \cdots & \underbrace{(\text{DP}_-,\text{DP}_+)}_{\text{center (CW)}} \\ \underbrace{(\text{DP}_+,\text{DP}_-)}_{\text{center (CCW)}} & \underbrace{(\text{UP}_-,\text{UP}_-)}_{\text{saddle }(-)} & \cdots & \underbrace{(\text{UP}_-,\text{UP}_-)}_{\text{saddle }(-)} \\ \vdots & \vdots & \vdots & \vdots \\ \underbrace{(\text{UP}_+,\text{UP}_+)}_{\text{saddle }(+)} & \underbrace{(\text{DP}_-,\text{DP}_+)}_{\text{center (CW)}} & \cdots & \underbrace{(\text{DP}_-,\text{DP}_+)}_{\text{center (CW)}} \end{Bmatrix}_{(2m_1+1)\times(2n_1)}. \tag{4.27}$$

In the network, the number of $(\text{DP}_+, \text{DP}_-)$*-counter-clockwise centers is* $m_1 \times n_1$; *the number of* $(\text{DP}_-, \text{DP}_+)$*-clockwise centers is* $(m_1+1) \times n_1$; *the number of* $(\text{UP}_+, \text{UP}_+)$*-positive saddles is* $(m_1+1) \times n_1$; *and the number of* $(\text{UP}_-, \text{UP}_-)$*-negative saddles is* $m_1 \times n_1$.

- *For* $a_{j_1j_20} < 0$ *and* $a_{j_2j_10} < 0$,

$$\bigcup_{s=1}^{2m_1+1}\bigcup_{l=1}^{2n_1}\underbrace{(a_{j_1j_2s}, a_{j_2j_1l})}_{\text{XX}}$$

$$
= \left\{ \begin{array}{cccc}
\underbrace{(\mathrm{DP}_+,\mathrm{DP}_-)}_{\text{center (CCW)}} & \underbrace{(\mathrm{UP}_-,\mathrm{UP}_-)}_{\text{saddle }(-)} & \cdots & \underbrace{(\mathrm{UP}_-,\mathrm{UP}_-)}_{\text{saddle }(-)} \\
\underbrace{(\mathrm{UP}_+,\mathrm{UP}_+)}_{\text{saddle }(+)} & \underbrace{(\mathrm{DP}_-,\mathrm{DP}_+)}_{\text{center (CW)}} & \cdots & \underbrace{(\mathrm{DP}_-,\mathrm{DP}_+)}_{\text{center (CW)}} \\
\vdots & \vdots & \vdots & \vdots \\
\underbrace{(\mathrm{DP}_+,\mathrm{DP}_-)}_{\text{center (CCW)}} & \underbrace{(\mathrm{UP}_-,\mathrm{UP}_-)}_{\text{saddle }(-)} & \cdots & \underbrace{(\mathrm{UP}_-,\mathrm{UP}_-)}_{\text{saddle }(-)}
\end{array} \right\}_{(2m_1+1)\times(2n_1)} . \tag{4.28}
$$

In the network, the number of $(\mathrm{DP}_+, \mathrm{DP}_-)$*-counter-clockwise centers is* $(m_1 + 1) \times n_1$*; the number of* $(\mathrm{DP}_-, \mathrm{DP}_+)$*-clockwise centers is* $m_1 \times n_1$*; the number of* $(\mathrm{UP}_+, \mathrm{UP}_+)$*-positive saddles is* $m_1 \times n_1$*; and the number of* $(\mathrm{UP}_-, \mathrm{UP}_-)$*-negative saddles is* $(m_1 + 1) \times n_1$.

(iii$_2$) *The numbers of positive and negative saddles, clockwise and counter-clockwise centers for* $(2m_1 + 1) \times (2n_1)$*-equilibriums are summarized in* Table 4.5.

(iii$_3$) Without global centers and saddles existing, the total numbers of limit cycles and saddles are obtained as follows.

- *For* $a_{j_1j_20} > 0$ *and* $a_{j_2j_10} > 0$,

$$
\begin{aligned}
N_{LC} &= (m_1 + 1) \times n_1 + m_1 \times n_1 = (2m_1 + 1)n_1; \\
N_{SD} &= (m_1 + 1) \times n_1 + m_1 \times n_1 = (2m_1 + 1)n_1.
\end{aligned} \tag{4.29}
$$

- *For* $a_{j_1j_20} < 0$ *and* $a_{j_2j_10} > 0$,

$$
\begin{aligned}
N_{LC} &= (m_1 + 1) \times n_1 + m_1 \times n_1 = (2m_1 + 1)n_1; \\
N_{SD} &= (m_1 + 1) \times n_1 + m_1 \times n_1 = (2m_1 + 1)n_1.
\end{aligned} \tag{4.30}
$$

- *For* $a_{j_1j_20} > 0$ *and* $a_{j_2j_10} < 0$,

$$
\begin{aligned}
N_{LC} &= (m_1 + 1) \times n_1 + m_1 \times n_1 = (2m_1 + 1)n_1; \\
N_{SD} &= (m_1 + 1) \times n_1 + m_1 \times n_1 = (2m_1 + 1)n_1.
\end{aligned} \tag{4.31}
$$

Table 4.5 Numbers of positive and negative saddles, and clockwise and counter-clockwise centers in network for $(2m_1 + 1) \times (2n_1)$-equilibriums

$(a_{j_1j_20}, a_{j_2j_10})$	Numbers			
	$(\mathrm{DP}_+, \mathrm{DP}_-)$-center (CCW)	$(\mathrm{DP}_-, \mathrm{DP}_+)$-center (CW)	$(\mathrm{UP}_+, \mathrm{UP}_+)$-saddle (+)	$(\mathrm{UP}_-, \mathrm{UP}_-)$-saddle (−)
(+, +)	$m_1 \times n_1$	$(m_1 + 1) \times n_1$	$(m_1 + 1) \times n_1$	$m_1 \times n_1$
(−, +)	$(m_1 + 1) \times n_1$	$m_1 \times n_1$	$m_1 \times n_1$	$(m_1 + 1) \times n_1$
(+, −)	$m_1 \times n_1$	$(m_1 + 1) \times n_1$	$(m_1 + 1) \times n_1$	$m_1 \times n_1$
(−, −)	$(m_1 + 1) \times n_1$	$m_1 \times n_1$	$m_1 \times n_1$	$(m_1 + 1) \times n_1$

- *For* $a_{j_1j_20} < 0$ *and* $a_{j_2j_10} < 0$,

$$N_{LC} = (m_1 + 1) \times n_1 + m_1 \times n_1 = (2m_1 + 1)n_1;$$
$$N_{SD} = (m_1 + 1) \times n_1 + m_1 \times n_1 = (2m_1 + 1)n_1. \quad (4.32)$$

The detailed descriptions are given in Table 4.6.

(iv$_1$) *For* $m = 2m_1$ and $n = 2n_1$, *the equilibrium networks with* $(2m_1) \times (2n_1)$ *equilibriums have the following properties.*

- *For* $a_{j_1j_20} > 0$ *and* $a_{j_2j_10} > 0$,

$$\bigcup_{s=1}^{2m_1}\bigcup_{l=1}^{2n_1}\underbrace{(a_{j_1j_2s}, a_{j_2j_1l})}_{XX} = \left\{\begin{matrix} \underbrace{(DP_-,DP_+)}_{\text{center (CW)}} & \underbrace{(UP_+,UP_+)}_{\text{saddle }(+)} & \cdots & \underbrace{(UP_+,UP_+)}_{\text{saddle }(+)} \\ \underbrace{(UP_-,UP_-)}_{\text{saddle }(-)} & \underbrace{(DP_+,DP_-)}_{\text{center (CCW)}} & \cdots & \underbrace{(DP_+,DP_-)}_{\text{center (CCW)}} \\ \vdots & \vdots & \vdots & \vdots \\ \underbrace{(UP_-,UP_-)}_{\text{saddle }(-)} & \underbrace{(DP_+,DP_-)}_{\text{center (CCW)}} & \cdots & \underbrace{(DP_+,DP_-)}_{\text{center (CCW)}} \end{matrix}\right\}_{(2m_1)\times(2n_1)}. \quad (4.33)$$

In the network, the number of (DP_+, DP_-)*-counter-clockwise centers is* $m_1 \times n_1$*; the number of* (DP_-, DP_+)*-clockwise centers is* $m_1 \times n_1$*; the number of* (UP_+, UP_+)*-positive saddles is* $m_1 \times n_1$*; and the number of* (UP_-, UP_-)*-negative saddles is* $m_1 \times n_1$.

- *For* $a_{j_1j_20} < 0$ *and* $a_{j_2j_10} > 0$,

Table 4.6 Total numbers of clockwise and counter-clockwise limit cycles in network for $(2m_1 + 1) \times (2n_1)$-equilibriums

$(a_{j_1j_20}, a_{j_2j_10})$	Total Numbers			
	Limit cycles (CCW)	Limit cycles (CW)	Saddles (+)	Saddles (−)
$(+,+)$	$m_1 \times n_1$	$(m_1+1) \times n_1$	$(m_1+1) \times n_1$	$m_1 \times n_1$
$(-,+)$	$(m_1+1) \times n_1$	$m_1 \times n_1$	$m_1 \times n_1$	$(m_1+1) \times n_1$
$(+,-)$	$m_1 \times n_1$	$(m_1+1) \times n_1$	$(m_1+1) \times n_1$	$m_1 \times n_1$
$(-,-)$	$(m_1+1) \times n_1$	$m_1 \times n_1$	$m_1 \times n_1$	$(m_1+1) \times n_1$

Notice: no global cycles or saddles but global parabola-saddle existence

$$\bigcup_{s=1}^{2m_1}\bigcup_{l=1}^{2n_1}\underbrace{(a_{j_1j_2s},a_{j_2j_1l})}_{XX}=\left\{\begin{array}{cccc}\underbrace{(UP_-,UP_-)}_{\text{saddle }(-)} & \underbrace{(DP_+,DP_-)}_{\text{center (CCW)}} & \cdots & \underbrace{(DP_+,DP_-)}_{\text{center (CCW)}}\\ \underbrace{(DP_-,DP_+)}_{\text{center (CW)}} & \underbrace{(UP_+,UP_+)}_{\text{saddle }(+)} & \cdots & \underbrace{(UP_+,UP_+)}_{\text{saddle }(+)}\\ \vdots & \vdots & \vdots & \vdots\\ \underbrace{(DP_-,DP_+)}_{\text{center (CW)}} & \underbrace{(UP_+,UP_+)}_{\text{saddle }(+)} & \cdots & \underbrace{(UP_+,UP_+)}_{\text{saddle }(+)}\end{array}\right\}_{(2m_1)\times(2n_1)}. \tag{4.34}$$

In the network, the number of (DP_+, DP_-)*-counter-clockwise centers is* $m_1 \times n_1$*; the number of* (DP_-, DP_+)*-clockwise centers is* $m_1 \times n_1$*; the number of* (UP_+, UP_+)*-positive saddles is* $m_1 \times n_1$*; and the number of* (UP_-, UP_-)*-negative saddles is* $m_1 \times n_1$*.*

- *For* $a_{j_1j_20} > 0$ *and* $a_{j_2j_10} < 0$,

$$\bigcup_{s=1}^{2m_1}\bigcup_{l=1}^{2n_1}\underbrace{(a_{j_1j_2s},a_{j_2j_1l})}_{XX}=\left\{\begin{array}{cccc}\underbrace{(UP_+,UP_+)}_{\text{saddle }(+)} & \underbrace{(DP_-,DP_+)}_{\text{center (CW)}} & \cdots & \underbrace{(DP_-,DP_+)}_{\text{center (CW)}}\\ \underbrace{(DP_+,DP_-)}_{\text{center (CCW)}} & \underbrace{(UP_-,UP_-)}_{\text{saddle }(-)} & \cdots & \underbrace{(UP_-,UP_-)}_{\text{saddle }(-)}\\ \vdots & \vdots & \vdots & \vdots\\ \underbrace{(DP_+,DP_-)}_{\text{center (CCW)}} & \underbrace{(UP_-,UP_-)}_{\text{saddle }(-)} & \cdots & \underbrace{(UP_-,UP_-)}_{\text{saddle }(-)}\end{array}\right\}_{(2m_1)\times(2n_1)}. \tag{4.35}$$

In the network, the number of (DP_+, DP_-)*-counter-clockwise centers is* $m_1 \times n_1$*; the number of* (DP_-, DP_+)*-clockwise centers is* $m_1 \times n_1$*; the number of* (UP_+, UP_+)*-positive saddles is* $m_1 \times n_1$*; and the number of* (UP_-, UP_-)*-negative saddles is* $m_1 \times n_1$*.*

- *For* $a_{j_1j_20} < 0$ *and* $a_{j_2j_10} < 0$,

$$\bigcup_{s=1}^{2m_1}\bigcup_{l=1}^{2n_1}\underbrace{(a_{j_1j_2s},a_{j_2j_1l})}_{XX}=\left\{\begin{array}{cccc}\underbrace{(DP_+,DP_-)}_{\text{center (CCW)}} & \underbrace{(UP_-,UP_-)}_{\text{saddle }(-)} & \cdots & \underbrace{(UP_-,UP_-)}_{\text{saddle }(-)}\\ \underbrace{(UP_+,UP_+)}_{\text{saddle }(+)} & \underbrace{(DP_-,DP_+)}_{\text{center (CW)}} & \cdots & \underbrace{(DP_-,DP_+)}_{\text{center (CW)}}\\ \vdots & \vdots & \vdots & \vdots\\ \underbrace{(UP_+,UP_+)}_{\text{saddle }(+)} & \underbrace{(DP_-,DP_+)}_{\text{center (CW)}} & \cdots & \underbrace{(DP_-,DP_+)}_{\text{center (CW)}}\end{array}\right\}_{(2m_1)\times(2n_1)}. \tag{4.36}$$

In the network, the number of (DP_+, DP_-)*-counter-clockwise centers is* $m_1 \times n_1$*; the number of* (DP_-, DP_+)*-clockwise centers is* $m_1 \times n_1$*; the number of*

(UP_+, UP_+)-positive saddles is $m_1 \times n_1$; and the number of (UP_-, UP_-)-negative saddles is $m_1 \times n_1$.

(iv$_2$) *The numbers of positive and negative saddles, clockwise and counter-clockwise centers for $(2m_1) \times (2n_1)$-equilibriums are summarized in* Table 4.7.

(iv$_3$) Without global centers and saddles existing, the total numbers of limit cycles and saddles are obtained as follows.

- *For $a_{j_1 j_2 0} > 0$ and $a_{j_2 j_1 0} > 0$,*

$$\begin{aligned} N_{LC} &= m_1 \times n_1 + m_1 \times n_1 = 2m_1 n_1; \\ N_{SD} &= m_1 \times n_1 + m_1 \times n_1 = 2m_1 n_1. \end{aligned} \tag{4.37}$$

- *For $a_{j_1 j_2 0} < 0$ and $a_{j_2 j_1 0} > 0$,*

$$\begin{aligned} N_{LC} &= m_1 \times n_1 + m_1 \times n_1 = 2m_1 n_1; \\ N_{SD} &= m_1 \times n_1 + m_1 \times n_1 = 2m_1 n_1. \end{aligned} \tag{4.38}$$

- *For $a_{j_1 j_2 0} > 0$ and $a_{j_2 j_1 0} < 0$,*

$$\begin{aligned} N_{LC} &= m_1 \times n_1 + m_1 \times n_1 = 2m_1 n_1; \\ N_{SD} &= m_1 \times n_1 + m_1 \times n_1 = 2m_1 n_1. \end{aligned} \tag{4.39}$$

- *For $a_{j_1 j_2 0} < 0$ and $a_{j_2 j_1 0} < 0$,*

$$\begin{aligned} N_{LC} &= m_1 \times n_1 + m_1 \times n_1 = 2m_1 n_1; \\ N_{SD} &= m_1 \times n_1 + m_1 \times n_1 = 2m_1 n_1. \end{aligned} \tag{4.40}$$

The detailed descriptions are given in Table 4.8.

Table 4.7 Numbers of positive and negative saddles, and clockwise and counter-clockwise centers in network for $(2m_1) \times (2n_1)$-equilibriums

$(a_{j_1 j_2 0}, a_{j_2 j_1 0})$	Numbers			
	(DP_+, DP_-)-center (CCW)	(DP_-, DP_+)-center (CW)	(UP_+, UP_+)-saddle (+)	(UP_-, UP_-)-saddle (−)
$(+, +)$	$m_1 \times n_1$	$m_1 \times n_1$	$m_1 \times n_1$	$m_1 \times n_1$
$(-, +)$	$m_1 \times n_1$	$m_1 \times n_1$	$m_1 \times n_1$	$m_1 \times n_1$
$(+, -)$	$m_1 \times n_1$	$m_1 \times n_1$	$m_1 \times n_1$	$m_1 \times n_1$
$(-, -)$	$m_1 \times n_1$	$m_1 \times n_1$	$m_1 \times n_1$	$m_1 \times n_1$

Table 4.8 Total numbers of clockwise and counter-clockwise limit cycles in network for $(2m_1) \times (2n_1)$-equilibriums

$(a_{j_1j_20}, a_{j_2j_10})$	Total numbers			
	Limit cycles (CCW)	Limit cycles (CW)	Saddles (+)	Saddles (−)
$(+, +)$	$m_1 \times n_1$	$m_1 \times n_1$	$m_1 \times n_1$	$m_1 \times n_1$
$(-, +)$	$m_1 \times n_1$	$m_1 \times n_1$	$m_1 \times n_1$	$m_1 \times n_1$
$(+, -)$	$m_1 \times n_1$	$m_1 \times n_1$	$m_1 \times n_1$	$m_1 \times n_1$
$(-, -)$	$m_1 \times n_1$	$m_1 \times n_1$	$m_1 \times n_1$	$m_1 \times n_1$

Notice: no global cycles or saddles but global inflection-saddle existence

4.2 Proof of Theorem 4.1

Consider a crossing-univariate polynomial system as

$$\begin{aligned}
\dot{x}_{j_1} &= a_{j_1j_20}\prod_{s_1=1}^{m}(x_{j_2}-a_{j_1j_2s_1}),\\
\dot{x}_{j_2} &= a_{j_2j_10}\prod_{l_1=1}^{n}(x_{j_1}-a_{j_2j_1l_1}),\\
& j_1, j_2 \in \{1,2\}; j_1 \neq j_2.
\end{aligned}$$

In phase space, for $a_{j_1j_2s_1} \neq a_{j_1j_2s_2}(s_1, s_2 \in \{1, 2, \cdots, m\}, s_1 \neq s_2)$ and $a_{j_1j_2l_1} \neq a_{j_2j_1l_2}$ $(l_1, l_2 \in \{1, 2, \cdots, n\}, l_1 \neq l_2)$,

$$\begin{aligned}
\frac{dx_{j_1}}{dx_{j_2}} &= \frac{a_{j_1j_20}}{a_{j_2j_10}}\frac{(x_{j_2}-a_{j_1j_2s_1})\prod_{s_2=1,s_2\neq s_1}^{m}(x_{j_2}-a_{j_1j_2s_2})}{(x_{j_1}-a_{j_2j_1l_1})\prod_{l_2=1,l_2\neq l_1}^{n}(x_{j_1}-a_{j_2j_1l_2})}\\
&= \frac{a_{j_1j_20}}{a_{j_2j_10}}\frac{(x_{j_2}-a_{j_1j_2s_1})\prod_{s_2=1,s_2\neq s_1}^{m}(x_{j_2}-a_{j_1j_2s_1}+a_{j_1j_2s_1}-a_{j_1j_2s_2})}{(x_{j_1}-a_{j_2j_1l_1})\prod_{l_2=1,l_2\neq l_1}^{n}(x_{j_1}-a_{j_2j_1l_1}+a_{j_2j_1l_1}-a_{j_2j_1l_2})},
\end{aligned}$$

thus, from the relations of coefficients and roots, we have

$$\begin{aligned}
& a_{j_2j_10}\Big[(x_{j_1}-a_{j_2j_1l_1})^{n}+\sum_{k=1}^{n-1}b_{j_2j_1k}(x_{j_1}-a_{j_2j_1l_1})^{n-k}\Big]dx_{j_1}\\
&= a_{j_1j_20}\Big[(x_{j_2}-a_{j_1j_2s_1})^{m}+\sum_{k=1}^{m-1}b_{j_1j_2k}(x_{j_2}-a_{j_1j_2s_1})^{m-k}\Big]dx_{j_2},
\end{aligned}$$

and

$$b_{j_2j_11} = \sum_{l_2=1,l_2\neq l_1}^{n}(a_{j_2j_1l_1}-a_{j_2j_1l_2}),$$

$$b_{j_2j_12} = \sum_{\substack{l_2,l_3=1;l_2,l_3\neq l_1\\(l_2<l_3)}}^{n} \prod_{r=2}^{3} (a_{j_2j_1l_1} - a_{j_2j_1l_r}), \cdots,$$

$$b_{j_2j_1k} = \sum_{\substack{l_2,l_3,\cdots,l_{k+1}=1;\\ l_2,l_3,\cdots,l_{k+1}\neq l_1\\(l_2<l_3<\cdots<l_{k+1})}}^{n} \prod_{r=2}^{k+1} (a_{j_2j_1l_1} - a_{j_2j_1l_r}), \cdots,$$

$$b_{j_2j_1(n-1)} = \prod_{l_2=1,l_2\neq l_1}^{n} (a_{j_2j_1l_1} - a_{j_2j_1l_2});$$

$$b_{j_1j_21} = \sum_{s_2=1,s_2\neq s_1}^{m} (a_{j_1j_2s_1} - a_{j_1j_2s_2}),$$

$$b_{j_1j_22} = \sum_{\substack{s_2,s_3=1;s_2,s_3\neq s_1\\(s_2,<_3)}}^{m} \prod_{r=2}^{3} (a_{j_1j_2s_1} - a_{j_1j_2s_r}), \cdots,$$

$$b_{j_1j_2k} = \sum_{\substack{s_2,s_3,\cdots,s_{k+1}=1;\\ s_2,s_3,\cdots,s_{k+1}\neq s_1\\(s_2<s_3<\cdots<s_{k+1})}}^{m} \prod_{r=2}^{k+1} (a_{j_1j_2s_1} - a_{j_1j_2s_r}), \cdots,$$

$$b_{j_1j_2(m-1)} = \prod_{s_1=1,s_2\neq s_1}^{m} (a_{j_1j_2s_1} - a_{j_1j_2s_2}).$$

With initial conditions of (x_{j_10}, x_{j_20}) at $t = t_0$, the integration of the above equations gives the first integral manifold as

$$\begin{aligned}
&a_{j_2j_10}\Bigg\{\frac{1}{n+1}\big[(x_{j_1} - a_{j_2j_1l_1})^{n+1} - (x_{j_10} - a_{j_2j_1l_1})^{n+1}\big] \\
&\quad + \sum_{k=1}^{n-1} \frac{1}{n-k+1} b_{j_2j_1k}\big[(x_{j_1} - a_{j_2j_1l_1})^{n-k+1} - (x_{j_10} - a_{j_2j_1l_1})^{n-k+1}\big]\Bigg\} \\
&= a_{j_1j_20}\Bigg\{\frac{1}{m+1}\big[(x_{j_2} - a_{j_1j_2s_1})^{m+1} - (x_{j_20} - a_{j_1j_2s_1})^{m+1}\big] \\
&\quad + \sum_{k=1}^{m-1} \frac{1}{m-k+1} b_{j_1j_2k}\big[(x_{j_1} - a_{j_1j_2s_1})^{m-k+1} - (x_{j_10} - a_{j_1j_2s_1})^{m-k+1}\big]\Bigg\}.
\end{aligned}$$

In phase space, at $(x^*_{j_2}, x^*_{j_1}) = (a_{j_1j_2s_1}, a_{j_2j_1l_1})$, two cases are discussed.

(I) In phase space, at $x^*_{j_2} = a_{j_1j_2s_1}(s_1, s_2 \in \{1, 2, \cdots, m\}, s_1 \neq s_2)$ and $\overline{x}_{j_1} \neq a_{j_2j_1l_1}$ $(l_1 = 1, 2, \cdots, n)$, we have

$$\frac{dx_{j_1}}{dx_{j_2}}\bigg|_{x^*_{j_2}=a_{j_1j_2s_1}} = \frac{a_{j_1j_20}}{a_{j_2j_10}} \frac{(x_{j_2} - a_{j_1j_2s_1}) \prod_{s_2=1,s_2\neq s_1}^{m} (x_{j_2} - a_{j_1j_2s_2})}{\prod_{l_1=1}^{n} (\overline{x}_{j_1} - a_{j_2j_1l_1})}\bigg|_{x^*_{j_2}=a_{j_1j_2s_1}} = 0.$$

If

$$\frac{d^2x_{j_1}}{dx_{j_2}^2}\bigg|_{x_{j_2}^*=a_{j_1j_2s_1}} = \frac{a_{j_1j_20}}{a_{j_2j_10}} \frac{\prod_{s_2=1,s_2\neq s_1}^{m}(a_{j_1j_2s_1}-a_{j_1j_2s_2})}{\prod_{l_1=1}^{n}(\overline{x}_{j_1}-a_{j_2j_1l_1})} > 0,$$

there is an up-parabola flow at $x_{j_2}^* = a_{j_1j_2s_1}$ in the x_{j_1}-direction. If

$$\frac{d^2x_{j_1}}{dx_{j_2}^2}\bigg|_{x_{j_2}^*=a_{j_1j_2s_1}} = \frac{a_{j_1j_20}}{a_{j_2j_10}} \frac{\prod_{s_2=1,s_2\neq s_1}^{m}(a_{j_1j_2s_1}-a_{j_1j_2s_2})}{\prod_{l_1=1}^{n}(\overline{x}_{j_1}-a_{j_2j_1l_1})} < 0,$$

there is a down-parabola flow at $x_{j_2}^* = a_{j_1j_2s_1}$ in the x_{j_1}-direction. Because of

$$\dot{x}_{j_2} = a_{j_2j_10}\prod_{l_1=1}^{n}(\overline{x}_{j_1}-a_{j_2j_1l_1}),$$

the flows at $x_{j_2}^* = a_{j_1j_2s_1}$ in the x_{j_2}-direction are positive and negative for $\dot{x}_{j_2} > 0$ and $\dot{x}_{j_2} < 0$, respectively.

Thus, the equilibrium of $x_{j_2}^* = a_{j_1j_2s_1}$ $(s_1, s_2 \in \{1, 2, \cdots, m\}, s_1 \neq s_2)$ on the x_{j_2}-direction with $\overline{x}_{j_1} \neq a_{j_2j_1l_1}(l_1 = 1, 2, \cdots, n)$ has the following properties.

- For $a_{j_1j_20}\prod_{s_2=1,s_2\neq s_1}^{m}(a_{j_1j_2s_1}-a_{j_1j_2s_2}) > 0$ and $a_{j_2j_10}\prod_{l_1=1}^{n}(\overline{x}_{j_1}-a_{j_2j_1l_1}) > 0$,

$$(a_{j_1j_2s_1}, \dot{x}_{j_2}) = \underbrace{(\text{UP, pF})}_{\text{up-parabola flow}(+)}.$$

- For $a_{j_1j_20}\prod_{s_2=1,s_2\neq s_1}^{m}(a_{j_1j_2s_1}-a_{j_1j_2s_2}) < 0$ and $a_{j_2j_10}\prod_{l_1=1}^{n}(\overline{x}_{j_1}-a_{j_2j_1l_1}) > 0$,

$$(a_{j_1j_2s_1}, \dot{x}_{j_2}) = \underbrace{(\text{DP, pF})}_{\text{down-parabola flow}(+)}.$$

- For $a_{j_1j_20}\prod_{s_2=1,s_2\neq s_1}^{m}(a_{j_1j_2s_1}-a_{j_1j_2s_2}) > 0$ and $a_{j_2j_10}\prod_{l_1=1}^{n}(\overline{x}_{j_1}-a_{j_2j_1l_1}) < 0$,

$$(a_{j_1j_2s_1}, \dot{x}_{j_2}) = \underbrace{(\text{DP,nF})}_{\text{down-parabola flow }(-)}.$$

- For $a_{j_1j_20}\prod_{s_2=1,s_2\neq s_1}^{m}(a_{j_1j_2s_1}-a_{j_1j_2s_2}) < 0$ and $a_{j_2j_10}\prod_{l_1=1}^{n}(\overline{x}_{j_1}-a_{j_2j_1l_1}) < 0$,

$$(a_{j_1j_2s_1}, \dot{x}_{j_2}) = \underbrace{(\text{UP,nF})}_{\text{up-parabola flow }(-)}.$$

(II) In phase space, at $x_{j_1}^* = a_{j_2j_1l_1}$ $(l_1, l_2 \in \{1, 2, \cdots, n\}, l_1 \neq l_2)$ and $\overline{x}_{j_2} \neq a_{j_1j_2s_1}$ $(s_1 = 1, 2, \cdots, m)$, we have

$$\frac{dx_{j_2}}{dx_{j_1}}\Big|_{x^*_{j_1}=a_{j_2j_1l_1}} = \frac{a_{j_2j_10}}{a_{j_1j_20}} \frac{(x_{j_1}-a_{j_2j_1l_1})\prod_{l_2=1,l_2\neq l_1}^{n}(x_{j_1}-a_{j_2j_1l_2})}{\prod_{s_1=1}^{m}(\bar{x}_{j_2}-a_{j_1j_2s_1})}\Big|_{x^*_{j_1}=a_{j_2j_1l_1}} = 0.$$

If

$$\frac{d^2x_{j_2}}{dx^2_{j_1}}\Big|_{x^*_{j_1}=a_{j_2j_1l_1}} = \frac{a_{j_2j_10}}{a_{j_1j_20}} \frac{\prod_{l_2=1,l_2\neq l_1}^{n}\left(a_{j_2j_1l_1}-a_{j_2j_1l_2}\right)}{\prod_{s_1=1}^{m}(\bar{x}_{j_2}-a_{j_1j_2s_1})} > 0,$$

there is an up-parabola flow at $x^*_{j_1} = a_{j_2j_1l_1}$ in the x_{j_2}-direction. If

$$\frac{d^2x_{j_2}}{dx^2_{j_1}}\Big|_{x^*_{j_1}=a_{j_2j_1l_1}} = \frac{a_{j_2j_10}}{a_{j_1j_20}} \frac{\prod_{l_2=1,l_2\neq l_1}^{n}\left(a_{j_2j_1l_1}-a_{j_2j_1l_2}\right)}{\prod_{s_1=1}^{m}(\bar{x}_{j_2}-a_{j_1j_2s_1})} < 0,$$

there is a down-parabola flow at $x^*_{j_1} = a_{j_2j_1l_1}$ in the x_{j_1}-direction. Because of

$$\dot{x}_{j_1} = a_{j_1j_20}\prod_{s_1=1}^{m}(\bar{x}_{j_2}-a_{j_1j_2s_1}),$$

the flows at $x^*_{j_1} = a_{j_2j_1l_1}$ in the x_{j_1}-direction are positive and negative for $\dot{x}_{j_1} > 0$ and $\dot{x}_{j_1} < 0$, respectively.

Thus, the equilibrium of $x^*_{j_1} = a_{j_2j_1l_1}(l_1, l_2 \in \{1, 2, \cdots, n\}, l_1 \neq l_2)$ and $\bar{x}_{j_2} \neq a_{j_1j_2s_1}$ $(s_1 = 1, 2, \cdots, m)$ on the x_{j_2}-direction has the following properties.

- For $a_{j_1j_20}\prod_{s_1=1}^{m}(\bar{x}_{j_2}-a_{j_1j_2s_1}) > 0$ and $a_{j_2j_10}\prod_{l_2=1,l_2\neq l_1}^{n}(a_{j_2j_1l_1}-a_{j_2j_1l_2}) > 0$,
$$(\dot{x}_{j_1}, a_{j_2j_1l_1}) = \underbrace{\text{(pF,UP)}}_{\text{up-parabola flow }(+)}.$$
- For $a_{j_1j_20}\prod_{s_1=1}^{m}(\bar{x}_{j_2}-a_{j_1j_2s_1}) < 0$ and $a_{j_2j_10}\prod_{l_2=1,l_2\neq l_1}^{3}\left(a_{j_2j_1l_1}-a_{j_2j_1l_2}\right) > 0$,
$$(\dot{x}_{j_1}, a_{j_2j_1l_1}) = \underbrace{\text{(nF,DP)}}_{\text{down-parabola flow }(-)}.$$
- For $a_{j_1j_20}\prod_{s_1=1}^{m}(\bar{x}_{j_2}-a_{j_1j_2s_1}) > 0$ and $a_{j_2j_10}\prod_{l_2=1,l_2\neq l_1}^{n}(a_{j_2j_1l_1}-a_{j_2j_1l_2}) < 0$,
$$(\dot{x}_{j_1}, a_{j_2j_1l_1}) = \underbrace{\text{(pF,DP)}}_{\text{down-parabola flow }(+)}.$$
- For $a_{j_1j_20}\prod_{s_1=1}^{m}(\bar{x}_{j_2}-a_{j_1j_2s_1}) < 0$ and $a_{j_2j_10}\prod_{l_2=1,l_2\neq l_1}^{n}(a_{j_2j_1l_1}-a_{j_2j_1l_2}) < 0$,
$$(\dot{x}_{j_1}, a_{j_2j_1l_1}) = \underbrace{\text{(nF,UP)}}_{\text{up-parabola flow }(-)}.$$

Therefore, from the two cases (I) and (II), the equilibrium of $(x^*_{j_2}, x^*_{j_1}) = (a_{j_1j_2s_1}, a_{j_2j_1l_1})$ $(s_1, s_2 \in \{1, 2, \cdots, m\},\ s_1 \neq s_2;\ l_1, l_2 \in \{1, 2, \cdots, n\}, l_1 \neq l_2)$ has the following properties as in Eqs. (4.7)–(4.10).

- For $a_{j_1j_20}\prod_{s_2=1,s_2\neq s_1}^{m}(a_{j_1j_2s_1} - a_{j_1j_2s_2}) > 0$ and $a_{j_2j_10}\prod_{l_2=1,l_2\neq l_1}^{n}(a_{j_2j_1l_1} - a_{j_2j_1l_2}) > 0$,

$$(a_{j_1j_2s_1}, a_{j_2j_1l_1}) = \underbrace{(\mathrm{UP}_+,\mathrm{UP}_+)}_{\text{positive saddle}}.$$

  The equilibrium of $(x^*_{j_2}, x^*_{j_1}) = (a_{j_1j_2s_1}, a_{j_2j_1l_1})$ is a $(\mathrm{UP}_+,\mathrm{UP}_+)$-positive saddle.
- For $a_{j_1j_20}\prod_{s_2=1,s_2\neq s_1}^{m}(a_{j_1j_2s_1} - a_{j_1j_2s_2}) < 0$ and $a_{j_2j_10}\prod_{l_2=1,l_2\neq l_1}^{n}(a_{j_2j_1l_1} - a_{j_2j_1l_2}) > 0$,

$$(a_{j_1j_2s_1}, a_{j_2j_1l_1}) = \underbrace{(\mathrm{DP}_+,\mathrm{DP}_-)}_{\text{CCW center}}.$$

  The equilibrium of $(x^*_{j_2}, x^*_{j_1}) = (a_{j_1j_2s_1}, a_{j_2j_1l_1})$ is a $(\mathrm{DP}_+,\mathrm{DP}_-)$-counter-clockwise center.
- For $a_{j_1j_20}\prod_{s_2=1,s_2\neq s_1}^{m}(a_{j_1j_2s_1} - a_{j_1j_2s_2}) > 0$ and $a_{j_2j_10}\prod_{l_2=1,l_2\neq l_1}^{n}(a_{j_2j_1l_1} - a_{j_2j_1l_2}) < 0$,

$$(a_{j_1j_2s_1}, a_{j_2j_1l_1}) = \underbrace{(\mathrm{DP}_-,\mathrm{DP}_+)}_{\text{CW center}}.$$

  The equilibrium of $(x^*_{j_2}, x^*_{j_1}) = (a_{j_1j_2s_1}, a_{j_2j_1l_1})$ is a $(\mathrm{DP}_-,\mathrm{DP}_+)$-clockwise center.
- For $a_{j_1j_20}\prod_{s_2=1,s_2\neq s_1}^{m}(a_{j_1j_2s_1} - a_{j_1j_2s_2}) < 0$ and $a_{j_2j_10}\prod_{l_2=1,l_2\neq l_1}^{n}(a_{j_2j_1l_1} - a_{j_2j_1l_2}) < 0$,

$$\left(a_{j_1j_2s_1}, a_{j_2j_1l_1}\right) = \underbrace{(\mathrm{UP}_-,\ \mathrm{UP}_-)}_{\text{negative saddle}}.$$

  The equilibrium of $(x^*_{j_2}, x^*_{j_1}) = (a_{j_1j_2s_1}, a_{j_2j_1l_1})$ is a $(\mathrm{UP}_-,\mathrm{UP}_-)$-negative saddle.

From the above equations, the equilibrium networks can be developed. Consider four cases:

- For $m = 2m_1 + 1$ and $n = 2n_1 + 1$, the corresponding equilibrium networks can be matrixed.
- For $m = 2m_1$ and $n = 2n_1 + 1$, the corresponding equilibrium networks can be matrixed.

- For $m = 2m_1 + 1$ and $n = 2n_1$, the corresponding equilibrium networks can be matrixed.
- For $m = 2m_1$ and $n = 2n_1$, the corresponding equilibrium networks can be matrixed.

The $(\mathrm{UP}_+,\mathrm{UP}_+)$-positive saddles, $(\mathrm{DP}_+,\mathrm{DP}_-)$-counter-clockwise centers, $(\mathrm{DP}_-,\mathrm{DP}_+)$ -clockwise centers, $(\mathrm{UP}_-,\mathrm{UP}_-)$-negative saddles are determined and counted for each row and column. Summations of numbers yield the total numbers of positive and negative saddles and counter-clockwise and clockwise centers.

Define a notation as

$$\bigcup_{s=1}^{m}\bigcup_{l=1}^{n}\underbrace{(a_{j_1j_2s},a_{j_2j_1l})}_{\mathrm{XX}} = \left\{\begin{array}{cccc} (a_{j_1j_2m},a_{j_2j_11}) & (a_{j_1j_2m},a_{j_2j_12}) & \cdots & (a_{j_1j_2m},a_{j_2j_1n}) \\ (a_{j_1j_2(m-1)},a_{j_2j_11}) & (a_{j_1j_2(m-1)},a_{j_2j_12}) & \cdots & (a_{j_1j_2(m-1)},a_{j_2j_1n}) \\ \vdots & \vdots & \vdots & \vdots \\ (a_{j_1j_21},a_{j_2j_21}) & (a_{j_1j_11},a_{j_2j_12}) & \cdots & (a_{j_1j_21},a_{j_2j_1n}) \end{array}\right\}_{m\times n}$$

(i_1) From the above conditions, for $m = 2m_1 + 1$ and $n = 2n_1 + 1$, the equilibrium networks with $(2m_1 + 1) \times (2n_1 + 1)$ equilibriums have the following properties as in Eqs. (4.9)–(4.12).

- For $a_{j_1j_20} > 0$ and $a_{j_2j_10} > 0$,

$$\bigcup_{s=1}^{2m_1+1}\bigcup_{l=1}^{2n_1+1}\underbrace{(a_{j_1j_2s},a_{j_2j_1l})}_{\mathrm{XX}} = \left\{\begin{array}{cccc} \underbrace{(\mathrm{UP}_+,\mathrm{UP}_+)}_{\text{saddle }(+)} & \underbrace{(\mathrm{DP}_-,\mathrm{DP}_+)}_{\text{center (CW)}} & \cdots & \underbrace{(\mathrm{UP}_+,\mathrm{UP}_+)}_{\text{saddle }(+)} \\ \underbrace{(\mathrm{DP}_+,\mathrm{DP}_-)}_{\text{center (CCW)}} & \underbrace{(\mathrm{UP}_-,\mathrm{UP}_-)}_{\text{saddle }(-)} & \cdots & \underbrace{(\mathrm{DP}_+,\mathrm{DP}_-)}_{\text{center (CCW)}} \\ \vdots & \vdots & \vdots & \vdots \\ \underbrace{(\mathrm{UP}_+,\mathrm{UP}_+)}_{\text{saddle }(+)} & \underbrace{(\mathrm{DP}_-,\mathrm{DP}_+)}_{\text{center (CW)}} & \cdots & \underbrace{(\mathrm{UP}_+,\mathrm{UP}_+)}_{\text{saddle }(+)} \end{array}\right\}_{(2m_1+1)\times(2n_1+1)}.$$

In the network, the number of $(\mathrm{DP}_+, \mathrm{DP}_-)$-counter-clockwise centers is $m_1 \times (n_1 + 1)$; the number of $(\mathrm{DP}_-, \mathrm{DP}_+)$-clockwise centers is $(m_1 + 1) \times n_1$; the number of $(\mathrm{UP}_+, \mathrm{UP}_+)$-positive saddles is $(m_1+1)\times(n_1+1)$; and the number of $(\mathrm{UP}_-, \mathrm{UP}_-)$-negative saddles is $m_1 \times n_1$.

- For $a_{j_1j_20} < 0$ and $a_{j_2j_10} > 0$,

$$\bigcup_{s=1}^{2m_1+1}\bigcup_{l=1}^{2n_1+1}\underbrace{(a_{j_1j_2s}, a_{j_2j_1l})}_{\text{XX}} = \left\{\begin{array}{cccc} \underbrace{(\text{DP}_+,\text{DP}_-)}_{\text{center (CCW)}} & \underbrace{(\text{UP}_-,\text{UP}_-)}_{\text{saddle }(-)} & \cdots & \underbrace{(\text{DP}_+,\text{DP}_-)}_{\text{center (CCW)}} \\ \underbrace{(\text{UP}_+,\text{UP}_+)}_{\text{saddle }(+)} & \underbrace{(\text{DP}_-,\text{DP}_+)}_{\text{center (CW)}} & \cdots & \underbrace{(\text{UP}_+,\text{UP}_+)}_{\text{saddle }(+)} \\ \vdots & \vdots & \vdots & \vdots \\ \underbrace{(\text{DP}_+,\text{DP}_-)}_{\text{center (CCW)}} & \underbrace{(\text{UP}_-,\text{UP}_-)}_{\text{saddle }(-)} & \cdots & \underbrace{(\text{DP}_+,\text{DP}_-)}_{\text{center (CCW)}} \end{array}\right\}_{(2m_1+1)\times(2n_1+1)}.$$

In the network, the number of $(\text{DP}_+, \text{DP}_-)$-counter-clockwise centers is $(m_1 + 1) \times (n_1 + 1)$; the number of $(\text{DP}_-, \text{DP}_+)$-clockwise centers is $m_1 \times n_1$; the number of $(\text{UP}_+, \text{UP}_+)$-positive saddles is $m_1 \times (n_1 + 1)$; and the number of $(\text{UP}_-, \text{UP}_-)$-negative saddles is $(m_1 + 1) \times n_1$.

- For $a_{j_1j_20} > 0$ and $a_{j_2j_10} < 0$,

$$\bigcup_{s=1}^{2m_1+1}\bigcup_{l=1}^{2n_1+1}\underbrace{(a_{j_1j_2s}, a_{j_2j_1l})}_{\text{XX}} = \left\{\begin{array}{cccc} \underbrace{(\text{DP}_-,\text{DP}_+)}_{\text{center (CW)}} & \underbrace{(\text{UP}_+,\text{UP}_+)}_{\text{saddle }(+)} & \cdots & \underbrace{(\text{DP}_-,\text{DP}_+)}_{\text{center (CW)}} \\ \underbrace{(\text{UP}_-,\text{UP}_-)}_{\text{saddle }(-)} & \underbrace{(\text{DP}_+,\text{DP}_-)}_{\text{center (CCW)}} & \cdots & \underbrace{(\text{UP}_-,\text{UP}_-)}_{\text{saddle }(-)} \\ \vdots & \vdots & \vdots & \vdots \\ \underbrace{(\text{DP}_-,\text{DP}_+)}_{\text{center (CW)}} & \underbrace{(\text{UP}_+,\text{UP}_+)}_{\text{saddle }(+)} & \cdots & \underbrace{(\text{DP}_-,\text{DP}_+)}_{\text{center (CW)}} \end{array}\right\}_{(2m_1+1)\times(2n_1+1)}.$$

In the network, the number of $(\text{DP}_+, \text{DP}_-)$-counter-clockwise centers is $m_1 \times n_1$; the number of $(\text{DP}_-, \text{DP}_+)$-clockwise centers is $(m_1 + 1) \times (n_1 + 1)$; the number of $(\text{UP}_+, \text{UP}_+)$-positive saddles is $(m_1 + 1) \times n_1$; and the number of $(\text{UP}_-, \text{UP}_-)$-negative saddles is $m_1 \times (n_1 + 1)$.

- For $a_{j_1j_20} < 0$ and $a_{j_2j_10} < 0$,

$$\bigcup_{s=1}^{2m_1+1}\bigcup_{l=1}^{2n_1+1}\underbrace{(a_{j_1j_2s},a_{j_2j_1l})}_{\text{XX}}$$

$$=\left\{\begin{array}{cccc}\underbrace{(\text{UP}_-,\text{UP}_-)}_{\text{saddle }(-)} & \underbrace{(\text{DP}_+,\text{DP}_-)}_{\text{center (CCW)}} & \cdots & \underbrace{(\text{UP}_-,\text{UP}_-)}_{\text{saddle }(-)}\\ \underbrace{(\text{DP}_-,\text{DP}_+)}_{\text{center (CW)}} & \underbrace{(\text{UP}_+,\text{UP}_+)}_{\text{saddle }(+)} & \cdots & \underbrace{(\text{DP}_-,\text{DP}_+)}_{\text{center (CW)}}\\ \vdots & \vdots & \vdots & \vdots\\ \underbrace{(\text{UP}_-,\text{UP}_-)}_{\text{saddle }(-)} & \underbrace{(\text{DP}_+,\text{DP}_-)}_{\text{center (CCW)}} & \cdots & \underbrace{(\text{UP}_-,\text{UP}_-)}_{\text{saddle }(-)}\end{array}\right\}_{(2m_1+1)\times(2n_1+1)}.$$

In the network, the number of (DP_+, DP_-)-counter-clockwise centers is $(m_1+1)\times n_1$; the number of (DP_-, DP_+)-clockwise centers is $m_1\times(n_1+1)$; the number of (UP_+, UP_+)-positive saddles is $m_1\times n_1$; and the number of (UP_-, UP_-)-negative saddles is $(m_1+1)\times(n_1+1)$.

(i_2) From case (i_1), the numbers of positive and negative saddles, clockwise and counter-clockwise centers for $(2m_1+1)\times(2n_1+1)$-equilibriums can be calculated and summarized as in Table 4.1.

(i_3) From case (i_2), with global centers and saddles, the total numbers of limit cycles and saddles are calculated and obtained as follows as in Eqs. (4.13)–(4.16).

- For $a_{j_1j_20}>0$ and $a_{j_2j_10}>0$,

$$N_{LC}=m_1\times(n_1+1)+(m_1+1)\times n_1=2m_1n_1+m_1+n_1;$$
$$N_{SD}=(m_1+1)\times(n_1+1)+m_1\times n_1+1=2m_1n_1+m_1+n_1+2.$$

- For $a_{j_1j_20}<0$ and $a_{j_2j_10}>0$,

$$N_{LC}=(m_1+1)\times(n_1+1)+m_1\times n_1+1=2m_1n_1+m_1+n_1+2;$$
$$N_{SD}=m_1\times(n_1+1)+(m_1+1)\times n_1=2m_1n_1+m_1+n_1.$$

- For $a_{j_1j_20}>0$ and $a_{j_2j_10}<0$,

$$N_{LC}=(m_1+1)\times(n_1+1)+m_1\times n_1+1=2m_1n_1+m_1+n_1+2;$$
$$N_{SD}=m_1\times(n_1+1)+(m_1+1)\times n_1=2m_1n_1+m_1+n_1.$$

- For $a_{j_1j_20}<0$ and $a_{j_2j_10}<0$,

$$N_{LC}=m_1\times(n_1+1)+(m_1+1)\times n_1=2m_1n_1+m_1+n_1;$$
$$N_{SD}=(m_1+1)\times(n_1+1)+m_1\times n_1+1=2m_1n_1+m_1+n_1+2.$$

From case (i$_2$), the detailed distributions are listed as in Table 4.2.

(ii$_1$) From the basic properties, for $m = 2m_1$ and $n = 2n_1 + 1$, the equilibrium networks with $(2m_1) \times (2n_1 + 1)$ equilibriums have the following properties as in Eqs. (4.17)–(4.20).

- For $a_{j_1 j_2 0} > 0$ and $a_{j_2 j_1 0} > 0$,

$$
\bigcup_{s=1}^{2m_1}\bigcup_{l=1}^{2n_1+1} \underbrace{(a_{j_1 j_2 s}, a_{j_2 j_1 l})}_{\mathrm{XX}}
= \left\{ \begin{array}{cccc}
\underbrace{(\mathrm{UP}_+,\mathrm{UP}_+)}_{\text{saddle }(+)} & \underbrace{(\mathrm{DP}_-,\mathrm{DP}_+)}_{\text{center (CW)}} & \cdots & \underbrace{(\mathrm{UP}_+,\mathrm{UP}_+)}_{\text{saddle }(+)} \\
\underbrace{(\mathrm{DP}_+,\mathrm{DP}_-)}_{\text{center (CCW)}} & \underbrace{(\mathrm{UP}_-,\mathrm{UP}_-)}_{\text{saddle }(-)} & \cdots & \underbrace{(\mathrm{DP}_+,\mathrm{DP}_-)}_{\text{center (CCW)}} \\
\vdots & \vdots & \vdots & \vdots \\
\underbrace{(\mathrm{DP}_+,\mathrm{DP}_-)}_{\text{center (CCW)}} & \underbrace{(\mathrm{UP}_-,\mathrm{UP}_-)}_{\text{saddle }(-)} & \cdots & \underbrace{(\mathrm{DP}_+,\mathrm{DP}_-)}_{\text{center (CCW)}}
\end{array} \right\}_{(2m_1)\times(2n_1+1)} .
$$

  In the network, the number of $(\mathrm{DP}_+, \mathrm{DP}_-)$-counter-clockwise centers is $m_1 \times (n_1+1)$; the number of $(\mathrm{DP}_-, \mathrm{DP}_+)$-clockwise centers is $m_1 \times n_1$; the number of $(\mathrm{UP}_+, \mathrm{UP}_+)$-positive saddles is $m_1 \times (n_1 + 1)$; and the number of $(\mathrm{UP}_-, \mathrm{UP}_-)$-negative saddles is $m_1 \times n_1$.
- For $a_{j_1 j_2 0} < 0$ and $a_{j_2 j_1 0} > 0$,

$$
\bigcup_{s=1}^{2m_1}\bigcup_{l=1}^{2n_1+1} \underbrace{(a_{j_1 j_2 s}, a_{j_2 j_1 l})}_{\mathrm{XX}}
= \left\{ \begin{array}{cccc}
\underbrace{(\mathrm{DP}_+,\mathrm{DP}_-)}_{\text{center (CCW)}} & \underbrace{(\mathrm{UP}_-,\mathrm{UP}_-)}_{\text{saddle }(-)} & \cdots & \underbrace{(\mathrm{DP}_+,\mathrm{DP}_-)}_{\text{center (CCW)}} \\
\underbrace{(\mathrm{UP}_+,\mathrm{UP}_+)}_{\text{saddle }(+)} & \underbrace{(\mathrm{DP}_-,\mathrm{DP}_+)}_{\text{center (CW)}} & \cdots & \underbrace{(\mathrm{UP}_+,\mathrm{UP}_+)}_{\text{saddle }(+)} \\
\vdots & \vdots & \vdots & \vdots \\
\underbrace{(\mathrm{UP}_+,\mathrm{UP}_+)}_{\text{saddle }(+)} & \underbrace{(\mathrm{DP}_-,\mathrm{DP}_+)}_{\text{center (CW)}} & \cdots & \underbrace{(\mathrm{UP}_+,\mathrm{UP}_+)}_{\text{saddle }(+)}
\end{array} \right\}_{(2m_1)\times(2n_1+1)} .
$$

  In the network, the number of $(\mathrm{DP}_+, \mathrm{DP}_-)$-counter-clockwise centers is $m_1 \times (n_1+1)$; the number of $(\mathrm{DP}_-, \mathrm{DP}_+)$-clockwise centers is $m_1 \times n_1$; the number of

$(\mathrm{UP}_+, \mathrm{UP}_+)$-positive saddles is $m_1 \times (n_1+1)$; and the number of $(\mathrm{UP}_-, \mathrm{UP}_-)$-negative saddles is $m_1 \times n_1$.

- For $a_{j_1 j_2 0} > 0$ and $a_{j_2 j_1 0} < 0$,

$$\bigcup_{s=1}^{2m_1} \bigcup_{l=1}^{2n_1+1} \underbrace{(a_{j_1 j_2 s}, a_{j_2 j_1 l})}_{\mathrm{XX}} = \left\{ \begin{array}{cccc} \underbrace{(\mathrm{DP}_-,\mathrm{DP}_+)}_{\text{center (CW)}} & \underbrace{(\mathrm{UP}_+,\mathrm{UP}_+)}_{\text{saddle }(+)} & \cdots & \underbrace{(\mathrm{DP}_-,\mathrm{DP}_+)}_{\text{center (CW)}} \\ \underbrace{(\mathrm{UP}_-,\mathrm{UP}_-)}_{\text{saddle }(-)} & \underbrace{(\mathrm{DP}_+,\mathrm{DP}_-)}_{\text{center (CCW)}} & \cdots & \underbrace{(\mathrm{UP}_-,\mathrm{UP}_-)}_{\text{saddle }(-)} \\ \vdots & \vdots & \vdots & \vdots \\ \underbrace{(\mathrm{UP}_-,\mathrm{UP}_-)}_{\text{saddle }(-)} & \underbrace{(\mathrm{DP}_+,\mathrm{DP}_-)}_{\text{center (CCW)}} & \cdots & \underbrace{(\mathrm{UP}_-,\mathrm{UP}_-)}_{\text{saddle }(-)} \end{array} \right\}_{(2m_1)\times(2n_1+1)}.$$

In the network, the number of $(\mathrm{DP}_+, \mathrm{DP}_-)$-counter-clockwise centers is $m_1 \times n_1$; the number of $(\mathrm{DP}_-, \mathrm{DP}_+)$-clockwise centers is $m_1 \times (n_1+1)$; the number of $(\mathrm{UP}_+, \mathrm{UP}_+)$-positive saddles is $m_1 \times n_1$; and the number of $(\mathrm{UP}_-, \mathrm{UP}_-)$-negative saddles is $m_1 \times (n_1+1)$.

- For $a_{j_1 j_2 0} < 0$ and $a_{j_2 j_1 0} < 0$,

$$\bigcup_{s=1}^{2m_1} \bigcup_{l=1}^{2n_1+1} \underbrace{(a_{j_1 j_2 s}, a_{j_2 j_1 l})}_{\mathrm{XX}} = \left\{ \begin{array}{cccc} \underbrace{(\mathrm{UP}_-,\mathrm{UP}_-)}_{\text{saddle }(-)} & \underbrace{(\mathrm{DP}_+,\mathrm{DP}_-)}_{\text{center (CCW)}} & \cdots & \underbrace{(\mathrm{UP}_-,\mathrm{UP}_-)}_{\text{saddle }(-)} \\ \underbrace{(\mathrm{DP}_-,\mathrm{DP}_+)}_{\text{center (CW)}} & \underbrace{(\mathrm{UP}_+,\mathrm{UP}_+)}_{\text{saddle }(+)} & \cdots & \underbrace{(\mathrm{DP}_-,\mathrm{DP}_+)}_{\text{center (CW)}} \\ \vdots & \vdots & \vdots & \vdots \\ \underbrace{(\mathrm{DP}_-,\mathrm{DP}_+)}_{\text{center (CW)}} & \underbrace{(\mathrm{UP}_+,\mathrm{UP}_+)}_{\text{saddle }(+)} & \cdots & \underbrace{(\mathrm{DP}_-,\mathrm{DP}_+)}_{\text{center (CW)}} \end{array} \right\}_{(2m_1)\times(2n_1+1)}.$$

In the network, the number of $(\mathrm{DP}_+, \mathrm{DP}_-)$-counter-clockwise centers is $m_1 \times n_1$; the number of $(\mathrm{DP}_-, \mathrm{DP}_+)$-clockwise centers is $m_1 \times (n_1+1)$; the number of $(\mathrm{UP}_+, \mathrm{UP}_+)$-positive saddles is $m_1 \times n_1$; and the number of $(\mathrm{UP}_-, \mathrm{UP}_-)$-negative saddles is $m_1 \times (n_1+1)$.

(ii_2) From case (ii_1), the numbers of positive and negative saddles, clockwise and counter-clockwise centers for $(2m_1) \times (2n_1+1)$-equilibriums are calculated and summarized as in Table 4.3.

(ii$_3$) From case (ii$_2$), due to without global centers and saddles existing, the total numbers of limit cycles and saddles are obtained as follows as in Eqs. (4.21)–(4.24).

- For $a_{j_1 j_2 0} > 0$ and $a_{j_2 j_1 0} > 0$,

$$\begin{aligned} N_{LC} &= m_1 \times (n_1 + 1) + m_1 \times n_1 = m_1(2n_1 + 1); \\ N_{SD} &= m_1 \times (n_1 + 1) + m_1 \times n_1 = m_1(2n_1 + 1). \end{aligned}$$

- For $a_{j_1 j_2 0} < 0$ and $a_{j_2 j_1 0} > 0$,

$$\begin{aligned} N_{LC} &= m_1 \times (n_1 + 1) + m_1 \times n_1 = m_1(2n_1 + 1); \\ N_{SD} &= m_1 \times (n_1 + 1) + m_1 \times n_1 = m_1(2n_1 + 1). \end{aligned}$$

- For $a_{j_1 j_2 0} > 0$ and $a_{j_2 j_1 0} < 0$,

$$\begin{aligned} N_{LC} &= m_1 \times (n_1 + 1) + m_1 \times n_1 = m_1(2n_1 + 1); \\ N_{SD} &= m_1 \times (n_1 + 1) + m_1 \times n_1 = m_1(2n_1 + 1). \end{aligned}$$

- For $a_{j_1 j_2 0} < 0$ and $a_{j_2 j_1 0} < 0$,

$$\begin{aligned} N_{LC} &= m_1 \times (n_1 + 1) + m_1 \times n_1 = m_1(2n_1 + 1); \\ N_{SD} &= m_1 \times (n_1 + 1) + m_1 \times n_1 = m_1(2n_1 + 1). \end{aligned}$$

Such a $(2m_1) \times (2n_1 + 1)$-polynomial systems have the same numbers of saddles and centers. From case (ii$_2$), the detailed descriptions are given as in Table 4.4.

(iii$_1$) From the basic properties, for $m = 2m_1 + 1$ and $n = 2n_1$, the equilibrium networks with $(2m_1 + 1) \times (2n_1)$ equilibriums have the following properties as in Eq. (4.25)–(4.28).

- For $a_{j_1 j_2 0} > 0$ and $a_{j_2 j_1 0} > 0$,

$$\bigcup_{s=1}^{2m_1+1} \bigcup_{l=1}^{2n_1} \underbrace{(a_{j_1 j_2 s}, a_{j_2 j_1 l})}_{XX} = \left\{ \begin{matrix} \underbrace{(\mathrm{DP}_-,\mathrm{DP}_+)}_{\text{center (CW)}} & \underbrace{(\mathrm{UP}_+,\mathrm{UP}_+)}_{\text{saddle }(+)} & \cdots & \underbrace{(\mathrm{UP}_+,\mathrm{UP}_+)}_{\text{saddle }(+)} \\ \underbrace{(\mathrm{UP}_-,\mathrm{UP}_-)}_{\text{saddle }(-)} & \underbrace{(\mathrm{DP}_+,\mathrm{DP}_-)}_{\text{center (CCW)}} & \cdots & \underbrace{(\mathrm{DP}_+,\mathrm{DP}_-)}_{\text{center (CCW)}} \\ \vdots & \vdots & \vdots & \vdots \\ \underbrace{(\mathrm{DP}_-,\mathrm{DP}_+)}_{\text{center (CW)}} & \underbrace{(\mathrm{UP}_+,\mathrm{UP}_+)}_{\text{saddle }(+)} & \cdots & \underbrace{(\mathrm{UP}_+,\mathrm{UP}_+)}_{\text{saddle }(+)} \end{matrix} \right\}_{(2m_1+1)\times(2n_1)}.$$

In the network, the number of $(\mathrm{DP}_+, \mathrm{DP}_-)$-counter-clockwise centers is $m_1 \times n_1$; the number of $(\mathrm{DP}_-, \mathrm{DP}_+)$-clockwise centers is $(m_1+1)\times n_1$; the number of $(\mathrm{UP}_+, \mathrm{UP}_+)$-positive saddles is $(m_1+1)\times n_1$; and the number of $(\mathrm{UP}_-, \mathrm{UP}_-)$-negative saddles is $m_1 \times n_1$.

- For $a_{j_1j_20} < 0$ and $a_{j_2j_10} > 0$,

$$\bigcup_{s=1}^{2m_1+1}\bigcup_{l=1}^{2n_1}\underbrace{(a_{j_1j_2s}, a_{j_2j_1l})}_{\mathrm{XX}} = \left\{\begin{matrix} \underbrace{(\mathrm{UP}_-,\mathrm{UP}_-)}_{\text{saddle }(-)} & \underbrace{(\mathrm{DP}_+,\mathrm{DP}_-)}_{\text{center (CCW)}} & \cdots & \underbrace{(\mathrm{DP}_+,\mathrm{DP}_-)}_{\text{center (CCW)}} \\ \underbrace{(\mathrm{DP}_-,\mathrm{DP}_+)}_{\text{center (CW)}} & \underbrace{(\mathrm{UP}_+,\mathrm{UP}_+)}_{\text{saddle }(+)} & \cdots & \underbrace{(\mathrm{UP}_+,\mathrm{UP}_+)}_{\text{saddle }(+)} \\ \vdots & \vdots & \vdots & \vdots \\ \underbrace{(\mathrm{UP}_-,\mathrm{UP}_-)}_{\text{saddle }(-)} & \underbrace{(\mathrm{DP}_+,\mathrm{DP}_-)}_{\text{center (CCW)}} & \cdots & \underbrace{(\mathrm{DP}_+,\mathrm{DP}_-)}_{\text{center (CCW)}} \end{matrix}\right\}_{(2m_1+1)\times(2n_1)}.$$

In the network, the number of $(\mathrm{DP}_+, \mathrm{DP}_-)$-counter-clockwise centers is $(m_1+1)\times n_1$; the number of $(\mathrm{DP}_-, \mathrm{DP}_+)$-clockwise centers is $m_1 \times n_1$; the number of $(\mathrm{UP}_+, \mathrm{UP}_+)$-positive saddles is $m_1 \times n_1$; and the number of $(\mathrm{UP}_-, \mathrm{UP}_-)$-negative saddles is $(m_1+1)\times n_1$.

- For $a_{j_1j_20} > 0$ and $a_{j_2j_10} < 0$,

$$\bigcup_{s=1}^{2m_1+1}\bigcup_{l=1}^{2n_1}\underbrace{(a_{j_1j_2s}, a_{j_2j_1l})}_{\mathrm{XX}} = \left\{\begin{matrix} \underbrace{(\mathrm{UP}_+,\mathrm{UP}_+)}_{\text{saddle }(+)} & \underbrace{(\mathrm{DP}_-,\mathrm{DP}_+)}_{\text{center (CW)}} & \cdots & \underbrace{(\mathrm{DP}_-,\mathrm{DP}_+)}_{\text{center (CW)}} \\ \underbrace{(\mathrm{DP}_+,\mathrm{DP}_-)}_{\text{center (CCW)}} & \underbrace{(\mathrm{UP}_-,\mathrm{UP}_-)}_{\text{saddle }(-)} & \cdots & \underbrace{(\mathrm{UP}_-,\mathrm{UP}_-)}_{\text{saddle }(-)} \\ \vdots & \vdots & \vdots & \vdots \\ \underbrace{(\mathrm{UP}_+,\mathrm{UP}_+)}_{\text{saddle }(+)} & \underbrace{(\mathrm{DP}_-,\mathrm{DP}_+)}_{\text{center (CW)}} & \cdots & \underbrace{(\mathrm{DP}_-,\mathrm{DP}_+)}_{\text{center (CW)}} \end{matrix}\right\}_{(2m_1+1)\times(2n_1)}.$$

In the network, the number of $(\mathrm{DP}_+, \mathrm{DP}_-)$-counter-clockwise centers is $m_1 \times n_1$; the number of $(\mathrm{DP}_-, \mathrm{DP}_+)$-clockwise centers is $(m_1+1)\times n_1$; the number of $(\mathrm{UP}_+, \mathrm{UP}_+)$-positive saddles is $(m_1+1)\times n_1$; and the number of $(\mathrm{UP}_-, \mathrm{UP}_-)$-negative saddles is $m_1 \times n_1$.

- For $a_{j_1j_20} < 0$ and $a_{j_2j_10} < 0$,

$$\bigcup_{s=1}^{2m_1+1}\bigcup_{l=1}^{2n_1}\underbrace{(a_{j_1j_2s},a_{j_2j_1l})}_{\text{XX}}$$
$$=\left\{\begin{matrix}\underbrace{(\text{DP}_+,\text{DP}_-)}_{\text{center (CCW)}} & \underbrace{(\text{UP}_-,\text{UP}_-)}_{\text{saddle }(-)} & \cdots & \underbrace{(\text{UP}_-,\text{UP}_-)}_{\text{saddle }(-)}\\ \underbrace{(\text{UP}_+,\text{UP}_+)}_{\text{saddle }(+)} & \underbrace{(\text{DP}_-,\text{DP}_+)}_{\text{center (CW)}} & \cdots & \underbrace{(\text{DP}_-,\text{DP}_+)}_{\text{center (CW)}}\\ \vdots & \vdots & \vdots & \vdots\\ \underbrace{(\text{DP}_+,\text{DP}_-)}_{\text{center (CCW)}} & \underbrace{(\text{UP}_-,\text{UP}_-)}_{\text{saddle }(-)} & \cdots & \underbrace{(\text{UP}_-,\text{UP}_-)}_{\text{saddle }(-)}\end{matrix}\right\}_{(2m_1+1)\times(2n_1)}.$$

In the network, the number of (DP_+, DP_-)-counter-clockwise centers is $(m_1+1)\times n_1$; the number of (DP_-, DP_+)-clockwise centers is $m_1\times n_1$; the number of (UP_+, UP_+)-positive saddles is $m_1\times n_1$; and the number of (UP_-, UP_-)-negative saddles is $(m_1+1)\times n_1$.

(iii_2) From case (iii_1), the numbers of positive and negative saddles, clockwise and counter-clockwise centers for $(2m_1+1)\times(2n_1)$-equilibriums are calculated and summarized as in Table 4.5.

(iii_3) From case (iii_2), due to without global centers and saddles existing, the total numbers of limit cycles and saddles are calculated and obtained as follows as in Eqs. (4.29)–(4.32).

- For $a_{j_1j_20}>0$ and $a_{j_2j_10}>0$,

$$N_{LC}=(m_1+1)\times n_1+m_1\times n_1=(2m_1+1)n_1;$$
$$N_{SD}=(m_1+1)\times n_1+m_1\times n_1=(2m_1+1)n_1.$$

- For $a_{j_1j_20}<0$ and $a_{j_2j_10}>0$,

$$N_{LC}=(m_1+1)\times n_1+m_1\times n_1=(2m_1+1)n_1;$$
$$N_{SD}=(m_1+1)\times n_1+m_1\times n_1=(2m_1+1)n_1.$$

- For $a_{j_1j_20}>0$ and $a_{j_2j_10}<0$,

$$N_{LC}=(m_1+1)\times n_1+m_1\times n_1=(2m_1+1)n_1;$$
$$N_{SD}=(m_1+1)\times n_1+m_1\times n_1=(2m_1+1)n_1.$$

- For $a_{j_1j_20}<0$ and $a_{j_2j_10}<0$,

$$N_{LC}=(m_1+1)\times n_1+m_1\times n_1=(2m_1+1)n_1;$$
$$N_{SD}=(m_1+1)\times n_1+m_1\times n_1=(2m_1+1)n_1.$$

From case (iii$_3$), the detailed descriptions are given as in Table 4.6.

(iv$_1$) From the basic properties, for $m = 2m_1$ and $n = 2n_1$, the equilibrium networks with $(2m_1) \times (2n_1)$ equilibriums have the following properties as in Eqs. (4.33)–(4.36).

- For $a_{j_1 j_2 0} > 0$ and $a_{j_2 j_1 0} > 0$,

$$\bigcup_{s=1}^{2m_1}\bigcup_{l=1}^{2n_1}\underbrace{(a_{j_1j_2s}, a_{j_2j_1l})}_{\text{XX}} = \left\{\begin{array}{cccc} \underbrace{(\text{DP}_-,\text{DP}_+)}_{\text{center (CW)}} & \underbrace{(\text{UP}_+,\text{UP}_+)}_{\text{saddle }(+)} & \cdots & \underbrace{(\text{UP}_+,\text{UP}_+)}_{\text{saddle }(+)} \\ \underbrace{(\text{UP}_-,\text{UP}_-)}_{\text{saddle }(-)} & \underbrace{(\text{DP}_+,\text{DP}_-)}_{\text{center (CCW)}} & \cdots & \underbrace{(\text{DP}_+,\text{DP}_-)}_{\text{center (CCW)}} \\ \vdots & \vdots & \vdots & \vdots \\ \underbrace{(\text{UP}_-,\text{UP}_-)}_{\text{saddle }(-)} & \underbrace{(\text{DP}_+,\text{DP}_-)}_{\text{center (CCW)}} & \cdots & \underbrace{(\text{DP}_+,\text{DP}_-)}_{\text{center (CCW)}} \end{array}\right\}_{(2m_1)\times(2n_1)}.$$

  In the network, the number of $(\text{DP}_+, \text{DP}_-)$-counter-clockwise centers is $m_1 \times n_1$; the number of $(\text{DP}_-, \text{DP}_+)$-clockwise centers is $m_1 \times n_1$; the number of $(\text{UP}_+, \text{UP}_+)$-positive saddles is $m_1 \times n_1$; and the number of $(\text{UP}_-, \text{UP}_-)$-negative saddles is $m_1 \times n_1$.
- For $a_{j_1 j_2 0} < 0$ and $a_{j_2 j_1 0} > 0$,

$$\bigcup_{s=1}^{2m_1}\bigcup_{l=1}^{2n_1}\underbrace{(a_{j_1j_2s}, a_{j_2j_1l})}_{\text{XX}} = \left\{\begin{array}{cccc} \underbrace{(\text{UP}_-,\text{UP}_-)}_{\text{saddle }(-)} & \underbrace{(\text{DP}_+,\text{DP}_-)}_{\text{center (CCW)}} & \cdots & \underbrace{(\text{DP}_+,\text{DP}_-)}_{\text{center (CCW)}} \\ \underbrace{(\text{DP}_-,\text{DP}_+)}_{\text{center (CW)}} & \underbrace{(\text{UP}_+,\text{UP}_+)}_{\text{saddle }(+)} & \cdots & \underbrace{(\text{UP}_+,\text{UP}_+)}_{\text{saddle }(+)} \\ \vdots & \vdots & \vdots & \vdots \\ \underbrace{(\text{DP}_-,\text{DP}_+)}_{\text{center (CW)}} & \underbrace{(\text{UP}_+,\text{UP}_+)}_{\text{saddle }(+)} & \cdots & \underbrace{(\text{UP}_+,\text{UP}_+)}_{\text{saddle }(+)} \end{array}\right\}_{(2m_1)\times(2n_1)}.$$

  In the network, the number of $(\text{DP}_+, \text{DP}_-)$-counter-clockwise centers is $m_1 \times n_1$; the number of $(\text{DP}_-, \text{DP}_+)$-clockwise centers is $m_1 \times n_1$; the number of $(\text{UP}_+, \text{UP}_+)$-positive saddles is $m_1 \times n_1$; and the number of $(\text{UP}_-, \text{UP}_-)$-negative saddles is $m_1 \times n_1$.
- For $a_{j_1 j_2 0} > 0$ and $a_{j_2 j_1 0} < 0$,

$$\bigcup_{s=1}^{2m_1}\bigcup_{l=1}^{2n_1}\underbrace{(a_{j_1j_2s},a_{j_2j_1l})}_{XX}=\left\{\begin{matrix}\underbrace{(UP_+,UP_+)}_{\text{saddle }(+)} & \underbrace{(DP_-,DP_+)}_{\text{center (CW)}} & \cdots & \underbrace{(DP_-,DP_+)}_{\text{center (CW)}}\\ \underbrace{(DP_+,DP_-)}_{\text{center (CCW)}} & \underbrace{(UP_-,UP_-)}_{\text{saddle }(-)} & \cdots & \underbrace{(UP_-,UP_-)}_{\text{saddle }(-)}\\ \vdots & \vdots & \vdots & \vdots\\ \underbrace{(DP_+,DP_-)}_{\text{center (CCW)}} & \underbrace{(UP_-,UP_-)}_{\text{saddle }(-)} & \cdots & \underbrace{(UP_-,UP_-)}_{\text{saddle }(-)}\end{matrix}\right\}_{(2m_1)\times(2n_1)}.$$

In the network, the number of (DP_+, DP_-)-counter-clockwise centers is $m_1 \times n_1$; the number of (DP_-, DP_+)-clockwise centers is $m_1 \times n_1$; the number of (UP_+, UP_+)-positive saddles is $m_1 \times n_1$; and the number of (UP_-, UP_-)-negative saddles is $m_1 \times n_1$.

- For $a_{j_1j_20} < 0$ and $a_{j_2j_10} < 0$,

$$\bigcup_{s=1}^{2m_1}\bigcup_{l=1}^{2n_1}\underbrace{(a_{j_1j_2s},a_{j_2j_1l})}_{XX}=\left\{\begin{matrix}\underbrace{(DP_+,DP_-)}_{\text{center (CCW)}} & \underbrace{(UP_-,UP_-)}_{\text{saddle }(-)} & \cdots & \underbrace{(UP_-,UP_-)}_{\text{saddle }(-)}\\ \underbrace{(UP_+,UP_+)}_{\text{saddle }(+)} & \underbrace{(DP_-,DP_+)}_{\text{center (CW)}} & \cdots & \underbrace{(DP_-,DP_+)}_{\text{center (CW)}}\\ \vdots & \vdots & \vdots & \vdots\\ \underbrace{(UP_+,UP_+)}_{\text{saddle }(+)} & \underbrace{(DP_-,DP_+)}_{\text{center (CW)}} & \cdots & \underbrace{(DP_-,DP_+)}_{\text{center (CW)}}\end{matrix}\right\}_{(2m_1)\times(2n_1)}.$$

In the network, the number of (DP_+, DP_-)-counter-clockwise centers is $m_1 \times n_1$; the number of (DP_-, DP_+)-clockwise centers is $m_1 \times n_1$; the number of (UP_+, UP_+)-positive saddles is $m_1 \times n_1$; and the number of (UP_-, UP_-)-negative saddles is $m_1 \times n_1$.

(iv_2) From case (iv_1), the numbers of positive and negative saddles, clockwise and counter-clockwise centers for $(2m_1) \times (2n_1)$-equilibriums are calculated and summarized as in Table 4.7.

(iv_3) From case (iv_2). Due to without global centers and saddles existing, the total numbers of limit cycles and saddles are obtained as follows as in Eqs. (4.37)–(4.40).

- For $a_{j_1j_20} > 0$ and $a_{j_2j_10} > 0$,

$$N_{LC} = m_1 \times n_1 + m_1 \times n_1 = 2m_1n_1;$$
$$N_{SD} = m_1 \times n_1 + m_1 \times n_1 = 2m_1n_1.$$

- For $a_{j_1j_20} < 0$ and $a_{j_2j_10} > 0$,

$$N_{LC} = m_1 \times n_1 + m_1 \times n_1 = 2m_1n_1;$$

$$N_{SD} = m_1 \times n_1 + m_1 \times n_1 = 2m_1n_1.$$

- For $a_{j_1j_20} > 0$ and $a_{j_2j_10} < 0$,

$$N_{LC} = m_1 \times n_1 + m_1 \times n_1 = 2m_1n_1;$$
$$N_{SD} = m_1 \times n_1 + m_1 \times n_1 = 2m_1n_1.$$

- For $a_{j_1j_20} < 0$ and $a_{j_2j_10} < 0$,

$$N_{LC} = m_1 \times n_1 + m_1 \times n_1 = 2m_1n_1;$$
$$N_{SD} = m_1 \times n_1 + m_1 \times n_1 = 2m_1n_1.$$

From case (iv$_2$), the detailed descriptions of limit cycles and saddles are given as in Table 4.8.

In the end, Theorem 4.1 is proved.

4.3 Homoclinic Networks with Centers

As in Luo [1], the networks for limit cycles and saddles in lower-degrees polynomial system with crossing-univariate vector fields are presented for a better understanding of the Hilbert's sixteenth problem.

(A) Consider a quadratic dynamical system with $(2m_1) \times (2n_1) = 2 \times 2$ equilibriums as

$$\begin{aligned} \dot{x}_1 &= a_{120}(x_2 - a_{121})(x_2 - a_{122}), \\ \dot{x}_2 &= a_{210}(x_1 - a_{211})(x_1 - a_{212}). \end{aligned} \tag{4.41}$$

and the first integral manifold is determined by

$$\begin{aligned} &a_{120}\Big\{\frac{1}{3}\big[(x_2 - a_{121})^3 - (x_{20} - a_{121})^3\big] \\ &\quad + \frac{1}{2}(a_{121} - a_{122})\big[(x_2 - a_{121})^2 - (x_{20} - a_{121})^2\big]\Big\} \\ &= a_{210}\Big\{\frac{1}{3}\big[(x_1 - a_{211})^3 - (x_{10} - a_{211})^3\big] \\ &\quad + \frac{1}{2}(a_{211} - a_{212})\big[(x_1 - a_{211})^2 - (x_{10} - a_{211})^2\big]\Big\}. \end{aligned} \tag{4.42}$$

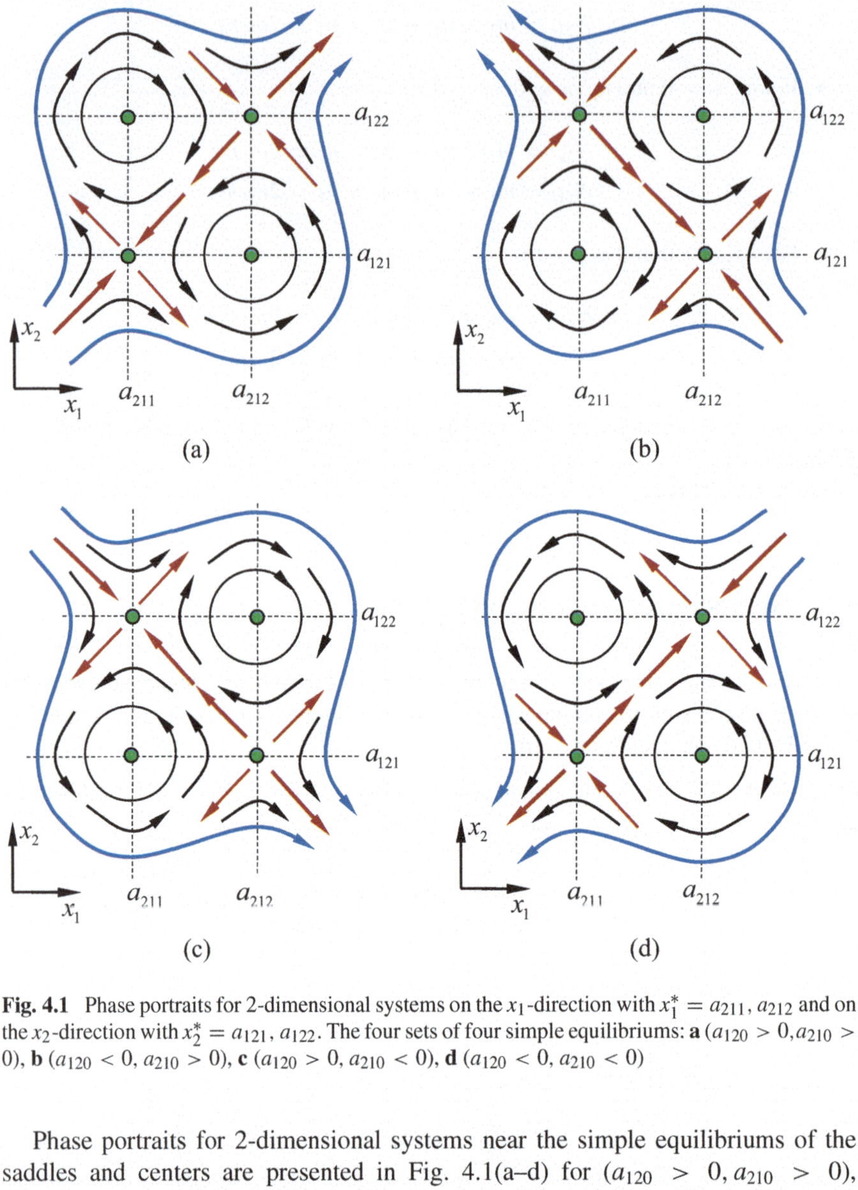

Fig. 4.1 Phase portraits for 2-dimensional systems on the x_1-direction with $x_1^* = a_{211}, a_{212}$ and on the x_2-direction with $x_2^* = a_{121}, a_{122}$. The four sets of four simple equilibriums: **a** ($a_{120} > 0, a_{210} > 0$), **b** ($a_{120} < 0$, $a_{210} > 0$), **c** ($a_{120} > 0$, $a_{210} < 0$), **d** ($a_{120} < 0$, $a_{210} < 0$)

Phase portraits for 2-dimensional systems near the simple equilibriums of the saddles and centers are presented in Fig. 4.1(a–d) for ($a_{120} > 0, a_{210} > 0$), ($a_{120} < 0, a_{210} > 0$), ($a_{120} > 0, a_{210} < 0$) and ($a_{120} < 0, a_{210} < 0$). The numbers of limit cycles and centers are tabulated in Table 4.9. The six equilibriums are three saddles and three centers. There are three limit cycles around the centers and three closed separatrices connected three saddles. The blue curves show the (2,2)-inflection-saddles, which will be discussed in the next chapter. The network matrices of saddle and centers are generated by the (2,2)-inflection-saddles, as presented as follows.

Table 4.9 Numbers of limit cycles and saddles in networks for 2 × 2-equilibriums

(a_{120}, a_{210})	Numbers			
	$(\mathrm{DP}_+, \mathrm{DP}_-)$-center (CCW)	$(\mathrm{DP}_-, \mathrm{DP}_+)$-center (CW)	$(\mathrm{DP}_+, \mathrm{DP}_+)$-saddle (+)	$(\mathrm{DP}_-, \mathrm{DP}_-)$-saddle (−)
$(+, +)$	$m_1 \times n_1 = 1$	$m_1 \times n_1 = 1$	$m_1 \times n_1 = 1$	$m_1 \times n_1 = 1$
$(-, +)$	$m_1 \times n_1 = 1$	$m_1 \times n_1 = 1$	$m_1 \times n_1 = 1$	$m_1 \times n_1 = 1$
$(+, -)$	$m_1 \times n_1 = 1$	$m_1 \times n_1 = 1$	$m_1 \times n_1 = 1$	$m_1 \times n_1 = 1$
$(-, -)$	$m_1 \times n_1 = 1$	$m_1 \times n_1 = 1$	$m_1 \times n_1 = 1$	$m_1 \times n_1 = 1$

- For $a_{210} > 0$ and $a_{120} > 0$,

$$\left\{\begin{array}{cc} (a_{122}, a_{211}) & (a_{122}, a_{212}) \\ (a_{121}, a_{211}) & (a_{121}, a_{212}) \end{array}\right\} = \left\{\begin{array}{cc} \underbrace{(\mathrm{DP}_-,\mathrm{DP}_+)}_{\text{center (CW)}} & \underbrace{(\mathrm{UP}_+,\mathrm{UP}_+)}_{\text{saddle (+)}} \\ \underbrace{(\mathrm{UP}_-,\mathrm{UP}_-)}_{\text{saddle (−)}} & \underbrace{(\mathrm{DP}_+,\mathrm{DP}_-)}_{\text{center (CCW)}} \end{array}\right\}. \tag{4.43}$$

- For $a_{210} < 0$ and $a_{120} > 0$,

$$\left\{\begin{array}{cc} (a_{122}, a_{211}) & (a_{122}, a_{212}) \\ (a_{121}, a_{211}) & (a_{121}, a_{212}) \end{array}\right\} = \left\{\begin{array}{cc} \underbrace{(\mathrm{UP}_-,\mathrm{UP}_-)}_{\text{saddle (−)}} & \underbrace{(\mathrm{DP}_+,\mathrm{DP}_-)}_{\text{center (CCW)}} \\ \underbrace{(\mathrm{DP}_-,\mathrm{DP}_+)}_{\text{center (CW)}} & \underbrace{(\mathrm{UP}_+,\mathrm{UP}_+)}_{\text{saddle (+)}} \end{array}\right\}. \tag{4.44}$$

- For $a_{210} > 0$ and $a_{120} < 0$,

$$\left\{\begin{array}{cc} (a_{122}, a_{211}) & (a_{122}, a_{212}) \\ (a_{121}, a_{211}) & (a_{121}, a_{212}) \end{array}\right\} = \left\{\begin{array}{cc} \underbrace{(\mathrm{UP}_+,\mathrm{UP}_+)}_{\text{saddle (+)}} & \underbrace{(\mathrm{DP}_-,\mathrm{DP}_+)}_{\text{center (CW)}} \\ \underbrace{(\mathrm{DP}_+,\mathrm{DP}_-)}_{\text{center (CCW)}} & \underbrace{(\mathrm{UP}_-,\mathrm{UP}_-)}_{\text{saddle (−)}} \end{array}\right\}. \tag{4.45}$$

- For $a_{210} < 0$ and $a_{120} < 0$,

$$\left\{\begin{array}{cc} (a_{122}, a_{211}) & (a_{122}, a_{212}) \\ (a_{121}, a_{211}) & (a_{121}, a_{212}) \end{array}\right\} = \left\{\begin{array}{cc} \underbrace{(\mathrm{DP}_+,\mathrm{DP}_-)}_{\text{center (CCW)}} & \underbrace{(\mathrm{UP}_-,\mathrm{UP}_-)}_{\text{saddle (−)}} \\ \underbrace{(\mathrm{UP}_+,\mathrm{UP}_+)}_{\text{saddle (+)}} & \underbrace{(\mathrm{DP}_-,\mathrm{DP}_+)}_{\text{center (CW)}} \end{array}\right\}. \tag{4.46}$$

(B) Consider a quadratic dynamical system with $(2m_1) \times (2n_1 + 1) = 2 \times 3$ equilibriums as

$$\begin{aligned}\dot{x}_1 &= a_{120}(x_2 - a_{121})(x_2 - a_{122}),\\ \dot{x}_2 &= a_{210}(x_1 - a_{211})(x_1 - a_{212})(x_1 - a_{213}).\end{aligned} \tag{4.47}$$

and the first integral manifold is determined by

$$\begin{aligned}&a_{120}\left\{\frac{1}{3}\left[(x_2 - a_{121})^3 - (x_{20} - a_{121})^3\right]\right.\\ &\quad\left.+\frac{1}{2}(a_{121} - a_{122})\left[(x_2 - a_{121})^2 - (x_{20} - a_{121})^2\right]\right\}\\ &= a_{210}\left\{\frac{1}{4}\left[(x_1 - a_{211})^4 - (x_{10} - a_{211})^4\right]\right.\\ &\quad+\frac{1}{3}(2a_{211} - a_{212} - a_{213})\left[(x_1 - a_{211})^3 - (x_{10} - a_{211})^3\right]\\ &\quad\left.+\frac{1}{2}(a_{211} - a_{212})(a_{211} - a_{213})\left[(x_1 - a_{211})^2 - (x_{10} - a_{211})^2\right]\right\}.\end{aligned} \tag{4.48}$$

Phase portraits for 2-dimensional systems with saddles and centers are presented in Fig. 4.2a–d for $(a_{120} > 0, a_{210} > 0)$, $(a_{120} < 0, a_{210} > 0)$, $(a_{120} > 0, a_{210} < 0)$ and $(a_{120} < 0, a_{210} < 0)$. The six equilibriums are three saddles and three centers. There are three limit cycles around the centers and three closed separatrices connected three saddles. The blue curves show the (2,3)-parabola-saddles, which will be discussed in the next chapter. The numbers of saddles and centers are tabulated in Table 4.10.

(i) For $a_{120} > 0$ and $a_{210} > 0$,

$$\begin{Bmatrix}(a_{122}, a_{211}) & (a_{122}, a_{212}) & (a_{122}, a_{213})\\ (a_{121}, a_{211}) & (a_{121}, a_{212}) & (a_{121}, a_{213})\end{Bmatrix} = \begin{Bmatrix}\underbrace{(\mathrm{UP}_+,\mathrm{UP}_+)}_{\text{saddle }(+)} & \underbrace{(\mathrm{DP}_-,\mathrm{DP}_+)}_{\text{center (CW)}} & \underbrace{(\mathrm{UP}_+,\mathrm{UP}_+)}_{\text{saddle }(+)}\\ \underbrace{(\mathrm{DP}_+,\mathrm{DP}_-)}_{\text{center (CCW)}} & \underbrace{(\mathrm{UP}_-,\mathrm{UP}_-)}_{\text{saddle }(-)} & \underbrace{(\mathrm{DP}_+,\mathrm{DP}_-)}_{\text{center (CCW)}}\end{Bmatrix}. \tag{4.49}$$

(ii) For $a_{120} < 0$ and $a_{210} > 0$,

$$\begin{Bmatrix}(a_{122}, a_{211}) & (a_{122}, a_{212}) & (a_{122}, a_{213})\\ (a_{121}, a_{211}) & (a_{121}, a_{212}) & (a_{121}, a_{213})\end{Bmatrix} = \begin{Bmatrix}\underbrace{(\mathrm{DP}_+,\mathrm{DP}_-)}_{\text{center (CCW)}} & \underbrace{(\mathrm{UP}_-,\mathrm{UP}_-)}_{\text{saddle }(-)} & \underbrace{(\mathrm{DP}_+,\mathrm{DP}_-)}_{\text{center (CCW)}}\\ \underbrace{(\mathrm{UP}_+,\mathrm{UP}_+)}_{\text{saddle }(+)} & \underbrace{(\mathrm{DP}_-,\mathrm{DP}_+)}_{\text{center (CW)}} & \underbrace{(\mathrm{UP}_+,\mathrm{UP}_+)}_{\text{saddle }(+)}\end{Bmatrix}. \tag{4.50}$$

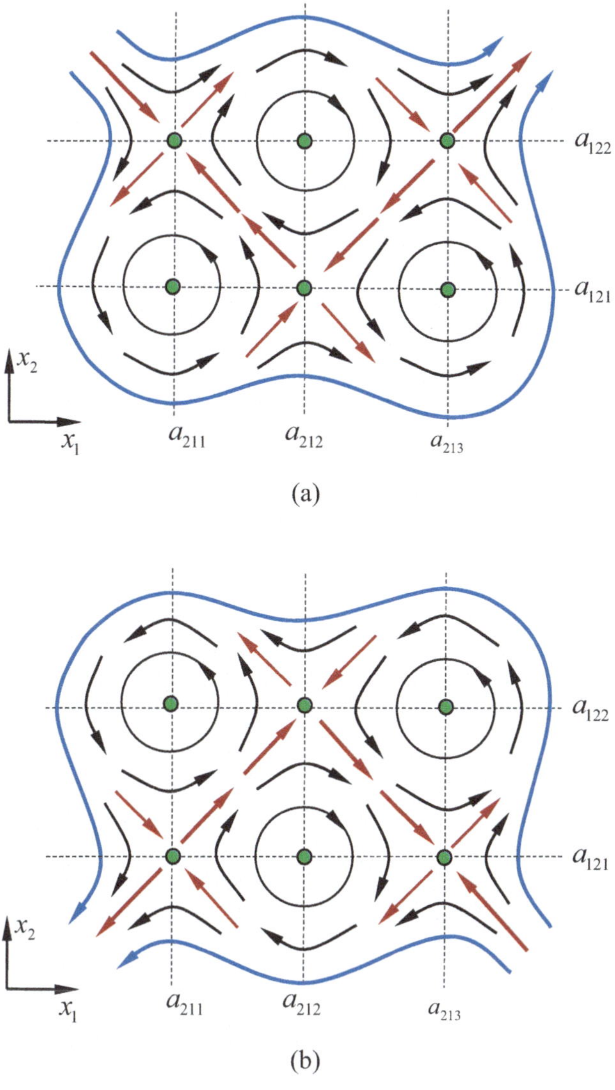

Fig. 4.2 Phase portraits for 2-dimensional systems with $x_1^* = a_{211}, a_{212}, a_{213}$ and $x_2^* = a_{121}, a_{122}$. The four sets of six simple saddle and center: **a** ($a_{120} > 0, a_{210} > 0$), **b** ($a_{120} < 0, a_{210} > 0$), **c** ($a_{120} > 0, a_{210} < 0$), **d** ($a_{120} < 0, a_{210} < 0$)

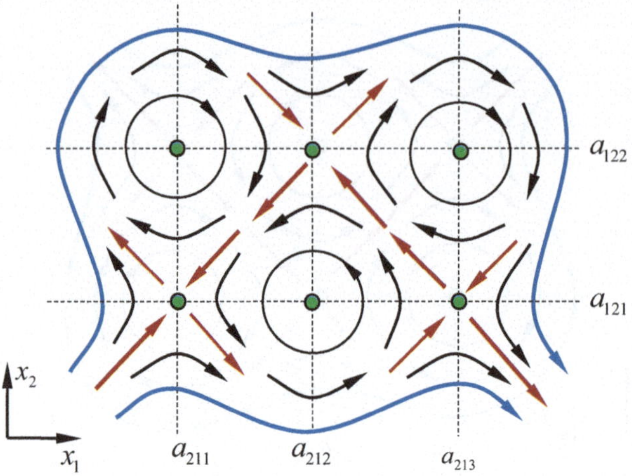

(c)

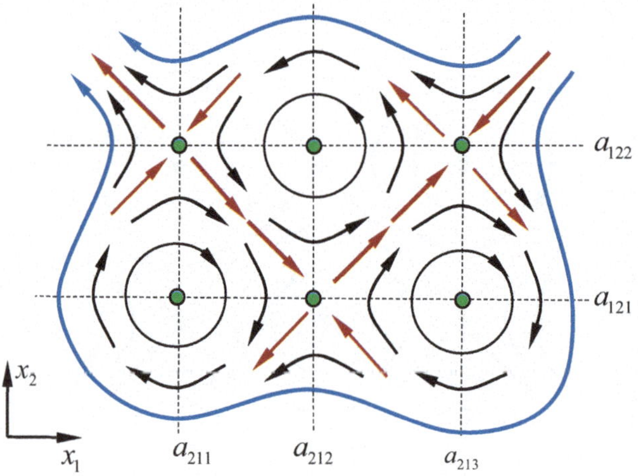

(d)

Fig. 4.2 (continued)

Table 4.10 Numbers of limit cycles and saddles in networks for 2×3-equilibriums

(a_{120}, a_{210})	Numbers			
	(DP_+, DP_-)-center (CCW)	(DP_-, DP_+)-center (CW)	(UP_+, UP_+)-saddle (+)	(UP_-, UP_-)-saddle (−)
$(+, +)$	$m_1 \times (n_1 + 1) = 2$	$m_1 \times n_1 = 1$	$m_1 \times (n_1 + 1) = 2$	$m_1 \times n_1 = 1$
$(-, +)$	$m_1 \times (n_1 + 1) = 2$	$m_1 \times n_1 = 1$	$m_1 \times (n_1 + 1) = 2$	$m_1 \times n_1 = 1$
$(+, -)$	$m_1 \times n_1 = 1$	$m_1 \times (n_1 + 1) = 2$	$m_1 \times n_1 = 1$	$m_1 \times (n_1 + 1) = 2$
$(-, -)$	$m_1 \times n_1 = 1$	$m_1 \times (n_1 + 1) = 2$	$m_1 \times n_1 = 1$	$m_1 \times (n_1 + 1) = 2$

(iii) For $a_{120} > 0$ and $a_{210} < 0$,

$$\begin{Bmatrix} (a_{122}, a_{211}) & (a_{122}, a_{212}) & (a_{122}, a_{213}) \\ (a_{121}, a_{211}) & (a_{121}, a_{212}) & (a_{121}, a_{213}) \end{Bmatrix} = \begin{Bmatrix} \underbrace{(\mathrm{DP}_-,\mathrm{DP}_+)}_{\text{center (CW)}} & \underbrace{(\mathrm{UP}_+,\mathrm{UP}_+)}_{\text{saddle }(+)} & \underbrace{(\mathrm{DP}_-,\mathrm{DP}_+)}_{\text{center (CW)}} \\ \underbrace{(\mathrm{UP}_-,\mathrm{UP}_-)}_{\text{saddle }(-)} & \underbrace{(\mathrm{DP}_+,\mathrm{DP}_-)}_{\text{center (CCW)}} & \underbrace{(\mathrm{UP}_-,\mathrm{UP}_-)}_{\text{saddle }(-)} \end{Bmatrix}. \tag{4.51}$$

(iv) For $a_{120} < 0$ and $a_{210} < 0$,

$$\begin{Bmatrix} (a_{122}, a_{211}) & (a_{122}, a_{212}) & (a_{122}, a_{213}) \\ (a_{121}, a_{211}) & (a_{121}, a_{212}) & (a_{121}, a_{213}) \end{Bmatrix} = \begin{Bmatrix} \underbrace{(\mathrm{UP}_-,\mathrm{UP}_-)}_{\text{saddle }(-)} & \underbrace{(\mathrm{DP}_+,\mathrm{DP}_-)}_{\text{center (CCW)}} & \underbrace{(\mathrm{UP}_-,\mathrm{UP}_-)}_{\text{saddle }(-)} \\ \underbrace{(\mathrm{DP}_-,\mathrm{DP}_+)}_{\text{center (CW)}} & \underbrace{(\mathrm{UP}_+,\mathrm{UP}_+)}_{\text{saddle }(+)} & \underbrace{(\mathrm{DP}_-,\mathrm{DP}_+)}_{\text{center (CW)}} \end{Bmatrix}. \tag{4.52}$$

(C) Consider a cubic dynamical system with $(2m_1 + 1) \times (2n_1) = 3 \times 2$ equilibriums as

$$\begin{aligned} \dot{x}_1 &= a_{120}(x_2 - a_{121})(x_1 - a_{122})(x_1 - a_{123}), \\ \dot{x}_2 &= a_{210}(x_1 - a_{211})(x_1 - a_{212}), \end{aligned} \tag{4.53}$$

and the first integral manifold is determined by

$$\begin{aligned} & a_{210}\Big\{\frac{1}{3}\big[(x_1 - a_{211})^3 - (x_{10} - a_{211})^3\big] \\ &\quad + \frac{1}{2}(a_{211} - a_{212})\big[(x_1 - a_{211})^2 - (x_{10} - a_{211})^2\big]\Big\} \\ &= a_{120}\Big\{\frac{1}{4}\big[(x_2 - a_{121})^4 - (x_{20} - a_{121})^4\big] \\ &\quad + \frac{1}{3}(2a_{121} - a_{122} - a_{123})\big[(x_2 - a_{121})^3 - (x_{20} - a_{121})^3\big] \\ &\quad + \frac{1}{2}(a_{121} - a_{122})(a_{121} - a_{123})\big[(x_2 - a_{121})^2 - (x_{20} - a_{121})^2\big]\Big\}. \end{aligned} \tag{4.54}$$

From the first integral manifold, the corresponding center and saddle networks are given as follows.

(i) For $a_{120} > 0$ and $a_{210} > 0$,

$$\left\{\begin{array}{l}(a_{123},a_{211})\ (a_{123},a_{212})\\(a_{122},a_{211})\ (a_{122},a_{212})\\(a_{121},a_{211})\ (a_{121},a_{212})\end{array}\right\}=\left\{\begin{array}{l}\underbrace{(\mathrm{DP}_-,\mathrm{DP}_+)}_{\text{center (CW)}}\ \underbrace{(\mathrm{UP}_+,\mathrm{UP}_+)}_{\text{saddle }(+)}\\\underbrace{(\mathrm{UP}_-,\mathrm{UP}_-)}_{\text{saddle }(-)}\ \underbrace{(\mathrm{DP}_+,\mathrm{DP}_-)}_{\text{center (CCW)}}\\\underbrace{(\mathrm{DP}_-,\mathrm{DP}_+)}_{\text{center (CW)}}\ \underbrace{(\mathrm{UP}_+,\mathrm{UP}_+)}_{\text{saddle }(+)}\end{array}\right\}. \tag{4.55}$$

(ii) For $a_{120} < 0$ and $a_{210} > 0$,

$$\left\{\begin{array}{l}(a_{123},a_{211})\ (a_{123},a_{212})\\(a_{122},a_{211})\ (a_{122},a_{212})\\(a_{121},a_{211})\ (a_{121},a_{212})\end{array}\right\}=\left\{\begin{array}{l}\underbrace{(\mathrm{UP}_-,\mathrm{UP}_-)}_{\text{saddle }(-)}\ \underbrace{(\mathrm{DP}_+,\mathrm{DP}_-)}_{\text{center (CCW)}}\\\underbrace{(\mathrm{DP}_-,\mathrm{DP}_+)}_{\text{center (CW)}}\ \underbrace{(\mathrm{UP}_+,\mathrm{UP}_+)}_{\text{saddle }(+)}\\\underbrace{(\mathrm{UP}_-,\mathrm{UP}_-)}_{\text{saddle }(-)}\ \underbrace{(\mathrm{DP}_+,\mathrm{DP}_-)}_{\text{center (CCW)}}\end{array}\right\}. \tag{4.56}$$

(iii) For $a_{120} > 0$ and $a_{210} < 0$,

$$\left\{\begin{array}{l}(a_{123},a_{211})\ (a_{123},a_{212})\\(a_{122},a_{211})\ (a_{122},a_{212})\\(a_{121},a_{211})\ (a_{121},a_{212})\end{array}\right\}=\left\{\begin{array}{l}\underbrace{(\mathrm{UP}_+,\mathrm{UP}_+)}_{\text{saddle }(+)}\ \underbrace{(\mathrm{DP}_-,\mathrm{DP}_+)}_{\text{center (CW)}}\\\underbrace{(\mathrm{DP}_+,\mathrm{DP}_-)}_{\text{center (CCW)}}\ \underbrace{(\mathrm{UP}_-,\mathrm{UP}_-)}_{\text{saddle }(-)}\\\underbrace{(\mathrm{UP}_+,\mathrm{UP}_+)}_{\text{saddle }(+)}\ \underbrace{(\mathrm{DP}_-,\mathrm{DP}_+)}_{\text{center (CW)}}\end{array}\right\}. \tag{4.57}$$

(iv) For $a_{120} < 0$ and $a_{210} < 0$,

$$\left\{\begin{array}{l}(a_{123},a_{211})\ (a_{123},a_{212})\\(a_{122},a_{211})\ (a_{122},a_{212})\\(a_{121},a_{211})\ (a_{121},a_{212})\end{array}\right\}=\left\{\begin{array}{l}\underbrace{(\mathrm{DP}_+,\mathrm{DP}_-)}_{\text{center (CCW)}}\ \underbrace{(\mathrm{UP}_-,\mathrm{UP}_-)}_{\text{saddle }(-)}\\\underbrace{(\mathrm{UP}_+,\mathrm{UP}_+)}_{\text{saddle }(+)}\ \underbrace{(\mathrm{DP}_-,\mathrm{DP}_+)}_{\text{center (CW)}}\\\underbrace{(\mathrm{DP}_+,\mathrm{DP}_-)}_{\text{center (CCW)}}\ \underbrace{(\mathrm{UP}_-,\mathrm{UP}_-)}_{\text{saddle }(-)}\end{array}\right\}. \tag{4.58}$$

Phase portraits for 2-dimensional systems with the saddles and centers are presented in Fig. 4.3a–d for ($a_{120} > 0, a_{210} > 0$), ($a_{120} < 0, a_{210} > 0$), ($a_{120} > 0$, $a_{210} < 0$) and ($a_{120} < 0, a_{210} < 0$). The six equilibriums are three saddles and three centers. There are three limit cycles around the centers and three closed separatrices connected three saddles. The blue curves show the (3,2)-parabola-saddles, which will be discussed in the next chapter. The numbers of saddles and centers are tabulated in Table 4.11.

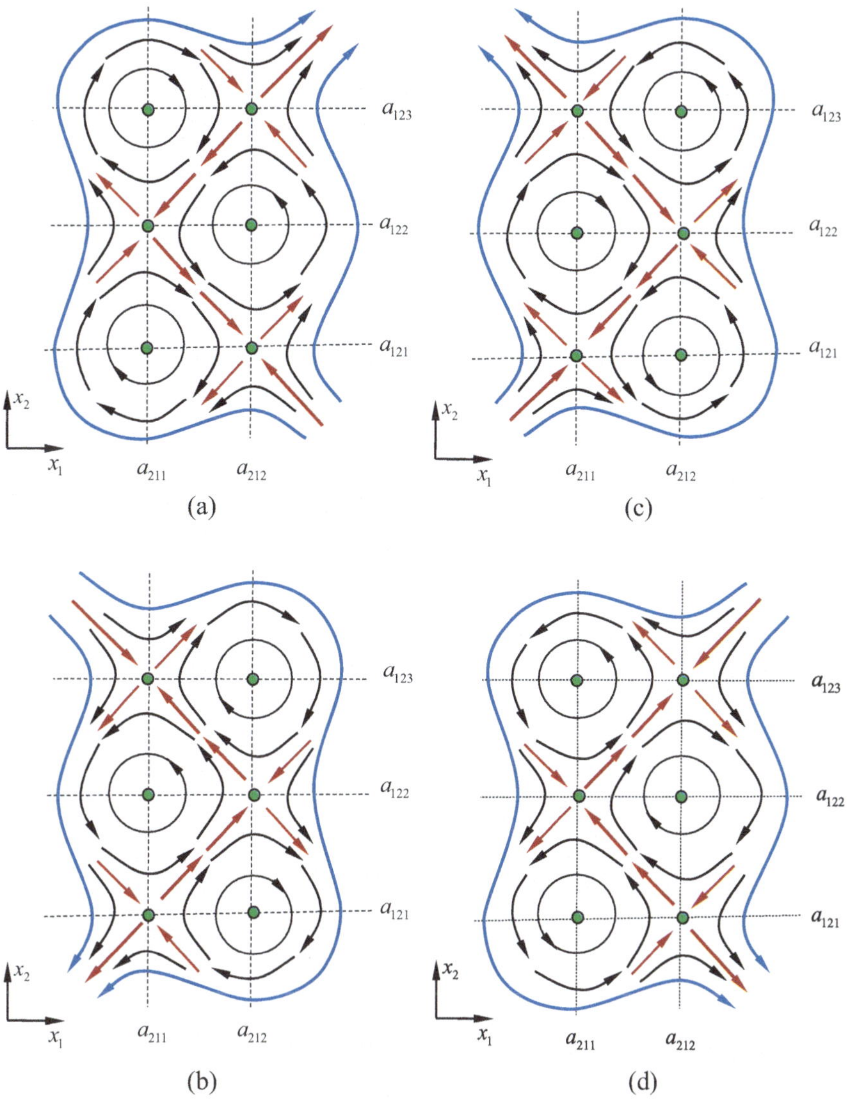

Fig. 4.3 Phase portraits for 2-dimensional systems with $x_1^* = a_{211}, a_{212}$ and $x_2^* = a_{121}, a_{122}, a_{123}$. The four sets of six simple saddle and center: **a** ($a_{120} > 0, a_{210} > 0$), **b** ($a_{120} < 0, a_{210} > 0$), **c** ($a_{120} > 0, a_{210} < 0$), **d** ($a_{120} < 0, a_{210} < 0$)

Table 4.11 Numbers of limit cycles and saddles in networks for 3×2-equilibriums

(a_{120}, a_{210})	Numbers			
	$(\mathrm{DP}_+, \mathrm{DP}_-)$-center (CCW)	$(\mathrm{DP}_-, \mathrm{DP}_+)$-center (CW)	$(\mathrm{UP}_+, \mathrm{UP}_+)$-saddle (+)	$(\mathrm{UP}_-, \mathrm{UP}_-)$-saddle (−)
$(+, +)$	$m_1 \times n_1 = 1$	$(m_1 + 1) \times n_1 = 2$	$(m_1 + 1) \times n_1 = 2$	$m_1 \times n_1 = 1$
$(-, +)$	$(m_1 + 1) \times n_1 = 2$	$m_1 \times n_1 = 1$	$m_1 \times n_1 = 1$	$(m_1 + 1) \times n_1 = 2$
$(+, -)$	$m_1 \times n_1 = 1$	$(m_1 + 1) \times n_1 = 2$	$(m_1 + 1) \times n_1 = 2$	$m_1 \times n_1 = 1$
$(-, -)$	$(m_1 + 1) \times n_1 = 2$	$m_1 \times n_1 = 1$	$m_1 \times n_1 = 1$	$(m_1 + 1) \times n_1 = 2$

(D) Consider a cubic dynamical system with $(2m_1 + 1) \times (2n_1 + 1) = 3 \times 3$ equilibriums as

$$\begin{aligned}\dot{x}_1 &= a_{120}(x_2 - a_{121})(x_2 - a_{122})(x_2 - a_{123}),\\ \dot{x}_2 &= a_{210}(x_1 - a_{211})(x_1 - a_{212})(x_1 - a_{213}),\end{aligned} \tag{4.59}$$

and the first integral manifold is determined by

$$\begin{aligned}
&a_{120}\Big\{\frac{1}{4}\big[(x_2 - a_{121})^4 - (x_{20} - a_{121})^4\big]\\
&\quad + \frac{1}{3}(2a_{121} - a_{122} - a_{123})\big[(x_2 - a_{121})^3 - (x_{20} - a_{121})^3\big]\\
&\quad + \frac{1}{2}(a_{121} - a_{122})(a_{121} - a_{123})\big[(x_2 - a_{121})^2 - (x_{20} - a_{121})^2\big]\Big\}\\
&= a_{210}\Big\{\frac{1}{4}\big[(x_1 - a_{211})^4 - (x_{10} - a_{211})^4\big]\\
&\quad + \frac{1}{3}(2a_{211} - a_{212} - a_{213})\big[(x_1 - a_{211})^3 - (x_{10} - a_{211})^3\big]\\
&\quad + \frac{1}{2}(a_{211} - a_{212})(a_{211} - a_{213})\big[(x_1 - a_{211})^2 - (x_{10} - a_{211})^2\big]\Big\}.
\end{aligned} \tag{4.60}$$

Phase portraits for 2-dimensional systems with the saddles and centers are presented in Fig. 4.4a–d for $(a_{120} > 0, a_{210} > 0)$, $(a_{120} < 0, a_{210} > 0)$, $(a_{120} > 0, a_{210} < 0)$ and $(a_{120} < 0, a_{210} < 0)$. The six equilibriums are three saddles and three centers. There are three limit cycles around the centers and three closed separatrices connected three saddles. The blue curves show the (3,3)-clockwise center, (3,3)-counter-clockwise centers, (3,3)-positive saddle, (3,3)-negative saddle. Such center and saddle bifurcations will be discussed in the next chapter. The numbers of saddles and centers are tabulated in Table 4.12. For the (3,3)-centers, there are six centers. For the (3,3)-saddles, there are six saddles. The equilibrium properties in the homoclinic networks are listed as follows.

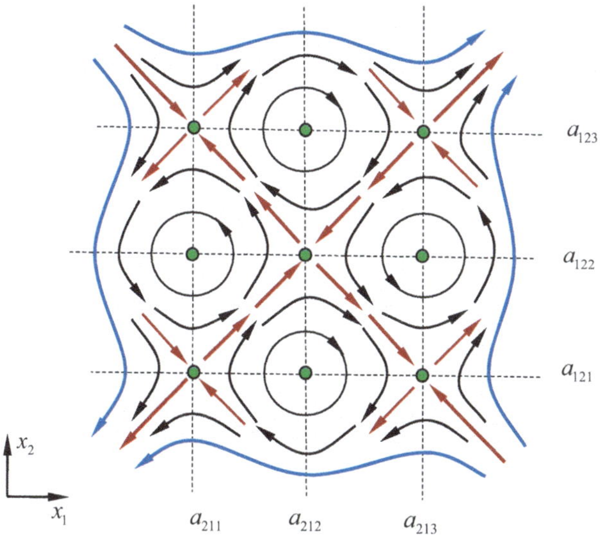

(a)

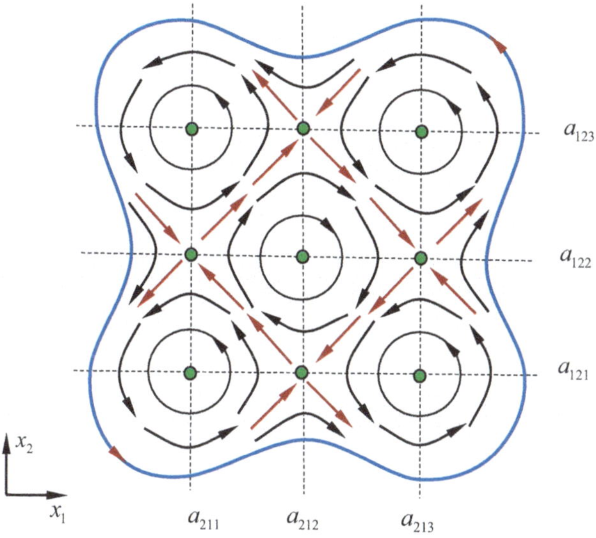

(b)

Fig. 4.4 Phase portraits for 2-dimensional systems with $x_1^* = a_{211}, a_{212}, a_{213}$ and $x_2^* = a_{121}, a_{122}, a_{123}$. The four sets of nine simple saddle and center: **a** ($a_{120} > 0, a_{210} > 0$), **b** ($a_{120} < 0, a_{210} > 0$), **c** ($a_{120} > 0, a_{210} < 0$), **d** ($a_{120} < 0, a_{210} < 0$)

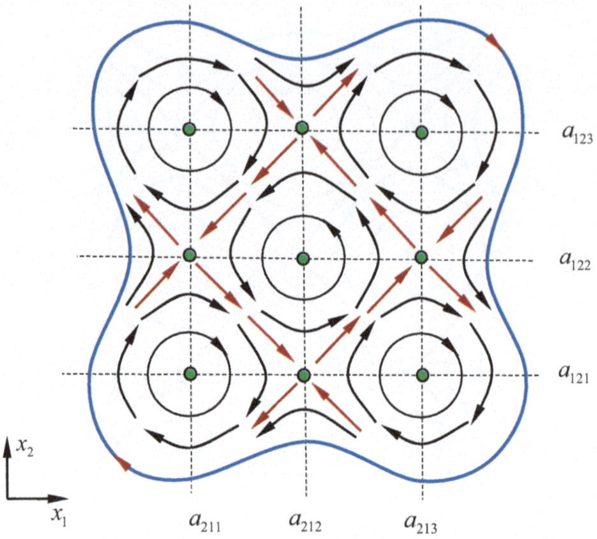

(c)

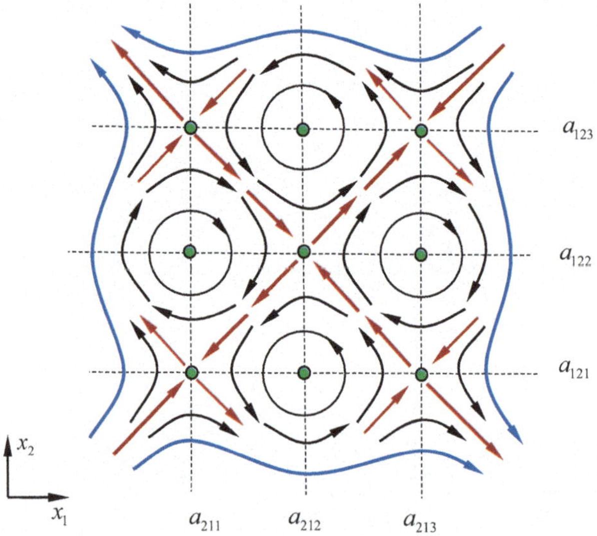

(d)

Fig. 4.4 (continued)

Table 4.12 Numbers of limit cycles and saddles in networks for 3×3-equilibriums

(a_{120}, a_{210})	Numbers			
	$(\mathrm{DP}_+, \mathrm{DP}_-)$-center (CCW)	$(\mathrm{DP}_-, \mathrm{DP}_+)$-center (CW)	$(\mathrm{UP}_+, \mathrm{UP}_+)$-saddle (+)	$(\mathrm{UP}_-, \mathrm{UP}_-)$-saddle (–)
$(+, +)$	$m_1 \times (n_1 + 1) = 2$	$(m_1 + 1) \times n_1 = 2$	$(m_1 + 1) \times (n_1 + 1) + 1 = 5$	$m_1 \times n_1 = 1$
$(-, +)$	$(m_1 + 1) \times (n_1 + 1) + 1 = 5$	$m_1 \times n_1 = 1$	$m_1 \times (n_1 + 1) = 2$	$(m_1 + 1) \times n_1 = 2$
$(+, -)$	$m_1 \times n_1 = 1$	$(m_1 + 1) \times (n_1 + 1) + 1 = 5$	$(m_1 + 1) \times n_1 = 2$	$m_1 \times (n_1 + 1) = 2$
$(-, -)$	$(m_1 + 1) \times n_1 = 2$	$m_1 \times (n_1 + 1) = 2$	$m_1 \times n_1 = 1$	$(m_1 + 1) \times (n_1 + 1) + 1 = 5$

(i) For $a_{120} > 0$ and $a_{210} > 0$,

$$\bigcup_{s=1}^{3}\bigcup_{l=1}^{3}\underbrace{(a_{j_1j_2s}, a_{j_2j_1l})}_{\mathrm{XX}} = \left\{\begin{array}{ccc} \underbrace{(\mathrm{UP}_+,\mathrm{UP}_+)}_{\text{saddle }(+)} & \underbrace{(\mathrm{DP}_-,\mathrm{DP}_+)}_{\text{center (CW)}} & \underbrace{(\mathrm{UP}_+,\mathrm{UP}_+)}_{\text{saddle }(+)} \\ \underbrace{(\mathrm{DP}_+,\mathrm{DP}_-)}_{\text{center (CCW)}} & \underbrace{(\mathrm{UP}_-,\mathrm{UP}_-)}_{\text{saddle }(-)} & \underbrace{(\mathrm{DP}_+,\mathrm{DP}_-)}_{\text{center (CCW)}} \\ \underbrace{(\mathrm{UP}_+,\mathrm{UP}_+)}_{\text{saddle }(+)} & \underbrace{(\mathrm{DP}_-,\mathrm{DP}_+)}_{\text{center (CW)}} & \underbrace{(\mathrm{UP}_+,\mathrm{UP}_+)}_{\text{saddle }(+)} \end{array}\right\} \tag{4.61}$$

(ii) For $a_{120} < 0$ and $a_{210} > 0$,

$$\bigcup_{s=1}^{3}\bigcup_{l=1}^{3}\underbrace{(a_{j_1j_2s}, a_{j_2j_1l})}_{\mathrm{XX}} = \left\{\begin{array}{ccc} \underbrace{(\mathrm{DP}_+,\mathrm{DP}_-)}_{\text{center (CCW)}} & \underbrace{(\mathrm{UP}_-,\mathrm{UP}_-)}_{\text{saddle }(-)} & \underbrace{(\mathrm{DP}_+,\mathrm{DP}_-)}_{\text{center (CCW)}} \\ \underbrace{(\mathrm{UP}_+,\mathrm{UP}_+)}_{\text{saddle }(+)} & \underbrace{(\mathrm{DP}_-,\mathrm{DP}_+)}_{\text{center (CW)}} & \underbrace{(\mathrm{UP}_+,\mathrm{UP}_+)}_{\text{saddle }(+)} \\ \underbrace{(\mathrm{DP}_+,\mathrm{DP}_-)}_{\text{center (CCW)}} & \underbrace{(\mathrm{UP}_-,\mathrm{UP}_-)}_{\text{saddle }(-)} & \underbrace{(\mathrm{DP}_+,\mathrm{DP}_-)}_{\text{center (CCW)}} \end{array}\right\}. \tag{4.62}$$

(iii) For $a_{120} > 0$ and $a_{210} < 0$,

$$\bigcup_{s=1}^{3}\bigcup_{l=1}^{3}\underbrace{(a_{j_1j_2s},a_{j_2j_1l})}_{\text{XX}}=\left\{\begin{array}{ccc}\underbrace{(\text{DP}_-,\text{DP}_+)}_{\text{center (CW)}} & \underbrace{(\text{UP}_+,\text{UP}_+)}_{\text{saddle (+)}} & \underbrace{(\text{DP}_-,\text{DP}_+)}_{\text{center (CW)}}\\ \underbrace{(\text{UP}_-,\text{UP}_-)}_{\text{saddle (−)}} & \underbrace{(\text{DP}_+,\text{DP}_-)}_{\text{center (CCW)}} & \underbrace{(\text{UP}_-,\text{UP}_-)}_{\text{saddle (−)}}\\ \underbrace{(\text{DP}_-,\text{DP}_+)}_{\text{center (CW)}} & \underbrace{(\text{UP}_+,\text{UP}_+)}_{\text{saddle (+)}} & \underbrace{(\text{DP}_-,\text{DP}_+)}_{\text{center (CW)}}\end{array}\right\}.$$

(iv) For $a_{120} < 0$ and $a_{210} < 0$,

$$\bigcup_{s=1}^{3}\bigcup_{l=1}^{3}\underbrace{(a_{j_1j_2s},a_{j_2j_1l})}_{\text{XX}}=\left\{\begin{array}{ccc}\underbrace{(\text{UP}_-,\text{UP}_-)}_{\text{saddle (−)}} & \underbrace{(\text{DP}_+,\text{DP}_-)}_{\text{center (CCW)}} & \underbrace{(\text{UP}_-,\text{UP}_-)}_{\text{saddle (−)}}\\ \underbrace{(\text{DP}_-,\text{DP}_+)}_{\text{center (CW)}} & \underbrace{(\text{UP}_+,\text{UP}_+)}_{\text{saddle (+)}} & \underbrace{(\text{DP}_-,\text{DP}_+)}_{\text{center (CW)}}\\ \underbrace{(\text{UP}_-,\text{UP}_-)}_{\text{saddle (−)}} & \underbrace{(\text{DP}_+,\text{DP}_-)}_{\text{center (CCW)}} & \underbrace{(\text{UP}_-,\text{UP}_-)}_{\text{saddle (−)}}\end{array}\right\}. \tag{4.63}$$

(E) Consider a dynamical system with $(2m_1+1)\times(2n_1)=3\times 4$ equilibriums as

$$\begin{aligned}\dot{x}_1&=a_{120}(x_2-a_{121})(x_2-a_{122})(x_2-a_{123}),\\ \dot{x}_2&=a_{210}(x_1-a_{211})(x_1-a_{212})(x_1-a_{213})(x_1-a_{214}),\end{aligned} \tag{4.64}$$

and the first integral manifold is determined by

$$\begin{aligned}&a_{120}\Big\{\frac{1}{4}\big[(x_2-a_{12s_1})^4-(x_{20}-a_{12s_1})^4\big]\\ &\quad+\frac{1}{3}\sum_{s_2=1,s_2\neq s_1}^{3}(a_{12s_1}-a_{12s_2})\big[(x_2-a_{121})^3-(x_{20}-a_{121})^3\big]\\ &\quad+\frac{1}{2}\prod\nolimits_{s_2=1,s_3\neq s_1}^{3}(a_{12s_1}-a_{12s_2})\big[(x_2-a_{12s_1})^2-(x_{20}-a_{12s_1})^2\big]\Big\}\\ &=a_{210}\Big\{\frac{1}{5}\big[(x_1-a_{21l_1})^5-(x_{10}-a_{21l_1})^5\big]\\ &\quad+\frac{1}{4}\sum_{l_2=2,l_2\neq l_1}^{4}(a_{21l_1}-a_{21l_2})\big[(x_1-a_{21l_1})^4-(x_{10}-a_{21l_1})^4\big]\\ &\quad+\frac{1}{3}\sum_{\substack{l_2,l_3=1,l_2,l_3\neq l_1\\(l_2<l_3)}}^{4}\prod\nolimits_{r=l_2}^{l_3}(a_{21l_1}-a_{12r})\big[(x_1-a_{21l_1})^3-(x_{10}-a_{21l_1})^3\big]\\ &\quad+\frac{1}{2}\prod\nolimits_{l_2=2,l_2\neq l_1}^{4}(a_{21l_1}-a_{21l_2})\big[(x_1-a_{21l_1})^2-(x_{10}-a_{21l_1})^2\big]\Big\}.\end{aligned} \tag{4.65}$$

The equilibrium properties in the homoclinic networks are listed as follows.

(i) For $a_{120} > 0$ and $a_{210} > 0$,

$$\bigcup_{s=1}^{4}\bigcup_{l=1}^{3}\underbrace{(a_{j_1j_2s}, a_{j_2j_1l})}_{\mathrm{XX}} = \left\{\begin{matrix} \underbrace{(\mathrm{DP}_-,\mathrm{DP}_+)}_{\text{center (CW)}} & \underbrace{(\mathrm{UP}_+,\mathrm{UP}_+)}_{\text{saddle }(+)} & \underbrace{(\mathrm{DP}_-,\mathrm{DP}_+)}_{\text{center (CW)}} & \underbrace{(\mathrm{UP}_+,\mathrm{UP}_+)}_{\text{saddle }(+)} \\ \underbrace{(\mathrm{UP}_-,\mathrm{UP}_-)}_{\text{saddle }(-)} & \underbrace{(\mathrm{DP}_+,\mathrm{DP}_-)}_{\text{center (CCW)}} & \underbrace{(\mathrm{UP}_-,\mathrm{UP}_-)}_{\text{saddle }(-)} & \underbrace{(\mathrm{DP}_+,\mathrm{DP}_-)}_{\text{center (CCW)}} \\ \underbrace{(\mathrm{DP}_-,\mathrm{DP}_+)}_{\text{center (CW)}} & \underbrace{(\mathrm{UP}_+,\mathrm{UP}_+)}_{\text{saddle }(+)} & \underbrace{(\mathrm{DP}_-,\mathrm{DP}_+)}_{\text{center (CW)}} & \underbrace{(\mathrm{UP}_+,\mathrm{UP}_+)}_{\text{saddle }(+)} \end{matrix}\right\}. \tag{4.66}$$

(ii) For $a_{120} < 0$ and $a_{210} > 0$,

$$\bigcup_{s=1}^{4}\bigcup_{l=1}^{3}\underbrace{(a_{j_1j_2s}, a_{j_2j_1l})}_{\mathrm{XX}} = \left\{\begin{matrix} \underbrace{(\mathrm{UP}_-,\mathrm{UP}_-)}_{\text{saddle }(-)} & \underbrace{(\mathrm{DP}_+,\mathrm{DP}_-)}_{\text{center (CCW)}} & \underbrace{(\mathrm{UP}_-,\mathrm{UP}_-)}_{\text{saddle }(-)} & \underbrace{(\mathrm{DP}_+,\mathrm{DP}_-)}_{\text{center (CCW)}} \\ \underbrace{(\mathrm{DP}_-,\mathrm{DP}_+)}_{\text{center (CW)}} & \underbrace{(\mathrm{UP}_+,\mathrm{UP}_+)}_{\text{saddle }(+)} & \underbrace{(\mathrm{DP}_-,\mathrm{DP}_+)}_{\text{center (CW)}} & \underbrace{(\mathrm{UP}_+,\mathrm{UP}_+)}_{\text{saddle }(+)} \\ \underbrace{(\mathrm{UP}_-,\mathrm{UP}_-)}_{\text{saddle }(-)} & \underbrace{(\mathrm{DP}_+,\mathrm{DP}_-)}_{\text{center (CCW)}} & \underbrace{(\mathrm{UP}_-,\mathrm{UP}_-)}_{\text{saddle }(-)} & \underbrace{(\mathrm{DP}_+,\mathrm{DP}_-)}_{\text{center (CCW)}} \end{matrix}\right\}. \tag{4.67}$$

(iii) For $a_{120} > 0$ and $a_{210} < 0$,

$$\bigcup_{s=1}^{4}\bigcup_{l=1}^{3}\underbrace{(a_{j_1j_2s}, a_{j_2j_1l})}_{\mathrm{XX}} = \left\{\begin{matrix} \underbrace{(\mathrm{UP}_+,\mathrm{UP}_+)}_{\text{saddle }(+)} & \underbrace{(\mathrm{DP}_-,\mathrm{DP}_+)}_{\text{center (CW)}} & \underbrace{(\mathrm{UP}_+,\mathrm{UP}_+)}_{\text{saddle }(+)} & \underbrace{(\mathrm{DP}_-,\mathrm{DP}_+)}_{\text{center (CW)}} \\ \underbrace{(\mathrm{DP}_+,\mathrm{DP}_-)}_{\text{center (CCW)}} & \underbrace{(\mathrm{UP}_-,\mathrm{UP}_-)}_{\text{saddle }(-)} & \underbrace{(\mathrm{DP}_+,\mathrm{DP}_-)}_{\text{center (CCW)}} & \underbrace{(\mathrm{UP}_-,\mathrm{UP}_-)}_{\text{saddle }(-)} \\ \underbrace{(\mathrm{UP}_+,\mathrm{UP}_+)}_{\text{saddle }(+)} & \underbrace{(\mathrm{DP}_-,\mathrm{DP}_+)}_{\text{center (CW)}} & \underbrace{(\mathrm{UP}_+,\mathrm{UP}_+)}_{\text{saddle }(+)} & \underbrace{(\mathrm{DP}_-,\mathrm{DP}_+)}_{\text{center (CW)}} \end{matrix}\right\}. \tag{4.68}$$

(iv) For $a_{120} < 0$ and $a_{210} < 0$,

$$\bigcup_{s=1}^{4}\bigcup_{l=1}^{3}\underbrace{(a_{j_1j_2s}, a_{j_2j_1l})}_{\mathrm{XX}} = \left\{\begin{matrix} \underbrace{(\mathrm{DP}_+,\mathrm{DP}_-)}_{\text{center (CCW)}} & \underbrace{(\mathrm{UP}_-,\mathrm{UP}_-)}_{\text{saddle }(-)} & \underbrace{(\mathrm{DP}_+,\mathrm{DP}_-)}_{\text{center (CCW)}} & \underbrace{(\mathrm{UP}_-,\mathrm{UP}_-)}_{\text{saddle }(-)} \\ \underbrace{(\mathrm{UP}_+,\mathrm{UP}_+)}_{\text{saddle }(+)} & \underbrace{(\mathrm{DP}_-,\mathrm{DP}_+)}_{\text{center (CW)}} & \underbrace{(\mathrm{UP}_+,\mathrm{UP}_+)}_{\text{saddle }(+)} & \underbrace{(\mathrm{DP}_-,\mathrm{DP}_+)}_{\text{center (CW)}} \\ \underbrace{(\mathrm{DP}_+,\mathrm{DP}_-)}_{\text{center (CCW)}} & \underbrace{(\mathrm{UP}_-,\mathrm{UP}_-)}_{\text{saddle }(-)} & \underbrace{(\mathrm{DP}_+,\mathrm{DP}_-)}_{\text{center (CCW)}} & \underbrace{(\mathrm{UP}_-,\mathrm{UP}_-)}_{\text{saddle }(-)} \end{matrix}\right\}. \tag{4.69}$$

Phase portraits for 2-dimensional systems with the saddles and centers are presented in Fig. 4.5a–d for ($a_{120} > 0, a_{210} > 0$), ($a_{120} < 0, a_{210} > 0$), ($a_{120} > 0$, $a_{210} < 0$) and ($a_{120} < 0, a_{210} < 0$). The twelve equilibriums are six saddles and six centers. There are six limit cycles around the centers and six closed separatrices connected six saddles. The blue curves show the (3,4)-parabola-saddles, which will be discussed in the next chapter. The numbers of saddles and centers are tabulated in Table 4.13.

(F) Consider a quartic dynamical system with $(2m_1) \times (2n_1 + 1) = 4 \times 3$ equilibriums as

$$\begin{aligned}\dot{x}_1 &= a_{120}(x_2 - a_{121})(x_2 - a_{122})(x_2 - a_{123})(x_2 - a_{124}),\\ \dot{x}_2 &= a_{210}(x_1 - a_{211})(x_1 - a_{212})(x_1 - a_{213}),\end{aligned} \tag{4.70}$$

and the first integral manifold is determined by

$$\begin{aligned}
&a_{120}\Big\{\frac{1}{5}\big[(x_2 - a_{12s_1})^5 - (x_{20} - a_{12s_1})^5\big]\\
&+\frac{1}{4}\sum_{s_2=1, s_2\neq s_1}^{4}(a_{12s_1} - a_{12s_2})\big[(x_2 - a_{121})^4 - (x_{20} - a_{121})^4\big]\\
&+\frac{1}{3}\sum_{\substack{s_2,s_3=1,s_2,s_3\neq s_1\\(s_2<s_3)}}^{4}\prod_{r=s_2}^{s_3}(a_{12s_1} - a_{12r})\big[(x_2 - a_{12s_1})^3 - (x_{20} - a_{12s_1})^3\big]\\
&+\frac{1}{2}\prod_{s_2=1,s_3\neq s_1}^{4}(a_{12s_1} - a_{12s_2})\big[(x_2 - a_{12s_1})^2 - (x_{20} - a_{12s_1})^2\big]\Big\}\\
&= a_{210}\Big\{\frac{1}{4}\big[(x_1 - a_{21l_1})^4 - (x_{10} - a_{21l_1})^4\big]\\
&+\frac{1}{3}\sum_{l_2=2,l_2\neq l_1}^{3}(a_{21l_1} - a_{21l_2})\big[(x_1 - a_{21l_1})^3 - (x_{10} - a_{21l_1})^3\big]\\
&+\frac{1}{2}\prod_{l_2=2,l_2\neq l_1}^{3}(a_{21l_1} - a_{21l_2})\big[(x_1 - a_{21l_1})^2 - (x_{10} - a_{21l_1})^2\big]\Big\}.
\end{aligned} \tag{4.71}$$

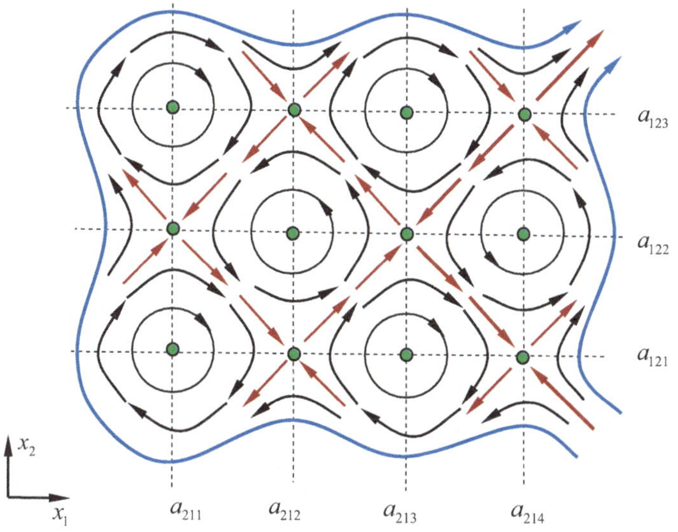

(a)

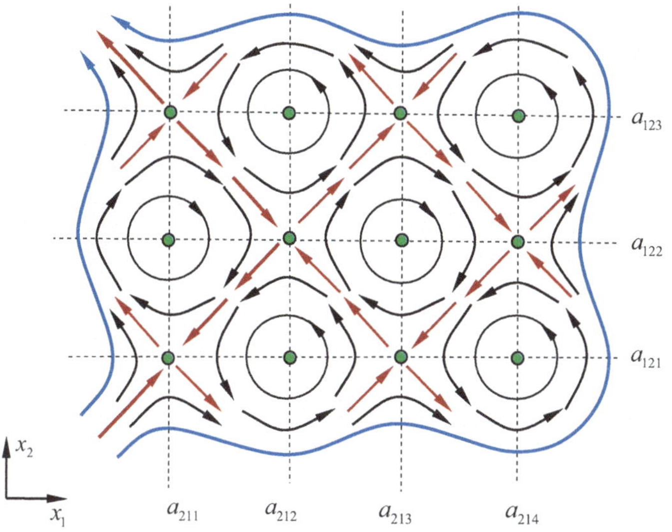

(b)

Fig. 4.5 Phase portraits for 2-dimensional systems with $x_1^* = a_{211}, a_{212}, a_{213}, a_{214}$ and $x_2^* = a_{121}, a_{122}, a_{123}$. The four sets of nine simple saddle and center: **a** ($a_{120} > 0, a_{210} > 0$), **b** ($a_{120} < 0, a_{210} > 0$), **c** ($a_{120} > 0, a_{210} < 0$), **d** ($a_{120} < 0, a_{210} < 0$)

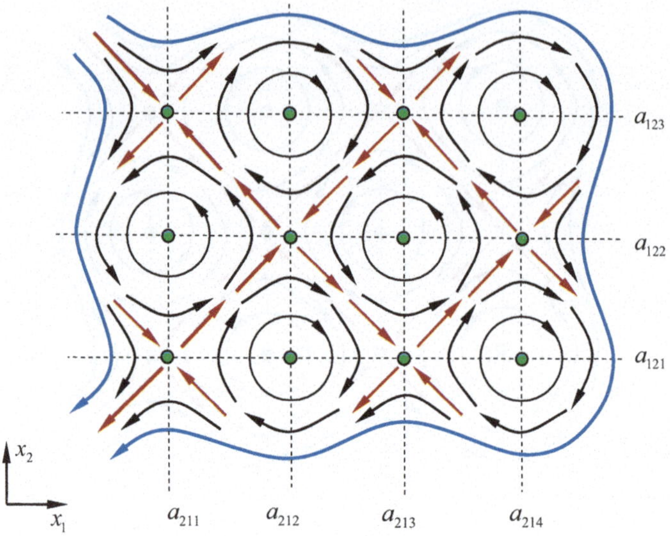

(c)

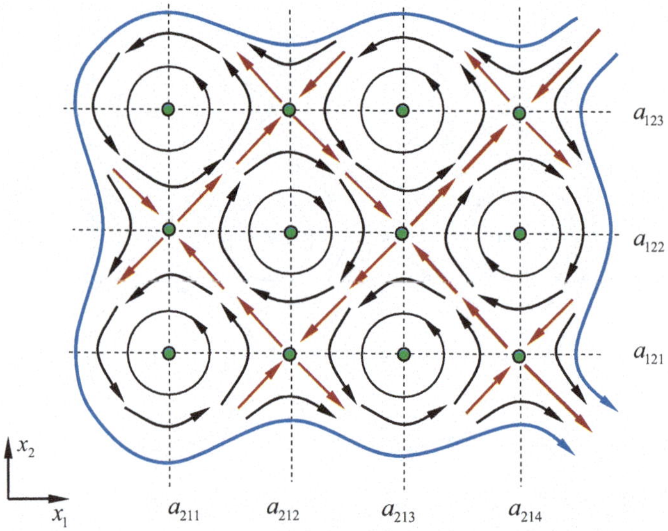

(d)

Fig. 4.5 (continued)

Table 4.13 Numbers of limit cycles and saddles in networks for 3 × 4-equilibriums

(a_{120}, a_{210})	Numbers			
	$(\mathrm{DP}_+, \mathrm{DP}_-)$-center (CCW)	$(\mathrm{DP}_-, \mathrm{DP}_+)$-center (CW)	$(\mathrm{UP}_+, \mathrm{UP}_+)$-saddle (+)	$(\mathrm{UP}_-, \mathrm{UP}_-)$-saddle (−)
$(+, +)$	$m_1 \times n_1 = 2$	$(m_1 + 1) \times n_1 = 4$	$(m_1 + 1) \times n_1 = 4$	$m_1 \times n_1 = 2$
$(-, +)$	$(m_1 + 1) \times n_1 = 4$	$m_1 \times n_1 = 2$	$m_1 \times n_1 = 2$	$(m_1 + 1) \times n_1 = 4$
$(+, -)$	$m_1 \times n_1 = 2$	$(m_1 + 1) \times n_1 = 4$	$(m_1 + 1) \times n_1 = 4$	$m_1 \times n_1 = 2$
$(-, -)$	$(m_1 + 1) \times n_1 = 4$	$m_1 \times n_1 = 2$	$m_1 \times n_1 = 2$	$(m_1 + 1) \times n_1 = 4$

The equilibrium properties in the homoclinic networks are listed as follows.

(i) For $a_{120} > 0$ and $a_{210} > 0$,

$$\bigcup_{s=1}^{3}\bigcup_{l=1}^{4}\underbrace{(a_{j_1j_2s}, a_{j_2j_1l})}_{\mathrm{XX}} = \left\{\begin{matrix} \underbrace{(\mathrm{UP}_+,\mathrm{UP}_+)}_{\text{saddle }(+)} & \underbrace{(\mathrm{DP}_-,\mathrm{DP}_+)}_{\text{center (CW)}} & \underbrace{(\mathrm{UP}_+,\mathrm{UP}_+)}_{\text{saddle }(+)} \\ \underbrace{(\mathrm{DP}_+,\mathrm{DP}_-)}_{\text{center (CCW)}} & \underbrace{(\mathrm{UP}_-,\mathrm{UP}_-)}_{\text{saddle }(-)} & \underbrace{(\mathrm{DP}_+,\mathrm{DP}_-)}_{\text{center (CCW)}} \\ \underbrace{(\mathrm{UP}_+,\mathrm{UP}_+)}_{\text{saddle }(+)} & \underbrace{(\mathrm{DP}_-,\mathrm{DP}_+)}_{\text{center (CW)}} & \underbrace{(\mathrm{UP}_+,\mathrm{UP}_+)}_{\text{saddle }(+)} \\ \underbrace{(\mathrm{DP}_+,\mathrm{DP}_-)}_{\text{center (CCW)}} & \underbrace{(\mathrm{UP}_-,\mathrm{UP}_-)}_{\text{saddle }(-)} & \underbrace{(\mathrm{DP}_+,\mathrm{DP}_-)}_{\text{center (CCW)}} \end{matrix}\right\}. \tag{4.72}$$

(ii) For $a_{120} < 0$ and $a_{210} > 0$,

$$\bigcup_{s=1}^{3}\bigcup_{l=1}^{4}\underbrace{(a_{j_1j_2s}, a_{j_2j_1l})}_{\mathrm{XX}} = \left\{\begin{matrix} \underbrace{(\mathrm{DP}_+,\mathrm{DP}_-)}_{\text{center (CCW)}} & \underbrace{(\mathrm{UP}_-,\mathrm{UP}_-)}_{\text{saddle }(-)} & \underbrace{(\mathrm{DP}_+,\mathrm{DP}_-)}_{\text{center (CCW)}} \\ \underbrace{(\mathrm{UP}_+,\mathrm{UP}_+)}_{\text{saddle }(+)} & \underbrace{(\mathrm{DP}_-,\mathrm{DP}_+)}_{\text{center (CW)}} & \underbrace{(\mathrm{UP}_+,\mathrm{UP}_+)}_{\text{saddle }(+)} \\ \underbrace{(\mathrm{DP}_+,\mathrm{DP}_-)}_{\text{center (CCW)}} & \underbrace{(\mathrm{UP}_-,\mathrm{UP}_-)}_{\text{saddle }(-)} & \underbrace{(\mathrm{DP}_+,\mathrm{DP}_-)}_{\text{center (CCW)}} \\ \underbrace{(\mathrm{UP}_+,\mathrm{UP}_+)}_{\text{saddle }(+)} & \underbrace{(\mathrm{DP}_-,\mathrm{DP}_+)}_{\text{center (CW)}} & \underbrace{(\mathrm{UP}_+,\mathrm{UP}_+)}_{\text{saddle }(+)} \end{matrix}\right\}. \tag{4.73}$$

(iii) For $a_{120} > 0$ and $a_{210} < 0$,

$$\bigcup_{s=1}^{3}\bigcup_{l=1}^{4}\underbrace{(a_{j_1j_2s},a_{j_2j_1l})}_{\text{XX}}=\left\{\begin{matrix}\underbrace{(\text{DP}_-,\text{DP}_+)}_{\text{center (CW)}} & \underbrace{(\text{UP}_+,\text{UP}_+)}_{\text{saddle }(+)} & \underbrace{(\text{DP}_-,\text{DP}_+)}_{\text{center (CW)}}\\ \underbrace{(\text{UP}_-,\text{UP}_-)}_{\text{saddle }(-)} & \underbrace{(\text{DP}_+,\text{DP}_-)}_{\text{center (CCW)}} & \underbrace{(\text{UP}_-,\text{UP}_-)}_{\text{saddle }(-)}\\ \underbrace{(\text{DP}_-,\text{DP}_+)}_{\text{center (CW)}} & \underbrace{(\text{UP}_+,\text{UP}_+)}_{\text{saddle }(+)} & \underbrace{(\text{DP}_-,\text{DP}_+)}_{\text{center (CW)}}\\ \underbrace{(\text{UP}_-,\text{UP}_-)}_{\text{saddle }(-)} & \underbrace{(\text{DP}_+,\text{DP}_-)}_{\text{center (CCW)}} & \underbrace{(\text{UP}_-,\text{UP}_-)}_{\text{saddle }(-)}\end{matrix}\right\}. \tag{4.74}$$

(iv) For $a_{120} < 0$ and $a_{210} < 0$,

$$\bigcup_{s=1}^{3}\bigcup_{l=1}^{4}\underbrace{(a_{j_1j_2s},a_{j_2j_1l})}_{\text{XX}}=\left\{\begin{matrix}\underbrace{(\text{UP}_-,\text{UP}_-)}_{\text{saddle }(-)} & \underbrace{(\text{DP}_+,\text{DP}_-)}_{\text{center (CCW)}} & \underbrace{(\text{UP}_-,\text{UP}_-)}_{\text{saddle }(-)}\\ \underbrace{(\text{DP}_-,\text{DP}_+)}_{\text{center (CW)}} & \underbrace{(\text{UP}_+,\text{UP}_+)}_{\text{saddle }(+)} & \underbrace{(\text{DP}_-,\text{DP}_+)}_{\text{center (CW)}}\\ \underbrace{(\text{UP}_-,\text{UP}_-)}_{\text{saddle }(-)} & \underbrace{(\text{DP}_+,\text{DP}_-)}_{\text{center (CCW)}} & \underbrace{(\text{UP}_-,\text{UP}_-)}_{\text{saddle }(-)}\\ \underbrace{(\text{DP}_-,\text{DP}_+)}_{\text{center (CW)}} & \underbrace{(\text{UP}_+,\text{UP}_+)}_{\text{saddle }(+)} & \underbrace{(\text{DP}_-,\text{DP}_+)}_{\text{center (CW)}}\end{matrix}\right\}. \tag{4.75}$$

Phase portraits for 2-dimensional systems with the saddles and centers are presented in Fig. 4.6a–d for $(a_{120} > 0, a_{210} > 0)$, $(a_{120} < 0, a_{210} > 0)$, $(a_{120} > 0, a_{210} < 0)$ and $(a_{120} < 0, a_{210} < 0)$. The twelve equilibriums are six saddles and six centers. There are six limit cycles around the centers and six closed separatrices connected six saddles. The blue curves show the (4,3)-parabola-saddles, which will be discussed in the next chapter. The numbers of saddles and centers are tabulated in Table 4.14.

(G) Consider a quartic dynamical system with $(2m_1)\times(2n_1)=4\times 4$ equilibriums as

$$\begin{aligned}\dot{x}_1&=a_{120}(x_2-a_{121})(x_2-a_{122})(x_2-a_{123})(x_2-a_{124}),\\ \dot{x}_2&=a_{210}(x_1-a_{211})(x_1-a_{212})(x_1-a_{213})(x_1-a_{214}),\end{aligned} \tag{4.76}$$

and the first integral manifold is determined by

$$\begin{aligned}&a_{120}\left\{\frac{1}{5}\left[(x_2-a_{12s_1})^5-(x_{20}-a_{12s_1})^5\right]\right.\\ &\quad+\frac{1}{4}\sum_{s_2=1,s_2\neq s_1}^{4}(a_{12s_1}-a_{12s_2})\left[(x_2-a_{121})^4-(x_{20}-a_{121})^4\right]\end{aligned}$$

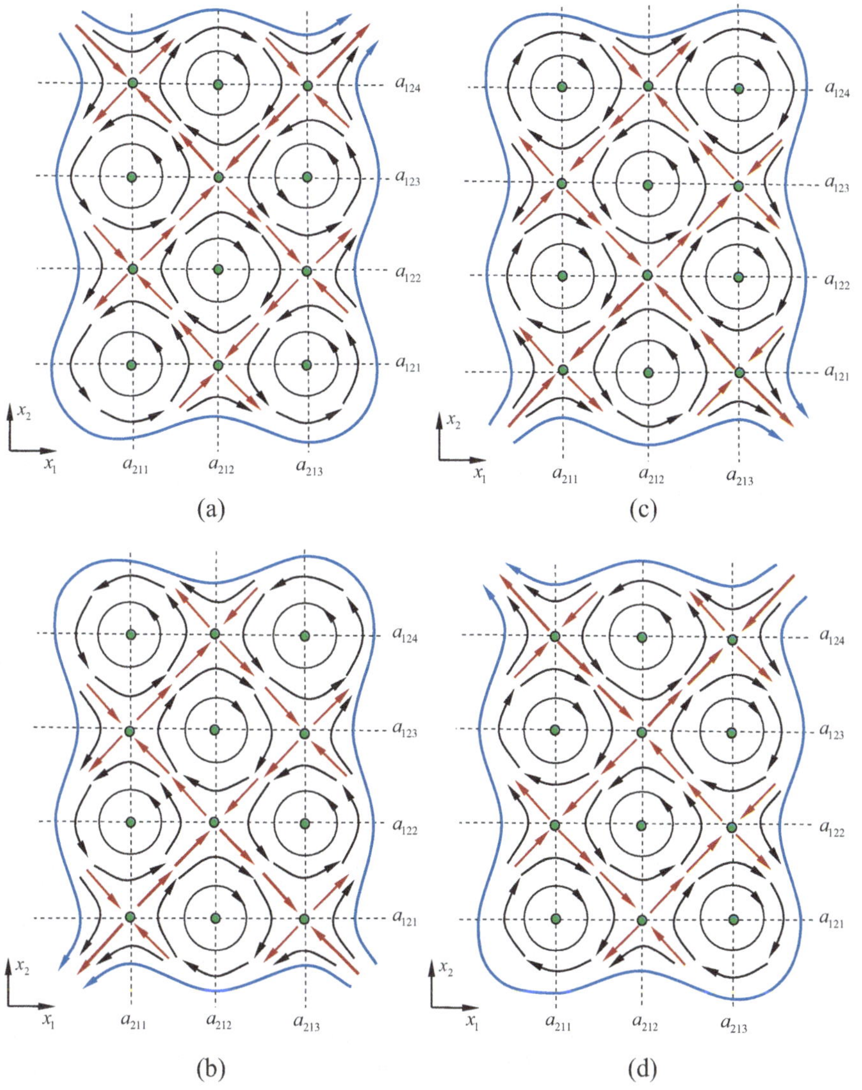

Fig. 4.6 Phase portraits for 2-dimensional systems with $x_1^* = a_{211}, a_{212}, a_{213}$ and $x_2^* = a_{121}, a_{122}, a_{123}, a_{124}$. The four sets of nine simple saddles and centers: **a** ($a_{120} > 0, a_{210} > 0$), **b** ($a_{120} < 0, a_{210} > 0$), **c** ($a_{120} > 0, a_{210} < 0$), **d** ($a_{120} < 0, a_{210} < 0$)

Table 4.14 Numbers of limit cycles and saddles in networks for 4 × 3-equilibriums

(a_{120}, a_{210})	Numbers			
	(DP_+, DP_-)-center (CCW)	(DP_-, DP_+)-center (CW)	(UP_+, UP_+)-saddle (+)	(UP_-, UP_-)-saddle (−)
(+, +)	$m_1 \times (n_1+1)=4$	$m_1 \times n_1 = 2$	$m_1 \times (n_1+1)=4$	$m_1 \times n_1 = 2$
(−, +)	$m_1 \times (n_1+1)=4$	$m_1 \times n_1 = 2$	$m_1 \times (n_1+1)=4$	$m_1 \times n_1 = 2$
(+, −)	$m_1 \times n_1 = 2$	$m_1 \times (n_1+1)=4$	$m_1 \times n_1 = 2$	$m_1 \times (n_1+1)=4$
(−, −)	$m_1 \times n_1 = 2$	$m_1 \times (n_1+1)=4$	$m_1 \times n_1 = 2$	$m_1 \times (n_1+1)=4$

$$
\begin{aligned}
&+\frac{1}{3}\sum_{\substack{s_2,s_3=1,s_2,s_3\neq s_1\\(s_2<s_3)}}^{4}\prod_{r=s_2}^{s_3}(a_{12s_1}-a_{12r})\big[(x_2-a_{12s_1})^3-(x_{20}-a_{12s_1})^3\big]\\
&+\frac{1}{2}\prod_{s_2=1,s_3\neq s_1}^{4}(a_{12s_1}-a_{12s_2})\big[(x_2-a_{12s_1})^2-(x_{20}-a_{12s_1})^2\big]\Big\}\\
&=a_{210}\Big\{\frac{1}{5}\big[(x_1-a_{21l_1})^5-(x_{10}-a_{21l_1})^5\big]\\
&+\frac{1}{4}\sum_{l_2=2,l_2\neq l_1}^{4}(a_{21l_1}-a_{21l_2})\big[(x_1-a_{21l_1})^4-(x_{10}-a_{21l_1})^4\big]\\
&+\frac{1}{3}\sum_{\substack{l_2,l_3=1,l_2,l_3\neq l_1\\(l_2<l_3)}}^{4}\prod_{r=l_2}^{l_3}(a_{21l_1}-a_{12r})\big[(x_1-a_{21l_1})^3-(x_{10}-a_{21l_1})^3\big]\\
&+\frac{1}{2}\prod_{l_2-2,l_2\neq l_1}^{4}(a_{21l_1}-a_{21l_2})\big[(x_1-a_{21l_1})^2-(x_{10}-a_{21l_1})^2\big]\Big\}. \qquad (4.77)
\end{aligned}
$$

The equilibrium properties in the homoclinic networks are listed as follows.

(i) For $a_{120} > 0$ and $a_{210} > 0$,

$$
\bigcup_{s=1}^{4}\bigcup_{l=1}^{4}\underbrace{(a_{j_1j_2s}, a_{j_2j_1l})}_{XX}=\left\{\begin{array}{cccc}
\underbrace{(DP_-,DP_+)}_{\text{center (CW)}} & \underbrace{(UP_+,UP_+)}_{\text{saddle (+)}} & \underbrace{(DP_-,DP_+)}_{\text{center (CW)}} & \underbrace{(UP_+,UP_+)}_{\text{saddle (+)}}\\
\underbrace{(UP_-,UP_-)}_{\text{saddle (−)}} & \underbrace{(DP_+,DP_-)}_{\text{center (CCW)}} & \underbrace{(UP_-,UP_-)}_{\text{saddle (−)}} & \underbrace{(DP_+,DP_-)}_{\text{center (CCW)}}\\
\underbrace{(DP_-,DP_+)}_{\text{center (CW)}} & \underbrace{(UP_+,UP_+)}_{\text{saddle (+)}} & \underbrace{(DP_-,DP_+)}_{\text{center (CW)}} & \underbrace{(UP_+,UP_+)}_{\text{saddle (+)}}\\
\underbrace{(UP_-,UP_-)}_{\text{saddle (−)}} & \underbrace{(DP_+,DP_-)}_{\text{center (CCW)}} & \underbrace{(UP_-,UP_-)}_{\text{saddle (−)}} & \underbrace{(DP_+,DP_-)}_{\text{center (CCW)}}
\end{array}\right\} \qquad (4.78)
$$

(ii) For $a_{120} < 0$ and $a_{210} > 0$,

$$\bigcup_{s=1}^{4}\bigcup_{l=1}^{4}\underbrace{(a_{j_1j_2s}, a_{j_2j_1l})}_{XX} = \left\{\begin{array}{cccc} \underbrace{(\mathrm{UP}_-,\mathrm{UP}_-)}_{\text{saddle }(-)} & \underbrace{(\mathrm{DP}_+,\mathrm{DP}_-)}_{\text{center (CCW)}} & \underbrace{(\mathrm{UP}_-,\mathrm{UP}_-)}_{\text{saddle }(-)} & \underbrace{(\mathrm{DP}_+,\mathrm{DP}_-)}_{\text{center (CCW)}} \\ \underbrace{(\mathrm{DP}_-,\mathrm{DP}_+)}_{\text{center (CW)}} & \underbrace{(\mathrm{UP}_+,\mathrm{UP}_+)}_{\text{saddle }(+)} & \underbrace{(\mathrm{DP}_-,\mathrm{DP}_+)}_{\text{center (CW)}} & \underbrace{(\mathrm{UP}_+,\mathrm{UP}_+)}_{\text{saddle }(+)} \\ \underbrace{(\mathrm{UP}_-,\mathrm{UP}_-)}_{\text{saddle }(-)} & \underbrace{(\mathrm{DP}_+,\mathrm{DP}_-)}_{\text{center (CCW)}} & \underbrace{(\mathrm{UP}_-,\mathrm{UP}_-)}_{\text{saddle }(-)} & \underbrace{(\mathrm{DP}_+,\mathrm{DP}_-)}_{\text{center (CCW)}} \\ \underbrace{(\mathrm{DP}_-,\mathrm{DP}_+)}_{\text{center (CW)}} & \underbrace{(\mathrm{UP}_+,\mathrm{UP}_+)}_{\text{saddle }(+)} & \underbrace{(\mathrm{DP}_-,\mathrm{DP}_+)}_{\text{center (CW)}} & \underbrace{(\mathrm{UP}_+,\mathrm{UP}_+)}_{\text{saddle }(+)} \end{array}\right\}. \tag{4.79}$$

(iii) For $a_{120} > 0$ and $a_{210} < 0$,

$$\bigcup_{s=1}^{4}\bigcup_{l=1}^{4}\underbrace{(a_{j_1j_2s}, a_{j_2j_1l})}_{XX} = \left\{\begin{array}{cccc} \underbrace{(\mathrm{UP}_+,\mathrm{UP}_+)}_{\text{saddle }(+)} & \underbrace{(\mathrm{DP}_-,\mathrm{DP}_+)}_{\text{center (CW)}} & \underbrace{(\mathrm{UP}_+,\mathrm{UP}_+)}_{\text{saddle }(+)} & \underbrace{(\mathrm{DP}_-,\mathrm{DP}_+)}_{\text{center (CW)}} \\ \underbrace{(\mathrm{DP}_+,\mathrm{DP}_-)}_{\text{center (CCW)}} & \underbrace{(\mathrm{UP}_-,\mathrm{UP}_-)}_{\text{saddle }(-)} & \underbrace{(\mathrm{DP}_+,\mathrm{DP}_-)}_{\text{center (CCW)}} & \underbrace{(\mathrm{UP}_-,\mathrm{UP}_-)}_{\text{saddle }(-)} \\ \underbrace{(\mathrm{UP}_+,\mathrm{UP}_+)}_{\text{saddle }(+)} & \underbrace{(\mathrm{DP}_-,\mathrm{DP}_+)}_{\text{center (CW)}} & \underbrace{(\mathrm{UP}_+,\mathrm{UP}_+)}_{\text{saddle }(+)} & \underbrace{(\mathrm{DP}_-,\mathrm{DP}_+)}_{\text{center (CW)}} \\ \underbrace{(\mathrm{DP}_+,\mathrm{DP}_-)}_{\text{center (CCW)}} & \underbrace{(\mathrm{UP}_-,\mathrm{UP}_-)}_{\text{saddle }(-)} & \underbrace{(\mathrm{DP}_+,\mathrm{DP}_-)}_{\text{center (CCW)}} & \underbrace{(\mathrm{UP}_-,\mathrm{UP}_-)}_{\text{saddle }(-)} \end{array}\right\}. \tag{4.80}$$

(iv) For $a_{120} > 0$ and $a_{210} < 0$,

$$\bigcup_{s=1}^{4}\bigcup_{l=1}^{4}\underbrace{(a_{j_1j_2s}, a_{j_2j_1l})}_{XX} = \left\{\begin{array}{cccc} \underbrace{(\mathrm{DP}_+,\mathrm{DP}_-)}_{\text{center (CCW)}} & \underbrace{(\mathrm{UP}_-,\mathrm{UP}_-)}_{\text{saddle }(-)} & \underbrace{(\mathrm{DP}_+,\mathrm{DP}_-)}_{\text{center (CCW)}} & \underbrace{(\mathrm{UP}_-,\mathrm{UP}_-)}_{\text{saddle }(-)} \\ \underbrace{(\mathrm{UP}_+,\mathrm{UP}_+)}_{\text{saddle }(+)} & \underbrace{(\mathrm{DP}_-,\mathrm{DP}_+)}_{\text{center (CW)}} & \underbrace{(\mathrm{UP}_+,\mathrm{UP}_+)}_{\text{saddle }(+)} & \underbrace{(\mathrm{DP}_-,\mathrm{DP}_+)}_{\text{center (CW)}} \\ \underbrace{(\mathrm{DP}_+,\mathrm{DP}_-)}_{\text{center (CCW)}} & \underbrace{(\mathrm{UP}_-,\mathrm{UP}_-)}_{\text{saddle }(-)} & \underbrace{(\mathrm{DP}_+,\mathrm{DP}_-)}_{\text{center (CCW)}} & \underbrace{(\mathrm{UP}_-,\mathrm{UP}_-)}_{\text{saddle }(-)} \\ \underbrace{(\mathrm{UP}_+,\mathrm{UP}_+)}_{\text{saddle }(+)} & \underbrace{(\mathrm{DP}_-,\mathrm{DP}_+)}_{\text{center (CW)}} & \underbrace{(\mathrm{UP}_+,\mathrm{UP}_+)}_{\text{saddle }(+)} & \underbrace{(\mathrm{DP}_-,\mathrm{DP}_+)}_{\text{center (CW)}} \end{array}\right\}. \tag{4.81}$$

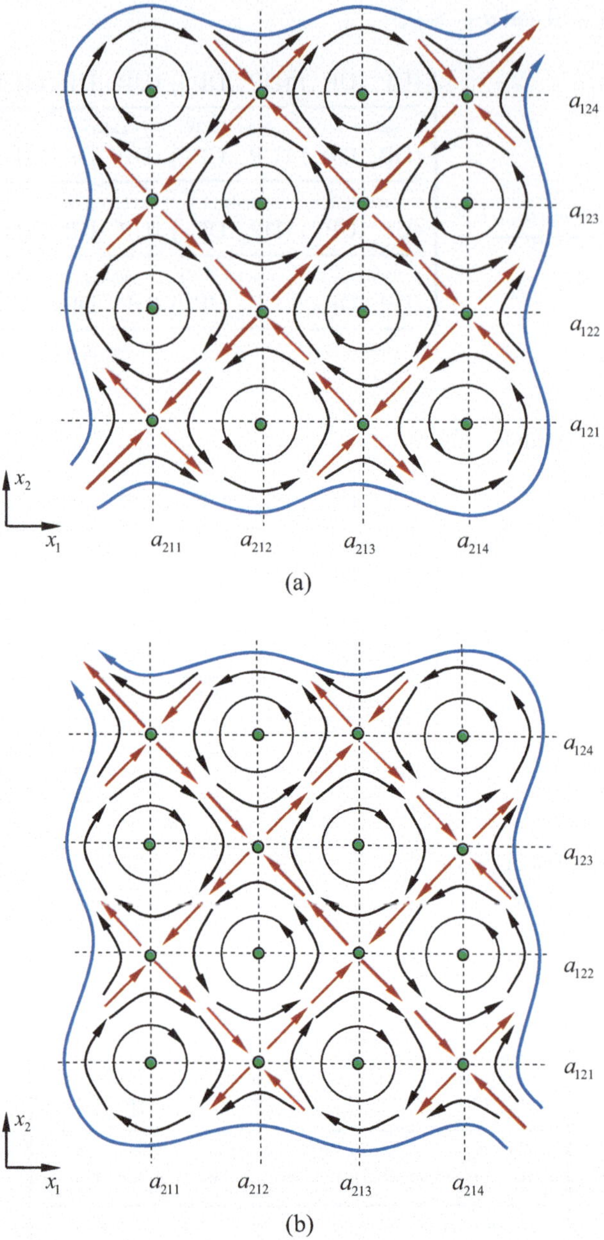

Fig. 4.7 Phase portraits for 2-dimensional systems with $x_1^* = a_{211}, a_{212}, a_{213}, a_{214}$ and $x_2^* = a_{121}, a_{122}, a_{123}, a_{124}$. The four sets of nine simple saddle and center: **a** ($a_{120} > 0, a_{210} > 0$), **b** ($a_{120} < 0, a_{210} > 0$), **c** ($a_{120} > 0, a_{210} < 0$), **d** ($a_{120} < 0, a_{210} < 0$)

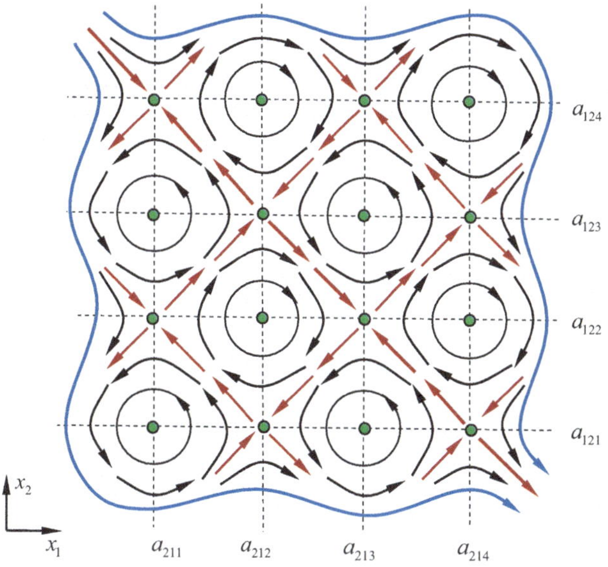

(c)

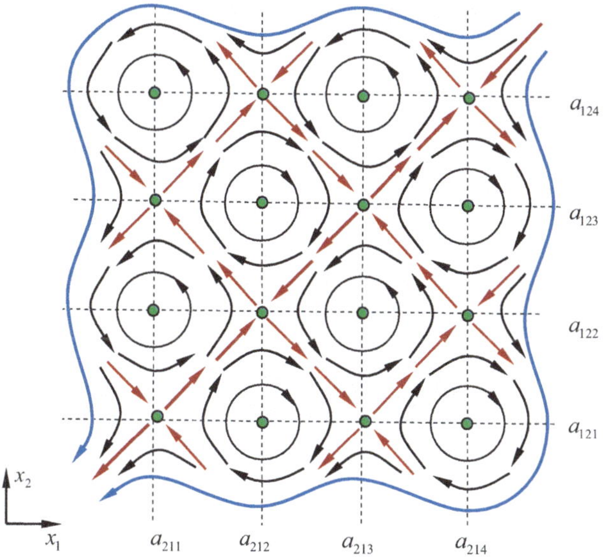

(d)

Fig. 4.7 (continued)

Phase portraits for 2-dimensional systems with the saddles and centers are presented in Fig. 4.7a–d for ($a_{120} > 0, a_{210} > 0$), ($a_{120} < 0, a_{210} > 0$), ($a_{120} > 0$, $a_{210} < 0$) and ($a_{120} < 0, a_{210} < 0$). The sixteen equilibriums are eight saddles and eight centers. There are eight limit cycles around the eight centers and eight closed separatrices connected eight saddles. The blue curves show the (4,4)-inflection-saddles, which will be discussed in the next chapter. The numbers of saddles and centers are tabulated in Table 4.15.

Table 4.15 Numbers of limit cycles and saddles in networks for 4 × 4-equilibriums

(a_{120}, a_{210})	Numbers			
	(DP_+, DP_-)-center (CCW)	(DP_-, DP_+)-center (CW)	(UP_+, UP_+)-saddle (+)	(UP_-, UP_-)-saddle (−)
$(+, +)$	$m_1 \times n_1 = 4$	$m_1 \times n_1 = 4$	$m_1 \times n_1 = 4$	$m_1 \times n_1 = 4$
$(-, +)$	$m_1 \times n_1 = 4$	$m_1 \times n_1 = 4$	$m_1 \times n_1 = 4$	$m_1 \times n_1 = 4$
$(+, -)$	$m_1 \times n_1 = 4$	$m_1 \times n_1 = 4$	$m_1 \times n_1 = 4$	$m_1 \times n_1 = 4$
$(-, -)$	$m_1 \times n_1 = 4$	$m_1 \times n_1 = 4$	$m_1 \times n_1 = 4$	$m_1 \times n_1 = 4$

Reference

1. Luo, A. C. J. (2023). Limit cycles and homoclinic network in 2-dimensional polynomial systems. *AIP Chaos*, 34, 022104 (66 pages).

Chapter 5
Bifurcations for Homoclinic Networks with Centers

In this chapter, the appearing and switching bifurcations are studied for homoclinic networks of singular and non-singular saddles and centers with singular parabola-saddles and double-inflection saddles in crossing-univariate polynomial systems, and the first integral manifolds of such homoclinic networks are determined through polynomial functions. The illustrations of singular equilibriums to networks of non-singular saddles and centers are given.

5.1 Higher-Order Singularity and Bifurcations

Consider singular equilibriums and bifurcations for homoclinic networks with centers in two-dimensional crossing-univariate polynomial nonlinear systems. To discuss bifurcations of equilibriums in polynomial systems, consider a polynomial system with crossing-univariate polynomial vector fields. The corresponding dynamical behaviors are presented through the following theorem.

Theorem 5.1 *(i) Consider a crossing-univariate polynomial dynamical system as*

$$\begin{aligned}&\dot{x}_{j_1} = a_{j_1 j_2 0}(x_{j_2} - a_{j_1 j_2 1})^m,\ \dot{x}_{j_2} = a_{j_2 j_1 0}(x_{j_1} - a_{j_2 j_1 1})^n,\\ &j_1, j_2 \in \{1, 2\}; j_1 \neq j_2.\end{aligned} \tag{5.1}$$

The first integral manifold is

$$\begin{aligned}&\frac{1}{n+1}\left[(x_{j_1} - a_{j_2 j_1 1})^{n+1} - (x_{j_1 0} - a_{j_2 j_1 1})^{n+1}\right]\\ &= \frac{a_{j_1 j_2 0}}{a_{j_2 j_1 0}} \frac{1}{m+1}\left[(x_{j_2} - a_{j_1 j_2 1})^{m+1} - (x_{j_2 0} - a_{j_1 j_2 1})^{m+1}\right],\\ &\text{for } m, n \geq 1.\end{aligned} \tag{5.2}$$

A. C. J. Luo, *Limit Cycles and Homoclinic Networks in Two-Dimensional Polynomial Systems*, https://doi.org/10.1007/978-981-97-2617-2_5

(i_1) For $m = 2m_1 + 1$ and $n = 2n_1 + 1$, *the equilibrium of* $(x^*_{j_2}, x^*_{j_1}) = (a_{j_1j_2 1}, a_{j_2j_1 1})$ *has the following properties.*

- *For* $a_{j_1j_2 0} > 0$ *and* $a_{j_2j_1 0} > 0$,

$$(a_{j_1j_2 1}, a_{j_2j_1 1}) = \underbrace{((2m_1{+}1)^{\text{th}}\text{UP}_+,(2n_1{+}1)^{\text{th}}\text{UP}_+)}_{((2m_1+1),(2n_1+1))\text{-positive saddle}}. \tag{5.3}$$

- *For* $a_{j_1j_2 0} < 0$ *and* $a_{j_2j_1 0} > 0$,

$$(a_{j_1j_2 1}, a_{j_2j_1 1}) = \underbrace{((2m_1{+}1)^{\text{th}}\text{DP}_+,(2n_1{+}1)^{\text{th}}\text{DP}_-)}_{((2m_1+1,)(2n_1+1))\text{-CCW center}}. \tag{5.4}$$

- *For* $a_{j_1j_2 0} > 0$ *and* $a_{j_2j_1 0} < 0$,

$$(a_{j_1j_2 1}, a_{j_2j_1 1}) = \underbrace{((2m_1{+}1)^{\text{th}}\text{DP}_-,(2n_1{+}1)^{\text{th}}\text{DP}_+)}_{((2m_1+1),(2n_1+1))\text{-CW center}}. \tag{5.5}$$

- *For* $a_{j_1j_2 0} < 0$ *and* $a_{j_2j_1 0} < 0$,

$$(a_{j_1j_2 1}, a_{j_2j_1 1}) = \underbrace{((2m_1{+}1)^{\text{th}}\text{UP}_-,(2n_1{+}1)^{\text{th}}\text{UP}_-)}_{((2m_1+1),(2n_1+1))\text{-negative saddle}}. \tag{5.6}$$

(i_2) For $m = 2m_1$ and $n = 2n_1 + 1$, *the equilibrium of* $(x^*_{j_2}, x^*_{j_1}) = (a_{j_1j_2 1}, a_{j_2j_1 1})$ *has the following properties.*

- *For* $a_{j_1j_2 0} > 0$ *and* $a_{j_2j_1 0} > 0$,

$$(a_{j_1j_2 1}, a_{j_2j_1 1}) = \underbrace{((2m_1)^{\text{th}}\text{US},(2n_1{+}1)^{\text{th}}\text{UP})}_{((2m_1),(2n_1+1))\text{-up-parabola upper-saddle}}. \tag{5.7}$$

- *For* $a_{j_1j_2 0} < 0$ *and* $a_{j_2j_1 0} > 0$,

$$(a_{j_1j_2 1}, a_{j_2j_1 1}) = \underbrace{((2m_1)^{\text{th}}\text{LS}, (2n_1 + 1)^{\text{th}}\text{DP})}_{((2m_1),(2n_1+1))\text{-down-parabola lower-saddle}}. \tag{5.8}$$

- *For* $a_{j_1j_2 0} > 0$ *and* $a_{j_2j_1 0} < 0$,

$$(a_{j_1j_2 1}, a_{j_2j_1 1}) = \underbrace{((2m_1)^{\text{th}}\text{US}, (2n_1 + 1)^{\text{th}}\text{DP})}_{((2m_1),(2n_1+1))\text{-down-parabola upper-saddle}}. \tag{5.9}$$

- *For* $a_{j_1j_2 0} < 0$ *and* $a_{j_2j_1 0} < 0$,

$$(a_{j_1j_2 1}, a_{j_2j_1 1}) = \underbrace{((2m_1)^{\text{th}}\text{LS}, (2n_1{+}1)^{\text{th}}\text{UP})}_{((2m_1),(2n_1+1))\text{-up-parabola lower-saddle}}. \tag{5.10}$$

(i₃) For $m = 2m_1 + 1$ and $n = 2n_1$, *the equilibrium of* $(x^*_{j_2}, x^*_{j_1}) = (a_{j_1j_2 1}, a_{j_2j_1 1})$ *has the following properties.*

- *For* $a_{j_1j_2 0} > 0$ *and* $a_{j_2j_1 0} > 0$,

$$(a_{j_1j_2 1}, a_{j_2j_1 1}) = \underbrace{((2m_1{+}1)^{\text{th}}\text{UP},(2n_1)^{\text{th}}\text{US})}_{((2m_1+1),(2n_1))\text{-up-parabola upper-saddle}}. \tag{5.11}$$

- *For* $a_{j_1j_2 0} < 0$ *and* $a_{j_2j_1 0} > 0$,

$$(a_{j_1j_2 1}, a_{j_2j_1 1}) = \underbrace{((2m_1{+}1)^{\text{th}}\text{DP},(2n_1)^{\text{th}}\text{US})}_{((2m_1+1),(2n_1))\text{-down-parabola upper-saddle}}. \tag{5.12}$$

- *For* $a_{j_1j_2 0} > 0$ *and* $a_{j_2j_1 0} < 0$,

$$(a_{j_1j_2 1}, a_{j_2j_1 1}) = \underbrace{((2m_1{+}1)^{\text{th}}\text{DP}, (2n_1)^{\text{th}}\text{LS})}_{((2m_1+1),(2n_1))\text{-down-parabola lower-saddle}}. \tag{5.13}$$

- *For* $a_{j_1j_2 0} < 0$ *and* $a_{j_2j_1 0} < 0$,

$$(a_{j_1j_2 1}, a_{j_2j_1 1}) = \underbrace{((2m_1{+}1)^{\text{th}}\text{UP}, (2n_1)^{\text{th}}\text{LS})}_{((2m_1+1),(2n_1))\text{-up-parabola lower-saddle}}. \tag{5.14}$$

(i₄) For $m = 2m_1$ and $n = 2n_1$, *the equilibrium of* $(x^*_{j_2}, x^*_{j_1}) = (a_{j_1j_2 1}, a_{j_2j_1 1})$ *has the following properties.*

- *For* $a_{j_1j_2 0} > 0$ *and* $a_{j_2j_1 0} > 0$,

$$(a_{j_1j_2 1}, a_{j_2j_1 1}) = \underbrace{((2m_1)^{\text{th}}\text{II}_{+},(2n_1)^{\text{th}}\text{II}_{+})}_{((2m_1),(2n_1))\text{-double-inflection saddle}}. \tag{5.15}$$

- *For* $a_{j_1j_2 0} < 0$ *and* $a_{j_2j_1 0} > 0$,

$$(a_{j_1j_2 1}, a_{j_2j_1 1}) = \underbrace{((2m_1)^{\text{th}}\text{DI}_{+}, (2n_1)^{\text{th}}\text{DI}_{-})}_{((2m_1),(2n_1))\text{-double-inflection -saddle}}. \tag{5.16}$$

- *For* $a_{j_1j_2 0} > 0$ *and* $a_{j_2j_1 0} < 0$,

$$(a_{j_1j_2 1}, a_{j_2j_1 1}) = \underbrace{((2m_1)^{\text{th}}\text{DI}_{-}, (2n_1)^{\text{th}}\text{DI}_{+})}_{((2m_1),(2n_1))\text{-double-inflection saddle}}. \tag{5.17}$$

- *For* $a_{j_1j_20}<0$ *and* $a_{j_2j_10}<0$,

$$(a_{j_1j_21},a_{j_2j_11})=\underbrace{((2m_1)^{\text{th}}\text{II}_-,(2n_1)^{\text{th}}\text{II}_-)}_{((2m_1),(2n_1))\text{-double-inflection saddle}}. \tag{5.18}$$

(ii) *Consider a crossing-univariate polynomial dynamical system as*

$$\begin{aligned}
\dot{x}_{j_1}&=a_{j_1j_20}\prod_{r=1}^{p}\left(x_{j_2}-a_{j_1j_2r}\right)^{m_r},\\
\dot{x}_{j_2}&=a_{j_2j_10}\prod_{s=1}^{q}\left(x_{j_1}-a_{j_2j_1s}\right)^{n_s},\\
\sum_{r=1}^{p}m_r&=m\text{ and }\sum_{s=1}^{q}n_s=n;\ j_1,j_2\in\{1,2\};j_1\neq j_2.
\end{aligned} \tag{5.19}$$

The first integral manifold is

$$\begin{aligned}
&a_{j_2j_10}\Big\{\frac{1}{n+1}[(x_{j_1}-a_{j_2j_1s_1})^{n+1}-(x_{j_10}-a_{j_2j_1s_1})^{n+1}]\\
&\quad+\sum_{\beta=1}^{n-n_{s_1}}\frac{1}{n-\beta+1}b_{j_2j_1\beta}\big[(x_{j_1}-a_{j_2j_1s_1})^{n-\beta+1}-(x_{j_10}-a_{j_2j_1s_1})^{n-\beta+1}\big]\Big\}\\
&=a_{j_1j_20}\Big\{\frac{1}{m+1}\big[(x_{j_2}-a_{j_1j_2r_1})^{m+1}-(x_{j_20}-a_{j_1j_2r_1})^{m+1}\big]\\
&\quad+\sum_{\alpha=1}^{m-m_{r_1}}\frac{1}{m-\alpha+1}b_{j_1j_2\alpha}\big[(x_{j_1}-a_{j_1j_2r_1})^{m-\alpha+1}-(x_{j_10}-a_{j_1j_2r_1})^{m-\alpha+1}\big]\Big\}
\end{aligned} \tag{5.20}$$

where

$$\begin{aligned}
b_{j_2j_1\beta}&=\sum_{l_{s_q},t_{s_q}=0}^{l_{s_q}+t_{s_q}=n_{s_q}}\cdots\sum_{l_{s_2},t_{s_2}=0}^{l_{s_2}+t_{s_2}=n_{s_2}}[\prod_{k=s_2}^{s_q}\frac{n_k!}{r_k!t_k!}(a_{j_2j_1l_1}-a_{j_2j_1k})^{r_k}]\delta_{t_{s_2}+\cdots+t_{s_q}}^{n-\beta},\\
b_{j_1j_2\alpha}&=\sum_{l_{r_p},t_{r_p}=0}^{l_{r_p}+t_{r_p}=m_{r_p}}\cdots\sum_{l_{r_2},t_{r_2}=0}^{l_{r_2}+t_{r_2}=m_{r_2}}[\prod_{k=r_2}^{r_p}\frac{m_k!}{l_k!t_k!}(a_{j_1j_2r_1}-a_{j_1j_2k})^{l_k}]\delta_{t_{r_2}+\cdots+t_{r_p}}^{m-\alpha};\\
b_{j_2j_10}&=1,b_{j_1j_20}=1.
\end{aligned} \tag{5.21}$$

The singular equilibrium network with $\sum_{r_i=1}^{p}m_{r_i}=m$ and $\sum_{r_i=1}^{q}n_{s_i}=n$ is defined as

$$
\bigcup_{r=1}^{p}\bigcup_{s=1}^{q}\underbrace{(a_{j_1j_2r},a_{j_2j_1s})}_{(m_r,n_s)\text{-XX}} \equiv \left\{\begin{matrix} (a_{j_1j_2p},a_{j_2j_11}) & \cdots & (a_{j_1j_2p},a_{j_2j_1q}) \\ \vdots & \cdots & \vdots \\ (a_{j_1j_21},a_{j_2j_11}) & \cdots & (a_{j_1j_21},a_{j_2j_1q}) \end{matrix}\right\}_{p\times q}
$$

$$
= \left\{\begin{matrix} \underbrace{((m_p)^{\text{th}}\text{XX},(n_q)^{\text{th}}\text{XX})}_{(m_1,n_q)\text{-XX}} & \cdots & \underbrace{((m_p)^{\text{th}}\text{XX},(n_q)^{\text{th}}\text{XX})}_{(m^{(i)}_{p_i},n^{(i)}_{q_i})\text{-XX}} \\ \vdots & \vdots & \vdots \\ \underbrace{((m_1)^{\text{th}}\text{XX},(n_1)^{\text{th}}\text{XX})}_{(m_1,n_1)\text{-XX}} & \cdots & \underbrace{((m_p)^{\text{th}}\text{XX},(n_q)^{\text{th}}\text{XX})}_{(m_p,n_1)\text{-XX}} \end{matrix}\right\}_{p\times q}. \tag{5.22}
$$

(ii$_1$) For $m_{r_1}=2m_{r_11}+1$ and $n_{s_1}=2n_{s_11}+1$, the equilibrium of $(x^*_{j_2},x^*_{j_1})=(a_{j_1j_2r_1},a_{j_2j_1s_1})$ has the following properties.

- For $a_{j_1j_20}\prod_{r_2=1,r_2\neq r_1}^{p}(a_{j_1j_2r_1}-a_{j_1j_2r_2})^{m_{r_2}}>0$ and $a_{j_2j_10}\prod_{s_2=1,s_2\neq s_1}^{q}(a_{j_2j_1s_1}-a_{j_2j_1s_2})^{n_{s_2}}>0$,

$$
(a_{j_1j_2r_1},a_{j_2j_1s_1})=\underbrace{((2m_{r_11}+1)^{\text{th}}\text{UP}_+,(2n_{s_11}+1)^{\text{th}}\text{UP}_+)}_{((2m_{r_11}+1),(2n_{s_11}+1))\text{-positive saddle}}. \tag{5.23}
$$

- For $a_{j_1j_20}\prod_{r_2=1,r_2\neq r_1}^{p}(a_{j_1j_2r_1}-a_{j_1j_2r_2})^{m_{r_2}}<0$ and $a_{j_2j_10}\prod_{s_2=1,s_2\neq s_1}^{q}(a_{j_2j_1s_1}-a_{j_2j_1s_2})^{n_{s_2}}>0$,

$$
(a_{j_1j_2r_1},a_{j_2j_1s_1})=\underbrace{((2m_{r_11}+1)^{\text{th}}\text{DP}_+,(2n_{s_11}+1)^{\text{th}}\text{DP}_-)}_{((2m_{r_11}+1),(2n_{s_11}+1))\text{-CCW center}}. \tag{5.24}
$$

- For $a_{j_1j_20}\prod_{r_2=1,r_2\neq r_1}^{p}(a_{j_1j_2r_1}-a_{j_1j_2r_2})^{m_{r_2}}>0$ and $a_{j_2j_10}\prod_{s_2=1,s_2\neq s_1}^{q}(a_{j_2j_1s_1}-a_{j_2j_1s_2})^{n_{s_2}}<0$,

$$
(a_{j_1j_2r_1},a_{j_2j_1s_1})=\underbrace{((2m_{r_11}+1)^{\text{th}}\text{DP}_-,(2n_{s_11}+1)^{\text{th}}\text{DP}_+)}_{((2m_{r_11}+1),(2n_{s_11}+1))\text{-CW center}}. \tag{5.25}
$$

- For $a_{j_1j_20}\prod_{r_2=1,r_2\neq r_1}^{p}(a_{j_1j_1r_1}-a_{j_1j_1r_2})^{m_{r_2}}<0$ and $a_{j_2j_10}\prod_{s_2=1,s_2\neq s_1}^{q}(a_{j_2j_2s_1}-a_{j_2j_2s_2})^{n_{s_2}}<0$,

$$
(a_{j_1j_2r_1},a_{j_2j_1s_1})=\underbrace{((2m_{r_11}+1)^{\text{th}}\text{UP}_-,(2n_{s_11}+1)^{\text{th}}\text{UP}_-)}_{((2m_{r_11}+1),(2n_{s_11}+1))\text{-negative saddle}}. \tag{5.26}
$$

(ii$_2$) For $m_{r_1}=2m_{r_1 1}$ and $n_{s_1}=2n_{s_1 1}+1$, the equilibrium of $(x^*_{j_2},x^*_{j_1})=(a_{j_1j_2r_1},a_{j_2j_1s_1})$ has the following properties.

- For $a_{j_1j_20}\prod_{r_2=1,r_2\neq r_1}^{p}(a_{j_1j_2r_1}-a_{j_1j_2r_2})^{m_{r_2}}>0$ and $a_{j_2j_10}\prod_{s_2=1,s_2\neq s_1}^{q}(a_{j_2j_1s_1}-a_{j_2j_1s_2})^{n_{s_2}}>0$,

$$(a_{j_1j_2r_1},a_{j_2j_1s_1})=\underbrace{((2m_{r_1 1})^{\text{th}}\text{US},(2n_{s_1 1}+1)^{\text{th}}\text{UP})}_{((2m_{r_1 1}),(2n_{s_1 1}+1))\text{-up-parabola upper-saddle}}. \tag{5.27}$$

- For $a_{j_1j_20}\prod_{r_2=1,r_2\neq r_1}^{p}(a_{j_1j_2r_1}-a_{j_1j_2r_2})^{m_{r_2}}<0$ and $a_{j_2j_10}\prod_{s_2=1,s_2\neq s_1}^{q}(a_{j_2j_1s_1}-a_{j_2j_1s_2})^{n_{s_2}}>0$,

$$(a_{j_1j_2r_1},a_{j_2j_1s_1})=\underbrace{((2m_{r_1 1})^{\text{th}}\text{LS},\ (2n_{s_1 1}+1)^{\text{th}}\text{DP})}_{((2m_{r_1 1}),(2n_{s_1 1}+1))\text{-down-parabola lower-saddle}}. \tag{5.28}$$

- For $a_{j_1j_20}\prod_{r_2=1,r_2\neq r_1}^{p}(a_{j_1j_2r_1}-a_{j_1j_2r_2})^{m_{r_2}}>0$ and $a_{j_2j_10}\prod_{s_2=1,s_2\neq s_1}^{q}(a_{j_2j_1s_1}-a_{j_2j_1s_2})^{n_{s_2}}<0$,

$$(a_{j_1j_2r_1},a_{j_2j_1s_1})=\underbrace{((2m_{r_1 1})^{\text{th}}\text{US},(2n_{s_1 1}+1)^{\text{th}}\text{DP})}_{((2m_{r_1 1}),(2n_{s_1 1}+1))\text{-down-parabola upper-saddle}}. \tag{5.29}$$

- For $a_{j_1j_20}\prod_{r_2=1,r_2\neq r_1}^{p}(a_{j_1j_2r_1}-a_{j_1j_2r_2})^{m_{r_2}}<0$ and $a_{j_2j_10}\prod_{s_2=1,s_2\neq s_1}^{q}(a_{j_2j_2s_1}-a_{j_2j_2s_2})^{n_{s_2}}<0$,

$$(a_{j_1j_2r_1},a_{j_2j_1s_1})=\underbrace{((2m_{r_1 1})^{\text{th}}\text{LS},(2n_{s_1 1}+1)^{\text{th}}\text{UP})}_{((2m_{r_1 1}),(2n_{s_1 1}+1))\text{-up-parabola lower-saddle}}. \tag{5.30}$$

(ii$_3$) For $m_{r_1}=2m_{r_1 1}+1$ and $n_{s_1}=2n_{s_1 1}$, the equilibrium of $(x^*_{j_2},x^*_{j_1})=(a_{j_1j_2r_1},a_{j_2j_1s_1})$ has the following properties.

- For $a_{j_1j_20}\prod_{r_2=1,r_2\neq r_1}^{p}(a_{j_1j_2r_1}-a_{j_1j_2r_2})^{m_{r_2}}>0$ and $a_{j_2j_10}\prod_{s_2=1,s_2\neq s_1}^{q}(a_{j_2j_1s_1}-a_{j_2j_1s_2})^{n_{s_2}}>0$,

$$(a_{j_1j_2r_1},a_{j_2j_1s_1})=\underbrace{((2m_{r_1 1}+1)^{\text{th}}\text{UP},(2n_{s_1 1})^{\text{th}}\text{US})}_{((2m_{r_1 1}+1),(2n_{s_1 1}))\text{-up-parabola upper-saddle}}. \tag{5.31}$$

- For $a_{j_1j_20}\prod_{r_2=1,r_2\neq r_1}^{p}(a_{j_1j_2r_1}-a_{j_1j_2r_2})^{m_{r_2}}<0$ and $a_{j_2j_10}\prod_{s_2=1,s_2\neq s_1}^{q}(a_{j_2j_1s_1}-a_{j_2j_1s_2})^{n_{s_2}}>0$,

$$(a_{j_1j_2r_1},a_{j_2j_1s_1})=\underbrace{((2m_{r_1 1}+1)^{\text{th}}\text{DP},(2n_{s_1 1})^{\text{th}}\text{US})}_{((2m_{r_1 1}+1),(2n_{s_1 1}))\text{-down-parabola upper-saddle}}. \tag{5.32}$$

- For $a_{j_1j_20}\prod_{r_2=1,r_2\neq r_1}^{p}(a_{j_1j_2r_1}-a_{j_1j_2r_2})^{m_{r_2}}>0$ and $a_{j_2j_10}\prod_{s_2=1,s_2\neq s_1}^{q}(a_{j_2j_1s_1}-a_{j_2j_1s_2})^{n_{s_2}}<0$,

$$(a_{j_1j_2r_1},a_{j_2j_1s_1})=\underbrace{((2m_{r_1 1}+1)^{\text{th}}\text{DP},(2n_{s_1 1})^{\text{th}}\text{LS})}_{((2m_{r_1 1}+1),(2n_{s_1 1}))\text{-down-parabola lower-saddle}}. \tag{5.33}$$

- For $a_{j_1j_20}\prod_{r_2=1,r_2\neq r_1}^{p}(a_{j_1j_1r_1}-a_{j_1j_1r_2})^{m_{r_2}}<0$ and $a_{j_2j_10}\prod_{s_2=1,s_2\neq s_1}^{q}(a_{j_2j_2s_1}-a_{j_2j_2s_2})^{n_{s_2}}<0$,

$$(a_{j_1j_2r_1},a_{j_2j_1s_1})=\underbrace{((2m_{r_1 1}+1)^{\text{th}}\text{UP},(2n_{s_1 1})^{\text{th}}\text{LS})}_{((2m_{r_1 1}+1),(2n_{s_1 1}))\text{-up-parabola lower-saddle}}. \tag{5.34}$$

(ii$_4$) For $m_{r_1}=2m_{r_1 1}$ and $n_{s_1}=2n_{s_1 1}$, the equilibrium of $(x_{j_2}^*,x_{j_1}^*)=(a_{j_1j_2r_1},a_{j_2j_1s_1})$ has the following properties.

- For $a_{j_1j_20}\prod_{r_2=1,r_2\neq r_1}^{p}(a_{j_1j_2r_1}-a_{j_1j_2r_2})^{m_{r_2}}>0$ and $a_{j_2j_10}\prod_{s_2=1,s_2\neq s_1}^{q}(a_{j_2j_1s_1}-a_{j_2j_1s_2})^{n_{s_2}}>0$,

$$(a_{j_1j_2r_1},a_{j_2j_1s_1})=\underbrace{((2m_{r_1 1})^{\text{th}}\text{II}_+,(2n_{s_1 1})^{\text{th}}\text{II}_+)}_{((2m_{r_1 1}),(2n_{s_1 1}))\text{-double-inflection saddle}}. \tag{5.35}$$

- For $a_{j_1j_20}\prod_{r_2=1,r_2\neq r_1}^{p}(a_{j_1j_2r_1}-a_{j_1j_2r_2})^{m_{r_2}}<0$ and $a_{j_2j_10}\prod_{s_2=1,s_2\neq s_1}^{q}(a_{j_2j_1s_1}-a_{j_2j_1s_2})^{n_{s_2}}>0$,

$$(a_{j_1j_2r_1},a_{j_2j_1s_1})=\underbrace{((2m_{r_1 1})^{\text{th}}\text{DI}_+,(2n_{s_1 1})^{\text{th}}\text{DI}_-)}_{((2m_{r_1 1}),(2n_{s_1 1}))\text{-double-inflection saddle}}. \tag{5.36}$$

- For $a_{j_1j_20}\prod_{r_2=1,r_2\neq r_1}^{p}(a_{j_1j_2r_1}-a_{j_1j_2r_2})^{m_{r_2}}>0$ and $a_{j_2j_10}\prod_{s_2=1,s_2\neq s_1}^{q}(a_{j_2j_1s_1}-a_{j_2j_1s_2})^{n_{s_2}}<0$,

$$(a_{j_1j_2r_1},a_{j_2j_1s_1})=\underbrace{((2m_{r_1 1})^{\text{th}}\text{DI}_-,(2n_{s_1 1})^{\text{th}}\text{DI}_+)}_{((2m_{r_1 1}),(2n_{s_1 1}))\text{-double-inflection saddle}}. \tag{5.37}$$

- For $a_{j_1j_20}\prod_{r_2=1,r_2\neq r_1}^{p}(a_{j_1j_1r_1}-a_{j_1j_1r_2})^{m_{r_2}}<0$ and $a_{j_2j_10}\prod_{s_2=1,s_2\neq s_1}^{q}(a_{j_2j_2s_1}-a_{j_2j_2s_2})^{n_{s_2}}<0$,

$$(a_{j_1j_2r_1},a_{j_2j_1s_1})=\underbrace{((2m_{r_1 1})^{\text{th}}\text{II}_-,(2n_{s_1 1})^{\text{th}}\text{II}_-)}_{((2m_{r_1 1}),(2n_{s_1 1}))\text{-double-inflection saddle}}. \tag{5.38}$$

(iii) *Consider a crossing-univariate polynomial system as*

$$\begin{aligned}
\dot{x}_{j_1} &= a_{j_1 j_2 0} \prod_{s_1=1}^{m} (x_{j_2} - a_{j_1 j_2 s_1}),\\
\dot{x}_{j_2} &= a_{j_2 j_1 0} \prod_{l_1=1}^{n} (x_{j_1} - a_{j_2 j_1 l_1}),\\
& j_1, j_2 \in \{1, 2\}; j_1 \neq j_2.
\end{aligned} \tag{5.39}$$

The first integral manifold is

$$\begin{aligned}
& a_{j_2 j_1 0} \Big\{ \frac{1}{n+1} \big[(x_{j_1} - a_{j_2 j_1 l_1})^{n+1} - (x_{j_1 0} - a_{j_2 j_1 l_1})^{n+1} \big]\\
&\quad + \sum_{k=1}^{n-1} \frac{1}{n-k+1} b_{j_2 j_1 k} \big[(x_{j_1} - a_{j_2 j_1 l_1})^{n-k+1} - (x_{j_1 0} - a_{j_2 j_1 l_1})^{n-k+1} \big] \Big\}\\
&= a_{j_1 j_2 0} \Big\{ \frac{1}{m+1} \big[(x_{j_2} - a_{j_1 j_2 s_1})^{m+1} - (x_{j_2 0} - a_{j_1 j_2 s_1})^{m+1} \big]\\
&\quad + \sum_{k=1}^{m-1} \frac{1}{m-k+1} b_{j_1 j_2 k} \big[(x_{j_1} - a_{j_1 j_2 s_1})^{m-k+1} - (x_{j_1 0} - a_{j_1 j_2 s_1})^{m-k+1} \big] \Big\}.
\end{aligned} \tag{5.40}$$

where

$$\begin{aligned}
b_{j_2 j_1 1} &= \sum_{l_2=1, l_2 \neq l_1}^{n} (a_{j_2 j_1 l_1} - a_{j_2 j_1 l_2}),\\
b_{j_2 j_1 2} &= \sum_{\substack{l_2, l_3=1; l_2, l_3 \neq l_1\\ (l_2 < l_3)}}^{n} \prod_{r=2}^{3} (a_{j_2 j_1 l_1} - a_{j_2 j_1 l_r}), \cdots,\\
b_{j_2 j_1 k} &= \sum_{\substack{l_2, l_3, \cdots, l_{k+1}=1;\\ l_2, l_3, \cdots, l_{k+1} \neq l_1\\ (l_2 < l_3 < \cdots < l_{k+1})}}^{n} \prod_{r=2}^{k+1} (a_{j_2 j_1 l_1} - a_{j_2 j_1 l_r}), \cdots,\\
b_{j_2 j_1 (n-1)} &= \prod_{l_2=1, l_2 \neq l_1}^{n} (a_{j_2 j_1 l_1} - a_{j_2 j_1 l_2});\\
b_{j_1 j_2 1} &= \sum_{s_2=1, s_2 \neq s_1}^{m} (a_{j_1 j_2 s_1} - a_{j_1 j_2 s_2}),\\
b_{j_1 j_2 2} &= \sum_{\substack{s_2, s_3=1; s_2, s_3 \neq s_1\\ (s_2 < s_3)}}^{m} \prod_{r=2}^{3} (a_{j_1 j_2 s_1} - a_{j_1 j_2 s_r}), \cdots,
\end{aligned}$$

$$b_{j_1j_2k} = \sum_{\substack{s_2,s_3,\cdots,s_{k+1}=1;\\ s_2,s_3,\cdots,s_{k+1}\neq s_1\\ (s_2<s_3<\cdots<s_{k+1})}}^{m} \prod_{r=2}^{k+1} (a_{j_1j_2s_1} - a_{j_1j_2s_r}), \cdots,$$

$$b_{j_1j_2(m-1)} = \prod_{s_1=1,s_2\neq s_1}^{m} (a_{j_1j_2s_1} - a_{j_1j_2s_2}). \tag{5.41}$$

The nonsingular equilibrium network with $m \times n$ is defined as

$$\bigcup_{s=1}^{m}\bigcup_{l=1}^{n}\underbrace{(a_{j_1j_1s}, a_{j_2j_2l})}_{\text{XX}}$$

$$\equiv \begin{Bmatrix} (a_{j_1j_2m}, a_{j_2j_1}1) & (a_{j_1j_2m}, a_{j_2j_1}2) & \cdots & (a_{j_1j_2m}, a_{j_2j_1n}) \\ (a_{j_1j_2(m-1)}, a_{j_2j_1}1) & (a_{j_1j_1(m-1)}, a_{j_2j_1}2) & \cdots & (a_{j_1j_2(m-1)}, a_{j_2j_1n}) \\ \vdots & \vdots & \vdots & \vdots \\ (a_{j_1j_2}1, a_{j_2j_2}1) & (a_{j_1j_1}1, a_{j_2j_1}2) & \cdots & (a_{j_1j_2}1, a_{j_2j_1n}) \end{Bmatrix}_{m\times n}. \tag{5.42}$$

The equilibrium of $(x_{j_2}^*, x_{j_1}^*) = (a_{j_1j_2s_1}, a_{j_2j_1l_1})$ $(s_1, s_2 \in \{1, 2, \cdots, m\}, s_1 \neq s_2; l_1, l_2 \in \{1, 2, \cdots, n\}, l_1 \neq l_2)$ *has the following properties.*

- For $a_{j_1j_2}0 \prod_{s_2=1,s_2\neq s_1}^{m} (a_{j_1j_2s_1} - a_{j_1j_2s_2}) > 0$ and $a_{j_2j_1}0 \prod_{l_2=1,l_2\neq l_1}^{n} (a_{j_2j_1l_1} - a_{j_2j_1l_2}) > 0$,

$$\left(a_{j_1j_2s_1}, a_{j_2j_1}1\right) = \underbrace{(\text{UP}_+, \text{UP}_+)}_{\text{positive saddle}}. \tag{5.43}$$

The equilibrium of $(x_{j_2}^*, x_{j_1}^*) = (a_{j_1j_2s_1}, a_{j_2j_1l_1})$ is a $(\text{UP}_+, \text{UP}_+)$-positive saddle.

- For $a_{j_1j_2}0 \prod_{s_2=1,s_2\neq s_1}^{m} (a_{j_1j_2s_1} - a_{j_1j_2s_2}) < 0$ and $a_{j_2j_1}0 \prod_{l_2=1,l_2\neq l_1}^{n} (a_{j_2j_1l_1} - a_{j_2j_1l_2}) > 0$,

$$\left(a_{j_1j_2s_1}, a_{j_2j_1}1\right) = \underbrace{(\text{DP}_+, \text{DP}_-)}_{\text{CCW center}} \tag{5.44}$$

The equilibrium of $(x_{j_2}^*, x_{j_1}^*) = (a_{j_1j_2s_1}, a_{j_2j_1l_1})$ is a $(\text{DP}_+, \text{DP}_-)$-counter-clockwise center.

- For $a_{j_1j_2}0 \prod_{s_2=1,s_2\neq s_1}^{m} \left(a_{j_1j_2s_1} - a_{j_1j_2s_2}\right) > 0$ and $a_{j_2j_1}0 \prod_{l_2=1,l_2\neq l_1}^{n} \left(a_{j_2j_1l_1} - a_{j_2j_1l_2}\right) < 0$,

$$\left(a_{j_1j_2s_1}, a_{j_2j_1l_1}\right) = \underbrace{(\text{DP}_-,\text{DP}_+)}_{\text{CW center}}. \tag{5.45}$$

The equilibrium of $(x_{j_2}^*, x_{j_1}^*) = (a_{j_1j_2s_1}, a_{j_2j_1l_1})$ is a $(\mathrm{DP}_-, \mathrm{DP}_+)$-clockwise center.

- For $a_{j_1j_20}\prod_{s_2=1, s_2\neq s_1}^{m}(a_{j_1j_2s_1} - a_{j_1j_2s_2}) < 0$ and $a_{j_2j_10}\prod_{l_2=1, l_2\neq l_1}^{n}(a_{j_2j_1l_1} - a_{j_2j_1l_2}) < 0$,

$$(a_{j_1j_2s_1}, a_{j_2j_1l_1}) = \underbrace{(\mathrm{UP}_-,\mathrm{UP}_-)}_{\text{negative saddle}}. \tag{5.46}$$

The equilibrium of $(x_{j_2}^*, x_{j_1}^*) = (a_{j_1j_2s_1}, a_{j_2j_1l_1})$ is a $(\mathrm{UP}_-, \mathrm{UP}_-)$-negative saddle.

(iv) For $m = 2m_1 + 1$ and $n = 2n_1 + 1$, the equilibrium of $(x_{j_2}^*, x_{j_1}^*) = (a_{j_1j_21}, a_{j_2j_11})$ has the following bifurcation properties.

(iv$_1$) For $a_{j_1j_20} > 0$ and $a_{j_2j_10} > 0$, there is a $((2m_1{+}1)^{\text{th}}\mathrm{UP}_+, (2n_1{+}1)^{\text{th}}\mathrm{UP}_+)$-positive saddle equilibrium as

$$(a_{j_1j_21}, a_{j_2j_11}) = \underbrace{((2m_1{+}1)^{\text{th}}\mathrm{UP}_+, (2n_1{+}1)^{\text{th}}\mathrm{UP}_+)}_{((2m_1+1),(2n_1+1))\text{-positive saddle}}. \tag{5.47}$$

There are three following $((2m_1{+}1)^{\text{th}}\mathrm{UP}_+, (2n_1{+}1)^{\text{th}}\mathrm{UP}_+)$-positive saddle appearing and switching bifurcations.

(iv$_{1a}$) The $((2m_1+1)^{\text{th}}\mathrm{UP}_+,(2n_1+1)^{\text{th}}\mathrm{UP}_+)$-positive saddle appearing bifurcation is from a $(\mathrm{UP}_+,\mathrm{UP}_+)$-positive saddle to a $(2m_1+1)\times(2n_1+1)$-equilibrium network as

$$\begin{aligned}\underbrace{(a_{j_1j_21}, a_{j_2j_11})}_{(\mathrm{UP}_+,\mathrm{UP}_+)\text{-positive saddle}} &\rightleftharpoons \underbrace{((2m_1{+}1)^{\text{th}}\mathrm{UP}_+, (2n_1{+}1)^{\text{th}}\mathrm{UP}_+)}_{((2m_1+1),(2n_1+1))\text{-positive saddle}}\\ &\rightleftharpoons \bigcup_{r=1}^{2m_1+1}\bigcup_{s=1}^{2n_1+1}\underbrace{(a_{j_1j_2r}, a_{j_2j_1s})}_{\mathrm{XX}},\end{aligned} \tag{5.48}$$

where

$$\bigcup_{r=1}^{2m_1+1}\bigcup_{s=1}^{2n_1+1}\underbrace{(a_{j_1j_2r}, a_{j_2j_1s})}_{\mathrm{XX}} = \left\{\begin{matrix} \underbrace{(\mathrm{UP}_+,\mathrm{UP}_+)}_{\text{positive saddle}} & \underbrace{(\mathrm{DP}_-,\mathrm{DP}_+)}_{\text{center (CW)}} & \cdots & \underbrace{(\mathrm{UP}_+,\mathrm{UP}_+)}_{\text{positive saddle}} \\ \underbrace{(\mathrm{DP}_+,\mathrm{DP}_-)}_{\text{center (CCW)}} & \underbrace{(\mathrm{UP}_-,\mathrm{UP}_-)}_{\text{negative saddle}} & \cdots & \underbrace{(\mathrm{DP}_+,\mathrm{DP}_-)}_{\text{center (CCW)}} \\ \vdots & \vdots & \vdots & \vdots \\ \underbrace{(\mathrm{UP}_+,\mathrm{UP}_+)}_{\text{positive saddle}} & \underbrace{(\mathrm{DP}_-,\mathrm{DP}_+)}_{\text{center (CW)}} & \cdots & \underbrace{(\mathrm{UP}_+,\mathrm{UP}_+)}_{\text{positive saddle}} \end{matrix}\right\}_{(2m_1+1)\times(2n_1+1)}. \tag{5.49}$$

(iv_{1b}) The $((2m_1+1)^{\text{th}}\mathrm{UP}_+,(2n_1+1)^{\text{th}}\mathrm{UP}_+)$-positive saddle appearing bifurcation is from a $p_1 \times q_1$ to $p_2 \times q_2$ equilibrium network as

$$\bigcup_{r_1=1}^{p_1}\bigcup_{s_1=1}^{q_1}\underbrace{(a_{j_1 j_2 r_1}, a_{j_2 j_1 s_1})}_{(m_{r_1}^{(1)},\, n_{s_1}^{(1)})\text{-XX}} \rightleftharpoons \underbrace{((2m_1+1)^{\text{th}}\mathrm{UP}_+, (2n_1+1)^{\text{th}}\mathrm{UP}_+)}_{((2m_1+1),(2n_1+1))\text{-positive saddle}}$$

$$\rightleftharpoons \bigcup_{r_2=1}^{p_2}\bigcup_{s_2=1}^{q_2}\underbrace{(a_{j_1 j_2 r_2}, a_{j_2 j_1 s_2})}_{(m_{r_2}^{(2)},n_{s_2}^{(2)})\text{-XX}} \tag{5.50}$$

where

$2m_1 - \Sigma_{r_1=1}^{p_1} m_{r_1}^{(1)} = 2\Sigma_{s=1}^{l_1} m_s$
with l_1-quadratic polynomials without real roots
and s^{th}-quadratic polynomial with power m_s,
$2n_1 + 1 - \Sigma_{s_1=1}^{q_1} n_{s_1}^{(1)} = 2\Sigma_{l=1}^{l_2} n_l$
with l_2-quadratic polynomials without real roots
and l^{th}-quadratic polynomial with power n_1;
$\Sigma_{s_2=1}^{p_2} m_{r_2}^{(2)} = 2m_1 + 1,\ \Sigma_{s_2=1}^{q_2} n_{s_2}^{(2)} = 2n_1 + 1;$ for $i = 1, 2,$

$$\underbrace{((m_{p_i}^{(i)})^{\text{th}}\,\mathrm{XX}, (n_{q_i}^{(i)})^{\text{th}}\,\mathrm{XX})}_{(m_{p_i}^{(i)},n_{q_i}^{(i)})\text{-XX}}$$

$$\in \left\{\begin{array}{l}\underbrace{((2m_{p_i1}^{(i)}+1)^{\text{th}}\mathrm{UP}_+, (2n_{q_i1}^{(i)}+1)^{\text{th}}\mathrm{UP}_+)}_{((2m_{p_i1}^{(i)}+1),(2n_{q_i1}^{(i)}+1))\text{-positive saddle}},\ \underbrace{((2m_{p_i1}^{(i)})^{\text{th}}\mathrm{II}_+, (2n_{q_i1}^{(i)})^{\text{th}}\mathrm{II}_+)}_{((2m_{p_i1}^{(i)}),(2n_{q_i1}^{(i)}))\text{-double-infection saddle}},\\ \underbrace{((2m_{p_i1}^{(i)})^{\text{th}}\mathrm{US}, (2n_{q_i1}^{(i)}+1)^{\text{th}}\mathrm{UP})}_{((2m_{p_i1}^{(i)}+1),(2n_{q_i1}^{(i)}+1))\text{-up-parabola upper-saddle}},\ \underbrace{((2m_{p_i1}^{(i)}+1)^{\text{th}}\mathrm{UP}, (2n_{q_i1}^{(i)})^{\text{th}}\mathrm{US})}_{((2m_{p_i1}^{(i)}+1),(2n_{q_i1}^{(i)}))\text{-up-parabola upper-saddle}}\end{array}\right\};$$

$$\underbrace{((m_{p_i}^{(i)})^{\text{th}}\,\mathrm{XX}, (n_{1}^{(i)})^{\text{th}}\,\mathrm{XX})}_{(m_{p_i}^{(i)},n_{1}^{(i)})\text{-XX}}$$

$$\in \left\{\begin{array}{l}\underbrace{((2m_{p_i1}^{(i)}+1)^{\text{th}}\mathrm{UP}_+, (2n_{11}^{(i)}+1)^{\text{th}}\mathrm{UP}_+)}_{((2m_{p_i1}^{(i)}+1),(2n_{11}^{(i)}+1))\text{-positive saddle}},\ \underbrace{((2m_{p_i1}^{(i)})^{\text{th}}\mathrm{DI}_-, (2n_{11}^{(i)})^{\text{th}}\mathrm{DI}_+)}_{((2m_{p_i1}^{(i)}),(2n_{11}^{(i)}))\text{-double-infection saddle}},\\ \underbrace{((2m_{p_i1}^{(i)})^{\text{th}}\mathrm{US}, (2n_{11}^{(i)}+1)^{\text{th}}\mathrm{UP})}_{((2m_{p_i1}^{(i)}),(2n_{11}^{(i)}+1))\text{-up-parabola upper-saddle}},\ \underbrace{((2m_{p_i1}^{(i)}+1)^{\text{th}}\mathrm{DP}, (2n_{11}^{(i)})^{\text{th}}\mathrm{LS})}_{((2m_{p_i1}^{(i)}+1),(2n_{11}^{(i)}))\text{-down-parabola lower-saddle}}\end{array}\right\};$$

$$
\underbrace{((m_1^{(i)})^{\text{th}}\,\text{XX},\,(n_{q_i}^{(i)})^{\text{th}}\,\text{XX})}_{(m_1^{(i)},n_{q_i}^{(i)})\text{-XX}}
\in \left\{ \begin{array}{l}
\underbrace{((2m_{11}{}^{(i)}+1)^{\text{th}}\text{UP}_+,\,(2n_{11}^{(i)}+1)^{\text{th}}\text{UP}_+)}_{((2m_{11}^{(i)}+1),(2n_{11}^{(i)}+1))\text{-positive saddle}},\ \underbrace{((2m_{p_i1}^{(i)})^{\text{th}}\text{DI}_-,\,(2n_{q_i1}^{(i)})^{\text{th}}\text{DI}_+)}_{((2m_{11}^{(i)}),(2n_{q_i1}^{(i)}))\text{-double-infection saddle}},\\
\underbrace{((2m_{11}^{(i)})^{\text{th}}\text{DP}_+,\,(2n_{q_i1}^{(i)}+1)^{\text{th}}\text{DP}_-)}_{((2m_{11}^{(i)}),(2n_{q_i1}^{(i)}+1))\text{-down-parabola lower-saddle}},\ \underbrace{((2m_{11}^{(i)}+1)^{\text{th}}\text{DP}_-,\,(2n_{q_i1}^{(i)})^{\text{th}}\text{US})}_{((2m_{11}^{(i)}+1),(2n_{q_i1}^{(i)}))\text{-up-parabola upper-saddle}}
\end{array} \right\};
$$

$$
\underbrace{((m_1^{(i)})^{\text{th}}\,\text{XX},\,(n_1^{(i)})^{\text{th}}\,\text{XX})}_{(m_1^{(i)},n_1^{(i)})\text{-XX}}
\in \left\{ \begin{array}{l}
\underbrace{((2m_{11}{}^{(i)}+1)^{\text{th}}\text{UP}_+,\,(2n_{q_i1}^{(i)}+1)^{\text{th}}\text{UP}_+)}_{((2m_{11}^{(i)}+1),(2n_{11}^{(i)}+1))\text{-positive saddle}},\ \underbrace{((2m_{11}^{(i)})^{\text{th}}\text{II}_-,\,(2n_{11}^{(i)})^{\text{th}}\text{II}_-)}_{((2m_{11}^{(i)}),(2n_{11}^{(i)}))\text{-double-infection saddle}},\\
\underbrace{((2m_{11}^{(i)})^{\text{th}}\text{LS},\,(2n_{11}^{(i)}+1)^{\text{th}}\text{DP})}_{((2m_{11}^{(i)}),(2n_{11}^{(i)}+1))\text{-down-parabola lower-saddle}},\ \underbrace{((2m_{11}^{(i)}+1)^{\text{th}}\text{UP},\,(2n_{11}^{(i)})^{\text{th}}\text{LS})}_{((2m_{11}^{(i)}+1),(2n_{11}^{(i)}))\text{-down-parabola lower-saddle}}
\end{array} \right\};
\tag{5.51}
$$

(iv$_{1c}$) The $((2m_1+1)^{\text{th}}SO,(2n_1+1)^{\text{th}}SO)$-source switching bifurcation is for the switching of two $p \times q$ equilibrium networks as

$$
\begin{aligned}
\bigcup_{r_1=1}^{p}\bigcup_{s_1=1}^{q}\underbrace{(a_{j_1j_2r_1},a_{j_2j_1s_1})}_{(m_{r_1}^{(1)},\,n_{s_1}^{(1)})\text{-XX}} &\rightleftharpoons \underbrace{((2m_1+1)^{\text{th}}\text{UP}_+,\,(2n_1+1)^{\text{th}}\text{UP}_+)}_{((2m_1+1),(2n_1+1))\text{-positive saddle}}\\
&\rightleftharpoons \bigcup_{r_2=1}^{p}\bigcup_{s_2=1}^{q}\underbrace{(a_{j_1j_2r_2},a_{j_2j_1s_2})}_{(m_{r_2}^{(2)},\,n_{s_2}^{(2)})\text{-XX}}
\end{aligned}
\tag{5.52}
$$

with $\Sigma_{r_i=1}^{p} m_{r_i} = 2m_1+1$ and $\Sigma_{s_i=1}^{q} n_{s_i} = 2n_1+1$ for $i = 1, 2$.

(iv$_2$) For $a_{j_1j_10} < 0$ and $a_{j_2j_20} > 0$, there is a $((2m_1+1)^{\text{th}}\text{DP}_+,\,(2n_1+1)^{\text{th}}\text{DP}_-)$-CCW center equilibrium as

$$
(a_{j_1j_21},a_{j_2j_11}) = \underbrace{((2m_1+1)^{\text{th}}\text{DP}_+,\,(2n_1+1)^{\text{th}}\text{DP}_-)}_{((2m_1+1),(2n_1+1))\text{-CCW center}}.
\tag{5.53}
$$

There are the following three $((2m_1+1)^{\text{th}}\text{DP}_+,\,(2n_1+1)^{\text{th}}\text{DP}_-)$-CCW center appearing and switching bifurcations.

(iv_{2a}) The $((2m_1+1)^{\text{th}}\text{DP}_+,(2n_1+1)^{\text{th}}\text{DP}_-)$-CCW center appearing bifurcation is from a $(\text{DP}_+, \text{DP}_-)$-CCW center to a $(2m_1+1)\times(2n_1+1)$ network as

$$\underbrace{(a_{j_1j_21}, a_{j_2j_11})}_{(\text{DP}_+,\text{DP}_-)\text{-CCW center}} \rightleftharpoons \underbrace{((2m_1+1)^{\text{th}}\text{DP}_+,(2n_1+1)^{\text{th}}\text{DP}_-)}_{((2m_1+1),(2n_1+1))\text{-CCW center}}$$
$$\rightleftharpoons \bigcup_{r=1}^{2m_1+1}\bigcup_{s=1}^{2n_1+1}\underbrace{(a_{j_1j_2r}, a_{j_2j_1s})}_{\text{XX}} \tag{5.54}$$

where

$$\bigcup_{r=1}^{2m_1+1}\bigcup_{s=1}^{2n_1+1}\underbrace{(a_{j_1j_2r}, a_{j_2j_1s})}_{\text{XX}} = \left\{\begin{matrix} \underbrace{(\text{DP}_+,\text{DP}_-)}_{\text{center (CCW)}} & \underbrace{(\text{UP}_-,\text{UP}_-)}_{\text{saddle }(-)} & \cdots & \underbrace{(\text{DP}_+,\text{DP}_-)}_{\text{center (CCW)}} \\ \underbrace{(\text{UP}_+,\text{UP}_+)}_{\text{saddle }(+)} & \underbrace{(\text{DP}_-,\text{DP}_+)}_{\text{center (CW)}} & \cdots & \underbrace{(\text{UP}_+,\text{UP}_+)}_{\text{saddle }(+)} \\ \vdots & \vdots & \vdots & \vdots \\ \underbrace{(\text{DP}_+,\text{DP}_-)}_{\text{center (CCW)}} & \underbrace{(\text{UP}_-,\text{UP}_-)}_{\text{saddle }(-)} & \cdots & \underbrace{(\text{DP}_+,\text{DP}_-)}_{\text{center (CCW)}} \end{matrix}\right\}_{(2m_1+1)\times(2n_1+1)}. \tag{5.55}$$

(iv_{2b}) The $((2m_1+1)^{\text{th}}\text{DP}_+,(2n_1+1)^{\text{th}}\text{DP}_-)$-CCW center appearing bifurcation is from $p_1 \times q_1$ to $p_2 \times q_2$ equilibrium networks as

$$\bigcup_{r_1=1}^{p_1}\bigcup_{s_1=1}^{q_1}\underbrace{(a_{j_1j_2r_1}, a_{j_2j_1s_1})}_{(m_{r_1}^{(1)},\, n_{s_1}^{(1)})\text{-XX}} \rightleftharpoons \underbrace{((2m_1+1)^{\text{th}}\text{DP}_+,(2n_1+1)^{\text{th}}\text{DP}_-)}_{((2m_1+1),(2n_1+1))\text{-CCW center}}$$
$$\rightleftharpoons \bigcup_{r_2=1}^{p_2}\bigcup_{s_2=1}^{q_2}\underbrace{(a_{j_1j_2r_2}, a_{j_2j_1s_2})}_{(m_{r_2}^{(2)},\, n_{s_2}^{(2)})\text{-XX}} \tag{5.56}$$

where

$2m_1+1-\Sigma_{r_1=1}^{p_1} m_{r_1}^{(1)} = 2\Sigma_{s=1}^{l_1} m_s$
with l_1-quadratic polynomials without real roots
and s^{th}-quadratic polynomial with power m_s,
$2n_1+1-\Sigma_{s_1=1}^{q_1} n_{s_1}^{(1)} = 2\Sigma_{l=1}^{l_2} n_l$
with l_2-quadraitc polynomials without real roots
and l^{th}-quadratic polynomial with power n_l;
$\Sigma_{r_2=1}^{p_2} m_{r_2}^{(2)} = 2m_1+1,\ \Sigma_{s_2=1}^{q_2} n_{s_2}^{(2)} = 2n_1+1$; for $i = 1, 2$

$$
\begin{aligned}
&\underbrace{((m_{p_i}^{(i)})^{\rm th}{\rm XX},(n_{q_i}^{(i)})^{\rm th}{\rm XX})}_{(m_{p_i}^{(i)},n_{q_i}^{(i)})\text{-XX}}\\
&\in\left\{\begin{array}{l}
\underbrace{((2m_{p_i1}^{(i)}+1)^{\rm th}{\rm DP}_+,\,(2n_{q_i1}^{(i)}+1)^{\rm th}{\rm DP}_-)}_{((2m_{p_i1}^{(i)}+1),(2n_{q_i1}^{(i)}+1))\text{-CCW center}},\ \underbrace{((2m_{p_i1}^{(i)})^{\rm th}{\rm DI}_+,\,(2n_{q_i1}^{(i)})^{\rm th}{\rm DI}_-)}_{((2m_{p_i1}^{(i)}),(2n_{q_i1}^{(i)}))\text{-double-inflection saddle}},\\
\underbrace{((2m_{p_i1}^{(i)})^{\rm th}{\rm LS},(2n_{q_i1}^{(i)}+1)^{\rm th}{\rm DP})}_{((2m_{p_i1}^{(i)}),(2n_{q_i1}^{(i)}+1))\text{-down-parabola lower-saddle}},\ \underbrace{((2m_{p_i1}^{(i)}+1)^{\rm th}{\rm DP},(2n_{q_i1}^{(i)})^{\rm th}{\rm US})}_{((2m_{p_i1}^{(i)}+1),(2n_{q_i1}^{(i)}))\text{-down-parabola upper-saddle}}
\end{array}\right\};\\
&\underbrace{((m_{p_i}^{(i)})^{\rm th}{\rm XX},(n_{1}^{(i)})^{\rm th}{\rm XX})}_{(m_{p_i}^{(i)},n_{1}^{(i)})\text{-XX}}\\
&\in\left\{\begin{array}{l}
\underbrace{((2m_{p_i1}^{(i)}+1)^{\rm th}{\rm DP}_+,\,(2n_{11}^{(i)}+1)^{\rm th}{\rm DP}_-)}_{((2m_{p_i1}^{(i)}+1),(2n_{11}^{(i)}+1))\text{-CCW center}},\ \underbrace{((2m_{p_i1}^{(i)})^{\rm th}{\rm II}_-,\,(2n_{11}^{(i)})^{\rm th}{\rm II}_-)}_{((2m_{p_i1}^{(i)}),(2n_{11}^{(i)}))\text{-double-inflection saddle}},\\
\underbrace{((2m_{p_i1}^{(i)})^{\rm th}{\rm LS},(2n_{11}^{(i)}+1)^{\rm th}{\rm DP})}_{((2m_{p_i1}^{(i)}),(2n_{11}^{(i)}+1))\text{-down-parabola lower-saddle}},\ \underbrace{((2m_{p_i1}^{(i)}+1)^{\rm th}{\rm UP},(2n_{11}^{(i)})^{\rm th}{\rm LS})}_{((2m_{p_i1}^{(i)}+1),(2n_{11}^{(i)}))\text{-up-parabola lower-saddle}}
\end{array}\right\};\\
&\underbrace{((m_{1}^{(i)})^{\rm th}{\rm XX},(n_{q_i}^{(i)})^{\rm th}{\rm XX})}_{(m_{1}^{(i)},n_{q_i}^{(i)})\text{-XX}}\\
&\in\left\{\begin{array}{l}
\underbrace{((2m_{11}^{(i)}+1)^{\rm th}{\rm DP}_+,\,(2n_{q_i1}^{(i)}+1)^{\rm th}{\rm DP}_-)}_{((2m_{11}^{(i)}+1),(2n_{q_i1}^{(i)}+1))\text{-CCW center}},\ \underbrace{((2m_{11}^{(i)})^{\rm th}{\rm II}_+,\,(2n_{q_i1}^{(i)})^{\rm th}{\rm II}_+)}_{((2m_{11}^{(i)}),(2n_{q_i1}^{(i)}))\text{-double-saddle}},\\
\underbrace{((2m_{11}^{(i)})^{\rm th}{\rm US},(2n_{q_i1}^{(i)}+1)^{\rm th}{\rm UP})}_{((2m_{11}^{(i)}),(2n_{q_i1}^{(i)}+1))\text{-up-parabola upper-saddle}},\ \underbrace{((2m_{11}^{(i)}+1)^{\rm th}{\rm DP},(2n_{q_i1}^{(i)})^{\rm th}{\rm US})}_{((2m_{11}^{(i)}+1),(2n_{q_i1}^{(i)}))\text{-down-parabola upper-saddle}}
\end{array}\right\};\\
&\underbrace{((m_{1}^{(i)})^{\rm th}{\rm XX},(n_{1}^{(i)})^{\rm th}{\rm XX})}_{(m_{1}^{(i)},n_{1}^{(i)})\text{-XX}}\\
&\in\left\{\begin{array}{l}
\underbrace{((2m_{11}^{(i)}+1)^{\rm th}{\rm DP}_+,\,(2n_{11}^{(i)}+1)^{\rm th}{\rm DP}_-)}_{((2m_{11}^{(i)}+1),(2n_{11}^{(i)}+1))\text{-CCW center}},\ \underbrace{((2m_{11}^{(i)})^{\rm th}{\rm DI}_-,\,(2n_{11}^{(i)})^{\rm th}{\rm DI}_+)}_{((2m_{11}^{(i)}),(2n_{11}^{(i)}))\text{-double-inflection saddle}},\\
\underbrace{((2m_{11}^{(i)})^{\rm th}{\rm US},(2n_{11}^{(i)}+1)^{\rm th}{\rm UP})}_{((2m_{11}^{(i)}),(2n_{11}^{(i)}+1))\text{-up-parabola upper-saddle}},\ \underbrace{((2m_{11}^{(i)}+1)^{\rm th}{\rm DP},(2n_{11}^{(i)})^{\rm th}{\rm LS})}_{((2m_{11}^{(i)}+1),(2n_{11}^{(i)}))\text{-down-parabola lower-saddle}}
\end{array}\right\}.
\end{aligned}
\tag{5.57}
$$

(iv_{2c}) The $((2m_1+1)^{\rm th}{\rm DP}_+,\,(2n_1+1)^{\rm th}{\rm DP}_-)$-CCW center switching bifurcations is for two $p\times q$ equilibrium networks as

$$\bigcup_{r_1=1}^{p}\bigcup_{s_1=1}^{q}\underbrace{(a_{j_1j_2r_1},a_{j_2j_1s_1})}_{(m_{r_1}^{(1)},n_{s_1}^{(1)})\text{-XX}} \rightleftharpoons \underbrace{((2m_1+1)^{\text{th}}\text{DP}_+,(2n_1+1)^{\text{th}}\text{DP}_-)}_{((2m_1+1),(2n_1+1))\text{-CCW center}}$$
$$\rightleftharpoons \bigcup_{r_2=1}^{p}\bigcup_{s_2=1}^{q}\underbrace{(a_{j_1j_2r_2},a_{j_2j_1s_2})}_{(m_{r_2}^{(2)},n_{s_2}^{(2)})\text{-XX}} \tag{5.58}$$

with $\Sigma_{r_i=1}^{p}m_{r_i}=2m_1+1$ and $\Sigma_{s_i=1}^{q}n_{s_i}=2n_1+1$ for $i=1,2$.

(iv$_3$) For $a_{j_1j_10}>0$ and $a_{j_2j_20}<0$, there is a $((2m_1+1)^{\text{th}}\text{DP}_-,(2n_1+1)^{\text{th}}\text{DP}_+)$-CW center equilibrium as

$$(a_{j_1j_21},a_{j_2j_11})=\underbrace{((2m_1+1)^{\text{th}}\text{DP}_-,(2n_1+1)^{\text{th}}\text{DP}_+)}_{((2m_1+1),(2n_1+1))\text{-CW center}}. \tag{5.59}$$

There are the following three $((2m_1+1)^{\text{th}}\text{DP}_-,(2n_1+1)^{\text{th}}\text{DP}_+)$-CW center appearing and switching bifurcations.

(iv$_{3a}$) The $((2m_1+1)^{\text{th}}\text{DP}_-,(2n_1+1)^{\text{th}}\text{DP}_+)$-CW center appearing bifurcation is from a $(\text{DP}_-,\text{DP}_+)$-CW center to a $(2m_1+1)\times(2n_1+1)$-equilibrium network as

$$\underbrace{(a_{j_1j_21},a_{j_2j_11})}_{(\text{DP}_-,\text{DP}_+)\text{-CW center}} \rightleftharpoons \underbrace{((2m_1+1)^{\text{th}}\text{DP}_-,(2n_1+1)^{\text{th}}\text{DP}_+)}_{((2m_1+1),(2n_1+1))\text{-CW center}}$$
$$\rightleftharpoons \bigcup_{r=1}^{2m_1+1}\bigcup_{s=1}^{2n_1+1}\underbrace{(a_{j_1j_1r},a_{j_2j_2s})}_{\text{XX}} \tag{5.60}$$

where

$$\bigcup_{r=1}^{2m_1+1}\bigcup_{s=1}^{2n_1+1}\underbrace{(a_{j_1j_2r},a_{j_2j_1s})}_{\text{XX}}$$
$$=\left\{\begin{matrix} \underbrace{(\text{DP}_-,\text{DP}_+)}_{\text{center (CW)}} & \underbrace{(\text{UP}_+,\text{UP}_+)}_{\text{saddle }(+)} & \cdots & \underbrace{(\text{DP}_-,\text{DP}_+)}_{\text{center (CW)}} \\ \underbrace{(\text{UP}_-,\text{UP}_-)}_{\text{saddle }(-)} & \underbrace{(\text{DP}_+,\text{DP}_-)}_{\text{center (CCW)}} & \cdots & \underbrace{(\text{UP}_-,\text{UP}_-)}_{\text{saddle }(-)} \\ \vdots & \vdots & \vdots & \vdots \\ \underbrace{(\text{DP}_-,\text{DP}_+)}_{\text{center (CW)}} & \underbrace{(\text{UP}_+,\text{UP}_+)}_{\text{saddle }(+)} & \cdots & \underbrace{(\text{DP}_-,\text{DP}_+)}_{\text{center (CW)}} \end{matrix}\right\}_{(2m_1+1)\times(2n_1+1)}. \tag{5.61}$$

(iv$_{3b}$) The $((2m_1+1)^{\text{th}}\text{DP}_-,(2n_1+1)^{\text{th}}\text{DP}_+)$-CW center appearing bifurcation is from a $p_1\times q_1$ to $p_2\times q_2$ equilibrium network as

$$\bigcup_{r_1=1}^{p_1}\bigcup_{s_1=1}^{q_1}\underbrace{(a_{j_1j_2r_1},a_{j_2j_1s_1})}_{(m^{(1)}_{r_1},n^{(1)}_{s_1})\text{-XX}} \rightleftharpoons \underbrace{((2m_1+1)^{\text{th}}\text{DP}_-,(2n_1+1)^{\text{th}}\text{DP}_+)}_{((2m_1+1),(2n_1+1))\text{-CW center}}$$
$$\rightleftharpoons \bigcup_{r_2=1}^{p_2}\bigcup_{s_2=1}^{q_2}\underbrace{(a_{j_1j_2r_2},a_{j_2j_1s_2})}_{(m^{(2)}_{r_2},n^{(2)}_{s_2})\text{-XX}} \tag{5.62}$$

where

$2m_1+1-\Sigma_{r_1=1}^{p_1}m^{(1)}_{r_1}=2\Sigma_{s=1}^{l_1}m_s$
with l_1-quadratic polynomials without real roots
and s^{th}-quadratic polynomial with power m_s,
$2n_1+1-\Sigma_{s_1=1}^{q_1}n^{(1)}_{s_1}=2\Sigma_{l=1}^{l_2}n_l$
with l_2-quadraitc polynomials without real roots
and l^{th}-quadratic polynomial with power n_l;
$\Sigma_{r_2=1}^{p_2}m^{(2)}_{r_2}=2m_1+1,\ \Sigma_{s_2=1}^{q_2}n^{(2)}_{s_2}=2n_1+1$; for $i=1,2$

$$\underbrace{((m^{(i)}_{p_i})^{\text{th}}\text{XX},(n^{(i)}_{q_i})^{\text{th}}\text{XX})}_{(m^{(i)}_{p_i},n^{(i)}_{q_i})\text{-XX}}
\in\left\{\begin{array}{l}\underbrace{((2m^{(i)}_{p_i1}+1)^{\text{th}}\text{DP}_-,(2n^{(i)}_{q_i1}+1)^{\text{th}}\text{DP}_+)}_{((2m^{(i)}_{p_i1}+1),(2n^{(i)}_{q_i1}+1))\text{-CW center}},\ \underbrace{((2m^{(i)}_{p_i1})^{\text{th}}\text{DI}_-,(2n^{(i)}_{q_i1})^{\text{th}}\text{DI}_+)}_{((2m^{(i)}_{p_i1}),(2n^{(i)}_{q_i1}))\text{-double-inflection saddle}},\\ \underbrace{((2m^{(i)}_{p_i1})^{\text{th}}\text{US},(2n^{(i)}_{q_i1}+1)^{\text{th}}\text{DP})}_{((2m^{(i)}_{p_i1}),(2n^{(i)}_{q_i1}+1))\text{-down-parabola upper-saddle}},\ \underbrace{((2m^{(i)}_{p_i1}+1)^{\text{th}}\text{DP},(2n^{(i)}_{q_i1})^{\text{th}}\text{LS})}_{((2m^{(i)}_{p_i1}+1),(2n^{(i)}_{q_i1}))\text{-down-parabola lower-saddle}}\end{array}\right\};$$

$$\underbrace{((m^{(i)}_{p_i})^{\text{th}}\text{XX},(n^{(i)}_{1})^{\text{th}}\text{XX})}_{(m^{(i)}_{p_i},n^{(i)}_{1})\text{-XX}}
\in\left\{\begin{array}{l}\underbrace{((2m^{(i)}_{p_i1}+1)^{\text{th}}\text{DP}_-,(2n^{(i)}_{11}+1)^{\text{th}}\text{DP}_+)}_{((2m^{(i)}_{p_i1}+1),(2n^{(i)}_{11}+1))\text{-CW center}},\ \underbrace{((2m^{(i)}_{p_i1})^{\text{th}}\text{II}_+,(2n^{(i)}_{11})^{\text{th}}\text{II}_+)}_{((2m^{(i)}_{p_i1}),(2n^{(i)}_{11}))\text{-double-inflection saddle}},\\ \underbrace{((2m^{(i)}_{p_i1})^{\text{th}}\text{US},(2n^{(i)}_{11}+1)^{\text{th}}DP)}_{((2m^{(i)}_{p_i1}),(2n^{(i)}_{11}+1))\text{-down-parabola upper-saddle}},\ \underbrace{((2m^{(i)}_{p_i1}+1)^{\text{th}}\text{UP},(2n^{(i)}_{11})^{\text{th}}\text{US})}_{((2m^{(i)}_{p_i1}+1),(2n^{(i)}_{11}))\text{-up-parabola upper-saddle}}\end{array}\right\};$$

$$\underbrace{((m_1^{(i)})^{\text{th}}\text{XX},(n_{q_i}^{(i)})^{\text{th}}\text{XX})}_{(m_1^{(i)},n_{q_i}^{(i)})\text{-XX}}$$
$$\in \left\{ \begin{array}{l} \underbrace{((2m_{11}^{(i)}+1)^{\text{th}}\text{DP}_-,\ (2n_{q_i1}^{(i)}+1)^{\text{th}}\text{DP}_+)}_{((2m_{11}^{(i)}+1),(2n_{q_i1}^{(i)}+1))\text{-CW center}},\ \underbrace{((2m_{11}^{(i)})^{\text{th}}\text{II}_-,\ (2n_{q_i1}^{(i)})^{\text{th}}\text{II}_-)}_{((2m_{11}^{(i)}),(2n_{q_i1}^{(i)}))\text{-double-inflection saddle}}, \\ \underbrace{((2m_{11}^{(i)})^{\text{th}}\text{LS},(2n_{q_i1}^{(i)}+1)^{\text{th}}\text{UP})}_{((2m_{11}^{(i)}),(2n_{q_i1}^{(i)}+1))\text{-up-parabola lower-saddle}},\ \underbrace{((2m_{11}^{(i)}+1)^{\text{th}}\text{DP},(2n_{q_i1}^{(i)})^{\text{th}}\text{LS})}_{((2m_{11}^{(i)}+1),(2n_{q_i1}^{(i)}))\text{-down-parabola lower-saddle}} \end{array} \right\};$$
$$\underbrace{((m_1^{(i)})^{\text{th}}\text{XX},(n_1^{(i)})^{\text{th}}\text{XX})}_{(m_1^{(i)},n_1^{(i)})\text{-XX}}$$
$$\in \left\{ \begin{array}{l} \underbrace{((2m_{11}^{(i)}+1)^{\text{th}}\text{DP}_-,\ (2n_{11}^{(i)}+1)^{\text{th}}\text{DP}_+)}_{((2m_{11}^{(i)}+1),(2n_{11}^{(i)}+1))\text{-CW center}},\ \underbrace{((2m_{11}^{(i)})^{\text{th}}\text{DI}_+,\ (2n_{11}^{(i)})^{\text{th}}\text{DI}_-)}_{((2m_{11}^{(i)}),(2n_{11}^{(i)}))\text{-double-inflection saddle}}, \\ \underbrace{((2m_{11}^{(i)})^{\text{th}}\text{LS},(2n_{11}^{(i)}+1)^{\text{th}}\text{UP})}_{((2m_{11}^{(i)}),(2n_{11}^{(i)}+1))\text{-up-paraobla lower-saddle}},\ \underbrace{((2m_{11}^{(i)}+1)^{\text{th}}\text{UP},(2n_{11}^{(i)})^{\text{th}}\text{US})}_{((m_{11}^{(i)}),(2n_{11}^{(i)}))\text{-down-paraobla upper-saddle}} \end{array} \right\}. \tag{5.63}$$

(iv_{3c}) The $((2m_1+1)^{\text{th}}\text{DP}_-,(2n_1+1)^{\text{th}}\text{DP}_+)$-CW center switching bifurcation is for the switching of two $p \times q$ equilibrium networks as

$$\begin{aligned} \bigcup_{r_1=1}^{p_1}\bigcup_{s_1=1}^{q_1}\underbrace{(a_{j_1j_2r_1},a_{j_2j_1s_1})}_{(m_{r_1}^{(1)},n_{s_1}^{(1)})\text{-XX}} &\rightleftharpoons \underbrace{((2m_1+1)^{\text{th}}\text{DP}_-,\ (2n_1+1)^{\text{th}}\text{DP}_+)}_{((2m_1+1),(2n_1+1))\text{-CW center}} \\ &\rightleftharpoons \bigcup_{r_2=1}^{p_2}\bigcup_{s_2=1}^{q_2}\underbrace{(a_{j_1j_2r_2},a_{j_2j_1s_2})}_{(m_{r_2}^{(2)},n_{s_2}^{(2)})\text{-XX}} \end{aligned} \tag{5.64}$$

with $\Sigma_{r_i=1}^{p} m_{r_i} = 2m_1+1$ and $\Sigma_{s_i=1}^{q} n_{s_i} = 2n_1+1$ for $i = 1, 2$.

(iv_4) For $a_{j_1j_10} < 0$ and $a_{j_2j_20} > 0$, there is a $((2m_1+1)^{\text{th}}\text{UP}_-,\ (2n_1+1)^{\text{th}}\text{UP}_-)$-negative saddle equilibrium as

$$\left(a_{j_1j_21},a_{j_2j_11}\right) = \underbrace{\left((2m_1+1)^{\text{th}}\text{UP}_-,\ (2n_1+1)^{\text{th}}\text{UP}_-\right)}_{\left((2m_1+1),(2n_1+1)\right)\text{-negative saddle}} \tag{5.65}$$

There are the following three $((2m_1+1)^{\text{th}}\text{UP}_-,(2n_1+1)^{\text{th}}\text{UP}_-)$-negative saddle appearing and switching bifurcations.

(iv_{4a}) The $((2m_1+1)^{\text{th}}\text{UP}_-,(2n_1+1)^{\text{th}}\text{UP}_-)$-negative saddle appearing bifurcation is from a (UP_-,UP_-)-negative saddle to a $(2m_1+1) \times (2n_1+1)$ network as

$$\begin{aligned}\underbrace{(a_{j_1 j_2 1}, a_{j_2 j_1 1})}_{(\mathrm{UP}_-,\mathrm{UP}_-)\text{-negative-saddle}} &\rightleftharpoons \underbrace{((2m_1+1)^{\mathrm{th}}\mathrm{UP}_-, (2n_1+1)^{\mathrm{th}}\mathrm{UP}_-)}_{((2m_1+1),(2n_1+1))\text{-negative saddle}} \\ &\rightleftharpoons \bigcup_{r=1}^{2m_1+1}\bigcup_{s=1}^{2n_1+1}\underbrace{(a_{j_1 j_2 r}, a_{j_2 j_1 s})}_{\mathrm{XX}}\end{aligned} \tag{5.66}$$

where

$$\bigcup_{s=1}^{2m_1+1}\bigcup_{l=1}^{2n_1+1}\underbrace{(a_{j_1 j_2 s}, a_{j_2 j_1 l})}_{\mathrm{XX}} = \left\{\begin{array}{cccc} \underbrace{(\mathrm{UP}_-, \mathrm{UP}_-)}_{\text{saddle }(-)} & \underbrace{(\mathrm{DP}_+, \mathrm{DP}_-)}_{\text{center (CCW)}} & \cdots & \underbrace{(\mathrm{UP}_-, \mathrm{UP}_-)}_{\text{saddle }(-)} \\ \underbrace{(\mathrm{DP}_-, \mathrm{DP}_+)}_{\text{center (CW)}} & \underbrace{(\mathrm{UP}_+, \mathrm{UP}_+)}_{\text{saddle }(+)} & \cdots & \underbrace{(\mathrm{DP}_-, \mathrm{DP}_+)}_{\text{center (CW)}} \\ \vdots & \vdots & \vdots & \vdots \\ \underbrace{(\mathrm{UP}_-, \mathrm{UP}_-)}_{\text{saddle }(-)} & \underbrace{(\mathrm{DP}_+, \mathrm{DP}_-)}_{\text{center (CCW)}} & \cdots & \underbrace{(\mathrm{UP}_-, \mathrm{UP}_-)}_{\text{saddle }(-)} \end{array}\right\}_{(2m_1+1)\times(2n_1+1)} \tag{5.67}$$

(iv_{4b}) The $((2m_1+1)^{\mathrm{th}}\mathrm{UP}_-, (2n_1+1)^{\mathrm{th}}\mathrm{UP}_-)$-negative saddle appearing bifurcation is from a $p_1 \times q_1$ to $p_2 \times q_2$ equilibrium network as

$$\begin{aligned}\bigcup_{r_1=1}^{p_1}\bigcup_{s_1=1}^{q_1}\underbrace{(a_{j_1 j_2 r_1}, a_{j_2 j_1 s_1})}_{(m_{r_1}^{(1)}, n_{s_1}^{(1)})\text{-XX}} &\rightleftharpoons \underbrace{((2m_1+1)^{\mathrm{th}}\mathrm{UP}_-, (2n_1+1)^{\mathrm{th}}\mathrm{UP}_-)}_{((2m_1+1),(2n_1+1))\text{-negative saddle}} \\ &\rightleftharpoons \bigcup_{r_2=1}^{p_2}\bigcup_{s_2=1}^{q_2}\underbrace{(a_{j_1 j_2 r_2}, a_{j_2 j_1 s_2})}_{(m_{r_2}^{(2)}, n_{s_2}^{(2)})\text{-XX}}\end{aligned} \tag{5.68}$$

where

$2m_1 + 1 - \Sigma_{r_1=1}^{p_1} m_{r_1}^{(1)} = 2\Sigma_{s=1}^{l_1} m_s$
with l_1-quadratic polynomials without real roots
and s^{th}-quadratic polynomial with power m_s,
$2n_1 + 1 - \Sigma_{s_1=1}^{q_1} n_{s_1}^{(1)} = 2\Sigma_{l=1}^{l_2} n_l$
with l_2-quadraitc polynomials without real roots
and l^{th}-quadratic polynomial with power n_l;

$$
\begin{array}{l}
\Sigma_{r_2=1}^{p_2} m_{r_2}^{(2)} = 2m_1+1,\ \Sigma_{s_2=1}^{q_2} n_{s_2}^{(2)} = 2n_1+1; \text{for } i=1,2,\\
\underbrace{((m_{p_i}^{(i)})^{\rm th}\mathrm{XX},(n_{q_i}^{(i)})^{\rm th}\mathrm{XX})}_{(m_{p_i}^{(i)},n_{q_i}^{(i)})\text{-XX}}\\
\in \left\{\begin{array}{c}
\underbrace{((2m_{p_i1}^{(i)}+1)^{\rm th}\mathrm{UP}_-,(2n_{q_i1}^{(i)}+1)^{\rm th}\mathrm{UP}_-)}_{((2m_{p_i1}^{(i)}+1),(2n_{q_i1}^{(i)}+1))\text{-nagative saddle}},\ \underbrace{((2m_{p_i1}^{(i)})^{\rm th}\mathrm{II}_-,(2n_{q_i1}^{(i)})^{\rm th}\mathrm{II}_-)}_{((2m_{p_i1}^{(i)}),(2n_{q_i1}^{(i)}))\text{-double-inflection saddle}},\\
\underbrace{((2m_{p_i1}^{(i)})^{\rm th}\mathrm{LS},(2n_{q_i1}^{(i)}+1)^{\rm th}UP)}_{((2m_{p_i1}^{(i)}),(2n_{q_i1}^{(i)}+1))\text{-up-parabola lower-saddle}},\ \underbrace{((2m_{p_i1}^{(i)}+1)^{\rm th}\mathrm{UP},(2n_{q_i1}^{(i)})^{\rm th}\mathrm{LS})}_{((2m_{p_i1}^{(i)}+1),(2n_{q_i1}^{(i)}))\text{-up-parabola lower-saddle}}
\end{array}\right\};\\
\underbrace{((m_{p_i}^{(i)})^{\rm th}\mathrm{XX},(n_{1}^{(i)})^{\rm th}\mathrm{XX})}_{(m_{p_i}^{(i)},n_{1}^{(i)})\text{-XX}}\\
\in \left\{\begin{array}{c}
\underbrace{((2m_{p_i1}^{(i)}+1)^{\rm th}\mathrm{UP}_-,(2n_{q_i1}^{(i)}+1)^{\rm th}\mathrm{UP}_-)}_{((2m_{p_i1}^{(i)}+1),(2n_{q_i1}^{(i)}+1))\text{-nagative saddle}},\ \underbrace{((2m_{p_i1}^{(i)})^{\rm th}\mathrm{DI}_+,(2n_{11}^{(i)})^{\rm th}\mathrm{DI}_-)}_{((2m_{p_i1}^{(i)}),(2n_{11}^{(i)}))\text{-double-inflection saddle}},\\
\underbrace{((2m_{p_i1}^{(i)})^{\rm th}\mathrm{LS},(2n_{11}^{(i)}+1)^{\rm th}\mathrm{UP})}_{((2m_{p_i1}^{(i)}),(2n_{11}^{(i)}+1))\text{-up-paraobla lower-saddle}},\ \underbrace{((2m_{p_i1}^{(i)}+1)^{\rm th}\mathrm{DP},(2n_{11}^{(i)})^{\rm th}\mathrm{US})}_{((2m_{p_i1}^{(i)}+1),(2n_{11}^{(i)}))\text{-down-parabpla upper-saddle}}
\end{array}\right\};\\
\underbrace{((m_{1}^{(i)})^{\rm th}\mathrm{XX},(n_{q_i}^{(i)})^{\rm th}\mathrm{XX})}_{(m_{1}^{(i)},n_{q_i}^{(i)})\text{-XX}}\\
\in \left\{\begin{array}{c}
\underbrace{((2m_{11}^{(i)}+1)^{\rm th}\mathrm{UP}_-,(2n_{q_i1}^{(i)}+1)^{\rm th}\mathrm{UP}_-)}_{((2m_{11}^{(i)}+1),(2n_{q_i1}^{(i)}+1))\text{-nagative saddle}},\ \underbrace{((2m_{11}^{(i)})^{\rm th}\mathrm{DI}_-,(2n_{q_i1}^{(i)})^{\rm th}\mathrm{DI}_+)}_{((2m_{11}^{(i)}),(2n_{q_i1}^{(i)}))\text{-double-inflection saddle}},\\
\underbrace{((2m_{11}^{(i)})^{\rm th}\mathrm{US},(2n_{q_i1}^{(i)}+1)^{\rm th}\mathrm{DP})}_{((2m_{11}^{(i)}),(2n_{q_i1}^{(i)}+1))\text{-down-parabola upper-saddle}},\ \underbrace{((2m_{11}^{(i)}+1)^{\rm th}\mathrm{UP},(2n_{q_i1}^{(i)})^{\rm th}\mathrm{LS})}_{((2m_{11}^{(i)}+1),(2n_{q_i1}^{(i)}))\text{-up-parobla lower-saddle}}
\end{array}\right\};\\
\underbrace{((m_{1}^{(i)})^{\rm th}\mathrm{XX},(n_{1}^{(i)})^{\rm th}\mathrm{XX})}_{(m_{1}^{(i)},n_{1}^{(i)})\text{-XX}}\\
\in \left\{\begin{array}{c}
\underbrace{((2m_{11}^{(i)}+1)^{\rm th}\mathrm{UP}_-,(2n_{11}^{(i)}+1)^{\rm th}\mathrm{UP}_-)}_{((2m_{11}^{(i)}+1),(2n_{q_i1}^{(i)}+1))\text{-nagative saddle}},\ \underbrace{((2m_{11}^{(i)})^{\rm th}\mathrm{II}_+,(2n_{11}^{(i)})^{\rm th}\mathrm{II}_+)}_{((2m_{11}^{(i)}),(2n_{11}^{(i)}))\text{-double-inflection saddle}},\\
\underbrace{((2m_{11}^{(i)})^{\rm th}\mathrm{US},(2n_{11}^{(i)}+1)^{\rm th}\mathrm{DP})}_{((2m_{11}^{(i)}),(2n_{11}^{(i)}+1))\text{-down-parabola upper-saddle}},\ \underbrace{((2m_{11}^{(i)}+1)^{\rm th}\mathrm{DP},(2n_{11}^{(i)})^{\rm th}\mathrm{US})}_{((2m_{11}^{(i)}+1),(2n_{11}^{(i)}))\text{-down-paraobla upper-saddle}}
\end{array}\right\}.
\end{array}
\tag{5.69}
$$

(iv$_{4c}$) The $((2m_1+1)^{\rm th}\mathrm{UP}_-,(2n_1+1)^{\rm th}\mathrm{UP}_-)$-negative saddle switching bifurcation is for two $p\times q$ equilibrium networks as

$$\bigcup_{r_1=1}^{p}\bigcup_{s_1=1}^{q}\underbrace{(a_{j_1j_2r_1},a_{j_2j_1s_1})}_{(m_{r_1}^{(1)},n_{s_1}^{(1)})\text{-XX}} \rightleftharpoons \underbrace{((2m_1+1)^{\text{th}}\text{UP}_-,(2n_1+1)^{\text{th}}\text{UP}_-)}_{((2m_1+1),(2n_1+1))\text{-negative saddle}}$$
$$\rightleftharpoons \bigcup_{r_2=1}^{p}\bigcup_{s_2=1}^{q}\underbrace{(a_{j_1j_2r_2},a_{j_2j_1s_2})}_{(m_{r_2}^{(2)},n_{s_2}^{(2)})\text{-XX}} \tag{5.70}$$

with $\Sigma_{r_i=1}^{p} m_{r_i} = 2m_1 + 1$ and $\Sigma_{s_i=1}^{q} n_{s_i} = 2n_1 + 1$ for $i = 1, 2$.

(v) For $m = 2m_1$ and $n = 2n_1 + 1$, the equilibrium of $(x_{j_2}^*, x_{j_1}^*) = (a_{j_1j_21}, a_{j_2j_11})$ has the following bifurcation properties.

(v_1) For $a_{j_1j_10} > 0$ and $a_{j_2j_20} > 0$, there is a $((2m_1)^{\text{th}}\text{US}, (2n_1+1)^{\text{th}}\text{UP})$-up-parabola upper-saddle equilibrium as

$$\left(a_{j_1j_21},a_{j_2j_11}\right) = \underbrace{((2m_1)^{\text{th}}\text{US},(2n_1+1)^{\text{th}}\text{UP})}_{((2m_1),(2n_1+1))\text{-up-parabola upper-saddle}}. \tag{5.71}$$

There are three following $((2m_1)^{\text{th}}\text{US}, (2n_1 + 1)^{\text{th}}\text{UP})$-up-parabola upper-saddle appearing and switching bifurcations.

(v_{1a}) The $((2m_1)^{\text{th}}\text{US}, (2n_1 + 1)^{\text{th}}\text{UP})$-up-paraonla upper-saddle appearing bifurcation is from a (pF,UP)-positive parabola flow to a $(2m_1) \times (2n_1 + 1)$-equilibrium network as

$$\underbrace{(\dot{x}_{j_1},a_{j_2j_11})}_{\text{(pF,UP)-up-parabola flow }(+)} \rightleftharpoons \underbrace{((2m_1)^{\text{th}}\text{US},(2n_1+1)^{\text{th}}\text{UP})}_{((2m_1),(2n_1+1))\text{-up-paraobla upper-saddle}}$$
$$\rightleftharpoons \bigcup_{r=1}^{2m_1}\bigcup_{s=1}^{2n_1+1}\underbrace{(a_{j_1j_2r},a_{j_2j_1s})}_{\text{XX}} \tag{5.72}$$

where

$$\bigcup_{s=1}^{2m_1}\bigcup_{l=1}^{2n_1+1}\underbrace{(a_{j_1j_2s},a_{j_2j_1l})}_{\text{XX}} = \begin{Bmatrix} \underbrace{(\text{UP}_+,\text{UP}_+)}_{\text{saddle }(+)} & \underbrace{(\text{DP}_-,\text{DP}_+)}_{\text{center (CW)}} & \cdots & \underbrace{(\text{UP}_+,\text{UP}_+)}_{\text{saddle }(+)} \\ \underbrace{(\text{DP}_+,\text{DP}_-)}_{\text{center (CCW)}} & \underbrace{(\text{UP}_-,\text{UP}_-)}_{\text{saddle }(-)} & \cdots & \underbrace{(\text{DP}_+,\text{DP}_-)}_{\text{center (CCW)}} \\ \vdots & \vdots & \vdots & \vdots \\ \underbrace{(\text{DP}_+,\text{DP}_-)}_{\text{center (CCW)}} & \underbrace{(\text{UP}_-,\text{UP}_-)}_{\text{saddle }(-)} & \cdots & \underbrace{(\text{DP}_+,\text{DP}_-)}_{\text{center (CCW)}} \end{Bmatrix}_{(2m_1)\times(2n_1+1)}. \tag{5.73}$$

(v_{1b}) The $((2m_1)^{\text{th}}\text{US}, (2n_1 + 1)^{\text{th}}\text{UP})$-up-paraonla upper-saddle appearing bifurcation is from a $p_1 \times q_1$ to $p_2 \times q_2$ equilibrium network as

$$\bigcup_{r_1=1}^{p_1}\bigcup_{s_1=1}^{q_1}\underbrace{(a_{j_1j_2r_1},a_{j_2j_1s_1})}_{(m_{r_1}^{(1)},n_{s_1}^{(1)})\text{-XX}} \rightleftharpoons \underbrace{((2m_1)^{\text{th}}\text{US},(2n_1+1)^{\text{th}}\text{UP})}_{((2m_1),(2n_1+1))\text{-up-paraobla upper-saddle}}$$
$$\rightleftharpoons \bigcup_{r_2=1}^{p_2}\bigcup_{s_2=1}^{q_2}\underbrace{(a_{j_1j_2r_2},a_{j_2j_1s_2})}_{(m_{r_2}^{(2)},n_{s_2}^{(2)})\text{-XX}} \tag{5.74}$$

where

$$\begin{aligned}
&2m_1-\Sigma_{r_1=1}^{p_1}m_{r_1}^{(1)}=2\Sigma_{s=1}^{l_1}m_s\\
&\text{with } l_1\text{-quadratic polynomials without real roots}\\
&\text{and } s^{\text{th}}\text{-quadratic polynomial with power } m_s,\\
&2n_1+1-\Sigma_{r_1=1}^{q_1}n_{s_1}^{(1)}=2\Sigma_{l=1}^{l_2}n_l\\
&\text{with } l_2\text{-quadratic polynomials without real roots}\\
&\text{and } l^{\text{th}}\text{-quadratic polynomial with power } n_1;\\
&\Sigma_{r_2=1}^{p_2}m_{r_2}^{(2)}=2m_1,\ \Sigma_{s_2=1}^{q_2}n_{s_2}^{(2)}=2n_1+1;\text{ for } i=1,2,
\end{aligned}$$

$$\underbrace{((m_{p_i}^{(i)})^{\text{th}}\,\text{XX},(n_{q_i}^{(i)})^{\text{th}}\,\text{XX})}_{(m_{p_i}^{(i)},n_{q_i}^{(i)})\text{-XX}}$$
$$\in\left\{\begin{array}{l}\underbrace{((2m_{p_i1}^{(i)}+1)^{\text{th}}\text{UP}_+,(2n_{q_i1}^{(i)}+1)^{\text{th}}\text{UP}_+)}_{((2m_{p_i1}^{(i)}+1),(2n_{q_i1}^{(i)}+1))\text{-positive saddle}},\ \underbrace{((2m_{p_i1}^{(i)})^{\text{th}}\text{II}_+,(2n_{q_i1}^{(i)})^{\text{th}}\text{II}_+)}_{((2m_{p_i1}^{(i)}),(2n_{q_i1}^{(i)}))\text{-double-infection saddle}},\\ \underbrace{((2m_{p_i1}^{(i)})^{\text{th}}\text{US},(2n_{q_i1}^{(i)}+1)^{\text{th}}\text{UP})}_{((2m_{p_i1}^{(i)}),(2n_{q_i1}^{(i)}+1))\text{-up-parabola upper-saddle}},\ \underbrace{((2m_{p_i1}^{(i)}+1)^{\text{th}}\text{UP},(2n_{q_i1}^{(i)})^{\text{th}}\text{US})}_{((2m_{p_i1}^{(i)}+1),(2n_{q_i1}^{(i)}))\text{-up-parabola upper-saddle}}\end{array}\right\};$$

$$\underbrace{((m_{p_i}^{(i)})^{\text{th}}\,\text{XX},(n_{q_i}^{(i)})^{\text{th}}\,\text{XX})}_{(m_{p_i}^{(i)},n_1^{(i)})\text{-XX}}$$
$$\in\left\{\begin{array}{l}\underbrace{((2m_{p_i1}^{(i)}+1)^{\text{th}}\text{UP}_+,(2n_{11}^{(i)}+1)^{\text{th}}\text{UP}_+)}_{((2m_{p_i1}^{(i)}+1),(2n_{11}^{(i)}+1))\text{-positive saddle}},\ \underbrace{((2m_{p_i1}^{(i)})^{\text{th}}\text{DI}_-,(2n_{11}^{(i)})^{\text{th}}\text{DI}_+)}_{((2m_{p_i1}^{(i)}),(2n_{11}^{(i)}))\text{-double -saddle}},\\ \underbrace{((2m_{p_i1}^{(i)})^{\text{th}}\text{US},(2n_{11}^{(i)}+1)^{\text{th}}\text{UP})}_{((2m_{p_i1}^{(i)}),(2n_{11}^{(i)}+1))\text{-up-parabola upper-saddle}},\ \underbrace{((2m_{p_i1}^{(i)}+1)^{\text{th}}\text{DP},(2n_{11}^{(i)})^{\text{th}}\text{LS})}_{((2m_{p_i1}^{(i)}+1),(2n_{11}^{(i)}))\text{-down-parabola lower-saddle}},\end{array}\right\};$$

$$\underbrace{((m_1^{(i)})^{\text{th}}\,\text{XX},\,(n_{q_i}^{(i)})^{\text{th}}\,\text{XX})}_{(m_1^{(i)},n_{q_i}^{(i)})\text{-XX}}$$

$$\in \left\{ \begin{array}{l} \underbrace{((2m_{11}{}^{(i)}+1)^{\text{th}}\text{US},\,(2n_{q_i1}^{(i)}+1)^{\text{th}}\text{UP})}_{((2m_{11}^{(i)}+1),(2n_{q_i1}^{(i)}+1))\text{-up-parabola upper-saddle}},\ \underbrace{((2m_{11}^{(i)})^{\text{th}}\text{II}_{+},\,(2n_{q_i1}^{(i)})^{\text{th}}\text{II}_{+})}_{((2m_{11}^{(i)}),(2n_{q_i1}^{(i)}))\text{-double-infection saddle}}, \\ \underbrace{((2m_{11}^{(i)})^{\text{th}}\text{DP}_{+},\,(2n_{q_i1}^{(i)}+1)^{\text{th}}\text{DP}_{-})}_{((2m_{11}^{(i)}),(2n_{q_i1}^{(i)}+1))\text{-CCW center}},\ \underbrace{((2m_{11}^{(i)}+1)^{\text{th}}\text{DP},\,(2n_{q_i1}^{(i)})^{\text{th}}\text{US})}_{((2m_{11}^{(i)}+1),(2n_{q_i1}^{(i)}))\text{-down-parabola upper-saddle}} \end{array} \right\};$$

$$\underbrace{((m_1^{(i)})^{\text{th}}\,\text{XX},\,(n_1^{(i)})^{\text{th}}\,\text{XX})}_{(m_1^{(i)},n_1^{(i)})\text{-XX}}$$

$$\in \left\{ \begin{array}{l} \underbrace{((2m_1 1^{(i)}+1)^{\text{th}}\text{US},\,(2n_{11}^{(i)}+1)^{\text{th}}\text{UP}_{+})}_{((2m_{11}^{(i)}+1),(2n_{11}^{(i)}+1))\text{-up-parabola upper-saddle}},\ \underbrace{((2m_{11}^{(i)})^{\text{th}}\text{DI}_{-},\,(2n_{11}^{(i)})^{\text{th}}\text{DI}_{+})}_{((2m_{11}^{(i)}),(2n_{11}^{(i)}))\text{-double-infection saddle}}, \\ \underbrace{((2m_{11}^{(i)})^{\text{th}}\text{DP}_{+},\,(2n_{11}^{(i)}+1)^{\text{th}}\text{DP}_{-})}_{((2m_{11}^{(i)}),(2n_{11}^{(i)}+1))\text{-CCW center}},\ \underbrace{((2m_{11}^{(i)}+1)^{\text{th}}\text{UP},\,(2n_{11}^{(i)})^{\text{th}}\text{LS})}_{((2m_{11}^{(i)}+1),(2n_{11}^{(i)}))\text{-up-parabola lower-saddle}} \end{array} \right\};$$
(5.75)

(v_{1c}) The $((2m_1)^{\text{th}}$ US, $(2n_1+1)^{\text{th}}$ UP)-up-paraonla upper-saddle switching bifurcation is for the switching of two $p \times q$ equilibrium networks as

$$\bigcup_{r_1=1}^{p}\bigcup_{s_1=1}^{q}\underbrace{(a_{j_1j_2r_1},a_{j_2j_1s_1})}_{(m_{r_1}^{(1)},\,n_{s_1}^{(1)})\text{-XX}} \rightleftharpoons \underbrace{((2m_1)^{\text{th}}\text{US},(2n_1+1)^{\text{th}}\text{UP})}_{((2m_1),(2n_1+1))\text{-up-paraobla upper-saddle}}$$
$$\rightleftharpoons \bigcup_{r_2=1}^{p}\bigcup_{s_2=1}^{q}\underbrace{(a_{j_1j_2r_2},a_{j_2j_1s_2})}_{(m_{r_2}^{(2)},\,n_{s_2}^{(2)})\text{-XX}} \tag{5.76}$$

with $\sum_{r_i=1}^{p} m_{r_i} = 2m_1$ and $\sum_{s_i=1}^{q} n_{s_i} = 2n_1+1$ for $i=1,2$.

(v_2) For $a_{j_1j_10} < 0$ and $a_{j_2j_20} > 0$, there is a $((2m_1)^{\text{th}}$LS, $(2n_1+1)^{\text{th}}$DP)-down-parabola lower-saddle equilibrium as

$$(a_{j_1j_21},a_{j_2j_11}) = \underbrace{((2m_1)^{\text{th}}\text{LS},(2n_1+1)^{\text{th}}\text{DP})}_{((2m_1),(2n_1+1))\text{-down-parabola lower-saddle}}. \tag{5.77}$$

There are the following three $((2m_1)^{\text{th}}$LS, $(2n_1+1)^{\text{th}}$UP)-up-paraonla lower-saddle appearing and switching bifurcations.

(v_{2a}) The $((2m_1)^{\text{th}}$LS, $(2n_1+1)^{\text{th}}$DP)-down-parabola lower-saddle appearing bifurcation is from an (nF,UP)-negative up-parabola flow to a $(2m_1) \times (2n_1+1)$

network as

$$\underbrace{(\dot{x}_{j_1}, a_{j_2 j_1 1})}_{\text{(nF,UP)-up-parabola flow }(-)} \rightleftharpoons \underbrace{((2m_1)^{\text{th}}\text{LS},(2n_1+1)^{\text{th}}\text{DP})}_{((2m_1),(2n_1+1))\text{-down-parabola lower-saddle}}$$
$$\rightleftharpoons \bigcup_{r=1}^{2m_1} \bigcup_{s=1}^{2n_1+1} \underbrace{(a_{j_1 j_1 r}, a_{j_2 j_2 s})}_{\text{XX}} \tag{5.78}$$

where

$$\bigcup_{s=1}^{2m_1} \bigcup_{l=1}^{2n_1+1} \underbrace{(a_{j_1 j_2 s}, a_{j_2 j_1 l})}_{\text{XX}} = \left\{ \begin{array}{cccc} \underbrace{(\text{DP}_+,\text{DP}_-)}_{\text{center (CCW)}} & \underbrace{(\text{UP}_-,\text{UP}_-)}_{\text{saddle }(-)} & \cdots & \underbrace{(\text{DP}_+,\text{DP}_-)}_{\text{center (CCW)}} \\ \underbrace{(\text{UP}_+,\text{UP}_+)}_{\text{saddle }(+)} & \underbrace{(\text{DP}_-,\text{DP}_+)}_{\text{center (CW)}} & \cdots & \underbrace{(\text{UP}_+,\text{UP}_+)}_{\text{saddle }(+)} \\ \vdots & \vdots & \vdots & \vdots \\ \underbrace{(\text{UP}_+,\text{UP}_+)}_{\text{saddle }(+)} & \underbrace{(\text{DP}_-,\text{DP}_+)}_{\text{center (CW)}} & \cdots & \underbrace{(\text{UP}_+,\text{UP}_+)}_{\text{saddle }(+)} \end{array} \right\}_{(2m_1)\times(2n_1+1)} . \tag{5.79}$$

(v_{2b}) The $((2m_1)^{\text{th}}\text{LS}, (2n_1+1)^{\text{th}}\text{DP})$-down-parabola lower-saddle appearing bifurcation is from a $p_1 \times q_1$ to $p_2 \times q_2$ equilibrium network as

$$\bigcup_{r_1=1}^{p_1} \bigcup_{s_1=1}^{q_1} \underbrace{(a_{j_1 j_2 r_1}, a_{j_2 j_1 s_1})}_{(m_{r_1}^{(1)}, n_{s_1}^{(1)})\text{-XX}} \rightleftharpoons \underbrace{((2m_1)^{\text{th}}\text{LS},(2n_1+1)^{\text{th}}\text{DP})}_{((2m_1),(2n_1+1))\text{-down-parabola lower-saddle}}$$
$$\rightleftharpoons \bigcup_{r_2=1}^{p_2} \bigcup_{s_2=1}^{q_2} \underbrace{(a_{j_1 j_2 r_2}, a_{j_2 j_1 s_2})}_{(m_{r_2}^{(2)}, n_{s_2}^{(2)})\text{-XX}} \tag{5.80}$$

where

$2m_1 - \Sigma_{r_1=1}^{p_1} m_{r_1}^{(1)} = 2\Sigma_{s=1}^{l_1} m_s$
with l_1-quadratic polynomials without real roots
and s^{th}-quadratic polynomial with power m_s,
$2n_1 + 1 - \Sigma_{r_1=1}^{q_1} n_{s_1}^{(1)} = 2\Sigma_{l=1}^{l_2} n_l$
with l_2-quadratic polynomials without real roots
and s^{th}-quadratic polynomial with power n_1;
$\Sigma_{r_2=1}^{p_2} m_{r_2}^{(2)} = 2m_1,\ \Sigma_{s_2=1}^{q_2} n_{s_2}^{(2)} = 2n_1 + 1$; for $i = 1, 2,$

$$
\begin{array}{l}
\underbrace{((m_{p_i}^{(i)})^{\rm th}\,\mathrm{XX},\,(n_{q_i}^{(i)})^{\rm th}\,\mathrm{XX})}_{(m_{p_i}^{(i)},n_{q_i}^{(i)})\text{-XX}}\\
\in\left\{\begin{array}{l}
\underbrace{((2m_{p_i1}^{(i)}+1)^{\rm th}\mathrm{DP}_{+},\,(2n_{q_i1}^{(i)}+1)^{\rm th}\mathrm{DP}_{-})}_{((2m_{p_i1}^{(i)}+1),(2n_{q_i1}^{(i)}+1))\text{-CCW center}},\;
\underbrace{((2m_{p_i1}^{(i)})^{\rm th}\mathrm{DI}_{+},\,(2n_{q_i1}^{(i)})^{\rm th}\mathrm{DI}_{-})}_{((2m_{p_i1}^{(i)}),(2n_{q_i1}^{(i)}))\text{-double-infection saddle}},\\
\underbrace{((2m_{p_i1}^{(i)})^{\rm th}\mathrm{LS},\,(2n_{q_i1}^{(i)}+1)^{\rm th}\mathrm{DP})}_{((2m_{p_i1}^{(i)}),(2n_{q_i1}^{(i)}+1))\text{-down-parabola lower-saddle}},\;
\underbrace{((2m_{p_i1}^{(i)}+1)^{\rm th}\mathrm{UP},\,(2n_{q_i1}^{(i)})^{\rm th}\mathrm{US})}_{((2m_{p_i1}^{(i)}+1),(2n_{q_i1}^{(i)}))\text{-down-parabola upper-saddle}}
\end{array}\right\};
\end{array}
$$

$$
\begin{array}{l}
\underbrace{((m_{p_i}^{(i)})^{\rm th}\,\mathrm{XX},\,(n_{1}^{(i)})^{\rm th}\,\mathrm{XX})}_{(m_{p_i}^{(i)},n_{1}^{(i)})\text{-XX}}\\
\in\left\{\begin{array}{l}
\underbrace{((2m_{p_i1}^{(i)}+1)^{\rm th}\mathrm{DP}_{+},\,(2n_{11}^{(i)}+1)^{\rm th}\mathrm{DP}_{-})}_{((2m_{p_i1}^{(i)}+1),(2n_{11}^{(i)}+1))\text{-CCW center}},\;
\underbrace{((2m_{p_i1}^{(i)})^{\rm th}\mathrm{II}_{-},\,(2n_{11}^{(i)})^{\rm th}\mathrm{II}_{-})}_{((2m_{p_i1}^{(i)}),(2n_{11}^{(i)}))\text{-double-infection saddle}},\\
\underbrace{((2m_{p_i1}^{(i)})^{\rm th}\mathrm{LS},\,(2n_{11}^{(i)}+1)^{\rm th}\mathrm{DP})}_{((2m_{p_i1}^{(i)}),(2n_{11}^{(i)}+1))\text{-down-parabola lower-saddle}},\;
\underbrace{((2m_{p_i1}^{(i)}+1)^{\rm th}\mathrm{UP},\,(2n_{11}^{(i)})^{\rm th}\mathrm{LS})}_{((2m_{p_i1}^{(i)}+1),(2n_{11}^{(i)}))\text{-up-parabola lower-saddle}},
\end{array}\right\};
\end{array}
$$

$$
\begin{array}{l}
\underbrace{((m_{1}^{(i)})^{\rm th}\,\mathrm{XX},\,(n_{q_i}^{(i)})^{\rm th}\,\mathrm{XX})}_{(m_{1}^{(i)},n_{q_i}^{(i)})\text{-XX}}\\
\in\left\{\begin{array}{l}
\underbrace{((2m_{11}^{(i)}+1)^{\rm th}\mathrm{UP}_{+},\,(2n_{q_i1}^{(i)}+1)^{\rm th}\mathrm{UP}_{+})}_{((2m_{11}^{(i)}+1),(2n_{q_i1}^{(i)}+1))\text{-positive saddle}},\;
\underbrace{((2m_{11}^{(i)})^{\rm th}\mathrm{DI}_{+},\,(2n_{q_i1}^{(i)})^{\rm th}\mathrm{DI}_{-})}_{((2m_{11}^{(i)}),(2n_{q_i1}^{(i)}))\text{-double-infection saddle}},\\
\underbrace{((2m_{11}^{(i)})^{\rm th}\mathrm{US},\,(2n_{q_i1}^{(i)}+1)^{\rm th}\mathrm{DP})}_{((2m_{11}^{(i)}),(2n_{q_i1}^{(i)}+1))\text{-down-parabola lower-saddle}},\;
\underbrace{((2m_{11}^{(i)}+1)^{\rm th}\mathrm{UP},\,(2n_{q_i1}^{(i)})^{\rm th}\mathrm{LS})}_{((2m_{11}^{(i)}+1),(2n_{q_i1}^{(i)}))\text{-up-parabola upper-saddle}}
\end{array}\right\};
\end{array}
$$

$$
\begin{array}{l}
\underbrace{((m_{1}^{(i)})^{\rm th}\mathrm{XX},(n_{1}^{(i)})^{\rm th}\mathrm{XX})}_{(m_{1}^{(i)},n_{1}^{(i)})\text{-XX}}\\
\in\left\{\begin{array}{l}
\underbrace{((2m_{11}^{(i)}+1)^{\rm th}\mathrm{UP}_{+},\,(2n_{11}^{(i)}+1)^{\rm th}\mathrm{UP}_{+})}_{((2m_{11}^{(i)}+1),(2n_{11}^{(i)}+1))\text{-positive saddle}},\;
\underbrace{((2m_{11}^{(i)})^{\rm th}\mathrm{II}_{-},\,(2n_{11}^{(i)})^{\rm th}\mathrm{II}_{-})}_{((2m_{11}^{(i)}),(2n_{11}^{(i)}))\text{-double-infelction saddle}},\\
\underbrace{((2m_{11}^{(i)})^{\rm th}\mathrm{LS},(2n_{11}^{(i)}+1)^{\rm th}\mathrm{DP})}_{((2m_{11}^{(i)}),(2n_{11}^{(i)}+1))\text{-down-parabola lower-saddle}},\;
\underbrace{((2m_{11}^{(i)}+1)^{\rm th}\mathrm{DP},(2n_{11}^{(i)})^{\rm th}\mathrm{LS})}_{((2m_{11}^{(i)}+1),(2n_{11}^{(i)}))\text{-down-parabola lower-saddle}}
\end{array}\right\}.
\end{array}
\tag{5.81}
$$

(v_{2c}) The $((2m_1)^{\rm th}\mathrm{LS},\ (2n_1+1)^{\rm th}\mathrm{UP})$-up-parabola lower-saddle switching bifurcation is for two $p \times q$ equilibrium networks as

$$\bigcup_{r_1=1}^{p}\bigcup_{s_1=1}^{q}\underbrace{(a_{j_1j_2r_1},a_{j_2j_1s_1})}_{(m_{r_1}^{(1)},n_{s_1}^{(1)})\text{-XX}} \rightleftharpoons \underbrace{((2m_1)^{\text{th}}\text{LS},(2n_1+1)^{\text{th}}\text{UP})}_{((2m_1),(2n_1+1))\text{-up-parabola lower-saddle}}$$
$$\rightleftharpoons \bigcup_{r_2=1}^{p}\bigcup_{s_2=1}^{q}\underbrace{(a_{j_1j_2r_2},a_{j_2j_1s_2})}_{(m_{r_2}^{(2)},n_{s_2}^{(2)})\text{-XX}} \tag{5.82}$$

with $\sum_{r_i=1}^{p} m_{r_i} = 2m_1$ and $\sum_{s_i=1}^{q} n_{s_i} = 2n_1 + 1$ for $i = 1, 2$.

(v_3) For $a_{j_1j_10} > 0$ and $a_{j_2j_20} < 0$, there is a $((2m_1)^{\text{th}}\text{US}, (2n_1{+}1)^{\text{th}}\text{DP})$-down-parabola upper-saddle equilibrium as

$$(a_{j_1j_21},a_{j_2j_11}) = \underbrace{((2m_1)^{\text{th}}\text{US},(2n_1+1)^{\text{th}}\text{DP})}_{((2m_1),(2n_1+1))\text{-down-parabola upper-saddle}}. \tag{5.83}$$

There are the following three $((2m_1)^{\text{th}}\text{US}, (2n_1{+}1)^{\text{th}}\text{DP})$-down-parabola upper-saddle appearing and switching bifurcations.

(v_{3a}) The $((2m_1)^{\text{th}}\text{US}, (2n_1+1)^{\text{th}}\text{DP})$-down-parabola upper-saddle appearing bifurcation is from a (pF, DP)-positive down-parabola flow to a $(2m_1) \times (2n_1 + 1)$-equilibrium network as

$$\underbrace{(\dot{x}_{j_1},a_{j_2j_11})}_{\text{(pF,DP)-down-parabola flow (+)}} \rightleftharpoons \underbrace{((2m_1)^{\text{th}}\text{US},(2n_1+1)^{\text{th}}\text{DP})}_{((2m_1),(2n_1+1))\text{-down-parabola upper-saddle}}$$
$$\rightleftharpoons \bigcup_{r=1}^{2m_1}\bigcup_{s=1}^{2n_1+1}\underbrace{(a_{j_1j_2r},a_{j_2j_1s})}_{\text{XX}} \tag{5.84}$$

where

$$\bigcup_{s=1}^{2m_1}\bigcup_{l=1}^{2n_1+1}\underbrace{(a_{j_1j_2s},a_{j_2j_1l})}_{\text{XX}} = \left\{\begin{matrix} \underbrace{(\text{DP}_-,\text{DP}_+)}_{\text{center (CW)}} & \underbrace{(\text{UP}_+,\text{UP}_+)}_{\text{saddle }(+)} & \cdots & \underbrace{(\text{DP}_-,\text{DP}_+)}_{\text{center (CW)}} \\ \underbrace{(\text{UP}_-,\text{UP}_-)}_{\text{saddle }(-)} & \underbrace{(\text{DP}_+,\text{DP}_-)}_{\text{center (CCW)}} & \cdots & \underbrace{(\text{UP}_-,\text{UP}_-)}_{\text{saddle }(-)} \\ \vdots & \vdots & \vdots & \vdots \\ \underbrace{(\text{UP}_-,\text{UP}_-)}_{\text{saddle }(-)} & \underbrace{(\text{DP}_+,\text{DP}_-)}_{\text{center (CCW)}} & \cdots & \underbrace{(\text{UP}_-,\text{UP}_-)}_{\text{saddle }(-)} \end{matrix}\right\}_{(2m_1)\times(2n_1+1)}. \tag{5.85}$$

(v_{3b}) The $((2m_1)^{\text{th}}\text{US}, (2n_1+1)^{\text{th}}\text{DP})$-down-parabola upper-saddle appearing bifurcation is from a $p_1 \times q_1$ to $p_2 \times q_2$ equilibrium network as

$$\bigcup_{r_1=1}^{p_1}\bigcup_{s_1=1}^{q_1}\underbrace{(a_{j_1j_2r_1},a_{j_2j_1s_1})}_{(m_{r_1}^{(1)},n_{s_1}^{(1)})\text{-XX}} \rightleftharpoons \underbrace{((2m_1)^{\text{th}}\text{US},(2n_1+1)^{\text{th}}\text{DP})}_{((2m_1),(2n_1+1))\text{-down-parabola upper-saddle}}$$
$$\rightleftharpoons \bigcup_{r_2=1}^{p_2}\bigcup_{s_2=1}^{q_2}\underbrace{(a_{j_1j_2r_2},a_{j_2j_1s_2})}_{(m_{r_2}^{(2)},n_{s_2}^{(2)})\text{-XX}} \tag{5.86}$$

where

$2m_1 - \Sigma_{r_1=1}^{p_1} m_{r_1}^{(1)} = 2\Sigma_{s=1}^{l_1} m_s$
with l_1-quadratic polynomials without real roots
and s^{th}-quadratic polynomial with power m_s,
$2n_1 + 1 - \Sigma_{s_1=1}^{q_1} n_{s_1}^{(1)} = 2\Sigma_{l=1}^{l_2} n_l$
with l_2-quadraitc polynomials without real roots
and l^{th}-quadratic polynomial with power n_l;
$\Sigma_{r_2=1}^{p_2} m_{r_2}^{(2)} = 2m_1, \ \Sigma_{s_2=1}^{q_2} n_{s_2}^{(2)} = 2n_1 + 1$; for $i = 1, 2$,

$$\underbrace{((m_{p_i}^{(i)})^{\text{th}}\text{XX},(n_{q_i}^{(i)})^{\text{th}}\text{XX})}_{(m_{p_i}^{(i)},n_{q_i}^{(i)})\text{-XX}}$$
$$\in \left\{\begin{array}{c} \underbrace{((2m_{p_i1}^{(i)}+1)^{\text{th}}\text{DP}_-,(2n_{q_i1}^{(i)}+1)^{\text{th}}\text{DP}_+)}_{((2m_{p_i1}^{(i)}+1),(2n_{q_i1}^{(i)}+1))\text{-CW center}},\ \underbrace{((2m_{p_i1}^{(i)})^{\text{th}}\text{DI}_-,(2n_{q_i1}^{(i)})^{\text{th}}\text{DI}_+)}_{((2m_{p_i1}^{(i)}),(2n_{q_i1}^{(i)}))\text{-double-inflection saddle}}, \\ \underbrace{((2m_{p_i1}^{(i)})^{\text{th}}\text{US},(2n_{q_i1}^{(i)}+1)^{\text{th}}\text{DP})}_{((2m_{p_i1}^{(i)}),(2n_{q_i1}^{(i)}+1))\text{-down-parabola upper-saddle}},\ \underbrace{((2m_{p_i1}^{(i)}+1)^{\text{th}}\text{DP},(2n_{q_i1}^{(i)})^{\text{th}}\text{LS})}_{((2m_{p_i1}^{(i)}+1),(2n_{q_i1}^{(i)}))\text{-down-parabola lower-saddle}} \end{array}\right\};$$

$$\underbrace{((m_{p_i}^{(i)})^{\text{th}}\text{XX},(n_1^{(i)})^{\text{th}}\text{XX})}_{(m_{p_i}^{(i)},n_1^{(i)})\text{-XX}}$$
$$\in \left\{\begin{array}{c} \underbrace{((2m_{p_i1}^{(i)}+1)^{\text{th}}\text{DP}_-,(2n_{11}^{(i)}+1)^{\text{th}}\text{DP}_+)}_{((2m_{p_i1}^{(i)}+1),(2n_{11}^{(i)}+1))\text{-CW center}},\ \underbrace{((2m_{p_i1}^{(i)})^{\text{th}}\text{II}_+,(2n_{11}^{(i)})^{\text{th}}\text{II}_+)}_{((2m_{p_i1}^{(i)}),(2n_{11}^{(i)}))\text{-double-inflection saddle}}, \\ \underbrace{((2m_{p_i1}^{(i)})^{\text{th}}\text{US},(2n_{11}^{(i)}+1)^{\text{th}}DP)}_{((2m_{p_i1}^{(i)}),(2n_{11}^{(i)}+1))\text{-down-paraobla upper-saddle}},\ \underbrace{((2m_{p_i1}^{(i)}+1)^{\text{th}}\text{UP},(2n_{11}^{(i)})^{\text{th}}\text{US})}_{((2m_{p_i1}^{(i)}+1),(2n_{11}^{(i)}))\text{-up-parabola upper-saddle}} \end{array}\right\};$$

$$\underbrace{((m_1^{(i)})^{\text{th}}\text{XX},(n_{q_i1}^{(i)})^{\text{th}}\text{XX})}_{(m_1^{(i)},n_{q_i}^{(i)})\text{-XX}}$$

$$\in\left\{\begin{array}{l}\underbrace{((2m_{11}^{(i)}+1)^{\text{th}}\text{UP}_-,(2n_{q_i1}^{(i)}+1)^{\text{th}}\text{UP}_-)}_{((2m_{11}^{(i)}+1),(2n_{q_i1}^{(i)}+1))\text{-negative saddle}},\ \underbrace{((2m_{11}^{(i)})^{\text{th}}\text{DI}_-,(2n_{q_i1}^{(i)})^{\text{th}}\text{DI}_+)}_{((2m_{11}^{(i)}),(2n_{q_i1}^{(i)}))\text{-double-inflection saddle}},\\ \underbrace{((2m_{11}^{(i)})^{\text{th}}\text{US},(2n_{q_i1}^{(i)}+1)^{\text{th}}\text{DP})}_{((2m_{11}^{(i)}),(2n_{q_i1}^{(i)}+1))\text{-down-paraobla upper-saddle}},\ \underbrace{((2m_{11}^{(i)}+1)^{\text{th}}\text{UP},(2n_{q_i1}^{(i)})^{\text{th}}\text{LS})}_{((2m_{11}^{(i)}+1),(2n_{q_i1}^{(i)}))\text{-up-paraobla lower-saddle}}\end{array}\right\};$$

$$\underbrace{((m_1^{(i)})^{\text{th}}\text{XX},(n_1^{(i)})^{\text{th}}\text{XX})}_{(m_1^{(i)},n_1^{(i)})\text{-XX}}$$

$$\in\left\{\begin{array}{l}\underbrace{((2m_{11}^{(i)}+1)^{\text{th}}\text{DP}_-,(2n_{11}^{(i)}+1)^{\text{th}}\text{DP}_-)}_{((2m_{11}^{(i)}+1),(2n_{11}^{(i)}+1))\text{-negtive saddle}},\ \underbrace{((2m_{11}^{(i)})^{\text{th}}\text{II}_-,(2n_{11}^{(i)})^{\text{th}}\text{II}_+)}_{((2m_{11}^{(i)}),(2n_{11}^{(i)}))\text{-double-inflection saddle}},\\ \underbrace{((2m_{11}^{(i)})^{\text{th}}\text{US},(2n_{11}^{(i)}+1)^{\text{th}}\text{DP})}_{((2m_{11}^{(i)}),(2n_{11}^{(i)}+1))\text{-down-parabola upper-saddle}},\ \underbrace{((2m_{11}^{(i)}+1)^{\text{th}}\text{DP},(2n_{11}^{(i)})^{\text{th}}\text{US})}_{((2m_{11}^{(i)}+1),(2n_{11}^{(i)}))\text{-down-parabola upper-saddle}}\end{array}\right\}. \tag{5.87}$$

(v_{3c}) The $((2m_1)^{\text{th}}\text{US},\ (2n_1+1)^{\text{th}}\text{DP})$-down-parabola upper-saddle switching bifurcation is for the switching of two $q \times p$ equilibrium networks as

$$\begin{aligned}\bigcup_{r_1=1}^{p}\bigcup_{s_1=1}^{q}\underbrace{(a_{j_1j_2r_1},a_{j_2j_1s_1})}_{(m_{r_1}^{(1)},n_{s_1}^{(1)})\text{-XX}} &\rightleftharpoons \underbrace{((2m_1)^{\text{th}}\text{US},(2n_1+1)^{\text{th}}\text{DP})}_{((2m_1),(2n_1+1))\text{-down-parabola upper-saddle}}\\ &\rightleftharpoons \bigcup_{r_2=1}^{p}\bigcup_{s_2=1}^{q}\underbrace{(a_{j_1j_2r_2},a_{j_2j_1s_2})}_{(m_{r_2}^{(2)},n_{s_2}^{(2)})\text{-XX}},\end{aligned} \tag{5.88}$$

where $\sum_{r_i=1}^{p} m_{r_i} = 2m_i$, $\sum_{s_i=1}^{q} n_{s_i} = 2n_i + 1$ for $i = 1, 2$.

(v_4) For $a_{j_1j_10} < 0$ and $a_{j_2j_20} < 0$, there is a $((2m_1)^{\text{th}}\text{LS},\ (2n_1+1)^{\text{th}}\text{UP})$-up-parabola lower-saddle equilibrium as

$$(a_{j_1j_21},a_{j_2j_11}) = \underbrace{((2m_1)^{\text{th}}\text{LS},\ (2n_1+1)^{\text{th}}\text{UP})}_{((2m_1),(2n_1+1))\text{-up-parabola lower-saddle}}. \tag{5.89}$$

There are the following three $((2m_1)^{\text{th}}\text{LS},\ (2n_1+1)^{\text{th}}\text{UP})$-up-parabola lower-saddle appearing and switching bifurcations.

(v_{4a}) The $((2m_1)^{\text{th}}\text{LS},\ (2n_1+1)^{\text{th}}\text{UP})$-up-parabola lower-saddle appearing bifurcation is from an (nF, DP)-negative down-parabola flow to a $(2m_1) \times (2n_1+1)$ network as

$$\underbrace{(\dot{x}_{j_1}, a_{j_2j_11})}_{\text{(nF,UP)-up-parabola flow }(-)} \rightleftharpoons \underbrace{((2m_1)^{\text{th}}\text{LS},(2n_1+1)^{\text{th}}\text{UP})}_{((2m_1),(2n_1+1))\text{-up-parabola lower-saddle}}$$
$$\rightleftharpoons \bigcup_{r=1}^{2m_1}\bigcup_{s=1}^{2n_1+1}\underbrace{(a_{j_1j_2r}, a_{j_2j_1s})}_{\text{XX}} \tag{5.90}$$

where

$$\bigcup_{s=1}^{2m_1}\bigcup_{l=1}^{2n_1+1}\underbrace{(a_{j_1j_2s}, a_{j_2j_1l})}_{\text{XX}} = \left\{\begin{array}{cccc} \underbrace{(\text{UP}_-,\text{UP}_-)}_{\text{saddle }(-)} & \underbrace{(\text{DP}_+,\text{DP}_-)}_{\text{center (CCW)}} & \cdots & \underbrace{(\text{UP}_-,\text{UP}_-)}_{\text{saddle }(-)} \\ \underbrace{(\text{DP}_-,\text{DP}_+)}_{\text{center (CW)}} & \underbrace{(\text{UP}_+,\text{UP}_+)}_{\text{saddle }(+)} & \cdots & \underbrace{(\text{DP}_-,\text{DP}_+)}_{\text{center (CW)}} \\ \vdots & \vdots & \vdots & \vdots \\ \underbrace{(\text{DP}_-,\text{DP}_+)}_{\text{center (CW)}} & \underbrace{(\text{UP}_+,\text{UP}_+)}_{\text{saddle }(+)} & \cdots & \underbrace{(\text{DP}_-,\text{DP}_+)}_{\text{center (CW)}} \end{array}\right\}_{(2m_1)\times(2n_1+1)}. \tag{5.91}$$

(v_{4b}) The $((2m_1)^{\text{th}}\text{LS},\ (2n_1+1)^{\text{th}}\text{UP})$-up-paraobla lower-saddle appearing bifurcation is from a $p_1 \times q_1$ to $p_2 \times q_2$ equilibrium network as

$$\bigcup_{r_1=1}^{p_1}\bigcup_{s_1=1}^{q_1}\underbrace{(a_{j_1j_2r_1}, a_{j_2j_1s_1})}_{(m_{r_1}^{(1)}, n_{s_1}^{(1)})\text{-XX}} \rightleftharpoons \underbrace{((2m_1)^{\text{th}}\text{LS},(2n_1+1)^{\text{th}}\text{UP})}_{((2m_1),(2n_1+1))\text{-up-parabola lower-saddle}}$$
$$\rightleftharpoons \bigcup_{r_2=1}^{p_2}\bigcup_{s_2=1}^{q_2}\underbrace{(a_{j_1j_2r_2}, a_{j_2j_1s_2})}_{(m_{r_2}^{(2)}, n_{s_2}^{(2)})\text{-XX}} \tag{5.92}$$

where

$$\begin{array}{l} 2m_1 - \Sigma_{r_1=1}^{p_1} m_{r_1}^{(1)} = 2\Sigma_{s=1}^{l_1} m_s \\ \text{with } l_1\text{-quadratic polynomials without real roots} \\ \text{and } s^{\text{th}}\text{-quadratic polynomial with power } m_s, \\ 2n_1 + 1 - \Sigma_{s_1=1}^{q_1} n_{s_1}^{(1)} = 2\Sigma_{l=1}^{l_2} n_l \\ \text{with } l_2\text{-quadraitc polynomials without real roots} \\ \text{and } l^{\text{th}}\text{-quadratic polynomial with power } n_l; \\ \Sigma_{r_2=1}^{p_2} m_{r_2}^{(2)} = 2m_1,\ \Sigma_{s_2=1}^{q_2} n_{s_2}^{(2)} = 2n_1 + 1;\ \text{for } i = 1, 2, \end{array}$$

$$
\begin{array}{l}
\underbrace{((m_{p_i}^{(i)})^{\text{th}}\text{XX},(n_{q_i}^{(i)})^{\text{th}}\text{XX})}_{(m_{p_i}^{(i)},n_{q_i}^{(i)})\text{-XX}} \\
\in \left\{\begin{array}{ll}
\underbrace{((2m_{p_i1}^{(i)}+1)^{\text{th}}\text{UP}_-,\,(2n_{11}^{(i)}+1)^{\text{th}}\text{UP}_-)}_{((2m_{p_i1}^{(i)}+1),(2n_{11}^{(i)}+1))\text{-negative saddle}}, & \underbrace{((2m_{p_i1}^{(i)})^{\text{th}}\text{II}_-,\,(2n_{q_i1}^{(i)})^{\text{th}}\text{II}_-)}_{((2m_{p_i1}^{(i)}),(2n_{q_i1}^{(i)}))\text{-double-inflection saddle}}, \\
\underbrace{((2m_{p_i1}^{(i)})^{\text{th}}\text{LS},(2n_{q_i1}^{(i)}+1)^{\text{th}}\text{UP})}_{((2m_{p_i1}^{(i)}+1),(2n_{q_i1}^{(i)}+1))\text{-up-parabola lower-saddle}}, & \underbrace{((2m_{p_i1}^{(i)}+1)^{\text{th}}\text{UP},(2n_{q_i1}^{(i)})^{\text{th}}\text{LS})}_{((2m_{p_i1}^{(i)}+1),(2n_{q_i1}^{(i)}))\text{-up-parabola lower-saddle}}
\end{array}\right\}; \\
\underbrace{((m_{p_i}^{(i)})^{\text{th}}\text{XX},(n_{1}^{(i)})^{\text{th}}\text{XX})}_{(m_{p_i}^{(i)},n_{1}^{(i)})\text{-XX}} \\
\in \left\{\begin{array}{ll}
\underbrace{((2m_{p_i1}^{(i)}+1)^{\text{th}}\text{UP}_-,\,(2n_{11}^{(i)}+1)^{\text{th}}\text{UP}_-)}_{((2m_{p_i1}^{(i)}+1),(2n_{11}^{(i)}+1))\text{-negative saddle}}, & \underbrace{((2m_{p_i1}^{(i)})^{\text{th}}\text{DI}_+,\,(2n_{11}^{(i)})^{\text{th}}\text{DI}_-)}_{((2m_{p_i1}^{(i)}),(2n_{11}^{(i)}))\text{-double-inflection saddle}}, \\
\underbrace{((2m_{p_i1}^{(i)})^{\text{th}}\text{LS},(2n_{11}^{(i)}+1)^{\text{th}}\text{UP})}_{((2m_{p_i1}^{(i)}),(2n_{11}^{(i)}+1))\text{-up-parabola lower-saddle}}, & \underbrace{((2m_{p_i1}^{(i)}+1)^{\text{th}}\text{DP},(2n_{11}^{(i)})^{\text{th}}\text{US})}_{((2m_{p_i1}^{(i)}+1),(2n_{11}^{(i)}))\text{-down-parabola upper-saddle}}
\end{array}\right\}; \\
\underbrace{((m_{1}^{(i)})^{\text{th}}\text{XX},(n_{q_i}^{(i)})^{\text{th}}\text{XX})}_{(m_{1}^{(i)},n_{q_i}^{(i)})\text{-XX}} \\
\in \left\{\begin{array}{ll}
\underbrace{((2m_{11}^{(i)}+1)^{\text{th}}\text{DP}_-,\,(2n_{q_i1}^{(i)}+1)^{\text{th}}\text{DP}_+)}_{((2m_{11}^{(i)}+1),(2n_{q_i1}^{(i)}+1))\text{-CW center}}, & \underbrace{((2m_{11}^{(i)})^{\text{th}}\text{II}_-,\,(2n_{q_i1}^{(i)})^{\text{th}}\text{II}_-)}_{((2m_{11}^{(i)}),(2n_{q_i1}^{(i)}))\text{-double-inflection saddle}}, \\
\underbrace{((2m_{11}^{(i)})^{\text{th}}\text{LS},(2n_{q_i1}^{(i)}+1)^{\text{th}}\text{UP})}_{((2m_{11}^{(i)}),(2n_{q_i1}^{(i)}+1))\text{-up-parabola lower-saddle}}, & \underbrace{((2m_{11}^{(i)}+1)^{\text{th}}\text{DP},(2n_{q_i1}^{(i)})^{\text{th}}\text{LS})}_{((2m_{11}^{(i)}+1),(2n_{q_i1}^{(i)}))\text{-down-paraobla lower-saddle}}
\end{array}\right\}; \\
\underbrace{((m_{1}^{(i)})^{\text{th}}\text{XX},(n_{1}^{(i)})^{\text{th}}\text{XX})}_{(m_{1}^{(i)},n_{1}^{(i)})\text{-XX}} \\
\in \left\{\begin{array}{ll}
\underbrace{((2m_{11}^{(i)}+1)^{\text{th}}\text{DP}_-,\,(2n_{11}^{(i)}+1)^{\text{th}}\text{DP}_+)}_{((2m_{11}^{(i)}+1),(2n_{11}^{(i)}+1))\text{-CW center}}, & \underbrace{((2m_{11}^{(i)})^{\text{th}}\text{DI}_+,\,(2n_{11}^{(i)})^{\text{th}}\text{DI}_-)}_{((2m_{11}^{(i)}),(2n_{11}^{(i)}))\text{-double-infelction saddle}}, \\
\underbrace{((2m_{11}^{(i)})^{\text{th}}\text{LS},(2n_{11}^{(i)}+1)^{\text{th}}\text{UP})}_{((2m_{11}^{(i)}),(2n_{11}^{(i)}+1))\text{-up-parabola lower-saddle}}, & \underbrace{((2m_{11}^{(i)}+1)^{\text{th}}\text{UP},(2n_{11}^{(i)})^{\text{th}}\text{US})}_{((2m_{11}^{(i)}+1),(2n_{11}^{(i)}))\text{-up-parabola upper-saddle}}
\end{array}\right\}.
\end{array}
\tag{5.93}
$$

(v_{4c}) The $((2m_1)^{\text{th}}\text{LS},\ (2n_1+1)^{\text{th}}\text{UP})$-up-paraobla lower-saddle switching bifurcation is for two $p \times q$ equilibrium networks as

$$
\bigcup_{r_1=1}^{p}\bigcup_{s_1=1}^{q}\underbrace{(a_{j_1j_2r_1},a_{j_2j_1s_1})}_{(m_{r_1}^{(1)},\,n_{s_1}^{(1)})\text{-XX}} \rightleftharpoons \underbrace{((2m_1)^{\text{th}}\text{LS},(2n_1+1)^{\text{th}}\text{DP})}_{((2m_1),(2n_1+1))\text{-down-parabola lower-saddle}}
$$

$$\rightleftharpoons \bigcup_{r_2=1}^{p} \bigcup_{s_2=1}^{q} \underbrace{(a_{j_1 j_2 r_2}, a_{j_2 j_1 s_2})}_{(m_{r_2}^{(2)}, n_{s_2}^{(2)})\text{-XX}} \tag{5.94}$$

where $\sum_{r_i=1}^{p} m_{r_i} = 2m_i$, $\sum_{s_i=1}^{q} n_{s_i} = 2n_i + 1$ for $i = 1, 2$.

(vi) For $m = 2m_1 + 1$ and $n = 2n_1$, the equilibrium of $(x_{j_2}^*, x_{j_1}^*) = (a_{j_1 j_2 1}, a_{j_2 j_1 1})$ has the following properties.

(vi$_1$) For $a_{j_1 j_1 0} > 0$ and $a_{j_2 j_2 0} > 0$, there is a $((2m_1{+}1)^{\text{th}}\text{UP},(2n_1)^{\text{th}}\text{US})$-up-parabola-upper-saddle equilibrium as

$$(a_{j_1 j_2 1}, a_{j_2 j_1 1}) = \underbrace{((2m_1{+}1)^{\text{th}}\text{UP},(2n_1)^{\text{th}}\text{US})}_{((2m_1+1),(2n_1))\text{-up-paraobla upper-saddle}}. \tag{5.95}$$

There are three following $((2m_1 + 1)^{\text{th}}\text{UP}, (2n_1)^{\text{th}}\text{US})$-up-parabola upper-saddle appearing and switching bifurcations.

(vi$_{1a}$) The $((2m_1 + 1)^{\text{th}}\text{UP}, (2n_1)^{\text{th}}\text{US})$-up-parabola upper-saddle appearing bifurcation is from an (UP, pF)-positive up-parabola flow to a $(2m_1 + 1)\times(2n_1)$-equilibrium network as

$$\begin{aligned}\underbrace{(a_{j_1 j_2 1}, \dot{x}_{j_2})}_{\text{(UP,pF)-up-parabola flow }(+)} &\rightleftharpoons \underbrace{((2m_1{+}1)^{\text{th}}\text{UP},(2n_1)^{\text{th}}\text{US})}_{((2m_1+1),(2n_1))\text{-up-paranola upper-saddle}}\\ &\rightleftharpoons \bigcup_{r=1}^{2m_1+1} \bigcup_{s=1}^{2n_1} \underbrace{(a_{j_1 j_1 r}, a_{j_2 j_2 s})}_{\text{XX}}\end{aligned} \tag{5.96}$$

where

$$\bigcup_{s=1}^{2m_1+1} \bigcup_{l=1}^{2n_1} \underbrace{(a_{j_1 j_2 s}, a_{j_2 j_1 l})}_{\text{XX}} = \left\{\begin{matrix} \underbrace{(\text{DP}_-,\text{DP}_+)}_{\text{center (CW)}} & \underbrace{(\text{UP}_+,\text{UP}_+)}_{\text{saddle }(+)} & \cdots & \underbrace{(\text{UP}_+,\text{UP}_+)}_{\text{saddle }(+)} \\ \underbrace{(\text{UP}_-,\text{UP}_-)}_{\text{saddle }(-)} & \underbrace{(\text{DP}_+,\text{DP}_-)}_{\text{center (CCW)}} & \cdots & \underbrace{(\text{DP}_+,\text{DP}_-)}_{\text{center (CCW)}} \\ \vdots & \vdots & \vdots & \vdots \\ \underbrace{(\text{DP}_-,\text{DP}_+)}_{\text{center (CW)}} & \underbrace{(\text{UP}_+,\text{UP}_+)}_{\text{saddle }(+)} & \cdots & \underbrace{(\text{UP}_+,\text{UP}_+)}_{\text{saddle }(+)} \end{matrix}\right\}_{(2m_1+1)\times(2n_1)}. \tag{5.97}$$

(vi$_{1b}$) The $((2m_1 + 1)^{\text{th}}$ UP, $(2n_1)^{\text{th}}$ US)-up-parabola upper-saddle appearing bifurcation is from a $p_1 \times q_1$ to $p_2 \times q_2$ equilibrium network as

$$\bigcup_{r_1=1}^{p_1}\bigcup_{s_1=1}^{q_1}\underbrace{(a_{j_1j_2r_1},a_{j_2j_1s_1})}_{(m_{r_1}^{(1)},\,n_{s_1}^{(1)})\text{-XX}} \rightleftharpoons \underbrace{((2m_1+1)^{\text{th}}\text{UP},(2n_1)^{\text{th}}\text{US})}_{((2m_1+1),\,(2n_1))\text{-up-parabola upper-saddle}}$$
$$\rightleftharpoons \bigcup_{r_2=1}^{p_2}\bigcup_{s_2=1}^{q_2}\underbrace{(a_{j_1j_2r_2},a_{j_2j_1s_2})}_{(m_{r_2}^{(2)},\,n_{s_2}^{(2)})\text{-XX}} \tag{5.98}$$

where

$2m_1-\Sigma_{r_1=1}^{p_1}m_{r_1}^{(1)}=2\Sigma_{s=1}^{l_1}m_s$
with l_1-quadratic polynomials without real roots
and s^{th}-quadratic polynomial with power m_s,
$2n_1+1-\Sigma_{s_1=1}^{q_1}n_{s_1}^{(1)}=2\Sigma_{l=1}^{l_2}n_l$
with l_2-quadratic polynomials without real roots
and l^{th}-quadratic polynomial with power n_1;
$\Sigma_{r_2=1}^{p_2}m_{r_2}^{(2)}=2m_1+1,\ \Sigma_{s_2=1}^{q_2}n_{s_2}^{(2)}=2n_1;$ for $i=1,2,$

$$\underbrace{((m_{p_i}^{(i)})^{\text{th}}\,\text{XX},(n_{q_i}^{(i)})^{\text{th}}\,\text{XX})}_{(m_{p_i}^{(i)},n_{q_i}^{(i)})\text{-XX}}$$
$$\in\left\{\begin{array}{l}\underbrace{((2m_{p_i1}^{(i)}+1)^{\text{th}}\text{UP}_+,(2n_{q_i1}^{(i)}+1)^{\text{th}}\text{UP}_+)}_{((2m_{p_i1}^{(i)}+1),(2n_{q_i1}^{(i)}+1))\text{-positive saddle}},\ \underbrace{((2m_{p_i1}^{(i)})^{\text{th}}\text{II}_+,(2n_{q_i1}^{(i)})^{\text{th}}\text{II}_+)}_{((2m_{p_i1}^{(i)}),(2n_{q_i1}^{(i)}))\text{-double-infection saddle}},\\ \underbrace{((2m_{p_i1}^{(i)})^{\text{th}}\text{US},(2n_{q_i1}^{(i)}+1)^{\text{th}}\text{UP})}_{((2m_{p_i1}^{(i)}),(2n_{q_i1}^{(i)}+1))\text{-up-parabola upper-saddle}},\ \underbrace{((2m_{p_i1}^{(i)}+1)^{\text{th}}\text{UP},(2n_{q_i1}^{(i)})^{\text{th}}\text{US})}_{((2m_{p_i1}^{(i)}+1),(2n_{q_i1}^{(i)}))\text{-up-parabola upper-saddle}}\end{array}\right\};$$

$$\underbrace{((m_{p_i}^{(i)})^{\text{th}}\,\text{XX},(n_1^{(i)})^{\text{th}}\,\text{XX})}_{(m_{p_i}^{(i)},n_1^{(i)})\text{-XX}}$$
$$\in\left\{\begin{array}{l}\underbrace{((2m_{p_i1}^{(i)}+1)^{\text{th}}\text{DP}_-,(2n_{11}^{(i)}+1)^{\text{th}}\text{DP})}_{((2m_{p_i1}^{(i)}+1),(2n_{11}^{(i)}+1))\text{-down-parabola upper-saddle}},\ \underbrace{((2m_{p_i1}^{(i)})^{\text{th}}\text{UP},(2n_{11}^{(i)})^{\text{th}}\text{US})}_{((2m_{p_i1}^{(i)}),(2n_{11}^{(i)}))\text{-up-parabola upper saddle}},\\ \underbrace{((2m_{p_i1}^{(i)})^{\text{th}}\text{US},(2n_{11}^{(i)}+1)^{\text{th}}\text{UP})}_{((2m_{p_i1}^{(i)}),(2n_{11}^{(i)}+1))\text{-down-parabola upper-saddle}},\ \underbrace{((2m_{p_i1}^{(i)}+1)^{\text{th}}\text{DP},(2n_{11}^{(i)})^{\text{th}}\text{US})}_{((2m_{p_i1}^{(i)}+1),(2n_{11}^{(i)}))\text{-up-parabola upper-saddle}},\end{array}\right\};$$

$$\underbrace{((m_1^{(i)})^{\text{th}}\,\text{XX},\,(n_{q_i}^{(i)})^{\text{th}}\,\text{XX})}_{(m_1^{(i)},n_{q_i}^{(i)})\text{-XX}}$$

$$\in\left\{\begin{array}{l}\underbrace{((2m_{11}^{(i)}+1)^{\text{th}}\text{UP}_+,\,(2n_{11}^{(i)}+1)^{\text{th}}\text{UP}_+)}_{((2m_{11}^{(i)}+1),(2n_{q_i1}^{(i)}+1))\text{-positive saddle}},\ \underbrace{((2m_{11}^{(i)})^{\text{th}}\text{DI}_+,\,(2n_{q_i1}^{(i)})^{\text{th}}\text{DI}_-)}_{((2m_{11}^{(i)}),(2n_{q_i1}^{(i)}))\text{-double-infection saddle}},\\ \underbrace{((2m_{11}^{(i)})^{\text{th}}\text{LS},\,(2n_{q_i1}^{(i)}+1)^{\text{th}}\text{DP})}_{((2m_{11}^{(i)}),(2n_{q_i1}^{(i)}+1))\text{-down-parabola lower-saddle}},\ \underbrace{((2m_{11}^{(i)}+1)^{\text{th}}\text{UP}_-,\,(2n_{q_i1}^{(i)})^{\text{th}}\text{US})}_{((2m_{11}^{(i)}+1),(2n_{q_i1}^{(i)}))\text{-up-parabola upper-saddle}}\end{array}\right\};$$

$$\underbrace{((m_1^{(i)})^{\text{th}}\,\text{XX},\,(n_1^{(i)})^{\text{th}}\,\text{XX})}_{(m_1^{(i)},n_1^{(i)})\text{-XX}}$$

$$\in\left\{\begin{array}{l}\underbrace{((2m_{11}^{(i)}+1)^{\text{th}}\text{DP}_-,\,(2n_{q_i1}^{(i)}+1)^{\text{th}}\text{DP}_+)}_{((2m_{11}^{(i)}+1),(2n_{11}^{(i)}+1))\text{-CCW center}},\ \underbrace{((2m_{11}^{(i)})^{\text{th}}\text{DI}_+,\,(2n_{11}^{(i)})^{\text{th}}\text{DI}_-)}_{((2m_{11}^{(i)}),(2n_{11}^{(i)}))\text{-double-infection saddle}},\\ \underbrace{((2m_{11}^{(i)})^{\text{th}}\text{LS},\,(2n_{11}^{(i)}+1)^{\text{th}}\text{UP})}_{((2m_{11}^{(i)}),(2n_{11}^{(i)}+1))\text{-up-parabola lower-saddle}},\ \underbrace{((2m_{11}^{(i)}+1)^{\text{th}}\text{UP},\,(2n_{11}^{(i)})^{\text{th}}\text{US})}_{((2m_{11}^{(i)}+1),(2n_{11}^{(i)}))\text{-up-parabola upper-saddle}}\end{array}\right\};\tag{5.99}$$

(vi_{1c}) The $((2m_1+1)^{\text{th}}$ UP, $(2n_1)^{\text{th}}$ US-up-parabola upper-saddle switching bifurcation is for the switching of two $p\times q$ equilibrium networks as

$$\begin{aligned}\bigcup_{r_1=1}^{p}\bigcup_{s_1=1}^{q}\underbrace{(a_{j_1j_2r_1},a_{j_2j_1s_1})}_{(m_{r_1}^{(1)},n_{s_1}^{(1)})\text{-XX}} &\rightleftharpoons \underbrace{((2m_1+1)^{\text{th}}\text{UP},(2n_1)^{\text{th}}\text{US})}_{((2m_1+1),(2n_1))\text{-up-parabola upper-saddle}}\\ &\rightleftharpoons \bigcup_{r_2=1}^{p}\bigcup_{s_2=1}^{q}\underbrace{(a_{j_1j_2r_2},a_{j_2j_1s_2})}_{(m_{r_2}^{(2)},n_{s_2}^{(2)})\text{-XX}}\end{aligned}\tag{5.100}$$

where $\sum_{r_i=1}^{p} m_{r_i}=2m_1+1$, $\sum_{s_i=1}^{q} n_{s_i}=2n_1$ for $i=1,2$.

(vi_2) For $a_{j_1j_10}<0$ and $a_{j_2j_20}>0$, there is a $((2m_1+1)^{\text{th}}\text{DP},\,(2n_1)^{\text{th}}\text{US})$-down-parabola upper-saddle equilibrium as

$$(a_{j_1j_21},a_{j_2j_11})=\underbrace{((2m_1+1)^{\text{th}}\text{DP},(2n_1)^{\text{th}}\text{US})}_{((2m_1+1),(2n_1))\text{-down-parabola upper-saddle}}.\tag{5.101}$$

There are the following three $((2m_1+1)^{\text{th}}\text{DP},\,(2n_1)^{\text{th}}\text{US})$-down-parabola upper-saddle appearing and switching bifurcations.

(vi_{2a}) The $((2m_1+1)^{\text{th}}$ DP, $(2n_1)^{\text{th}}$ US)-down-parabola upper-saddle appearing bifurcation is from a (DP, pF)-positive down-parabola flow to a $(2m_1+1)\times(2n_1)$ network as

$$\underbrace{(a_{j_1j_21}, \dot{x}_{j_2})}_{\text{(DP,pF)-down-parabola flow }(+)} \rightleftharpoons \underbrace{((2m_1+1)^{\text{th}}\text{DP},(2n_1)^{\text{th}}\text{US})}_{((2m_1+1),(2n_1))\text{-down-parabola upper-saddle}}$$
$$\rightleftharpoons \bigcup_{r=1}^{2m_1+1}\bigcup_{s=1}^{2n_1}\underbrace{(a_{j_1j_2r}, a_{j_2j_1s})}_{\text{XX}} \tag{5.102}$$

where

$$\bigcup_{s=1}^{2m_1+1}\bigcup_{l=1}^{2n_1}\underbrace{(a_{j_1j_2s}, a_{j_2j_1l})}_{\text{XX}} = \left\{\begin{array}{cccc} \underbrace{(\text{UP}_-,\text{UP}_-)}_{\text{saddle }(-)} & \underbrace{(\text{DP}_+,\text{DP}_-)}_{\text{center (CCW)}} & \cdots & \underbrace{(\text{DP}_+,\text{DP}_-)}_{\text{center (CCW)}} \\ \underbrace{(\text{DP}_-,\text{DP}_+)}_{\text{center (CW)}} & \underbrace{(\text{UP}_+,\text{UP}_+)}_{\text{saddle }(+)} & \cdots & \underbrace{(\text{UP}_+,\text{UP}_+)}_{\text{saddle }(+)} \\ \vdots & \vdots & \vdots & \vdots \\ \underbrace{(\text{UP}_-,\text{UP}_-)}_{\text{saddle }(-)} & \underbrace{(\text{DP}_+,\text{DP}_-)}_{\text{center (CCW)}} & \cdots & \underbrace{(\text{DP}_+,\text{DP}_-)}_{\text{center (CCW)}} \end{array}\right\}_{(2m_1+1)\times(2n_1)}. \tag{5.103}$$

($\text{vi}_{2\text{b}}$) The $((2m_1+1)^{\text{th}}\text{DP}, (2n_1)^{\text{th}}\text{US})$-down-parabola upper-saddle appearing bifurcation is from a $p_1 \times q_1$ to $p_2 \times q_2$ equilibrium network as

$$\bigcup_{r_1=1}^{p_1}\bigcup_{s_1=1}^{q_1}\underbrace{(a_{j_1j_2r_1}, a_{j_2j_1s_1})}_{(m_{r_1}^{(1)}, n_{s_1}^{(1)})\text{-XX}} \rightleftharpoons \underbrace{((2m_1+1)^{\text{th}}\text{DP}, (2n_1)^{\text{th}}\text{US})}_{((2m_1+1),(2n_1))\text{-down-parabola upper-saddle}}$$
$$\rightleftharpoons \bigcup_{r_2=1}^{p_2}\bigcup_{s_2=1}^{q_2}\underbrace{(a_{j_1j_2r_2}, a_{j_2j_1s_2})}_{(m_{r_2}^{(2)}, n_{s_2}^{(2)})\text{-XX}} \tag{5.104}$$

where

$2m_1+1-\Sigma_{r_1=1}^{p_1}m_{r_1}^{(1)} = 2\Sigma_{s=1}^{l_1}m_s$
with l_1-quadratic polynomials without real roots
and s^{th}-quadratic polynomial with power m_s,
$2n_1-\Sigma_{s_1=1}^{q_1}n_{s_1}^{(1)} = 2\Sigma_{l=1}^{l_2}n_l$
with l_2-quadraitc polynomials without real roots
and l^{th}-quadratic polynomial with power n_l;
$\Sigma_{r_2=1}^{p_2}m_{r_2}^{(2)} = 2m_1+1,\ \Sigma_{s_2=1}^{q_2}n_{s_2}^{(2)} = 2n_1$; for $i=1,2$,

$$\underbrace{((m_{p_i}^{(i)})^{\text{th}}\text{XX},(n_{q_i}^{(i)})^{\text{th}}\text{XX})}_{(m_{p_i}^{(i)},n_{q_i}^{(i)})\text{-XX}}$$
$$\in\left\{\begin{array}{c}\underbrace{((2m_{p_i1}^{(i)}+1)^{\text{th}}\text{DP}_+,\,(2n_{q_i1}^{(i)}+1)^{\text{th}}\text{DP}_-)}_{((2m_{p_i1}^{(i)}+1),(2n_{q_i1}^{(i)}+1))\text{-CCW center}},\;\underbrace{((2m_{p_i1}^{(i)})^{\text{th}}\text{DI}_+,\,(2n_{q_i1}^{(i)})^{\text{th}}\text{DI}_-)}_{((2m_{p_i1}^{(i)}),(2n_{q_i1}^{(i)}))\text{-double-infelction saddle}},\\ \underbrace{((2m_{p_i1}^{(i)})^{\text{th}}\text{LS},(2n_{q_i1}^{(i)}+1)^{\text{th}}\text{DP})}_{((2m_{p_i1}^{(i)}),(2n_{q_i1}^{(i)}+1))\text{-down-parabola lower-saddle}},\;\underbrace{((2m_{p_i1}^{(i)}+1)^{\text{th}}\text{DP},(2n_{q_i1}^{(i)})^{\text{th}}\text{US})}_{((2m_{p_i1}^{(i)}+1),(2n_{q_i1}^{(i)}))\text{-down-parabola upper-saddle}}\end{array}\right\};$$

$$\underbrace{((m_{p_i}^{(i)})^{\text{th}}\text{XX},(n_{1}^{(i)})^{\text{th}}\text{XX})}_{(m_{p_i}^{(i)},n_{1}^{(i)})\text{-XX}}$$
$$\in\left\{\begin{array}{c}\underbrace{((2m_{p_i1}^{(i)}+1)^{\text{th}}\text{UP}_-,\,(2n_{11}^{(i)}+1)^{\text{th}}\text{UP}_-)}_{((2m_{p_i1}^{(i)}+1),(2n_{11}^{(i)}+1))\text{-negative saddle}},\;\underbrace{((2m_{p_i1}^{(i)})^{\text{th}}\text{DI}_+,\,(2n_{11}^{(i)})^{\text{th}}\text{DI}_-)}_{((2m_{p_i1}^{(i)}),(2n_{11}^{(i)}))\text{-double-inflection saddle}},\\ \underbrace{((2m_{p_i1}^{(i)})^{\text{th}}\text{LS},(2n_{11}^{(i)}+1)^{\text{th}}\text{UP})}_{((2m_{p_i1}^{(i)}),(2n_{11}^{(i)}+1))\text{-up-parabola lower-saddle}},\;\underbrace{((2m_{p_i1}^{(i)}+1)^{\text{th}}\text{DP},(2n_{11}^{(i)})^{\text{th}}\text{US})}_{((2m_{p_i1}^{(i)}+1),(2n_{11}^{(i)}))\text{-down-parabola upper-saddle}}\end{array}\right\};$$

$$\underbrace{((m_{1}^{(i)})^{\text{th}}\text{XX},(n_{q_i}^{(i)})^{\text{th}}\text{XX})}_{(m_{1}^{(i)},n_{q_i}^{(i)})\text{-XX}}$$
$$\in\left\{\begin{array}{c}\underbrace{((2m_{11}^{(i)}+1)^{\text{th}}\text{DP}_+,\,(2n_{q_i1}^{(i)}+1)^{\text{th}}\text{DP}_-)}_{((2m_{11}^{(i)}+1),(2n_{q_i1}^{(i)}+1))\text{-CCW center}},\;\underbrace{((2m_{11}^{(i)})^{\text{th}}\text{II}_+,\,(2n_{q_i1}^{(i)})^{\text{th}}\text{II}_+)}_{((2m_{11}^{(i)}),(2n_{q_i1}^{(i)}))\text{-double-inflection saddle}},\\ \underbrace{((2m_{11}^{(i)})^{\text{th}}\text{US},(2n_{q_i1}^{(i)}+1)^{\text{th}}\text{UP})}_{((2m_{11}^{(i)}),(2n_{q_i1}^{(i)}+1))\text{-up-parabola upper-saddle}},\;\underbrace{((2m_{11}^{(i)}+1)^{\text{th}}\text{DP},(2n_{q_i1}^{(i)})^{\text{th}}\text{US})}_{((2m_{11}^{(i)}+1),(2n_{q_i1}^{(i)}))\text{-down-parabola upper-saddle}}\end{array}\right\};$$

$$\underbrace{((m_{1}^{(i)})^{\text{th}}\text{XX},(n_{1}^{(i)})^{\text{th}}\text{XX})}_{(m_{1}^{(i)},n_{1}^{(i)})\text{-XX}}$$
$$\in\left\{\begin{array}{c}\underbrace{((2m_{11}^{(i)}+1)^{\text{th}}\text{UP}_-,\,(2n_{11}^{(i)}+1)^{\text{th}}\text{UP}_-)}_{((2m_{11}^{(i)}+1),(2n_{11}^{(i)}+1))\text{-negative saddle}},\;\underbrace{((2m_{11}^{(i)})^{\text{th}}\text{II}_+,\,(2n_{11}^{(i)})^{\text{th}}\text{II}_+)}_{((2m_{11}^{(i)}),(2n_{q_i1}^{(i)}))\text{-double-inflection saddle}},\\ \underbrace{((2m_{11}^{(i)})^{\text{th}}\text{US},(2n_{11}^{(i)}+1)^{\text{th}}\text{DP})}_{((2m_{11}^{(i)}),(2n_{11}^{(i)}+1))\text{-down-parabola upper-saddle}},\;\underbrace{((2m_{11}^{(i)}+1)^{\text{th}}\text{DP},(2n_{11}^{(i)})^{\text{th}}\text{US})}_{((2m_{11}^{(i)}+1),(2n_{q_i1}^{(i)}))\text{-down-parabola upper-saddle}}\end{array}\right\}.\tag{5.105}$$

(vi_{2c}) The $((2m_1+1)^{\text{th}}\text{DP},\,(2n_1)^{\text{th}}\text{US})$-down-parabola upper-saddle switching bifurcation $((2m_1+1)^{\text{th}}\text{DP},\,(2n_1)^{\text{th}}\text{US})$-down-parabola upper-saddle is for two

$p \times q$ equilibrium networks as

$$\bigcup_{r_1=1}^{p}\bigcup_{s_1=1}^{q}\underbrace{(a_{j_1j_2r_1},a_{j_2j_1s_1})}_{(m_{r_1}^{(1)},n_{s_1}^{(1)})\text{-XX}} \rightleftharpoons \underbrace{((2m_1+1)^{\text{th}}\text{DP},(2n_1)^{\text{th}}\text{US})}_{((2m_1+1),(2n_1))\text{-down-parabola upper-saddle}} \rightleftharpoons \bigcup_{r_2=1}^{p}\bigcup_{s_2=1}^{q}\underbrace{(a_{j_1j_2r_2},a_{j_2j_1s_2})}_{(m_{r_2}^{(2)},n_{s_2}^{(2)})\text{-XX}} \tag{5.106}$$

where $\sum_{r_i=1}^{p} m_{r_i} = 2m_i + 1,\ \sum_{s_i=1}^{q} n_{s_i} = 2n_i$ for $i = 1, 2$.

(vi$_3$) For $a_{j_1j_10} > 0$ and $a_{j_2j_20} < 0$, there is a $((2m_1+1)^{\text{th}}\text{DP}, (2n_1)^{\text{th}}\text{LS})$-down-parabola lower-saddle equilibrium as

$$(a_{j_1j_21},a_{j_2j_11}) = \underbrace{((2m_1+1)^{\text{th}}\text{DP},(2n_1)^{\text{th}}\text{LS})}_{((2m_1+1),(2n_1))\text{-down-parabola lower-saddle}}. \tag{5.107}$$

There are the following three $((2m_1+1)^{\text{th}}\text{DP}, (2n_1)^{\text{th}}\text{LS})$-down-parabola lower-saddle appearing and switching bifurcations.

(vi$_{3a}$) The $((2m_1+1)^{\text{th}}\text{DP}, (2n_1)^{\text{th}}\text{LS})$-down-parabola lower-saddle appearing bifurcation are from an (UP, nF)-negative up-parabola flow to a $(2m_1+1) \times (2n_1)$-equilibrium network as

$$\underbrace{(a_{j_1j_21},\dot{x}_{j_2})}_{\text{(SO,nF)-negative source flow}} \rightleftharpoons \underbrace{((2m_1+1)^{\text{th}}\text{DP},(2n_1)^{\text{th}}\text{LS})}_{((2m_1+1),(2n_1))\text{-down-parabola lower-saddle}} \rightleftharpoons \bigcup_{r=1}^{2m_1+1}\bigcup_{s=1}^{2n_1}\underbrace{(a_{j_1j_2r},a_{j_2j_1s})}_{\text{XX}} \tag{5.108}$$

where

$$\bigcup_{s=1}^{2m_1+1}\bigcup_{l=1}^{2n_1}\underbrace{(a_{j_1j_2s},a_{j_2j_1l})}_{\text{XX}} = \left\{\begin{matrix} \underbrace{(\text{UP}_+,\text{UP}_+)}_{\text{saddle }(+)} & \underbrace{(\text{DP}_-,\text{DP}_+)}_{\text{center (CW)}} & \cdots & \underbrace{(\text{DP}_-,\text{DP}_+)}_{\text{center (CW)}} \\ \underbrace{(\text{DP}_+,\text{DP}_-)}_{\text{center (CCW)}} & \underbrace{(\text{UP}_-,\text{UP}_-)}_{\text{saddle }(-)} & \cdots & \underbrace{(\text{UP}_-,\text{UP}_-)}_{\text{saddle }(-)} \\ \vdots & \vdots & \vdots & \vdots \\ \underbrace{(\text{UP}_+,\text{UP}_+)}_{\text{saddle }(+)} & \underbrace{(\text{DP}_-,\text{DP}_+)}_{\text{center (CW)}} & \cdots & \underbrace{(\text{DP}_-,\text{DP}_+)}_{\text{center (CW)}} \end{matrix}\right\}_{(2m_1+1)\times(2n_1)}. \tag{5.109}$$

(vi$_{3b}$) The $((2m_1+1)^{\rm th}{\rm DP},\ (2n_1)^{\rm th}{\rm LS})$-down-parabola lower-saddle appearing bifurcation is from a $p_1 \times q_1$ to $p_2 \times q_2$ equilibrium network as

$$\bigcup_{r_1=1}^{p_1}\bigcup_{s_1=1}^{q_1}\underbrace{(a_{j_1 j_1 r_1}, a_{j_2 j_2 s_1})}_{(m_{r_1}^{(1)},\, n_{s_1}^{(1)})\text{-XX}} \rightleftharpoons \underbrace{((2m_1+1)^{\rm th}{\rm DP},(2n_1)^{\rm th}{\rm LS})}_{((2m_1+1),(2n_1))\text{-down-parabola lower-saddle}}$$
$$\rightleftharpoons \bigcup_{r_2=1}^{p_2}\bigcup_{s_2=1}^{q_2}\underbrace{(a_{j_1 j_1 r_2}, a_{j_2 j_2 s_2})}_{(m_{r_2}^{(2)},\, n_{s_2}^{(2)})\text{-XX}} \tag{5.110}$$

where

$2m_1 + 1 - \Sigma_{r_1=1}^{p_1} m_{r_1}^{(1)} = 2\Sigma_{s=1}^{l_1} m_s$
with l_1-quadratic polynomials without real roots
and $s^{\rm th}$-quadratic polynomial with power m_s,
$2n_1 - \Sigma_{s_1=1}^{q_1} n_{s_1}^{(1)} = 2\Sigma_{l=1}^{l_2} n_l$
with l_2-quadraitc polynomials without real roots
and $l^{\rm th}$-quadratic polynomial with power n_l;
$\Sigma_{r_2=1}^{p_2} m_{r_2}^{(2)} = 2m_1 + 1,\ \Sigma_{s_2=1}^{q_2} n_{s_2}^{(2)} = 2n_1$; for $i = 1, 2,$

$$\underbrace{((m_{p_i}^{(i)})^{\rm th}{\rm XX},(n_{q_i}^{(i)})^{\rm th}{\rm XX})}_{(m_{p_i}^{(i)},n_{q_i}^{(i)})\text{-XX}}$$
$$\in \left\{\begin{array}{l} \underbrace{((2m_{p_i1}^{(i)}+1)^{\rm th}{\rm DP}_-,\ (2n_{q_i1}^{(i)}+1)^{\rm th}{\rm DP}_+)}_{((2m_{p_i1}^{(i)}+1),(2n_{q_i1}^{(i)}+1))\text{-CW center}},\ \underbrace{((2m_{p_i1}^{(i)})^{\rm th}{\rm DI}_-,\ (2n_{q_i1}^{(i)})^{\rm th}{\rm DI}_+)}_{((2m_{p_i1}^{(i)}),(2n_{q_i1}^{(i)}))\text{-double-inflection saddle}},\\ \underbrace{((2m_{p_i1}^{(i)})^{\rm th}{\rm US},(2n_{q_i1}^{(i)}+1)^{\rm th}{\rm DP})}_{((2m_{p_i1}^{(i)}),(2n_{q_i1}^{(i)}+1))\text{-down-parabola upper-saddle}},\ \underbrace{((2m_{p_i1}^{(i)}+1)^{\rm th}{\rm DP},(2n_{q_i1}^{(i)})^{\rm th}{\rm LS})}_{((2m_{p_i1}^{(i)}+1),(2n_{q_i1}^{(i)}))\text{-down-parabola lower-saddle}} \end{array}\right\};$$

$$\underbrace{((m_{p_i}^{(i)})^{\rm th}{\rm XX},(n_1^{(i)})^{\rm th}{\rm XX})}_{(m_{p_i}^{(i)},n_1^{(i)})\text{-XX}}$$
$$\in \left\{\begin{array}{l} \underbrace{((2m_{p_i1}^{(i)}+1)^{\rm th}{\rm UP}_+,\ (2n_{11}^{(i)}+1)^{\rm th}{\rm UP}_+)}_{((2m_{p_i1}^{(i)}+1),(2n_{11}^{(i)}+1))\text{-positive saddle}},\ \underbrace{((2m_{p_i1}^{(i)})^{\rm th}{\rm DI}_-,\ (2n_{11}^{(i)})^{\rm th}{\rm DI}_+)}_{((2m_{p_i1}^{(i)}),(2n_{11}^{(i)}))\text{-double-inflection saddle}},\\ \underbrace{((2m_{p_i1}^{(i)})^{\rm th}{\rm US},(2n_{11}^{(i)}+1)^{\rm th}{\rm UP})}_{((2m_{p_i1}^{(i)}),(2n_{11}^{(i)}+1))\text{-up-parabola upper-saddle}},\ \underbrace{((2m_{p_i1}^{(i)}+1)^{\rm th}{\rm DP},(2n_{11}^{(i)})^{\rm th}{\rm LS})}_{((2m_{p_i1}^{(i)}+1),(2n_{11}^{(i)}))\text{-down-parabola lower-saddle}} \end{array}\right\};$$

$$\underbrace{((m_1^{(i)})^{\text{th}}\text{XX},(n_{q_i}^{(i)})^{\text{th}}\text{XX})}_{(m_1^{(i)},n_{q_i}^{(i)})\text{-XX}} \in \left\{\begin{array}{l} \underbrace{((2m_{11}^{(i)}+1)^{\text{th}}\text{DP}_{-},\ (2n_{q_i1}^{(i)}+1)^{\text{th}}\text{DP}_{+})}_{((2m_{11}^{(i)}+1),(2n_{q_i1}^{(i)}+1))\text{-CW center}},\ \underbrace{((2m_{11}^{(i)})^{\text{th}}\text{II}_{-},\ (2n_{q_i1}^{(i)})^{\text{th}}\text{II}_{-})}_{((2m_{11}^{(i)}),(2n_{q_i1}^{(i)}))\text{-double-infelction center}},\\ \underbrace{((2m_{11}^{(i)})^{\text{th}}\text{LS},(2n_{q_i1}^{(i)}+1)^{\text{th}}\text{UP})}_{((2m_{11}^{(i)}),(2n_{q_i1}^{(i)}+1))\text{-up-parabola lower-saddle}},\ \underbrace{((2m_{11}^{(i)}+1)^{\text{th}}\text{DP},(2n_{q_i1}^{(i)})^{\text{th}}\text{LS})}_{((2m_{11}^{(i)}+1),(2n_{q_i1}^{(i)}))\text{-down-parabola lower-saddle}} \end{array}\right\};$$

$$\underbrace{((m_1^{(i)})^{\text{th}}\text{XX},(n_1^{(i)})^{\text{th}}\text{XX})}_{(m_1^{(i)},n_1^{(i)})\text{-XX}} \in \left\{\begin{array}{l} \underbrace{((2m_{11}^{(i)}+1)^{\text{th}}\text{UP}_{+},\ (2n_{11}^{(i)}+1)^{\text{th}}\text{UP}_{+})}_{((2m_{11}^{(i)}+1),(2n_{11}^{(i)}+1))\text{-positive saddle}},\ \underbrace{((2m_{11}^{(i)})^{\text{th}}\text{II}_{-},\ (2n_{11}^{(i)})^{\text{th}}\text{II}_{-})}_{((2m_{11}^{(i)}),(2n_{11}^{(i)}))\text{-double-inflection saddle}},\\ \underbrace{((2m_{11}^{(i)})^{\text{th}}\text{LS},(2n_{11}^{(i)}+1)^{\text{th}}\text{DP})}_{((2m_{11}^{(i)}),(2n_{11}^{(i)}+1))\text{-down-parabola lower-saddle}},\ \underbrace{((2m_{11}^{(i)}+1)^{\text{th}}\text{DP},(2n_{11}^{(i)})^{\text{th}}\text{LS})}_{((2m_{11}^{(i)}+1),(2n_{11}^{(i)}))\text{-down-parabola lower-saddle}} \end{array}\right\}. \tag{5.111}$$

(vi_{3c}) The $((2m_1+1)^{\text{th}}\text{DP},\ (2n_1)^{\text{th}}\text{LS})$-down-parabola lower-saddle switching bifurcation is for the switching of two $p \times q$ equilibrium networks as

$$\bigcup_{r_1=1}^{p_1}\bigcup_{s_1=1}^{q_1}\underbrace{(a_{j_1j_2r_1},a_{j_2j_1s_1})}_{(m_{r_1}^{(1)},n_{s_1}^{(1)})\text{-XX}} \rightleftharpoons \underbrace{((2m_1+1)^{\text{th}}\text{DP},(2n_1)^{\text{th}}\text{LS})}_{((2m_1+1),(2n_1)\text{-down-parabola lower-saddle}} \rightleftharpoons \bigcup_{r_2=1}^{p_2}\bigcup_{s_2=1}^{q_2}\underbrace{(a_{j_1j_2r_2},a_{j_2j_1s_2})}_{(m_{r_2}^{(2)},n_{s_2}^{(2)})\text{-XX}}, \tag{5.112}$$

where $\sum_{r_i=1}^{p} m_{r_i} = 2m_i+1$, $\sum_{s_i=1}^{q} n_{s_i} = 2n_i$ for $i=1,2$.

(vi_4) For $a_{j_1j_10} < 0$ and $a_{j_2j_20} < 0$, there is a $((2m_1+1)^{\text{th}}\text{UP},(2n_1)^{\text{th}}\text{LS})$-up-parabola lower-saddle equilibrium as

$$(a_{j_1j_21},a_{j_2j_11}) = \underbrace{((2m_1+1)^{\text{th}}\text{UP},\ (2n_1)^{\text{th}}\text{LS})}_{((2m_1+1),(2n_1))\text{-up-parabola lower-saddle}} \tag{5.113}$$

There are the following three $((2m_1+1)^{\text{th}}\text{UP},\ (2n_1)^{\text{th}}\text{LS})$-up-parabola lower-saddle appearing and switching bifurcations.

(vi_{4a}) The $((2m_1+1)^{\mathrm{th}}\mathrm{UP}, (2n_1)^{\mathrm{th}}\mathrm{LS})$-up-parabola lower-saddle appearing bifurcation is from a (UP, nF)-negative up-parabola flow to a $(2m_1+1)\times(2n_1)$ network as

$$\underbrace{(a_{j_1j_21}, \dot{x}_{j_2})}_{\text{(DP,nF)-negative down-parabola flow}} \rightleftharpoons \underbrace{((2m_1+1)^{\mathrm{th}}\mathrm{UP}, (2n_1)^{\mathrm{th}}\mathrm{LS})}_{((2m_1+1),(2n_1))\text{-up-parabola lower-saddle}}$$
$$\rightleftharpoons \bigcup_{r=1}^{2m_1+1}\bigcup_{s=1}^{2n_1}\underbrace{(a_{j_1j_2r}, a_{j_2j_1s})}_{\mathrm{XX}} \tag{5.114}$$

where

$$\bigcup_{s=1}^{2m_1+1}\bigcup_{l=1}^{2n_1}\underbrace{(a_{j_1j_2s}, a_{j_2j_1l})}_{\mathrm{XX}} = \left\{\begin{array}{cccc} \underbrace{(\mathrm{DP}_+,\mathrm{DP}_-)}_{\text{center (CCW)}} & \underbrace{(\mathrm{UP}_-,\mathrm{UP}_-)}_{\text{saddle }(-)} & \cdots & \underbrace{(\mathrm{UP}_-,\mathrm{UP}_-)}_{\text{saddle }(-)} \\ \underbrace{(\mathrm{UP}_+,\mathrm{UP}_+)}_{\text{saddle }(+)} & \underbrace{(\mathrm{DP}_-,\mathrm{DP}_+)}_{\text{center (CW)}} & \cdots & \underbrace{(\mathrm{DP}_-,\mathrm{DP}_+)}_{\text{center (CW)}} \\ \vdots & \vdots & \vdots & \vdots \\ \underbrace{(\mathrm{DP}_+,\mathrm{DP}_-)}_{\text{center (CCW)}} & \underbrace{(\mathrm{UP}_-,\mathrm{UP}_-)}_{\text{saddle }(-)} & \cdots & \underbrace{(\mathrm{UP}_-,\mathrm{UP}_-)}_{\text{saddle }(-)} \end{array}\right\}_{(2m_1+1)\times(2n_1)}. \tag{5.115}$$

(vi_{4b}) The $((2m_1+1)^{\mathrm{th}}\mathrm{UP}, (2n_1)^{\mathrm{th}}\mathrm{LS})$-up-parabola lower-saddle appearing bifurcation is from $p_1\times q_1$ to $p_2\times q_2$ equilibrium networks as

$$\bigcup_{r_1=1}^{p_1}\bigcup_{s_1=1}^{q_1}\underbrace{(a_{j_1j_2r_1}, a_{j_2j_1s_1})}_{(m_{r_1}^{(1)}, n_{s_1}^{(1)})\text{-XX}} \rightleftharpoons \underbrace{((2m_1+1)^{\mathrm{th}}\mathrm{UP}, (2n_1)^{\mathrm{th}}\mathrm{LS})}_{((2m_1+1),(2n_1))\text{-up-parabola lower-saddle}}$$
$$\rightleftharpoons \bigcup_{r_2=1}^{p_2}\bigcup_{s_2=1}^{q_2}\underbrace{(a_{j_1j_2r_2}, a_{j_2j_1s_2})}_{(m_{r_2}^{(2)}, n_{s_2}^{(2)})\text{-XX}} \tag{5.116}$$

where

$2m_1+1-\Sigma_{r_1=1}^{p_1}m_{r_1}^{(1)} = 2\Sigma_{s=1}^{l_1}m_s$
with l_1-quadratic polynomials without real roots
and s^{th}-quadratic polynomial with power m_s,
$2n_1-\Sigma_{s_1=1}^{q_1}n_{s_1}^{(1)} = 2\Sigma_{l=1}^{l_2}n_l$
with l_2-quadraitc polynomials without real roots
and l^{th}-quadratic polynomial with power n_l;

$$\Sigma_{r_2=1}^{p_2} m_{r_2}^{(2)} = 2m_1 + 1,\ \ \Sigma_{s_2=1}^{q_2} n_{s_2}^{(2)} = 2n_1;\ \text{ for } i = 1, 2,$$

$$\underbrace{((m_{p_i}^{(i)})^{\text{th}}\text{XX}, (n_{q_i}^{(i)})^{\text{th}}\text{XX})}_{(m_{p_i}^{(i)}, n_{q_i}^{(i)})\text{-XX}}
\in \left\{ \begin{array}{c}
\underbrace{((2m_{p_i1}^{(i)}+1)^{\text{th}}\text{UP}_-, (2n_{q_i1}^{(i)}+1)^{\text{th}}\text{UP}_-)}_{((2m_{p_i1}^{(i)}+1),(2n_{q_i1}^{(i)}+1))\text{-negartive saddle}},\ \underbrace{((2m_{p_i1}^{(i)})^{\text{th}}\text{II}_-, (2n_{q_i1}^{(i)})^{\text{th}}\text{II}_-)}_{((2m_{p_i1}^{(i)}),(2n_{q_i1}^{(i)}))\text{-double-inflection saddle}},\\
\underbrace{((2m_{p_i1}^{(i)})^{\text{th}}\text{LS},(2n_{q_i1}^{(i)}+1)^{\text{th}}\text{UP})}_{((2m_{p_i1}^{(i)}),(2n_{q_i1}^{(i)}+1))\text{-up-parabola lower-saddle}},\ \underbrace{((2m_{p_i1}^{(i)}+1)^{\text{th}}\text{UP},(2n_{q_i1}^{(i)})^{\text{th}}\text{LS})}_{((2m_{p_i1}^{(i)}+1),(2n_{q_i1}^{(i)}))\text{-up-parabola lower-saddle}}
\end{array} \right\};$$

$$\underbrace{((m_{p_i}^{(i)})^{\text{th}}\text{XX},(n_1^{(i)})^{\text{th}}\text{XX})}_{(m_{p_i}^{(i)}, n_1^{(i)})\text{-XX}}
\in \left\{ \begin{array}{c}
\underbrace{((2m_{p_i1}^{(i)}+1)^{\text{th}}\text{DP}_+, (2n_{11}^{(i)}+1)^{\text{th}}\text{DP}_-)}_{((2m_{p_i1}^{(i)}+1),(2n_{11}^{(i)}+1))\text{-CCW center}},\ \underbrace{((2m_{p_i1}^{(i)})^{\text{th}}\text{II}_-, (2n_{11}^{(i)})^{\text{th}}\text{II}_-)}_{((2m_{p_i1}^{(i)}),(2n_{11}^{(i)}))\text{-double-inflection saddle}},\\
\underbrace{((2m_{p_i1}^{(i)})^{\text{th}}\text{LS},(2n_{11}^{(i)}+1)^{\text{th}}\text{DP})}_{((2m_{p_i1}^{(i)}),(2n_{11}^{(i)}+1))\text{-down-parabola lower-saddle}},\ \underbrace{((2m_{p_i1}^{(i)}+1)^{\text{th}}\text{UP},(2n_{11}^{(i)})^{\text{th}}\text{LS})}_{((2m_{p_i1}^{(i)}+1),(2n_{11}^{(i)}))\text{-up-parabola lower-saddle}}
\end{array} \right\};$$

$$\underbrace{((m_1^{(i)})^{\text{th}}\text{XX},(n_{q_i}^{(i)})^{\text{th}}\text{XX})}_{(m_1^{(i)}, n_{q_i}^{(i)})\text{-XX}}
\in \left\{ \begin{array}{c}
\underbrace{((2m_{11}^{(i)}+1)^{\text{th}}\text{UP}_-, (2n_{q_i1}^{(i)}+1)^{\text{th}}\text{UP}_-)}_{((2m_{11}^{(i)}+1),(2n_{q_i1}^{(i)}+1))\text{-nagative saddle}},\ \underbrace{((2m_{11}^{(i)})^{\text{th}}\text{DI}_-, (2n_{q_i1}^{(i)})^{\text{th}}\text{DI}_+)}_{((2m_{11}^{(i)}),(2n_{q_i1}^{(i)}))\text{-double-inflection saddle}},\\
\underbrace{((2m_{11}^{(i)})^{\text{th}}\text{US},(2n_{q_i1}^{(i)}+1)^{\text{th}}\text{DP})}_{((2m_{11}^{(i)}),(2n_{q_i1}^{(i)}+1))\text{-down-parabola upper-saddle}},\ \underbrace{((2m_{11}^{(i)}+1)^{\text{th}}\text{UP},(2n_{q_i1}^{(i)})^{\text{th}}\text{LS})}_{((2m_{11}^{(i)}+1),(2n_{q_i1}^{(i)}))\text{-up-parabola lower-saddle}}
\end{array} \right\};$$

$$\underbrace{((m_1^{(i)})^{\text{th}}\text{XX},(n_1^{(i)})^{\text{th}}\text{XX})}_{(m_1^{(i)}, n_1^{(i)})\text{-XX}}
\in \left\{ \begin{array}{c}
\underbrace{((2m_{11}^{(i)}+1)^{\text{th}}\text{DP}_+, (2n_{11}^{(i)}+1)^{\text{th}}\text{DP}_-)}_{((2m_{11}^{(i)}+1),(2n_{11}^{(i)}+1))\text{-CCW center}},\ \underbrace{((2m_{11}^{(i)})^{\text{th}}\text{DI}_-, (2n_{11}^{(i)})^{\text{th}}\text{DI}_+)}_{((2m_{11}^{(i)}),(2n_{11}^{(i)}))\text{-double-inflection saddle}},\\
\underbrace{((2m_{11}^{(i)})^{\text{th}}\text{US},(2n_{11}^{(i)}+1)^{\text{th}}\text{UP})}_{((2m_{11}^{(i)}),(2n_{11}^{(i)}+1))\text{-up-parabola upper-saddle}},\ \underbrace{((2m_{11}^{(i)}+1)^{\text{th}}\text{UP},(2n_{11}^{(i)})^{\text{th}}\text{LS})}_{((2m_{11}^{(i)}+1),(2n_{11}^{(i)}))\text{-up-parabola lower-saddle}}
\end{array} \right\}. \tag{5.117}$$

(vi_{4c}) The $((2m_1+1)^{\text{th}}\text{UP}, (2n_1)^{\text{th}}\text{LS})$-up-parabola lower-saddle switching bifurcation is for two $p \times q$ equilibrium networks as

$$\bigcup_{r_1=1}^{p}\bigcup_{s_1=1}^{q}\underbrace{(a_{j_1j_2r_1}, a_{j_2j_1s_1})}_{(m_{r_1}^{(1)}, n_{s_1}^{(1)})\text{-XX}} \rightleftharpoons \underbrace{((2m_1+1)^{\text{th}}\text{UP}, (2n_1)^{\text{th}}\text{LS})}_{((2m_1+1),(2n_1))\text{-up-parabola lower-saddle}}$$
$$\rightleftharpoons \bigcup_{r_2=1}^{p}\bigcup_{s_2=1}^{q}\underbrace{(a_{j_1j_2r_2}, a_{j_2j_1s_2})}_{(m_{r_2}^{(2)}, n_{s_2}^{(2)})\text{-XX}} \tag{5.118}$$

where $\sum_{r_i=1}^{p} m_{r_i} = 2m_1+1, \sum_{s_i=1}^{q} n_{s_i} = 2n_i$ for $i = 1, 2$.

(vii) For $m = 2m_1$ and $n = 2n_1$, the equilibrium of $(x_{j_2}^*, x_{j_1}^*) = (a_{j_1j_21}, a_{j_2j_11})$ has the following properties.

(vii_1) For $a_{j_1j_10} > 0$ and $a_{j_2j_20} > 0$, there is a $((2m_1)^{\text{th}}\text{II}_+, (2n_1)^{\text{th}}\text{II}_+)$-double-inflection saddle equilibrium as

$$(a_{j_1j_21}, a_{j_2j_11}) = \underbrace{((2m_1)^{\text{th}}\text{II}_+, (2n_1)^{\text{th}}\text{II}_+)}_{((2m_1),(2n_1))\text{-double-inflection saddle}}. \tag{5.119}$$

There are three following $((2m_1)^{\text{th}}\text{II}_+, (2n_1)^{\text{th}}\text{II}_+)$-double-inflection saddle appearing and switching bifurcations.

(vii_{1a}) The $((2m_1)^{\text{th}}\text{II}_+, (2n_1)^{\text{th}}\text{II}_+)$-double-inflection saddle appearing bifurcation are from a (pF, pF)-positive flow to a $((2m_1) \times (2n_1))$-equilibrium network as

$$\underbrace{(\dot{x}_{j_1}, \dot{x}_{j_2})}_{(\text{pF},\text{pF})\text{-flow}} \rightleftharpoons \underbrace{((2m_1)^{\text{th}}\text{II}_+, (2n_1)^{\text{th}}\text{II}_+)}_{((2m_1),(2n_1))\text{-double-inflection saddle}} \rightleftharpoons \bigcup_{r=1}^{2m_1}\bigcup_{s=1}^{2n_1}\underbrace{(a_{j_1j_2r}, a_{j_2j_1s})}_{\text{XX}} \tag{5.120}$$

where

$$\bigcup_{s=1}^{2m_1}\bigcup_{l=1}^{2n_1}\underbrace{(a_{j_1j_2s}, a_{j_2j_1l})}_{\text{XX}} = \left\{\begin{array}{cccc} \underbrace{(\text{DP}_-,\text{DP}_+)}_{\text{center (CW)}} & \underbrace{(\text{UP}_+,\text{UP}_+)}_{\text{saddle (+)}} & \cdots & \underbrace{(\text{UP}_+,\text{UP}_+)}_{\text{saddle (+)}} \\ \underbrace{(\text{UP}_-,\text{UP}_-)}_{\text{saddle }(-)} & \underbrace{(\text{DP}_+,\text{DP}_-)}_{\text{center (CCW)}} & \cdots & \underbrace{(\text{DP}_+,\text{DP}_-)}_{\text{center (CCW)}} \\ \vdots & \vdots & \vdots & \vdots \\ \underbrace{(\text{UP}_-,\text{UP}_-)}_{\text{saddle }(-)} & \underbrace{(\text{DP}_+,\text{DP}_-)}_{\text{center (CCW)}} & \cdots & \underbrace{(\text{DP}_+,\text{DP}_-)}_{\text{center (CCW)}} \end{array}\right\}_{(2m_1)\times(2n_1)}. \tag{5.121}$$

(vii_{1b}) The $((2m_1)^{\text{th}}\text{II}_+, (2n_1)^{\text{th}}\text{II}_+)$-double-inflection saddle appearing bifurcation is from a $p_1 \times q_1$ to $p_2 \times q_2$ equilibrium network as

$$\bigcup_{r_1=1}^{p_1}\bigcup_{s_1=1}^{q_1}\underbrace{(a_{j_1j_2r_1}, a_{j_2j_1s_1})}_{(m_{r_1}^{(1)},\, n_{s_1}^{(1)})\text{-XX}} \rightleftharpoons \underbrace{((2m_1)^{\text{th}}\text{II}_+, (2n_1)^{\text{th}}\text{II}_+)}_{((2m_1),(2n_1))\text{-double-inflection saddle}}$$
$$\rightleftharpoons \bigcup_{r_2=1}^{p_2}\bigcup_{s_2=1}^{q_2}\underbrace{(a_{j_1j_2r_2}, a_{j_2j_1s_2})}_{(m_{r_2}^{(2)},\, n_{s_2}^{(2)})\text{-XX}} \tag{5.122}$$

where

$$\begin{aligned}
&2m_1 - \Sigma_{r_1=1}^{p_1} m_{r_1}^{(1)} = 2\Sigma_{s=1}^{l_1} m_s\\
&\text{with } l_1\text{-quadratic polynomials without real roots}\\
&\text{and } s^{\text{th}}\text{-quadratic polynomial with power } m_s,\\
&2n_1 - \Sigma_{s_1=1}^{q_1} n_{s_1}^{(1)} = 2\Sigma_{l=1}^{l_2} n_l\\
&\text{with } l_2\text{-quadraitc polynomials without real roots}\\
&\text{and } l^{\text{th}}\text{-quadratic polynomial with power } n_l;\\
&\Sigma_{r_2=1}^{p_2} m_{r_2}^{(2)} = 2m_1,\ \Sigma_{s_2=1}^{q_2} n_{s_2}^{(2)} = 2n_1;\ \text{for } i = 1, 2,
\end{aligned}$$

$$\underbrace{((m_{p_i}^{(i)})^{\text{th}}\text{XX}, (n_{q_i}^{(i)})^{\text{th}}\text{XX})}_{(m_{p_i}^{(i)},n_{q_i}^{(i)})\text{-XX}}$$
$$\in \left\{\begin{array}{c}
\underbrace{((2m_{p_i1}^{(i)}+1)^{\text{th}}\text{UP}_+, (2n_{q_i1}^{(i)}+1)^{\text{th}}\text{UP}_+)}_{((2m_{p_i1}^{(i)}+1),(2n_{q_i1}^{(i)}+1))\text{-positive saddle}},\ \underbrace{((2m_{p_i1}^{(i)})^{\text{th}}\text{II}_+, (2n_{q_i1}^{(i)})^{\text{th}}\text{II}_+)}_{((2m_{p_i1}^{(i)}),(2n_{q_i1}^{(i)}))\text{-double-infection saddle}},\\
\underbrace{((2m_{p_i1}^{(i)})^{\text{th}}\text{US}, (2n_{q_i1}^{(i)}+1)^{\text{th}}\text{UP})}_{((2m_{p_i1}^{(i)}),(2n_{q_i1}^{(i)}+1))\text{-up-parabola upper-saddle}},\ \underbrace{((2m_{p_i1}^{(i)}+1)^{\text{th}}\text{UP}, (2n_{q_i1}^{(i)})^{\text{th}}\text{US})}_{((2m_{p_i1}^{(i)}+1),(2n_{q_i1}^{(i)}))\text{-up-parabola upper-saddle}}
\end{array}\right\};$$

$$\underbrace{((m_{p_i}^{(i)})^{\text{th}}\text{XX}, (n_1^{(i)})^{\text{th}}\text{XX})}_{(m_{p_i}^{(i)},n_1^{(i)})\text{-XX}}$$
$$\in \left\{\begin{array}{c}
\underbrace{((2m_{p_i1}^{(i)}+1)^{\text{th}}\text{DP}_-, (2n_{11}^{(i)}+1)^{\text{th}}\text{DP}_+)}_{((2m_{p_i1}^{(i)}+1),(2n_{11}^{(i)}+1))\text{-CW center}},\ \underbrace{((2m_{p_i1}^{(i)})^{\text{th}}\text{II}_+, (2n_{11}^{(i)})^{\text{th}}\text{II}_+)}_{((2m_{p_i1}^{(i)}),(2n_{11}^{(i)}))\text{-double-infection saddle}},\\
\underbrace{((2m_{p_i1}^{(i)})^{\text{th}}\text{US}, (2n_{11}^{(i)}+1)^{\text{th}}\text{DP})}_{((2m_{p_i1}^{(i)}),(2n_{11}^{(i)}+1))\text{-down-parabola upper-saddle}},\ \underbrace{((2m_{p_i1}^{(i)}+1)^{\text{th}}\text{UP}, (2n_{11}^{(i)})^{\text{th}}\text{US})}_{((2m_{p_i1}^{(i)}+1),(2n_{11}^{(i)}))\text{-up-parabola upper-saddle}},
\end{array}\right\};$$

$$\underbrace{((m_1^{(i)})^{\text{th}}\,\text{XX},(n_{q_i}^{(i)})^{\text{th}}\,\text{XX})}_{(m_1^{(i)},n_{q_i}^{(i)})\text{-XX}}$$
$$\in \left\{ \begin{array}{l} \underbrace{((2m_1 1^{(i)}+1)^{\text{th}}\text{DP}_+,(2n_{q_i 1}^{(i)}+1)^{\text{th}}\text{DP}_-)}_{((2m_{p_i 1}^{(i)}+1),(2n_{q_i 1}^{(i)}+1))\text{-CCW center}},\ \underbrace{((2m_{11}^{(i)})^{\text{th}}\text{II}_+,(2n_{q_i 1}^{(i)})^{\text{th}}\text{II}_+)}_{((2m_{11}^{(i)}),(2n_{q_i 1}^{(i)}))\text{-double-infection saddle}},\\ \underbrace{((2m_{11}^{(i)})^{\text{th}}\text{US},(2n_{q_i 1}^{(i)}+1)^{\text{th}}\text{UP})}_{((2m_{11}^{(i)}),(2n_{q_i 1}^{(i)}+1))\text{-up-parabola upper-saddle}},\ \underbrace{((2m_{11}^{(i)}+1)^{\text{th}}\text{UP},(2n_{q_i 1}^{(i)})^{\text{th}}\text{US})}_{((2m_{11}^{(i)}+1),(2n_{q_i 1}^{(i)}))\text{-down-parabola upper-saddle}} \end{array} \right\};$$
$$\underbrace{((m_1^{(i)})^{\text{th}}\,\text{XX},(n_1^{(i)})^{\text{th}}\,\text{XX})}_{(m_1^{(i)},n_1^{(i)})\text{-XX}}$$
$$\in \left\{ \begin{array}{l} \underbrace{((2m_{11}{}^{(i)}+1)^{\text{th}}\text{UP}_-,(2n_{11}^{(i)}+1)^{\text{th}}\text{UP}_-)}_{((2m_{11}^{(i)}+1),(2n_{11}^{(i)}+1))\text{-negative saddle}},\ \underbrace{((2m_{11}^{(i)})^{\text{th}}\text{DI}_+,(2n_{11}^{(i)})^{\text{th}}\text{DI}_-)}_{((2m_{11}^{(i)}),(2n_{11}^{(i)}))\text{-double-infection saddle}},\\ \underbrace{((2m_{11}^{(i)})^{\text{th}}\text{US},(2n_{11}^{(i)}+1)^{\text{th}}\text{DP})}_{((2m_{11}^{(i)}),(2n_{11}^{(i)}+1))\text{-down-parabola upper-saddle}},\ \underbrace{((2m_{11}^{(i)}+1)^{\text{th}}\text{DP},(2n_{11}^{(i)})^{\text{th}}\text{US})}_{((2m_{11}^{(i)}+1),(2n_{11}^{(i)}))\text{-down-parabola upper-saddle}} \end{array} \right\};$$
(5.123)

(vii$_{1c}$) The $((2m_1)^{\text{th}}\text{II}_+,(2n_1)^{\text{th}}\text{II}_+)$-double-inflection saddle switching bifurcation is for the switching of two $p \times q$ equilibrium networks as

$$\bigcup_{r_1=1}^{p}\bigcup_{s_1=1}^{q}\underbrace{(a_{j_1j_2r_1},a_{j_2j_1s_1})}_{(m_{r_1}^{(1)},n_{s_1}^{(1)})\text{-XX}} \rightleftharpoons \underbrace{((2m_1)^{\text{th}}\text{II}_+,(2n_1)^{\text{th}}\text{II}_+)}_{((2m_1),(2n_1))\text{-double-inflection saddle}}$$
$$\rightleftharpoons \bigcup_{r_2=1}^{p}\bigcup_{s_2=1}^{q}\underbrace{(a_{j_1j_2r_2},a_{j_2j_1s_2})}_{(m_{r_2}^{(2)},n_{s_2}^{(2)})\text{-XX}} \quad (5.124)$$

where $\sum_{r_i=1}^{p} m_{r_i} = 2m_1$, $\sum_{s_i=1}^{q} n_{s_i} = 2n_1$ for $i = 1, 2$.

(vii$_2$) For $a_{j_1j_10} < 0$ and $a_{j_2j_20} > 0$, there is a $((2m_1)^{\text{th}}\text{DI}_+,(2n_1)^{\text{th}}\text{DI}_-)$ double-inflection saddle equilibrium as

$$(a_{j_1j_21},a_{j_2j_11}) = \underbrace{((2m_1)^{\text{th}}\text{DI}_+,(2n_1)^{\text{th}}\text{DI}_-)}_{((2m_1),(2n_1))\text{-double-inflection saddle}}. \quad (5.125)$$

There are the following three $((2m_1)^{\text{th}}\text{DI}_+,(2n_1)^{\text{th}}\text{DI}_-)$-double-inflection saddle appearing and switching bifurcations.

(vii$_{2a}$) The $((2m_1)^{\text{th}}\text{DI}_+,(2n_1)^{\text{th}}\text{DI}_-)$-double-inflection saddle appearing bifurcation is from an (nF, pF)-flow to a $(2m_1) \times (2n_1)$ network as

$$\underbrace{(\dot{x}_{j_1}, \dot{x}_{j_2})}_{\text{(nF,pF)-flow}} \rightleftharpoons \underbrace{((2m_1)^{\text{th}}\text{DI}_+, (2n_1)^{\text{th}}\text{DI}_-)}_{((2m_1),(2n_1))\text{-double-inflection saddle}} \rightleftharpoons \bigcup_{r=1}^{2m_1}\bigcup_{s=1}^{2n_1}\underbrace{(a_{j_1j_2r}, a_{j_2j_1s})}_{\text{XX}} \tag{5.126}$$

where

$$\bigcup_{s=1}^{2m_1}\bigcup_{l=1}^{2n_1}\underbrace{(a_{j_1j_2s}, a_{j_2j_1l})}_{\text{XX}} = \left\{\begin{matrix} \underbrace{(\text{UP}_-,\text{UP}_-)}_{\text{saddle }(-)} & \underbrace{(\text{DP}_+,\text{DP}_-)}_{\text{center (CCW)}} & \cdots & \underbrace{(\text{DP}_+,\text{DP}_-)}_{\text{center (CCW)}} \\ \underbrace{(\text{DP}_-,\text{DP}_+)}_{\text{center (CW)}} & \underbrace{(\text{UP}_+,\text{UP}_+)}_{\text{saddle }(+)} & \cdots & \underbrace{(\text{UP}_+,\text{UP}_+)}_{\text{saddle }(+)} \\ \vdots & \vdots & \vdots & \vdots \\ \underbrace{(\text{DP}_-,\text{DP}_+)}_{\text{center (CW)}} & \underbrace{(\text{UP}_+,\text{UP}_+)}_{\text{saddle }(+)} & \cdots & \underbrace{(\text{UP}_+,\text{UP}_+)}_{\text{saddle }(+)} \end{matrix}\right\}_{(2m_1)\times(2n_1)}. \tag{5.127}$$

(vii$_{2b}$) The $(2m_1)^{\text{th}}$ DI$_+$, $(2n_1)^{\text{th}}$ DI$_-$)-double-inflection saddle appearing bifurcation is from $p_1 \times q_1$ to $p_2 \times q_2$ equilibrium networks as

$$\begin{aligned} \bigcup_{r_1=1}^{p_1}\bigcup_{s_1=1}^{q_1}\underbrace{(a_{j_1j_2r_1}, a_{j_2j_1s_1})}_{(m_{r_1}^{(1)}, n_{s_1}^{(1)})\text{-XX}} &\rightleftharpoons \underbrace{((2m_1)^{\text{th}}\text{DI}_+, (2n_1)^{\text{th}}\text{DI}_-)}_{((2m_1),(2n_1))\text{-double-inflection saddle}} \\ &\rightleftharpoons \bigcup_{r_2=1}^{p_2}\bigcup_{s_2=1}^{q_2}\underbrace{(a_{j_1j_2r_2}, a_{j_2j_1s_2})}_{(m_{r_2}^{(2)}, n_{s_2}^{(2)})\text{-XX}} \end{aligned} \tag{5.128}$$

where

$$\begin{aligned} &2m_1 - \Sigma_{r_1=1}^{p_1} m_{r_1}^{(1)} = 2\Sigma_{s=1}^{l_1} m_s \\ &\text{with } l_1\text{-quadratic polynomials without real roots} \\ &\text{and } s^{\text{th}}\text{-quadratic polynomial with power } m_s, \\ &2n_1 - \Sigma_{s_1=1}^{q_1} n_{s_1}^{(1)} = 2\Sigma_{l=1}^{l_2} n_l \\ &\text{with } l_2\text{-quadraitc polynomials without real roots} \\ &\text{and } l^{\text{th}}\text{-quadratic polynomial with power } n_l; \\ &\Sigma_{r_2=1}^{p_2} m_{r_2}^{(2)} = 2m_1,\ \Sigma_{s_2=1}^{q_2} n_{s_2}^{(2)} = 2n_1; \text{ for } i = 1, 2, \end{aligned}$$

$$\underbrace{((m_{p_i}^{(i)})^{\text{th}}\text{XX},(n_{q_i}^{(i)})^{\text{th}}\text{XX})}_{(m_{p_i}^{(i)},n_{q_i}^{(i)})\text{-XX}}$$

$$
\in \left\{ \begin{array}{c} \underbrace{((2m_{p_i1}^{(i)}+1)^{\text{th}}\text{DP}_+,(2n_{q_i1}^{(i)}+1)^{\text{th}}\text{DP}_-)}_{((2m_{p_i1}^{(i)}+1),(2n_{q_i1}^{(i)}+1))\text{-CCW center}}, \underbrace{((2m_{p_i1}^{(i)}+1)^{\text{th}}\text{DI}_+,(2n_{q_i1}^{(i)})^{\text{th}}\text{DI}_-)}_{((2m_{p_i1}^{(i)}),(2n_{q_i1}^{(i)}))\text{-double-inflection saddle}}, \\ \underbrace{((2m_{p_i1}^{(i)})^{\text{th}}\text{LS},(2n_{q_i1}^{(i)}+1)^{\text{th}}\text{DP})}_{((2m_{p_i1}^{(i)}),(2n_{q_i1}^{(i)}+1))\text{-down-parabola lower-saddle}}, \underbrace{((2m_{p_i1}^{(i)}+1)^{\text{th}}\text{UP},(2n_{q_i1}^{(i)})^{\text{th}}\text{US})}_{((2m_{p_i1}^{(i)}+1),(2n_{q_i1}^{(i)}))\text{-down-parabola upper-saddle}} \end{array} \right\};
$$

$$
\underbrace{((m_{p_i}^{(i)})^{\text{th}}\text{XX},(n_1^{(i)})^{\text{th}}\text{XX})}_{(m_{p_i}^{(i)},n_1^{(i)})\text{-XX}}
$$

$$
\in \left\{ \begin{array}{c} \underbrace{((2m_{p_i1}^{(i)}+1)^{\text{th}}\text{UP}_-,(2n_{11}^{(i)}+1)^{\text{th}}\text{UP}_-)}_{((2m_{p_i1}^{(i)}+1),(2n_{11}^{(i)}+1))\text{-negative saddle}}, \underbrace{((2m_{p_i1}^{(i)})^{\text{th}}\text{DI}_+,(2n_{11}^{(i)})^{\text{th}}\text{DI}_-)}_{((2m_{p_i1}^{(i)}),(2n_{11}^{(i)}))\text{-double-inflection saddle}}, \\ \underbrace{((2m_{p_i1}^{(i)})^{\text{th}}\text{LS},(2n_{11}^{(i)}+1)^{\text{th}}\text{UP})}_{((2m_{p_i1}^{(i)}),(2n_{11}^{(i)}+1))\text{-up-parabola lower-saddle}}, \underbrace{((2m_{p_i1}^{(i)}+1)^{\text{th}}\text{UP},(2n_{11}^{(i)}+1)^{\text{th}}\text{US})}_{((2m_{p_i1}^{(i)}+1),(2n_{11}^{(i)}))\text{-down-parabola lower-saddle}} \end{array} \right\};
$$

$$
\underbrace{((m_1^{(i)})^{\text{th}}\text{XX},(n_{q_i}^{(i)})^{\text{th}}\text{XX})}_{(m_1^{(i)},n_{q_i}^{(i)})\text{-XX}}
$$

$$
\in \left\{ \begin{array}{c} \underbrace{((2m_{11}^{(i)}+1)^{\text{th}}\text{UP}_+,(2n_{q_i1}^{(i)}+1)^{\text{th}}\text{UP}_+)}_{((2m_{11}^{(i)}+1),(2n_{q_i1}^{(i)}+1))\text{-positive saddle}}, \underbrace{((2m_{11}^{(i)})^{\text{th}}\text{DI}_+,(2n_{q_i1}^{(i)})^{\text{th}}\text{DI}_-)}_{((2m_{11}^{(i)}),(2n_{q_i1}^{(i)}))\text{-double-infection-saddle}}, \\ \underbrace{((2m_{11}^{(i)})^{\text{th}}\text{LS},(2n_{q_i1}^{(i)}+1)^{\text{th}}\text{DP})}_{((2m_{11}^{(i)}),(2n_{q_i1}^{(i)}+1))\text{-down-parabola lower-saddle}}, \underbrace{((2m_{11}^{(i)}+1)^{\text{th}}\text{DP},(2n_{q_i1}^{(i)})^{\text{th}}\text{US})}_{((2m_{11}^{(i)}+1),(2n_{q_i1}^{(i)}))\text{-up-parabola lower-saddle}} \end{array} \right\};
$$

$$
\underbrace{((m_1^{(i)})^{\text{th}}\text{XX},(n_1^{(i)})^{\text{th}}\text{XX})}_{(m_1^{(i)},n_1^{(i)})\text{-XX}}
$$

$$
\in \left\{ \begin{array}{c} \underbrace{((2m_{11}^{(i)}+1)^{\text{th}}\text{DP}_-,(2n_{11}^{(i)})^{\text{th}}\text{DP}_+)}_{((2m_{11}^{(i)}+1),(2n_{11}^{(i)}+1))\text{-CW center}}, \underbrace{((2m_{11}^{(i)})^{\text{th}}\text{DI}_+,(2n_{11}^{(i)})^{\text{th}}\text{DI}_-)}_{((2m_{11}^{(i)}),(2n_{11}^{(i)}))\text{-double-inflection saddle}}, \\ \underbrace{((2m_{11}^{(i)})^{\text{th}}\text{LS},(2n_{11}^{(i)}+1)^{\text{th}}\text{UP}_+)}_{((2m_{11}^{(i)}),(2n_{11}^{(i)}+1))\text{-up-parabola lower-saddle}}, \underbrace{((2m_{11}^{(i)}+1)^{\text{th}}\text{DP},(2n_{11}^{(i)})^{\text{th}}\text{LS})}_{((2m_{11}^{(i)}+1),(2n_{11}^{(i)}))\text{-down-parabola lower-saddle}} \end{array} \right\}. \tag{5.129}
$$

(vii$_{2c}$) The $(2m_1)^{\text{th}}$ DI$_+$, $(2n_1)^{\text{th}}$ DI$_-$)-double-inflection saddle switching bifurcation is for two $p \times q$ equilibrium networks as

$$
\bigcup_{r_1=1}^{p}\bigcup_{s_1=1}^{q} \underbrace{(a_{j_1j_2r_1},a_{j_2j_1s_1})}_{(m_{r_1}^{(1)},n_{s_1}^{(1)})\text{-XX}} \rightleftharpoons \underbrace{((2m_1)^{\text{th}}\text{DI}_+,(2n_1)^{\text{th}}\text{DI}_-)}_{((2m_1),(2n_1))\text{-double-inflection saddle}}
$$

$$\rightleftharpoons \bigcup_{r_2=1}^{p}\bigcup_{s_2=1}^{q}\underbrace{(a_{j_1j_2r_2}, a_{j_2j_1s_2})}_{(m_{r_2}^{(2)}, n_{s_2}^{(2)})\text{-XX}} \tag{5.130}$$

where $\sum_{r_i=1}^{p} m_{r_i} = 2m_i + 1$, $\sum_{s_i=1}^{q} n_{s_i} = 2n_i$ for $i = 1, 2$.

(vii$_3$) For $a_{j_1j_10} > 0$ and $a_{j_2j_20} < 0$, there is a $((2m_1)^{\text{th}}\text{DI}_-, (2n_1)^{\text{th}}\text{DI}_+)$ double-inflection saddle equilibrium as

$$(a_{j_1j_11}, a_{j_2j_21}) = \underbrace{((2m_1)^{\text{th}}\text{DI}_-, (2n_1)^{\text{th}}\text{DI}_+)}_{((2m_1),(2n_1))\text{-double-inflection saddle}}. \tag{5.131}$$

There are the following three $((2m_1)^{\text{th}}\text{DI}_-, (2n_1)^{\text{th}}\text{DI}_+)$-double-inflection saddle appearing and switching bifurcations.

(vii$_{3a}$) The $((2m_1)^{\text{th}}\text{DI}_-, (2n_1)^{\text{th}}\text{DI}_+)$-double-inflection saddle appearing bifurcation are from an (pF, nF)-flow to a $(2m_1) \times (2n_1)$-equilibrium network as

$$\underbrace{(\dot{x}_{j_1}, \dot{x}_{j_2})}_{(\text{pF,nF})\text{-flow}} \rightleftharpoons \underbrace{((2m_1)^{\text{th}}\text{DI}_-, (2n_1)^{\text{th}}\text{DI}_+)}_{((2m_1),(2n_1))\text{-double-inflection saddle}} \rightleftharpoons \bigcup_{r=1}^{2m_1}\bigcup_{s=1}^{2n_1}\underbrace{(a_{j_1j_2r}, a_{j_2j_1s})}_{\text{XX}} \tag{5.132}$$

where

$$\bigcup_{s=1}^{2m_1}\bigcup_{l=1}^{2n_1}\underbrace{(a_{j_1j_2s}, a_{j_2j_1l})}_{\text{XX}} = \begin{Bmatrix} \underbrace{(\text{UP}_+,\text{UP}_+)}_{\text{saddle }(+)} & \underbrace{(\text{DP}_-,\text{DP}_+)}_{\text{center (CW)}} & \cdots & \underbrace{(\text{DP}_-,\text{DP}_+)}_{\text{center (CW)}} \\ \underbrace{(\text{DP}_+,\text{DP}_-)}_{\text{center (CCW)}} & \underbrace{(\text{UP}_-,\text{UP}_-)}_{\text{saddle }(-)} & \cdots & \underbrace{(\text{UP}_-,\text{UP}_-)}_{\text{saddle }(-)} \\ \vdots & \vdots & \vdots & \vdots \\ \underbrace{(\text{DP}_+,\text{DP}_-)}_{\text{center (CCW)}} & \underbrace{(\text{UP}_-,\text{UP}_-)}_{\text{saddle }(-)} & \cdots & \underbrace{(\text{UP}_-,\text{UP}_-)}_{\text{saddle }(-)} \end{Bmatrix}_{(2m_1)\times(2n_1)}. \tag{5.133}$$

(vii$_{3b}$) The $((2m_1)^{\text{th}}\text{DI}_-, (2n_1)^{\text{th}}\text{DI}_+)$-double-inflection saddle appearing bifurcation is from a $p_1 \times q_1$ to $p_2 \times q_2$ equilibrium network as

$$\bigcup_{r_1=1}^{p_1}\bigcup_{s_1=1}^{q_1}\underbrace{(a_{j_1j_2r_1}, a_{j_2j_1s_1})}_{(m_{r_1}^{(1)}, n_{s_1}^{(1)})\text{-XX}} \rightleftharpoons \underbrace{((2m_1)^{\text{th}}\text{DI}_-, (2n_1)^{\text{th}}\text{DI}_+)}_{((2m_1),(2n_1))\text{-double-inflection saddle}}$$

$$\rightleftharpoons \bigcup_{r_2=1}^{p_2}\bigcup_{s_2=1}^{q_2}\underbrace{(a_{j_1j_2r_2}, a_{j_2j_1s_2})}_{(m_{r_2}^{(2)}, n_{s_2}^{(2)})\text{-XX}} \tag{5.134}$$

where

$$\begin{array}{l} 2m_1-\Sigma_{r_1=1}^{p_1}m_{r_1}^{(1)}=2\Sigma_{s=1}^{l_1}m_s \\ \text{with } l_1\text{-quadratic polynomials without real roots} \\ \text{and } s^{\text{th}}\text{-quadratic polynomial with power } m_s, \\ 2n_1-\Sigma_{s_1=1}^{q_1}n_{s_1}^{(1)}=2\Sigma_{l=1}^{l_2}n_l \\ \text{with } l_2\text{-quadraitc polynomials without real roots} \\ \text{and } l^{\text{th}}\text{-quadratic polynomial with power } n_l; \\ \Sigma_{r_2=1}^{p_2}m_{r_2}^{(2)}=2m_1,\ \Sigma_{s_2=1}^{q_2}n_{s_2}^{(2)}=2n_1;\ \text{for } i=1,2, \end{array}$$

$$\underbrace{((m_{p_i}^{(i)})^{\text{th}}\text{XX},(n_{q_i}^{(i)})^{\text{th}}\text{XX})}_{(m_{p_i}^{(i)},n_{q_i}^{(i)})\text{-XX}} \in \left\{\begin{array}{l} \underbrace{((2m_{p_i1}^{(i)}+1)^{\text{th}}\text{DP}_-,(2n_{q_i1}^{(i)}+1)^{\text{th}}\text{DP}_+)}_{((2m_{p_i1}^{(i)}+1),(2n_{q_i1}^{(i)}+1))\text{-CW center}},\ \underbrace{((2m_{p_i1}^{(i)})^{\text{th}}\text{DI}_-,(2n_{q_i1}^{(i)})^{\text{th}}\text{DI}_-)}_{((2m_{p_i1}^{(i)}),(2n_{q_i1}^{(i)}))\text{-double-inflection saddle}}, \\ \underbrace{((2m_{p_i1}^{(i)})^{\text{th}}\text{US},(2n_{q_i1}^{(i)}+1)^{\text{th}}\text{DP})}_{((2m_{p_i1}^{(i)}),(2n_{q_i1}^{(i)}+1))\text{-down-parabola upper-saddle}},\ \underbrace{((2m_{p_i1}^{(i)}+1)^{\text{th}}\text{UP},(2n_{q_i1}^{(i)})^{\text{th}}\text{US})}_{((2m_{p_i1}^{(i)}+1),(2n_{q_i1}^{(i)}))\text{-down-parabola lower-saddle}} \end{array}\right\};$$

$$\underbrace{((m_{p_i}^{(i)})^{\text{th}}\text{XX},(n_1^{(i)})^{\text{th}}\text{XX})}_{(m_{p_i}^{(i)},n_1^{(i)})\text{-XX}} \in \left\{\begin{array}{l} \underbrace{((2m_{p_i1}^{(i)}+1)^{\text{th}}\text{UP}_+,(2n_{11}^{(i)}+1)^{\text{th}}\text{UP}_+)}_{((2m_{p_i1}^{(i)}+1),(2n_{11}^{(i)}+1))\text{-positive saddle}},\ \underbrace{((2m_{p_i1}^{(i)})^{\text{th}}\text{DI}_-,(2n_{11}^{(i)})^{\text{th}}\text{DI}_+)}_{((2m_{p_i1}^{(i)}),(2n_{11}^{(i)}))\text{-double-inflection saddle}}, \\ \underbrace{((2m_{p_i1}^{(i)})^{\text{th}}\text{US},(2n_{11}^{(i)}+1)^{\text{th}}\text{UP})}_{((2m_{p_i1}^{(i)}),(2n_{11}^{(i)}+1))\text{-up-parabola upper-saddle}},\ \underbrace{((2m_{p_i1}^{(i)}+1)^{\text{th}}\text{DP},(2n_{11}^{(i)})^{\text{th}}\text{LS})}_{((2m_{p_i1}^{(i)}+1),(2n_{11}^{(i)}))\text{-down-parabola lower-saddle}} \end{array}\right\};$$

$$\underbrace{((m_1^{(i)})^{\text{th}}\text{XX},(n_{q_i}^{(i)})^{\text{th}}\text{XX})}_{(m_1^{(i)},n_{q_i}^{(i)})\text{-XX}} \in \left\{\begin{array}{l} \underbrace{((2m_{11}^{(i)}+1)^{\text{th}}\text{UP}_-,(2n_{q_i1}^{(i)}+1)^{\text{th}}\text{UP}_-)}_{((2m_{11}^{(i)}+1),(2n_{q_i1}^{(i)}+1))\text{-negative saddle}},\ \underbrace{((2m_{11}^{(i)})^{\text{th}}\text{DI}_-,(2n_{q_i1}^{(i)})^{\text{th}}\text{DI}_+)}_{((2m_{11}^{(i)}),(2n_{q_i1}^{(i)}))\text{-double-infection-saddle}}, \\ \underbrace{((2m_{11}^{(i)})^{\text{th}}\text{US},(2n_{q_i1}^{(i)}+1)^{\text{th}}\text{DP})}_{((2m_{11}^{(i)}),(2n_{q_i1}^{(i)}+1))\text{-down-parabola upper-saddle}},\ \underbrace{((2m_{11}^{(i)}+1)^{\text{th}}\text{UP},(2n_{q_i1}^{(i)})^{\text{th}}\text{LS})}_{((2m_{11}^{(i)}+1),(2n_{q_i1}^{(i)}))\text{-up-parabola lower-saddle}} \end{array}\right\};$$

$$\underbrace{((m_1^{(i)})^{\text{th}}\text{XX},(n_1^{(i)})^{\text{th}}\text{XX})}_{(m_1^{(i)},n_1^{(i)})\text{-XX}}$$

$$\in \left\{ \begin{array}{l} \underbrace{((2m_{11}^{(i)}+1)^{\text{th}}\text{DP}_{+}, (2n_{11}^{(i)}+1)^{\text{th}}\text{DP}_{-})}_{((2m_{11}^{(i)}+1),(2n_{11}^{(i)}+1))\text{-CCW center}}, \underbrace{((2m_{11}^{(i)})^{\text{th}}\text{DI}_{-}, (2n_{11}^{(i)})^{\text{th}}\text{DI}_{+})}_{((2m_{11}^{(i)}),(2n_{11}^{(i)}))\text{-double-inflection saddle}}, \\ \underbrace{((2m_{11}^{(i)})^{\text{th}}\text{US},(2n_{11}^{(i)}+1)^{\text{th}}\text{UP})}_{((2m_{11}^{(i)}),(2n_{11}^{(i)}+1))\text{-up-parabola upper-saddle}}, \underbrace{((2m_{11}^{(i)}+1)^{\text{th}}\text{UP},(2n_{11}^{(i)})^{\text{th}}\text{LS})}_{((2m_{11}^{(i)}+1),(2n_{11}^{(i)}))\text{-up-parabola lower-saddle}} \end{array} \right\}. \tag{5.135}$$

(vii$_{3c}$) The $((2m_1)^{\text{th}}\text{DI}_{-},\ (2n_1)^{\text{th}}\text{DI}_{+})$-double-inflection saddle switching bifurcation is for the switching of two $p \times q$ equilibrium networks as

$$\begin{aligned} \bigcup_{r_1=1}^{p_1}\bigcup_{s_1=1}^{q_1} \underbrace{(a_{j_1j_2r_1}, a_{j_2j_1s_1})}_{(m_{r_1}^{(1)}, n_{s_1}^{(1)})\text{-XX}} &\rightleftharpoons \underbrace{((2m_1)^{\text{th}}\text{DI}_{-}, (2n_1)^{\text{th}}\text{DI}_{+})}_{((2m_1),(2n_1))\text{-double-inflection saddle}} \\ &\rightleftharpoons \bigcup_{r_2=1}^{p_2}\bigcup_{s_2=1}^{q_2} \underbrace{(a_{j_1j_2r_2}, a_{j_2j_1s_2})}_{(m_{r_2}^{(2)}, n_{s_2}^{(2)})\text{-XX}} \end{aligned} \tag{5.136}$$

where $\sum_{r_i=1}^{p} m_{r_i} = 2m_1, \sum_{s_i=1}^{q} n_{s_i} = 2n_1$ for $i = 1, 2$.

(vii$_4$) For $a_{j_1j_10} < 0$ and $a_{j_2j_20} < 0$, there is a $((2m_1)^{\text{th}}\text{II}_{-}, (2n_1)^{\text{th}}\text{II}_{-})$-double-inflection saddle equilibrium as

$$(a_{j_1j_11}, a_{j_2j_21}) = \underbrace{((2m_1)^{\text{th}}\text{II}_{-}, (2n_1)^{\text{th}}\text{II}_{-})}_{((2m_1),\,(2n_1))\text{-double-inflection saddle}} \tag{5.137}$$

There are the following three $((2m_1)^{\text{th}}\text{II}_{-}, (2n_1)^{\text{th}}\text{II}_{-})$-double-inflection saddle appearing and switching bifurcations.

(vii$_{4a}$) The $((2m_1)^{\text{th}}\text{II}_{-}, (2n_1)^{\text{th}}\text{II}_{-})$-double-inflection saddle appearing bifurcation is from an (nF, nF)-flow to a $(2m_1) \times (2n_1)$ network as

$$\underbrace{(\dot{x}_{j_1}, \dot{x}_{j_2})}_{(\text{nF,nF})\text{-flow}} \rightleftharpoons \underbrace{((2m_1)^{\text{th}}\text{II}_{-}, (2n_1)^{\text{th}}\text{II}_{-})}_{((2m_1),(2n_1))\text{-double-inflection saddle}} \rightleftharpoons \bigcup_{r=1}^{2m_1}\bigcup_{s=1}^{2n_1} \underbrace{(a_{j_1j_2r}, a_{j_2j_1s})}_{\text{XX}} \tag{5.138}$$

where

$$\bigcup_{s=1}^{2m_1}\bigcup_{l=1}^{2n_1}\underbrace{(a_{j_1j_2s},a_{j_2j_1l})}_{\text{XX}}=\left\{\begin{matrix}\underbrace{(\text{DP}_+,\text{DP}_-)}_{\text{center (CCW)}} & \underbrace{(\text{UP}_-,\text{UP}_-)}_{\text{saddle }(-)} & \cdots & \underbrace{(\text{UP}_-,\text{UP}_-)}_{\text{saddle }(-)}\\ \underbrace{(\text{UP}_+,\text{UP}_+)}_{\text{saddle }(+)} & \underbrace{(\text{DP}_-,\text{DP}_+)}_{\text{center (CW)}} & \cdots & \underbrace{(\text{DP}_-,\text{DP}_+)}_{\text{center (CW)}}\\ \vdots & \vdots & \vdots & \vdots\\ \underbrace{(\text{UP}_+,\text{UP}_+)}_{\text{saddle }(+)} & \underbrace{(\text{DP}_-,\text{DP}_+)}_{\text{center (CW)}} & \cdots & \underbrace{(\text{DP}_-,\text{DP}_+)}_{\text{center (CW)}}\end{matrix}\right\}_{(2m_1)\times(2n_1)}. \tag{5.139}$$

(vii_{4b}) The $((2m_1)^{\text{th}}\text{II}_-,(2n_1)^{\text{th}}\text{II}_-)$-double-inflection saddle appearing bifurcation is from a $p_1\times q_1$ to $p_2\times q_2$ equilibrium network as

$$\begin{aligned}\bigcup_{r_1=1}^{p_1}\bigcup_{s_1=1}^{q_1}\underbrace{(a_{j_1j_2r_1},a_{j_2j_1s_1})}_{(m^{(1)}_{r_1},\,n^{(1)}_{s_1})\text{-XX}}&\rightleftharpoons\underbrace{((2m_1)^{\text{th}}\text{II}_-,\ (2n_1)^{\text{th}}\text{II}_-)}_{((2m_1),(2n_1))\text{-double-inflection saddle}}\\&\rightleftharpoons\bigcup_{r_2=1}^{p_2}\bigcup_{s_2=1}^{q_2}\underbrace{(a_{j_1j_2r_2},a_{j_2j_1s_2})}_{(m^{(2)}_{r_2},\,n^{(2)}_{s_2})\text{-XX}}\end{aligned} \tag{5.140}$$

where

$$\begin{aligned}&2m_1-\Sigma_{r_1=1}^{p_1}m^{(1)}_{r_1}=2\Sigma_{s=1}^{l_1}m_s\\&\text{with } l_1\text{-quadratic polynomials without real roots}\\&\text{and } s^{\text{th}}\text{-quadratic polynomial with power } m_s,\\&2n_1-\Sigma_{s_1-1}^{q_1}n^{(1)}_{s_1}=2\Sigma_{l-1}^{l_2}n_l\\&\text{with } l_2\text{-quadraitc polynomials without real roots}\\&\text{and } l^{\text{th}}\text{-quadratic polynomial with power } n_l;\\&\Sigma_{r_2=1}^{p_2}m^{(2)}_{r_2}=2m_1,\ \Sigma_{s_2=1}^{q_2}n^{(2)}_{s_2}=2n_1;\ \text{for } i=1,2,\end{aligned}$$

$$\underbrace{((m^{(i)}_{p_i})^{\text{th}}\text{XX},(n^{(i)}_{q_i})^{\text{th}}\text{XX})}_{(m^{(i)}_{p_i},n^{(i)}_{q_i})\text{-XX}}$$

$$\in\left\{\begin{matrix}\underbrace{((2m^{(i)}_{p_i1}+1)^{\text{th}}\text{UP}_-,\ (2n^{(i)}_{q_i1}+1)^{\text{th}}\text{UP}_-)}_{((2m^{(i)}_{p_i1}+1),(2n^{(i)}_{q_i1}+1))\text{-negative saddle}}, & \underbrace{((2m^{(i)}_{p_i1}+1)^{\text{th}}\text{II}_-,\ (2n^{(i)}_{q_i1}+1)^{\text{th}}\text{II}_-)}_{((2m^{(i)}_{p_i1}),(2n^{(i)}_{q_i1}))\text{-double-inflection saddle}},\\ \underbrace{((2m^{(i)}_{p_i1})^{\text{th}}\text{LS},(2n^{(i)}_{q_i1}+1)^{\text{th}}\text{UP})}_{((2m^{(i)}_{p_i1}),(2n^{(i)}_{q_i1}+1))\text{-up-parabola lower-saddle}}, & \underbrace{((2m^{(i)}_{p_i1}+1)^{\text{th}}\text{UP},(2n^{(i)}_{q_i1})^{\text{th}}\text{LS})}_{((2m^{(i)}_{p_i1}+1),(2n^{(i)}_{q_i1}))\text{-up-parabola lower-saddle}}\end{matrix}\right\};$$

$$
\underbrace{((m_{p_i}^{(i)})^{\text{th}}\text{XX},(n_1^{(i)})^{\text{th}}\text{XX})}_{(m_{p_i}^{(i)},n_1^{(i)})\text{-XX}}
\in \left\{ \begin{array}{l}
\underbrace{((2m_{p_i1}^{(i)}+1)^{\text{th}}\text{DP}_+,(2n_{11}^{(i)}+1)^{\text{th}}\text{DP}_-)}_{((2m_{p_i1}^{(i)}+1),(2n_{11}^{(i)}+1))\text{-CCW center}},
\underbrace{((2m_{p_i1}^{(i)}+1)^{\text{th}}\text{II}_-,(2n_{11}^{(i)})^{\text{th}}\text{II}_-)}_{((2m_{p_i1}^{(i)}),(2n_{11}^{(i)}))\text{-double-inflection saddle}}, \\
\underbrace{((2m_{p_i1}^{(i)})^{\text{th}}\text{LS},(2n_{11}^{(i)}+1)^{\text{th}}\text{DP})}_{((2m_{p_i1}^{(i)}),(2n_{11}^{(i)}+1))\text{-down-parabola lower-saddle}},
\underbrace{((2m_{p_i1}^{(i)}+1)^{\text{th}}\text{UP},(2n_{11}^{(i)}+1)^{\text{th}}\text{LS})}_{((2m_{p_i1}^{(i)}+1),(2n_{11}^{(i)}))\text{-up-parabola lower-saddle}}
\end{array} \right\};
$$

$$
\underbrace{((m_1^{(i)})^{\text{th}}\text{XX},(n_{q_i}^{(i)})^{\text{th}}\text{XX})}_{(m_1^{(i)},n_{q_i}^{(i)})\text{-XX}}
\in \left\{ \begin{array}{l}
\underbrace{((2m_{11}^{(i)}+1)^{\text{th}}\text{DP}_-,(2n_{q_i1}^{(i)}+1)^{\text{th}}\text{DP}_+)}_{((2m_{11}^{(i)}+1),(2n_{q_i1}^{(i)}+1))\text{-CW center}},
\underbrace{((2m_{11}^{(i)})^{\text{th}}\text{II}_-,(2n_{q_i1}^{(i)})^{\text{th}}\text{II}_-)}_{((2m_{11}^{(i)}),(2n_{q_i1}^{(i)}))\text{-double-infection-saddle}}, \\
\underbrace{((2m_{11}^{(i)})^{\text{th}}\text{LS},(2n_{q_i1}^{(i)}+1)^{\text{th}}\text{UP})}_{((2m_{11}^{(i)}),(2n_{q_i1}^{(i)}+1))\text{-up-parabola lower-saddle}},
\underbrace{((2m_{11}^{(i)}+1)^{\text{th}}\text{DP},(2n_{q_i1}^{(i)})^{\text{th}}\text{LS})}_{((2m_{11}^{(i)}+1),(2n_{q_i1}^{(i)}))\text{-down-parabola lower-saddle}}
\end{array} \right\};
$$

$$
\underbrace{((m_1^{(i)})^{\text{th}}\text{XX},(n_1^{(i)})^{\text{th}}\text{XX})}_{(m_1^{(i)},n_1^{(i)})\text{-XX}}
\in \left\{ \begin{array}{l}
\underbrace{((2m_{11}^{(i)}+1)^{\text{th}}\text{UP}_+,(2n_{11}^{(i)})^{\text{th}}\text{UP}_+)}_{((2m_{11}^{(i)}+1),(2n_{11}^{(i)}+1))\text{-positive saddle}},
\underbrace{((2m_{11}^{(i)})^{\text{th}}\text{II}_-,(2n_{11}^{(i)})^{\text{th}}\text{II}_-)}_{((2m_{11}^{(i)}),(2n_{11}^{(i)}))\text{-double-saddle}}, \\
\underbrace{((2m_{11}^{(i)})^{\text{th}}\text{LS},(2n_{11}^{(i)}+1)^{\text{th}}\text{UP}_+)}_{((2m_{11}^{(i)}),(2n_{11}^{(i)}+1))\text{-down-parabola lower-saddle}},
\underbrace{((2m_{11}^{(i)}+1)^{\text{th}}\text{DP},(2n_{11}^{(i)})^{\text{th}}\text{LS})}_{((2m_{11}^{(i)}+1),(2n_{11}^{(i)}))\text{-down-parabola lower-saddle}}
\end{array} \right\}.
\tag{5.141}
$$

(vii$_{4c}$) The $((2m_1)^{\text{th}}\text{II}_-,(2n_1)^{\text{th}}\text{II}_-)$-double-inflection saddle switching bifurcation is for two $p \times q$ equilibrium networks as

$$
\begin{aligned}
\bigcup_{r_1=1}^{p}\bigcup_{s_1=1}^{q}\underbrace{(a_{j_1j_2r_1},a_{j_2j_1s_1})}_{(m_{r_1}^{(1)},n_{s_1}^{(1)})\text{-XX}} &\rightleftharpoons \underbrace{((2m_1)^{\text{th}}\text{II}_-,(2n_1)^{\text{th}}\text{II}_-)}_{((2m_1),(2n_1))\text{-double-inflection saddle}} \\
&\rightleftharpoons \bigcup_{r_2=1}^{p}\bigcup_{s_2=1}^{q}\underbrace{(a_{j_1j_2r_2},a_{j_2j_1s_2})}_{(m_{r_2}^{(2)},n_{s_2}^{(2)})\text{-XX}}
\end{aligned}
\tag{5.142}
$$

where $\sum_{r_i=1}^{p} m_{r_i} = 2m_1,\ \sum_{s_i=1}^{q} n_{s_i} = 2n_1$ for $i = 1, 2$.

5.2 Proof of Theorem 5.1

(i) Consider a crossing-univariate polynomial dynamical system as

$$\dot{x}_{j_1} = a_{j_1 j_2 0}(x_{j_2} - a_{j_1 j_2 1})^m,\ \dot{x}_{j_2} = a_{j_2 j_1 0}(x_{j_1} - a_{j_2 j_1 1})^n,\\ j_1, j_2 \in \{1, 2\}; j_1 \neq j_2.$$

In phase space,

$$\frac{dx_{j_2}}{dx_{j_1}} = \frac{a_{j_2 j_1 0}}{a_{j_1 j_2 0}} \frac{(x_{j_1} - a_{j_2 j_1 1})^n}{(x_{j_2} - a_{j_1 j_2 1})^m}$$

gives

$$(x_{j_1} - a_{j_2 j_1 1})^n dx_{j_1} = \frac{a_{j_1 j_2 0}}{a_{j_2 j_1 0}} (x_{j_2} - a_{j_1 j_2 1})^m dx_{j_2}.$$

With initial condition $(x_{j_1 0}, x_{j_2 0})$, the integration of the foregoing equation yields the first integral manifold as

$$\frac{1}{n+1}\left[(x_{j_1} - a_{j_2 j_1 1})^{n+1} - (x_{j_1 0} - a_{j_2 j_1 1})^{n+1}\right]\\ = \frac{a_{j_1 j_2 0}}{a_{j_1 j_2 0}} \frac{1}{m+1}\left[(x_{j_2} - a_{j_1 j_2 1})^{m+1} - (x_{j_2 0} - a_{j_1 j_2 1})^{m+1}\right], \quad \text{for } m, n \geq 1.$$

At $(x^*_{j_2}, x^*_{j_1}) = (a_{j_1 j_2 1}, a_{j_2 j_1 1})$ for $\overline{x}_{j_2} \neq a_{j_1 j_2 1}$, in phase space,

$$\left.\frac{dx_{j_2}}{dx_{j_1}}\right|_{x^*_{j_1} = a_{j_2 j_1 1}} = \frac{a_{j_2 j_1 0}}{a_{j_1 j_2 0}} \frac{(x_{j_1} - a_{j_2 j_1 1})^n}{\left(\overline{x}_{j_2} - a_{j_1 j_2 1}\right)^m} = 0,$$

$$\vdots$$

$$\left.\frac{d^n x_{j_2}}{dx^n_{j_1}}\right|_{x^*_{j_1} = a_{j_2 j_1 1}} = 0.$$

If

$$\left.\frac{d^{n+1} x_{j_2}}{dx^{n+1}_{j_1}}\right|_{x^*_{j_1} = a_{j_2 j_1 1}} = \frac{a_{j_2 j_1 0}}{a_{j_1 j_2 0}} \frac{n!}{\left(\overline{x}_{j_2} - a_{j_1 j_2 1}\right)^m} > 0,$$

there is a $(2n_1 + 1)^{\text{th}}$-order up-parabola flow for $n = 2n_1 + 1$ or a $(2n_1)^{\text{th}}$-order increasing-inflection flow for $n = 2n_1$ in the x_{j_2}-direction. If

$$\frac{d^{n+1}x_{j_2}}{dx_{j_1}^{n+1}}\bigg|_{x_{j_1}^*=a_{j_2j_11}} = \frac{a_{j_2j_10}}{a_{j_1j_20}}\frac{n!}{(\overline{x}_{j_2}-a_{j_1j_21})^m} < 0,$$

there is a $(2n_1+1)^{\text{th}}$-order down-parabola flow for $n=2n_1+1$ or a $(2n_1)^{\text{th}}$-order decreasing-inflection flow for $n=2n_1$ in the x_{j_2}-direction. Let

$$\dot{x}_{j_1} = a_{j_1j_20}\left(\overline{x}_{j_2}-a_{j_1j_21}\right)^m.$$

If $a_{j_1j_20}\left(\overline{x}_{j_2}-a_{j_1j_21}\right)^m > 0$, the $(2n_1+1)^{\text{th}}$-order parabola flow is positive for $n=2n_1+1$ and the $(2n_1)^{\text{th}}$-order inflection flow is positive for $n=2n_1$ in the x_{j_1}-direction. If $a_{j_1j_20}\left(\overline{x}_{j_2}-a_{j_1j_21}\right)^m < 0$, the $(2n_1+1)^{\text{th}}$-order parabola flow is negative for $n=2n_1+1$ and the $(2n_1)^{\text{th}}$-order inflection flow is negative for $n=2n_1$ in the x_{j_1}-direction.

Similarly, at $(x_{j_2}^*, x_{j_1}^*) = (a_{j_1j_21}, a_{j_2j_11})$ for $\overline{x}_{j_1} \neq a_{j_2j_11}$, in phase space,

$$\frac{dx_{j_1}}{dx_{j_2}}\bigg|_{x_{j_2}^*=a_{j_1j_21}} = \frac{a_{j_1j_20}\left(\overline{x}_{j_2}-a_{j_1j_21}\right)^m}{a_{j_2j_10}\left(x_{j_1}-a_{j_2j_11}\right)^n} = 0,$$

$$\vdots$$

$$\frac{d^m x_{j_1}}{dx_{j_2}^m}\bigg|_{x_{j_2}^*=a_{j_1j_21}} = 0.$$

If

$$\frac{d^{m+1}x_{j_1}}{dx_{j_2}^{m+1}}\bigg|_{x_{j_2}^*=a_{j_1j_21}} = \frac{a_{j_1j_20}}{a_{j_2j_10}}\frac{m!}{\left(\overline{x}_{j_1}-a_{j_2j_11}\right)^n} > 0,$$

there is a $(2m_1+1)^{\text{th}}$-order up-parabola flow for $m=2m_1+1$ or a $(2m_1)^{\text{th}}$-order increasing-inflection flow for $m=2m_1$ in the x_{j_1}-direction. If

$$\frac{d^{m+1}x_{j_1}}{dx_{j_2}^{m+1}}\bigg|_{x_{j_2}^*=a_{j_1j_21}} = \frac{a_{j_1j_20}}{a_{j_2j_10}}\frac{m!}{\left(\overline{x}_{j_1}-a_{j_2j_11}\right)^n} < 0,$$

there is a $(2m_1+1)^{\text{th}}$-order down-parabola flow for $m=2m_1+1$ or a $(2m_1)^{\text{th}}$-order decreasing-inflection flow for $m=2m_1$ in the x_{j_1}-direction.

Let

$$\dot{x}_{j_2} = a_{j_2j_10}\left(\overline{x}_{j_1}-a_{j_2j_11}\right)^n.$$

If $a_{j_2j_10}\left(\overline{x}_{j_1}-a_{j_2j_11}\right)^n > 0$, the $(2m_1+1)^{\text{th}}$-order parabola flow is positive for $m=2m_1+1$ and the $(2m_1)^{\text{th}}$-order inflection flow is positive for $m=2m_1$ in the x_{j_2}-direction. If $a_{j_2j_10}\left(\overline{x}_{j_1}-a_{j_2j_11}\right)^n < 0$, the $(2m_1+1)^{\text{th}}$-order parabola flow is negative for $m = 2m_1 + 1$ and a $(2n_1)^{\text{th}}$-order inflection flow is negative for $m = 2m_1$ in the x_{j_2}-direction.

Therefore, for $\overline{x}_{j_1} = a_{j_2 j_1 1}$ and $\overline{x}_{j_2} = a_{j_1 j_2 1}$, four cases can be summarized as follows:

- Case I: The singular equilibriums of $(x^*_{j_2}, x^*_{j_1}) = (a_{j_1 j_2 1}, a_{j_2 j_1 1})$ for $m = 2m_1 + 1$ and $n = 2n_1 + 1$ are given as in case (i_1) through Eqs. (5.3)–(5.6).
- Case II: The singular equilibriums of $(x^*_{j_2}, x^*_{j_1}) = (a_{j_1 j_2 1}, a_{j_2 j_1 1})$ for $m = 2m_1$ and $n = 2n_1 + 1$ are given as in case (i_2) through Eqs. (5.7)–(5.10).
- Case III: The singular equilibriums of $(x^*_{j_2}, x^*_{j_1}) = (a_{j_1 j_2 1}, a_{j_2 j_1 1})$ for $m = 2m_1 + 1$ and $n = 2n_1$ are given as in case (i_3) through Eqs. (5.11)–(5.14).
- Case IV: The singular equilibriums of $(x^*_{j_2}, x^*_{j_1}) = (a_{j_1 j_2 1}, a_{j_2 j_1 1})$ for $m = 2m_1$ and $n = 2n_1$ are given as in case (i_4) through Eqs. (5.15)–(5.18).

(i_1) For $m = 2m_1 + 1$ and $n = 2n_1 + 1$, the equilibrium of $(x^*_{j_2}, x^*_{j_1}) = (a_{j_1 j_2 1}, a_{j_2 j_1 1})$ has the following properties.

- For $a_{j_1 j_2 0} > 0$ and $a_{j_2 j_1 0} > 0$,

$$\left(a_{j_1 j_2 1}, a_{j_2 j_1 1}\right) = \underbrace{((2m_1 + 1)^{\text{th}}\text{UP}_+, (2n_1 + 1)^{\text{th}}\text{UP}_+)}_{((2m_1+1), (2n_1+1))\text{-positive saddle}}.$$

- For $a_{j_1 j_2 0} < 0$ and $a_{j_2 j_1 0} > 0$,

$$\left(a_{j_1 j_2 1}, a_{j_2 j_1 1}\right) = \underbrace{((2m_1 + 1)^{\text{th}}\text{DP}_+, (2n_1 + 1)^{\text{th}}\text{DP}_-)}_{((2m_1+1), (2n_1+1))\text{-CCW center}}.$$

- For $a_{j_1 j_2 0} > 0$ and $a_{j_2 j_1 0} < 0$,

$$\left(a_{j_1 j_1 1}, a_{j_2 j_2 1}\right) = \underbrace{((2m_1 + 1)^{\text{th}}\text{DP}_-, (2n_1 + 1)^{\text{th}}\text{DP}_+)}_{((2m_1+1), (2n_1+1))\text{-CCW center}}.$$

- For $a_{j_1 j_2 0} < 0$ and $a_{j_2 j_1 0} < 0$,

$$\left(a_{j_1 j_2 1}, a_{j_2 j_1 1}\right) = \underbrace{((2m_1 + 1)^{\text{th}}\text{UP\_}, (2n_1 + 1)^{\text{th}}\text{UP\_})}_{((2m_1+1), (2n_1+1))\text{-negative saddle}}.$$

(i_2) For $m = 2m_1$ and $n = 2n_1 + 1$, the equilibrium of $(x^*_{j_2}, x^*_{j_1}) = (a_{j_1 j_2 1}, a_{j_2 j_1 1})$ has the following properties.

- For $a_{j_1 j_2 0} > 0$ and $a_{j_2 j_1 0} > 0$,

$$(a_{j_1 j_2 1}, a_{j_2 j_1 1}) = \underbrace{((2m_1)^{\text{th}}\text{US}, (2n_1 + 1)^{\text{th}}\text{UP})}_{((2m_1),(2n_1+1))\text{-up-parabola upper-saddle}}.$$

- For $a_{j_1j_20} < 0$ and $a_{j_2j_10} > 0$,

$$(a_{j_1j_21}, a_{j_2j_11}) = \underbrace{((2m_1)^{\text{th}}\text{LS},(2n_1+1)^{\text{th}}\text{DP})}_{((2m_1),(2n_1+1))\text{-down-parabola lower-saddle}}.$$

- For $a_{j_1j_20} > 0$ and $a_{j_2j_10} < 0$,

$$(a_{j_1j_21}, a_{j_2j_11}) = \underbrace{((2m_1)^{\text{th}}\text{US},(2n_1+1)^{\text{th}}\text{DP})}_{((2m_1),(2n_1+1))\text{-down-parabola upper-saddle}}.$$

- For $a_{j_1j_20} < 0$ and $a_{j_2j_10} < 0$,

$$(a_{j_1j_21}, a_{j_2j_11}) = \underbrace{((2m_1)^{\text{th}}\text{LS},(2n_1+1)^{\text{th}}\text{UP})}_{((2m_1),(2n_1+1))\text{-up-parabola lower-saddle}}.$$

(i_3) For $m = 2m_1 + 1$ and $n = 2n_1$, the equilibrium of $(x^*_{j_2}, x^*_{j_1}) = (a_{j_1j_21}, a_{j_2j_11})$ has the following properties.

- For $a_{j_1j_20} > 0$ and $a_{j_2j_10} > 0$,

$$(a_{j_1j_21}, a_{j_2j_11}) = \underbrace{((2m_1+1)^{\text{th}}\text{UP},(2n_1)^{\text{th}}\text{US})}_{((2m_1+1),(2n_1))\text{-up-parabola upper-saddle}}.$$

- For $a_{j_1j_20} < 0$ and $a_{j_2j_10} > 0$,

$$(a_{j_1j_21}, a_{j_2j_11}) = \underbrace{((2m_1+1)^{\text{th}}\text{DP},(2n_1)^{\text{th}}\text{US})}_{((2m_1+1),(2n_1))\text{-down-parabola upper-saddle}}.$$

- For $a_{j_1j_20} > 0$ and $a_{j_2j_10} < 0$,

$$(a_{j_1j_21}, a_{j_2j_11}) = \underbrace{((2m_1+1)^{\text{th}}\text{DP},(2n_1)^{\text{th}}\text{LS})}_{((2m_1+1),(2n_1))\text{-down-parabola lower-saddle}}.$$

- For $a_{j_1j_20} < 0$ and $a_{j_2j_10} < 0$,

$$(a_{j_1j_21}, a_{j_2j_11}) = \underbrace{((2m_1+1)^{\text{th}}\text{UP},(2n_1)^{\text{th}}\text{LS})}_{((2m_1+1),(2n_1))\text{-up-parabola lower-saddle}}.$$

(i_4) For $m = 2m_1$ and $n = 2n_1$, the equilibrium of $(x^*_{j_2}, x^*_{j_1}) = (a_{j_1j_21}, a_{j_2j_11})$ has the following properties.

- For $a_{j_1j_20} > 0$ and $a_{j_2j_10} > 0$,

$$(a_{j_1j_21}, a_{j_2j_11}) = \underbrace{((2m_1)^{\text{th}}\text{II}_+, (2n_1)^{\text{th}}\text{II}_+)}_{((2m_1),(2n_1))\text{-double-inflection saddle}}.$$

- For $a_{j_1j_20} < 0$ and $a_{j_2j_10} > 0$,

$$(a_{j_1j_21}, a_{j_2j_11}) = \underbrace{((2m_1)^{\text{th}}\text{DI}_+, (2n_1)^{\text{th}}\text{DI}_-)}_{((2m_1),(2n_1))\text{-double-inflection saddle}}.$$

- For $a_{j_1j_20} > 0$ and $a_{j_2j_10} < 0$,

$$(a_{j_1j_21}, a_{j_2j_11}) = \underbrace{((2m_1)^{\text{th}}\,\text{DI}_+, (2n_1)^{\text{th}}\,\text{DI}_-)}_{((2m_1),(2m_1))\text{-double-inflection saddle}}.$$

- For $a_{j_1j_20} < 0$ and $a_{j_2j_10} < 0$,

$$(a_{j_1j_21}, a_{j_2j_11}) = \underbrace{((2m_1)^{\text{th}}\text{II}_-, (2n_1)^{\text{th}}\text{II}_-)}_{((2m_1),(2n_1))\text{-double-inflection saddle}}.$$

(ii) *Consider a crossing-univariate polynomial dynamical system for* $\Sigma_{r=1}^{p} m_r = m$ and $\Sigma_{s=1}^{q} n_s = n$ *as*

$$\begin{aligned}
\dot{x}_{j_1} &= a_{j_1j_20}\prod_{r=1}^{p}(x_{j_2} - a_{j_1j_2r})^{m_r},\\
\dot{x}_{j_2} &= a_{j_2j_10}\prod_{s=1}^{q}(x_{j_1} - a_{j_2j_1s})^{n_s},\\
\sum_{r=1}^{p} m_r &= m \text{ and } \sum_{s=1}^{q} n_s = n; j_1, j_2 \in \{1, 2\}; j_1 \neq j_2.
\end{aligned}$$

In phase space, with $\sum_{r=1}^{p} m_r = m$ and $\sum_{s=1}^{q} n_s = n$,

$$\frac{dx_{j_2}}{dx_{j_1}} = \frac{a_{j_2j_10}\prod_{s=1}^{q}(x_{j_1} - a_{j_2j_1s})^{n_s}}{a_{j_1j_20}\prod_{r=1}^{p}(x_{j_2} - a_{j_1j_2r})^{m_r}}.$$

The deformation of the above equation gives

$$\begin{aligned}
&a_{j_1j_20}(x_{j_2} - a_{j_1j_2r_1})^{m_{r_1}} \prod_{r_2=1, r_2\neq r_1}^{r_p} (x_{j_2} - a_{j_1j_2r_1} + a_{j_1j_2r_1} - a_{j_1j_2r_2})^{m_{r_2}} dx_{j_2}\\
&= a_{j_2j_10}(x_{j_1} - a_{j_2j_1s_1})^{n_{s_1}} \prod_{s_2=1, s_2\neq s_1}^{s_q} (x_{j_1} - a_{j_2j_1s_1} + a_{j_2j_1s_1} - a_{j_2j_1s_2})^{n_{s_2}} dx_{j_1}.
\end{aligned}$$

$$r_1, r_2 \cdots, r_p \in \{1, 2, \cdots, p\} \text{ and } s_1, s_2 \cdots, s_q \in \{1, 2, \cdots, q\}.$$

Using expansion, the foregoing equation becomes

$$\begin{aligned}
& a_{j_1j_20}(x_{j_2} - a_{j_1j_2r_1})^{m_{r_1}} \\
& \quad \times \prod_{r_2=1, r_2\neq r_1}^{p} \left[\sum_{l_{r_2}, t_{r_2}=0}^{l_{r_2}+t_{r_2}=m_{s_2}} \frac{m_{r_2}!}{l_{r_2}! t_{r_2}!} (x_{j_2} - a_{j_1j_2r_1})^{t_{r_2}} (a_{j_1j_2r_1} - a_{j_1j_2r_2})^{l_{r_2}} \right] dx_{j_2} \\
& = a_{j_2j_10}(x_{j_1} - a_{j_2j_1s_1})^{n_{t_1}} \\
& \quad \times \prod_{s_2=1, s_2\neq s_1}^{q} \left[\sum_{l_{s_2}, t_{s_2}=0}^{l_{s_2}+t_{s_2}=n_{s_2}} \frac{n_{s_2}!}{l_{s_2}! t_{s_2}!} (x_{j_1} - a_{j_2j_1s_1})^{t_{s_2}} (a_{j_2j_1s_1} - a_{j_2j_1s_2})^{r_{l_2}} \right] dx_{j_1};
\end{aligned}$$

and exchanging summations and continuous multiplications yields

$$\begin{aligned}
& a_{j_1j_20}(x_{j_2} - a_{j_1j_2r_1})^{m_{r_1}} \sum_{l_{r_2}, t_{r_2}=0}^{l_{r_2}+t_{r_2}=m_{r_2}} \cdots \sum_{l_{r_p}, t_{r_p}=0}^{l_{r_p}+t_{r_p}=m_{r_p}} \left[\prod_{k=r_2}^{r_p} \frac{m_k!}{l_k! t_k!} (a_{j_1j_2r_1} - a_{j_1j_2r_2})^{l_k} \right] \\
& \times (x_{j_2} - a_{j_1j_2r_1})^{t_{r_2}+\cdots+t_{r_p}} dx_{j_2} \\
& = a_{j_2j_10}(x_{j_1} - a_{j_2j_1s_1})^{n_{s_1}} \sum_{l_{s_2}, t_{s_2}=0}^{l_{s_2}+t_{s_2}=n_{s_2}} \cdots \sum_{l_{s_q}, t_{s_q}=0}^{l_{s_q}+t_{s_q}=n_{s_q}} \left[\prod_{k=s_2}^{s_q} \frac{n_k!}{l_k! t_k!} (a_{j_1j_2s_1} - a_{j_1j_2s_2})^{l_k} \right] \\
& \times (x_{j_2} - a_{j_2j_1s_1})^{t_{s_2}+\cdots+t_{s_p}} dx_{j_1};
\end{aligned}$$

which is equivalent to

$$\begin{aligned}
& a_{j_1j_20}\{(x_{j_2} - a_{j_1j_2r_1})^{m} \\
& \quad + \sum_{\alpha=1}^{m-m_{s_1}} \left[\sum_{l_{r_2}, t_{r_2}=0}^{l_{r_2}+t_{r_2}=m_{r_2}} \cdots \sum_{l_{r_p}, t_{r_p}=0}^{l_{r_p}+t_{r_p}=m_{r_p}} \prod_{k=s_2}^{r_p} \frac{m_k!}{l_k! t_k!} (a_{j_1j_2r_1} - a_{j_1j_2r_2})^{l_k} \delta_{t_{r_2}+\cdots+t_{r_p}}^{m-\alpha} \right] \\
& \quad \times (x_{j_2} - a_{j_1j_2r_1})^{m-\alpha}\} dx_{j_2} \\
& = a_{j_2j_10}\Bigg\{ (x_{j_1} - a_{j_2j_1s_1})^{n} \\
& \quad + \sum_{\beta=1}^{n-n_{t_1}} \left[\sum_{l_{s_2}, t_{s_2}=0}^{l_{s_2}+t_{s_2}=n_{s_2}} \cdots \sum_{l_{s_q}, t_{s_q}=0}^{l_{s_q}+t_{s_q}=n_{s_q}} \prod_{k=s_2}^{s_q} \frac{n_k!}{l_k! t_k!} (a_{j_2j_1s_1} - a_{j_2j_1s_2})^{l_k} \delta_{t_{s_2}+\cdots+t_{s_q}}^{n-\beta} \right] \\
& \quad \times (x_{j_1} - a_{j_2j_1s_2})^{n-\beta} \Bigg\} dx_{j_1}.
\end{aligned}$$

Let

$$
\begin{aligned}
b_{j_2j_1\beta} &= \sum_{l_{s_q},t_{s_q}=0}^{l_{s_q}+t_{s_q}=n_{s_q}} \cdots \sum_{l_{s_2},t_{s_2}=0}^{l_{s_2}+t_{s_2}=n_2} \left[\prod_{k=s_2}^{s_q} \frac{n_k!}{l_k!t_k!}(a_{j_2j_1s_1}-a_{j_2j_1k})^{l_k}\right]\delta^{n-\beta}_{t_{s_2}+\cdots+t_{s_q}},\\
b_{j_1j_2\alpha} &= \sum_{l_{r_p},t_{r_p}=0}^{l_{r_p}+t_{r_p}=m_{r_p}} \cdots \sum_{l_{r_2},t_{r_2}=0}^{l_{r_2}+t_{r_2}=m_{r_2}} \left[\prod_{k=r_2}^{r_p} \frac{m_k!}{l_k!t_k!}(a_{j_1j_2r_1}-a_{j_1j_2k})^{l_k}\right]\delta^{m-\alpha}_{t_{r_2}+\cdots+t_{r_p}};\\
&\alpha = 0,1,\cdots,m-m_{r_1};\ \beta = 0,1,\cdots,n-n_{s_1};\\
&b_{j_2j_10} = 1,\ b_{j_1j_20} = 1.
\end{aligned}
$$

Thus,

$$
\begin{aligned}
&a_{j_2j_10}\left\{(x_{j_1}-a_{j_2j_1s_1})^n + \sum_{\beta=1}^{n-n_{l_1}} b_{j_2j_1\beta}(x_{j_1}-a_{j_2j_1s_1})^{n-\beta}\right\}dx_{j_1}\\
&\quad = a_{j_1j_20}\left\{(x_{j_2}-a_{j_1j_2r_1})^m + \sum_{\alpha=1}^{m-m_{s_1}} b_{j_1j_2\alpha}(x_{j_1}-a_{j_1j_2r_1})^{m-\alpha}\right\}dx_{j_2}.
\end{aligned}
$$

With initial condition (x_{j_10}, x_{j_20}), the integration of the foregoing equation yields the first integral manifold as

$$
\begin{aligned}
&a_{j_2j_10}\left\{\frac{1}{n+1}\left[(x_{j_1}-a_{j_2j_1s_1})^{n+1}-(x_{j_10}-a_{j_2j_1s_1})^{n+1}\right]\right.\\
&\quad\left.+\sum_{\beta=1}^{n-n_{s_1}}\frac{1}{n-\beta+1}b_{j_2j_1\beta}\left[(x_{j_1}-a_{j_2j_1s_1})^{n-\beta+1}-(x_{j_10}-a_{j_2j_1s_1})^{n-\beta+1}\right]\right\}\\
&= a_{j_1j_20}\left\{\frac{1}{m+1}\left[(x_{j_2}-a_{j_1j_2r_1})^{m+1}-(x_{j_20}-a_{j_1j_2r_1})^{m+1}\right]\right.\\
&\quad\left.+\sum_{\alpha=1}^{m-m_{r_1}}\frac{1}{m-\alpha+1}b_{j_1j_2\alpha}\left[(x_{j_1}-a_{j_1j_2r_1})^{m-\alpha+1}-(x_{j_10}-a_{j_1j_2r_1})^{m-\alpha+1}\right]\right\}.
\end{aligned}
$$

The singular equilibrium network with $\sum_{r_i=1}^{p} m_{r_i} = m$ and $\sum_{s_i=1}^{q} n_{s_i} = n$ is defined as

$$
\bigcup_{r=1}^{p}\bigcup_{s=1}^{q}\underbrace{(a_{j_1j_2r}, a_{j_2j_1s})}_{(m_r,n_s)\text{-XX}} \equiv \left\{\begin{matrix}(a_{j_1j_2p}, a_{j_2j_11}) & \cdots & (a_{j_1j_2p}, a_{j_2j_1q})\\ \vdots & \vdots & \vdots\\ (a_{j_1j_21}, a_{j_2j_11}) & \cdots & (a_{j_1j_21}, a_{j_2j_1q})\end{matrix}\right\}_{p\times q}
$$

$$= \left\{ \begin{array}{ccc} \underbrace{((m_p)^{\text{th}}\text{XX},(n_q)^{\text{th}}\text{XX})}_{(m_1,n_q)\text{-XX}} & \cdots & \underbrace{((m_p)^{\text{th}}\text{XX},(n_q)^{\text{th}}\text{XX})}_{(m_{p_i}^{(i)},n_{q_i}^{(i)})\text{-XX}} \\ \vdots & \vdots & \vdots \\ \underbrace{((m_1)^{\text{th}}\text{XX},(n_1)^{\text{th}}\text{XX})}_{(m_1,n_1)\text{-XX}} & \cdots & \underbrace{((m_1)^{\text{th}}\text{XX},(n_q)^{\text{th}}\text{XX})}_{(m_p,n_1)\text{-XX}} \end{array} \right\}.$$

At $(x_{j_2}^*, x_{j_1}^*) = (a_{j_1j_2r_1}, a_{j_2j_1s_1})$ for $\overline{x}_{j_2} \neq a_{j_1j_2r_1}$, in phase space,

$$\frac{dx_{j_2}}{dx_{j_1}}\bigg|_{x_{j_1}^*=a_{j_2j_1s_1}} = \frac{a_{j_2j_10}}{a_{j_1j_20}} \frac{(x_{j_1} - a_{j_2j_1s_1})^{n_{s_1}} \prod_{s_2=1,s_2\neq s_1}^{q} (x_{j_1} - a_{j_2j_1s_2})^{n_{s_2}}}{(\overline{x}_{j_2} - a_{j_1j_2s_1})^{m_{r_1}} \prod_{r_2=1,r_2\neq r_1}^{p} (\overline{x}_{j_2} - a_{j_1j_2r_2})^{m_{sr}}}\bigg|_{x_{j_1}^*=a_{j_2j_1s_1}} = 0,$$

$$\vdots$$

$$\frac{d^{n_{s_1}} x_{j_2}}{dx_{j_1}^{n_{s_1}}}\bigg|_{x_{j_1}^*=a_{j_2j_1s_1}} = 0.$$

If

$$\frac{d^{n_{s_1}+1} x_{j_2}}{dx_{j_1}^{n_{s_1}+1}}\bigg|_{x_{j_1}^*=a_{j_2j_1s_1}} = \frac{a_{j_2j_10}}{a_{j_1j_20}} \frac{(n_{s_1})! \prod_{s_2=1,s_2\neq s_1}^{q} (a_{j_2j_1s_1} - a_{j_2j_1s_2})^{n_{s_2}}}{(\overline{x}_{j_2} - a_{j_1j_2r_1})^{m_{r_1}} \prod_{r_2=1,r_2\neq r_1}^{p} (\overline{x}_{j_2} - a_{j_1j_2r_2})^{m_{r_2}}} > 0,$$

there is a $(2n_{s_1 1} + 1)^{\text{th}}$-order up-parabola flow for $n_{s_1} = 2n_{s_1 1} + 1$ or a $(2n_{s_1 1})^{\text{th}}$-order increasing-inflection flow for $n_{s_1} = 2n_{s_1 1}$ in the x_{j_2}-direction. If

$$\frac{d^{n_{s_1}+1} x_{j_2}}{dx_{j_1}^{n_{s_1}+1}}\bigg|_{x_{j_1}^*=a_{j_2j_1s_1}} = \frac{a_{j_2j_10}}{a_{j_1j_20}} \frac{(n_{s_1})! \prod_{s_2=1,s_2\neq s_1}^{q} (a_{j_2j_1s_1} - a_{j_2j_1s_2})^{n_{s_2}}}{(\overline{x}_{j_2} - a_{j_1j_2r_1})^{m_{r_1}} \prod_{r_2=1,r_2\neq r_1}^{p} (\overline{x}_{j_2} - a_{j_1j_2r_2})^{m_{r_2}}} < 0,$$

there is a $(2n_{s_1 1}+1)^{\text{th}}$-order down-parabola flow for $n_{s_1} = 2n_{s_1 1}+1$ or a $(2n_{s_1 1})^{\text{th}}$-order decreasing-inflection flow for $n_{s_1} = 2n_{s_1 1}$ in the x_{j_2}-direction.

Let

$$\dot{x}_{j_1} = a_{j_1j_20}(\overline{x}_{j_2} - a_{j_1j_2r_1})^{m_{r_1}} \prod_{r_2=1,r_2\neq r_1}^{p} (\overline{x}_{j_2} - a_{j_1j_2r_2})^{m_{r_2}} = a_{j_1j_20} \prod_{r_1=1}^{p} (\overline{x}_{j_2} - a_{j_1j_2r_1})^{m_{r_1}}$$

If $a_{j_1j_20} \prod_{r_1=1}^{p} (\overline{x}_{j_2} - a_{j_1j_2r_1})^{m_{r_1}} > 0$, the $(2n_{s_1 1} + 1)^{\text{th}}$-order parabola flow is positive for $n_{s_1} = 2n_{s_1 1}+1$ and the $(2n_{s_1 1})^{\text{th}}$-order inflection flow is positive for $n_{s_1} = 2n_{s_1 1}$ in the x_{j_1}-direction. If $a_{j_1j_20} \prod_{r_1=1}^{p} (\overline{x}_{j_2} - a_{j_1j_2r_1})^{m_{r_1}} < 0$, the $(2n_{s_1 1} + 1)^{\text{th}}$-order parabola flow is negative for $n_{s_1} = 2n_{s_1 1} + 1$ and the $(2n_{s_1 1})^{\text{th}}$-order inflection flow is negative for $n_{s_1} = 2n_{s_1 1}$ in the x_{j_1}-direction.

Similarly, at $(x_{j_2}^*, x_{j_1}^*) = (a_{j_1j_2r_1}, a_{j_2j_1s_1})$ for $\overline{x}_{j_1} \neq a_{j_2j_1s_1}$, in phase space,

$$\left.\frac{dx_{j_1}}{dx_{j_2}}\right|_{x^*_{j_2}=a_{j_1j_2r_1}} = \left.\frac{a_{j_1j_20}}{a_{j_2j_10}} \frac{(x_{j_2}-a_{j_1j_2r_1})^{m_{r_1}} \prod_{r_2=1,r_2\neq r_1}^{p} (x_{j_2}-a_{j_1j_2r_2})^{m_{r_2}}}{(\overline{x}_{j_1}-a_{j_2j_1s_1})^{n_{s_1}} \prod_{s_2=1,s_2\neq s_1}^{q} (\overline{x}_{j_1}-a_{j_2j_1s_2})^{n_{s_2}}}\right|_{x^*_{j_2}=a_{j_1j_2r_1}} = 0,$$

$$\vdots$$

$$\left.\frac{d^{m_{r_1}}x_{j_1}}{dx_{j_2}^{m_{r_1}}}\right|_{x^*_{j_2}=a_{j_1j_2r_1}} = 0.$$

If

$$\left.\frac{d^{m_{r_1}+1}x_{j_1}}{dx_{j_2}^{m_{r_1}+1}}\right|_{x^*_{j_2}=a_{j_1j_2r_1}} = \frac{a_{j_1j_20}}{a_{j_2j_10}} \frac{(m_{r_1})! \prod_{r_2=1,r_2\neq r_1}^{p} (a_{j_1j_2r_1}-a_{j_1j_2r_2})^{m_{r_2}}}{(\overline{x}_{j_1}-a_{j_2j_1s_1})^{n_{s_1}} \prod_{s_2=1,s_2\neq s_1}^{q} (\overline{x}_{j_1}-a_{j_2j_1s_2})^{n_{s_2}}} > 0,$$

there is a $(2m_{r_11}+1)^{\text{th}}$-order up-parabola flow for $m_{r_1} = 2m_{r_11}+1$ or a $(2m_{r_11})^{\text{th}}$-order increasing-inflection flow for $m_{r_1} = 2m_{r_11}$ in the x_{j_1}-direction. If

$$\left.\frac{d^{m_{r_1}+1}x_{j_1}}{dx_{j_2}^{m_{r_1}+1}}\right|_{x^*_{j_2}=a_{j_1j_2r_1}} = \frac{a_{j_1j_20}}{a_{j_2j_10}} \frac{(m_{r_1})! \prod_{r_2=1,r_2\neq r_1}^{p} (a_{j_1j_2r_1}-a_{j_1j_2r_2})^{m_{r_2}}}{(\overline{x}_{j_1}-a_{j_2j_1s_1})^{n_{s_1}} \prod_{s_2=1,s_2\neq s_1}^{q} (\overline{x}_{j_1}-a_{j_2j_1s_2})^{n_{s_2}}} < 0,$$

there is a $(2m_{r_11}+1)^{\text{th}}$-order down-parabola flow for $m_{r_1} = 2m_{r_11}+1$ or a $(2m_{r_11})^{\text{th}}$-order decreasing-inflection flow for $m_{r_1} = 2m_{r_11}$ in the x_{j_1}-direction. Let

$$\dot{x}_{j_2} = a_{j_2j_10}(\overline{x}_{j_1}-a_{j_2j_1s_1})^{n_{s_1}} \prod_{s_2=1,s_2\neq s_1}^{q} (\overline{x}_{j_1}-a_{j_2j_1s_2})^{n_{s_2}} = a_{j_2j_10} \prod_{s_1=1}^{q} (\overline{x}_{j_1}-a_{j_2j_1s_1})^{n_{s_1}}.$$

If $a_{j_2j_10}\prod_{s_1=1}^{q}(\overline{x}_{j_1}-a_{j_2j_1s_1})^{n_{s_1}} > 0$, the $(2m_{r_11}+1)^{\text{th}}$-order parabola flow is positive for $m_{r_1} = 2m_{r_11}+1$ and a $(2m_{r_11})^{\text{th}}$-order inflection flow is positive for $m_{r_1} - 2m_{r_11}$ in the x_{j_2}-direction. If $a_{j_2j_10}\prod_{s_1=1}^{q}(\overline{x}_{j_1}-a_{j_2j_1s_1})^{n_{s_1}} < 0$, the $(2m_{r_11}+1)^{\text{th}}$-order parabola flow is negative for $m_{r_1} = 2m_{r_11}+1$ and a $(2m_{r_11})^{\text{th}}$-order inflection flow is negative for $m_{r_1} = 2m_{r_11}$ in the x_{j_2}-direction.

Therefore, for $\overline{x}_{j_2} = a_{j_1j_2r_1}$ and $\overline{x}_{j_1} = a_{j_2j_1s_1}$ the following four cases exist:

- Case I: The singular equilibriums of $(x^*_{j_2}, x^*_{j_1}) = (a_{j_1j_2r_1}, a_{j_2j_1s_1})$ for $m_{r_1} = 2m_{r_11}+1$ and $n_{s_1} = 2n_{s_11}+1$ are given as in case (ii$_1$) through Eqs. (5.23)–(5.27).
- Case II: The singular equilibriums of $(x^*_{j_2}, x^*_{j_1}) = (a_{j_1j_2r_1}, a_{j_2j_1s_1})$ for $m_{r_1} = 2m_{r_11}$ and $n_{s_1} = 2n_{s_11}+1$ are given as in case (ii$_2$) through Eqs. (5.28)–(5.31).
- Case III: The singular equilibriums of $(x^*_{j_2}, x^*_{j_1}) = (a_{j_1j_2r_1}, a_{j_2j_1s_1})$ for $m_{r_1} = 2m_{r_11}+1$ and $n_{s_1} = 2n_{s_11}$ are given as in case (ii$_3$) through Eqs. (5.32)–(5.35).
- Case IV: The singular equilibriums of $(x^*_{j_2}, x^*_{j_1}) = (a_{j_1j_2r_1}, a_{j_2j_1s_1})$ for $m_{r_1} = 2m_{r_11}$ and $n_{s_1} = 2n_{s_11}$ are given as in case (ii$_4$) through Eqs. (5.36)–(5.39).

(ii$_1$) For $m_{r_1} = 2m_{r_11}+1$ and $n_{s_1} = 2n_{s_11}+1$, the equilibrium of $(x^*_{j_2}, x^*_{j_1}) = (a_{j_1j_2r_1}, a_{j_2j_1s_1})$ has the following properties.

- For $a_{j_1j_20}\prod_{r_2=1,r_2\neq s_1}^{p}(a_{j_1j_2r_1}-a_{j_1j_2r_2})^{m_{r_2}}>0$ and $a_{j_2j_10}\prod_{s_2=1,s_2\neq s_1}^{q}(a_{j_2j_1s_1}-a_{j_2j_1s_2})^{n_{s_2}}>0$,

$$(a_{j_1j_2r_1},a_{j_2j_1s_1})=\underbrace{((2m_{r_1 1}+1)^{\text{th}}\text{UP}_+,\,(2n_{s_1 1}+1)^{\text{th}}\text{UP}_+)}_{((2m_{r_1 1}+1),(2n_{s_1 1}+1))\text{-positive saddle}}.$$

- For $a_{j_1j_20}\prod_{r_2=1,r_2\neq s_1}^{p}\left(a_{j_1j_2r_1}-a_{j_1j_2r_2}\right)^{m_{r_2}}<0$ and $a_{j_2j_10}\prod_{s_2=1,s_2\neq s_1}^{q}\left(a_{j_2j_1s_1}-a_{j_2j_1s_2}\right)^{n_{s_2}}>0$,

$$(a_{j_1j_2r_1},a_{j_2j_1s_1})=\underbrace{((2m_{r_1 1}+1)^{\text{th}}\text{DP}_+,\,(2n_{s_1 1}+1)^{\text{th}}\text{DP}_-)}_{((2m_{r_1 1}+1),(2n_{s_1 1}+1))\text{-CCW center}}.$$

- For $a_{j_1j_20}\prod_{r_2=1,r_2\neq s_1}^{p}(a_{j_1j_2r_1}-a_{j_1j_2r_2})^{m_{r_2}}>0$ and $a_{j_2j_10}\prod_{s_2=1,s_2\neq s_1}^{q}(a_{j_2j_1s_1}-a_{j_2j_1s_2})^{n_{s_2}}<0$,

$$(a_{j_1j_2r_1},a_{j_2j_1s_1})=\underbrace{((2m_{r_1 1}+1)^{\text{th}}\text{DP}_-,\,(2n_{s_1 1}+1)^{\text{th}}\text{DP}_+)}_{((2m_{r_1 1}+1),(2n_{s_1 1}+1))\text{-CW center}}.$$

- For $a_{j_1j_20}\prod_{r_2=1,r_2\neq s_1}^{p}(a_{j_1j_1r_1}-a_{j_1j_1r_2})^{m_{r_2}}<0$ and $a_{j_2j_10}\prod_{s_2=1,s_2\neq s_1}^{q}(a_{j_2j_2s_1}-a_{j_2j_2s_2})^{n_{s_2}}<0$,

$$(a_{j_1j_2r_1},a_{j_2j_1s_1})=\underbrace{((2m_{r_1 1}+1)^{\text{th}}\text{UP}_-,\,(2n_{s_1 1}+1)^{\text{th}}\text{UP}_-)}_{((2m_{r_1 1}+1),(2n_{s_1 1}+1))\text{-negative saddle}}.$$

(ii$_2$) For $m_{r_1}=2m_{r_1 1}$ and $n_{s_1}=2n_{s_1 1}+1$, the equilibrium of $(x_{j_2}^*,x_{j_1}^*)=(a_{j_1j_2r_1},a_{j_2j_1s_1})$ has the following properties.

- For $a_{j_1j_20}\prod_{r_2=1,r_2\neq s_1}^{p}(a_{j_1j_2r_1}-a_{j_1j_2r_2})^{m_{r_2}}>0$ and $a_{j_2j_10}\prod_{s_2=1,s_2\neq s_1}^{q}(a_{j_2j_1s_1}-a_{j_2j_1s_2})^{n_{s_2}}>0$,

$$(a_{j_1j_2r_1},a_{j_2j_1s_1})=\underbrace{((2m_{r_1 1})^{\text{th}}\text{US},(2n_{s_1 1}+1)^{\text{th}}\text{UP})}_{((2m_{r_1 1}),(2n_{s_1 1}+1))\text{-up-parabola upper-saddle}}.$$

- For $a_{j_1j_20}\prod_{r_2=1,r_2\neq s_1}^{p}(a_{j_1j_2r_1}-a_{j_1j_2r_2})^{m_{r_2}}<0$ and $a_{j_2j_10}\prod_{s_2=1,s_2\neq s_1}^{q}(a_{j_2j_1s_1}-a_{j_2j_1s_2})^{n_{s_2}}>0$,

$$(a_{j_1j_2r_1},a_{j_2j_1s_1})=\underbrace{((2m_{r_1 1})^{\text{th}}\text{LS},(2n_{s_1 1}+1)^{\text{th}}\text{DP})}_{((2m_{r_1 1}),(2n_{s_1 1}+1))\text{-down-parabola lower-saddle}}.$$

- For $a_{j_1j_20}\prod_{r_2=1,r_2\neq s_1}^{p}(a_{j_1j_2r_1}-a_{j_1j_2r_2})^{m_{r_2}}>0$ and

$a_{j_2 j_1 0} \prod_{s_2=1, s_2 \neq s_1}^{q} (a_{j_2 j_1 s_1} - a_{j_2 j_1 s_2})^{n_{s_2}} < 0$,

$$(a_{j_1 j_2 r_1}, a_{j_2 j_1 s_1}) = \underbrace{((2m_{r_1 1})^{\text{th}}\text{US},(2n_{s_1 1}+1)^{\text{th}}\text{DP})}_{((2m_{r_1 1}),(2n_{s_1 1}+1))\text{-down-parabola upper-saddle}}.$$

- For $a_{j_1 j_2 0} \prod_{r_2=1, r_2 \neq s_1}^{p} (a_{j_1 j_2 r_1} - a_{j_1 j_2 r_2})^{m_{r_2}} < 0$ and $a_{j_2 j_1 0} \prod_{s_2=1, s_2 \neq s_1}^{q} (a_{j_2 j_2 s_1} - a_{j_2 j_2 s_2})^{n_{s_2}} < 0$,

$$(a_{j_1 j_2 r_1}, a_{j_2 j_1 s_1}) = \underbrace{((2m_{r_1 1})^{\text{th}}\text{LS},(2n_{s_1 1}+1)^{\text{th}}\text{UP})}_{((2m_{r_1 1}),(2n_{s_1 1}+1))\text{-up-parabola lower-saddle}}.$$

(ii$_3$) For $m_{r_1} = 2m_{r_1 1}+1$ and $n_{s_1} = 2n_{s_1 1}$, the equilibrium of $(x_{j_2}^*, x_{j_1}^*) = (a_{j_1 j_2 r_1}, a_{j_2 j_1 s_1})$ has the following properties.

- For $a_{j_1 j_2 0} \prod_{r_2=1, r_2 \neq s_1}^{p} (a_{j_1 j_2 r_1} - a_{j_1 j_2 r_2})^{m_{r_2}} > 0$ and $a_{j_2 j_1 0} \prod_{s_2=1, s_2 \neq s_1}^{q} (a_{j_2 j_1 s_1} - a_{j_2 j_1 s_2})^{n_{s_2}} > 0$,

$$(a_{j_1 j_2 r_1}, a_{j_2 j_1 s_1}) = \underbrace{((2m_{r_1 1}+1)^{\text{th}}\text{UP},(2n_{s_1 1})^{\text{th}}\text{US})}_{((2m_{r_1 1}+1),(2n_{s_1 1}))\text{-up-parabola upper-saddle}}.$$

- For $a_{j_1 j_2 0} \prod_{r_2=1, r_2 \neq s_1}^{p} (a_{j_1 j_2 r_1} - a_{j_1 j_2 r_2})^{m_{r_2}} < 0$ and $a_{j_2 j_1 0} \prod_{s_2=1, s_2 \neq s_1}^{q} (a_{j_2 j_1 s_1} - a_{j_2 j_1 s_2})^{n_{s_2}} > 0$,

$$(a_{j_1 j_2 r_1}, a_{j_2 j_1 s_1}) = \underbrace{((2m_{r_1 1}+1)^{\text{th}}\text{DP},(2n_{s_1 1})^{\text{th}}\text{US})}_{((2m_{r_1 1}+1),(2n_{s_1 1}))\text{-down-parabola upper-saddle}}.$$

- For $a_{j_1 j_2 0} \prod_{r_2=1, r_2 \neq s_1}^{p} (a_{j_1 j_2 r_1} - a_{j_1 j_2 r_2})^{m_{r_2}} > 0$ and $a_{j_2 j_1 0} \prod_{s_2=1, s_2 \neq s_1}^{q} (a_{j_2 j_1 s_1} - a_{j_2 j_1 s_2})^{n_{s_2}} < 0$,

$$(a_{j_1 j_2 r_1}, a_{j_2 j_1 s_1}) = \underbrace{((2m_{r_1 1}+1)^{\text{th}}\text{DP},(2n_{s_1 1})^{\text{th}}\text{LS})}_{((2m_{r_1 1}+1),(2n_{s_1 1}))\text{-down-parabola lower-saddle}}.$$

- For $a_{j_1 j_2 0} \prod_{r_2=1, r_2 \neq s_1}^{p} (a_{j_1 j_1 r_1} - a_{j_1 j_1 r_2})^{m_{r_2}} < 0$ and $a_{j_2 j_1 0} \prod_{s_2=1, s_2 \neq s_1}^{q} (a_{j_2 j_2 s_1} - a_{j_2 j_2 s_2})^{n_{s_2}} < 0$,

$$(a_{j_1 j_2 r_1}, a_{j_2 j_1 s_1}) = \underbrace{((2m_{r_1 1}+1)^{\text{th}}\text{UP},(2n_{s_1 1})^{\text{th}}\text{LS})}_{((2m_{r_1 1}+1),(2n_{s_1 1}))\text{-up-parabola lower-saddle}}.$$

(ii$_4$) For $m_{r_1} = 2m_{r_1 1}$ and $n_{s_1} = 2n_{s_1 1}$, the equilibrium of $(x_{j_2}^*, x_{j_1}^*) = (a_{j_1 j_2 r_1}, a_{j_2 j_1 s_1})$ has the following properties.

- For $a_{j_1j_20}\prod_{r_2=1,r_2\neq s_1}^{p}(a_{j_1j_2r_1}-a_{j_1j_2r_2})^{m_{r_2}}>0$ and $a_{j_2j_10}\prod_{s_2=1,s_2\neq s_1}^{q}(a_{j_2j_1s_1}-a_{j_2j_1s_2})^{n_{s_2}}>0$,

$$(a_{j_1j_2r_1},a_{j_2j_1s_1})=\underbrace{((2m_{r_1 1})^{\text{th}}\text{II}_+,(2n_{s_1 1})^{\text{th}}\text{II}_+)}_{((2m_{r_1 1}),(2n_{s_1 1}))\text{-double-inflection saddle}}.$$

- For $a_{j_1j_20}\prod_{r_2=1,r_2\neq s_1}^{p}(a_{j_1j_2r_1}-a_{j_1j_2r_2})^{m_{r_2}}<0$ and $a_{j_2j_10}\prod_{s_2=1,s_2\neq s_1}^{q}(a_{j_2j_1s_1}-a_{j_2j_1s_2})^{n_{s_2}}>0$,

$$(a_{j_1j_2r_1},a_{j_2j_1s_1})=\underbrace{((2m_{r_1 1})^{\text{th}}\text{DI}_+,(2n_{s_1 1})^{\text{th}}\text{DI}_-)}_{((2m_{r_1 1}),(2n_{s_1 1}))\text{-double-inflection saddle}}.$$

- For $a_{j_1j_20}\prod_{r_2=1,r_2\neq s_1}^{p}(a_{j_1j_2r_1}-a_{j_1j_2r_2})^{m_{r_2}}>0$ and $a_{j_2j_10}\prod_{s_2=1,s_2\neq s_1}^{q}(a_{j_2j_1s_1}-a_{j_2j_1s_2})^{n_{s_2}}<0$,

$$(a_{j_1j_2r_1},a_{j_2j_1s_1})=\underbrace{((2m_{r_1 1})^{\text{th}}\text{DI}_-,(2n_{s_1 1})^{\text{th}}\text{DI}_+)}_{((2m_{r_1 1}),(2n_{s_1 1}))\text{-double-inflection saddle}}.$$

- For $a_{j_1j_20}\prod_{r_2=1,r_2\neq s_1}^{p}(a_{j_1j_1r_1}-a_{j_1j_1r_2})^{m_{r_2}}<0$ and $a_{j_2j_10}\prod_{s_2=1,s_2\neq s_1}^{q}(a_{j_2j_2s_1}-a_{j_2j_2s_2})^{n_{s_2}}<0$,

$$(a_{j_1j_2r_1},a_{j_2j_1s_1})=\underbrace{((2m_{r_1 1})^{\text{th}}\text{II}_-,(2n_{s_1 1})^{\text{th}}\text{II}_-)}_{((2m_{r_1 1}),(2n_{s_1 1}))\text{-double-inflection saddle}}.$$

(iii) Consider a crossing-univariate polynomial system as

$$\begin{aligned}\dot{x}_{j_1}&=a_{j_1j_20}\prod_{s_1=1}^{m}(x_{j_2}-a_{j_1j_2s_1}),\\ \dot{x}_{j_2}&=a_{j_2j_10}\prod_{l_1=1}^{n}(x_{j_1}-a_{j_2j_1l_1}),\\ &j_1,j_2\in\{1,2\};j_1\neq j_2.\end{aligned}$$

In phase space, for $a_{j_1j_2s_1}\neq a_{j_1j_2s_2}(s_1,s_2\in\{1,2,3\},s_1\neq s_2)$ and $a_{j_2j_1l_1}\neq a_{j_2j_1l_2}(l_1,l_2\in\{1,2,3\},l_1\neq l_2\cdot)$,

$$\begin{aligned}\frac{dx_{j_1}}{dx_{j_2}}&=\frac{a_{j_1j_20}}{a_{j_2j_10}}\frac{(x_{j_2}-a_{j_1j_2s_1})\prod_{s_2=1,s_2\neq s_1}^{m}(x_{j_2}-a_{j_1j_2s_2})}{(x_{j_1}-a_{j_2j_1l_1})\prod_{l_2=1,l_2\neq l_1}^{n}(x_{j_1}-a_{j_2j_1l_2})}\\ &=\frac{a_{j_1j_20}}{a_{j_2j_10}}\frac{(x_{j_2}-a_{j_1j_2s_1})\prod_{s_2=1,s_2\neq s_1}^{m}(x_{j_2}-a_{j_1j_2s_1}+a_{j_1j_2s_1}-a_{j_1j_2s_2})}{(x_{j_1}-a_{j_2j_1l_1})\prod_{l_2=1,l_2\neq l_1}^{n}(x_{j_1}-a_{j_2j_1l_1}+a_{j_2j_1l_1}-a_{j_2j_1l_2})},\end{aligned}$$

thus, from the relations of coefficients and roots, we have

$$a_{j_2j_10}\big[(x_{j_1}-a_{j_2j_1l_1})^n+\sum_{k=1}^{n-1}b_{j_2j_1k}(x_{j_1}-a_{j_2j_1l_1})^{n-k}\big]dx_{j_1}$$
$$=a_{j_1j_20}\big[(x_{j_2}-a_{j_1j_2s_1})^m+\sum_{k=1}^{m-1}b_{j_1j_2k}(x_{j_1}-a_{j_1j_2s_1})^{m-k}\big]dx_{j_2},$$

and

$$b_{j_2j_11}=\sum_{l_2=1,l_2\neq l_1}^{n}(a_{j_2j_1l_1}-a_{j_2j_1l_2}),$$
$$b_{j_2j_12}=\sum_{\substack{l_2,l_3=1;l_2,l_3\neq l_1\\(l_2<l_3)}}^{n}\prod_{r=2}^{3}(a_{j_2j_1l_1}-a_{j_2j_1l_r}),\cdots,$$
$$b_{j_2j_1k}=\sum_{\substack{l_2,l_3,\cdots,l_{k+1}=1;\\ l_2,l_3,\cdots,l_{k+1}\neq l_1\\(l_2<l_3<\cdots<l_{k+1})}}^{n}\prod_{r=2}^{k+1}(a_{j_2j_1l_1}-a_{j_2j_1l_r}),\cdots,$$
$$b_{j_2j_1(n-1)}=\prod_{l_2=1,l_2\neq l_1}^{n}(a_{j_2j_1l_1}-a_{j_2j_1l_2});$$
$$b_{j_1j_21}=\sum_{s_2=1,s_2\neq s_1}^{m}(a_{j_1j_2s_1}-a_{j_1j_2s_2}),$$
$$b_{j_1j_22}=\sum_{\substack{s_2,s_3=1;s_2,s_3\neq s_1\\(s_2,<_3)}}^{m}\prod_{r=2}^{3}(a_{j_1j_2s_1}-a_{j_1j_2s_r}),\cdots,$$
$$b_{j_1j_2k}=\sum_{\substack{s_2,s_3,\cdots,s_{k+1}=1;\\ s_2,s_3,\cdots,s_{k+1}\neq s_1\\(s_2<s_3<\cdots<s_{k+1})}}^{m}\prod_{r=2}^{k+1}(a_{j_1j_2s_1}-a_{j_1j_2s_r}),\cdots,$$
$$b_{j_1j_2(m-1)}=\prod_{s_1=1,s_2\neq s_1}^{m}(a_{j_1j_2s_1}-a_{j_1j_2s_2}).$$

With initial conditions of $\left(x_{j_10},x_{j_20}\right)$ at $t=t_0$, the integration of the above equations gives the first integral manifold as

$$a_{j_2j_10}\Big\{\frac{1}{n+1}\big[(x_{j_1}-a_{j_2j_1l_1})^{n+1}-(x_{j_10}-a_{j_2j_1l_1})^{n+1}\big]$$

$$+\sum_{k=1}^{n-1}\frac{1}{n-k+1}b_{j_2j_1k}\big[(x_{j_1}-a_{j_2j_1l_1})^{n-k+1}-(x_{j_10}-a_{j_2j_1l_1})^{n-k+1}\big]\Big\}$$

$$=a_{j_1j_20}\Big\{\frac{1}{m+1}\big[(x_{j_2}-a_{j_1j_2s_1})^{m+1}-(x_{j_20}-a_{j_1j_2s_1})^{m+1}\big]$$

$$+\sum_{k=1}^{m-1}\frac{1}{m-k+1}b_{j_1j_2k}\big[(x_{j_1}-a_{j_1j_2s_1})^{m-k+1}-(x_{j_10}-a_{j_1j_2s_1})^{m-k+1}\big]\Big\}.$$

The nonsingular equilibrium network with $m\times n$ is defined as

$$\bigcup_{s=1}^{m}\bigcup_{l=1}^{n}\underbrace{(a_{j_1j_1s},a_{j_2j_2l})}_{\text{XX}}$$

$$\equiv\left\{\begin{matrix}(a_{j_1j_2m},a_{j_2j_11}) & (a_{j_1j_2m},a_{j_2j_12}) & \cdots & (a_{j_1j_2m},a_{j_2j_1n})\\ (a_{j_1j_2(m-1)},a_{j_2j_11}) & (a_{j_1j_1(m-1)},a_{j_2j_12}) & \cdots & (a_{j_1j_2(m-1)},a_{j_2j_1n})\\ \vdots & \vdots & \vdots & \vdots\\ (a_{j_1j_21},a_{j_2j_21}) & (a_{j_1j_11},a_{j_2j_12}) & \cdots & (a_{j_1j_21},a_{j_2j_1n})\end{matrix}\right\}_{m\times n}.$$

In phase space, at $(x_{j_2}^*,x_{j_1}^*)=(a_{j_1j_2s_1},a_{j_2j_ll_1})$, two cases are discussed.

(I) In phase space, at $x_{j_2}^*=a_{j_1j_2s_1}(s_1,s_2\in\{1,2,\cdots,m\},s_1\neq s_2)$ and $\overline{x}_{j_1}\neq a_{j_2j_1l_1}$ $(l_1=1,2,\ \cdots,n)$, we have

$$\frac{dx_{j_1}}{dx_{j_2}}\Big|_{x_{j_2}^*=a_{j_1j_2s_1}}=\frac{a_{j_1j_20}}{a_{j_2j_10}}\frac{(x_{j_2}-a_{j_1j_2s_1})\prod_{s_2=1,s_2\neq s_1}^{m}(x_{j_2}-a_{j_1j_2s_2})}{\prod_{l_1=1}^{n}(\overline{x}_{j_1}-a_{j_2j_1l_1})}\Big|_{x_{j_2}^*=a_{j_1j_2s_1}}=0.$$

If

$$\frac{d^2x_{j_1}}{dx_{j_2}^2}\Big|_{x_{j_2}^*=a_{j_1j_2s_1}}=\frac{a_{j_1j_20}}{a_{j_2j_10}}\frac{\prod_{s_2=1,s_2\neq s_1}^{m}(a_{j_1j_2s_1}-a_{j_1j_2s_2})}{\prod_{l_1=1}^{n}(\overline{x}_{j_1}-a_{j_2j_1l_1})}>0,$$

there is an up-parabola flow at $x_{j_2}^*=a_{j_1j_2s_1}$ in the x_{j_1}-direction. If

$$\frac{d^2x_{j_1}}{dx_{j_2}^2}\Big|_{x_{j_2}^*=a_{j_1j_2s_1}}=\frac{a_{j_1j_20}}{a_{j_2j_10}}\frac{\prod_{s_2=1,s_2\neq s_1}^{m}(a_{j_1j_2s_1}-a_{j_1j_2s_2})}{\prod_{l_1=1}^{n}(\overline{x}_{j_1}-a_{j_2j_1l_1})}<0,$$

there is a down-parabola flow at $x_{j_2}^*=a_{j_1j_2s_1}$ in the x_{j_1}-direction. Because of

$$\dot{x}_{j_2}=a_{j_2j_10}\prod_{l_1=1}^{n}(\overline{x}_{j_1}-a_{j_2j_1l_1}),$$

the flows at $x_{j_2}^* = a_{j_1 j_2 s_1}$ in the $x_{j_2}^*$-direction are positive and negative for $x_{j_2}^* > 0$ and $\dot{x}_{j_2} < 0$, respectively.

Thus, the equilibrium of $x_{j_2}^* = a_{j_1 j_2 s_1} (s_1, s_2 \in \{1, 2, \cdots, m\}, s_1 \neq s_2)$ on the $x_{j_2}^*$-direction with $\overline{x}_{j_1} \neq a_{j_2 j_1 l_1} (l_1 = 1, 2, \cdots, n)$ has the following properties.

- For $a_{j_1 j_2 0} \prod_{s_2=1, s_2 \neq s_1}^{m} (a_{j_1 j_2 s_1} - a_{j_1 j_2 s_2}) > 0$ and $a_{j_2 j_1 0} \prod_{l_1=1}^{n} (\overline{x}_{j_1} - a_{j_2 j_1 l_1}) > 0$,

$$(a_{j_1 j_2 s_1}, \dot{x}_{j_2}) = \underbrace{(\text{UP}, \text{pF})}_{\text{up-parabola flow}(+)}.$$

- For $a_{j_1 j_2 0} \prod_{s_2=1, s_2 \neq s_1}^{m} (a_{j_1 j_2 s_1} - a_{j_1 j_2 s_2}) < 0$ and $a_{j_2 j_1 0} \prod_{l_1=1}^{n} (\overline{x}_{j_1} - a_{j_2 j_1 l_1}) > 0$,

$$(a_{j_1 j_2 s_1}, \dot{x}_{j_2}) = \underbrace{(\text{DP}, \text{pF})}_{\text{down-parabola flow}(+)}.$$

- For $a_{j_1 j_2 0} \prod_{s_2=1, s_2 \neq s_1}^{m} (a_{j_1 j_2 s_1} - a_{j_1 j_2 s_2}) > 0$ and $a_{j_2 j_1 0} \prod_{l_1=1}^{n} (\overline{x}_{j_1} - a_{j_2 j_1 l_1}) < 0$,

$$(a_{j_1 j_2 s_1}, \dot{x}_{j_2}) = \underbrace{(\text{DP}, \text{nF})}_{\text{down-parabola flow}(-)}.$$

- For $a_{j_1 j_2 0} \prod_{s_2=1, s_2 \neq s_1}^{m} (a_{j_1 j_2 s_1} - a_{j_1 j_2 s_2}) < 0$ and $a_{j_2 j_1 0} \prod_{l_1=1}^{n} (\overline{x}_{j_1} - a_{j_2 j_1 l_1}) < 0$,

$$(a_{j_1 j_2 s_1}, \dot{x}_{j_2}) = \underbrace{(\text{UP}, \text{nF})}_{\text{up-parabola flow }(-)}.$$

(II) In phase space, at $x_{j_1}^* = a_{j_2 l_1} (l_1, l_2 \in \{1, 2, \cdots, n\}, l_1 \neq l_2)$ and $\overline{x}_{j_2} \neq a_{j_1 j_2 s_1}$ $(s_1 = 1, 2, \cdots, m)$, we have

$$\left.\frac{dx_{j_2}}{dx_{j_1}}\right|_{x_{j_1}^* = a_{j_2 j_1 l_1}} = \left.\frac{a_{j_2 j_1 0}}{a_{j_1 j_2 0}} \frac{(x_{j_1} - a_{j_2 j_1 l_1}) \prod_{l_2=1, l_2 \neq l_1}^{n} (x_{j_1} - a_{j_2 j_1 l_2})}{\prod_{s_1=1}^{m} (\overline{x}_{j_2} - a_{j_1 j_2 s_1})}\right|_{x_{j_1}^* = a_{j_2 j_1 l_1}} = 0.$$

If

$$\left.\frac{d^2 x_{j_2}}{dx_{j_1}^2}\right|_{x_{j_1}^* = a_{j_2 j_1 l_1}} = \frac{a_{j_2 j_1 0}}{a_{j_1 j_2 0}} \frac{\prod_{l_2=1, l_2 \neq l_1}^{n} (a_{j_2 j_1 l_1} - a_{j_2 j_1 l_2})}{\prod_{s_1=1}^{m} (\overline{x}_{j_2} - a_{j_1 j_2 s_1})} > 0,$$

there is an up-parabola flow at $x_{j_1}^* = a_{j_2 j_1 l_1}$ in the x_{j_2}-direction. If

$$\left.\frac{d^2 x_{j_2}}{dx_{j_1}^2}\right|_{x_{j_1}^* = a_{j_2 j_1 l_1}} = \frac{a_{j_2 j_1 0}}{a_{j_1 j_2 0}} \frac{\prod_{l_2=1, l_2 \neq l_1}^{n} (a_{j_2 j_1 l_1} - a_{j_2 j_1 l_2})}{\prod_{s_1=1}^{m} (\overline{x}_{j_2} - a_{j_1 j_2 s_1})} < 0,$$

there is a down-parabola flow at $x_{j_1}^* = a_{j_2 j_1 l_1}$ in the x_{j_1}-direction. Because of

$$\dot{x}_{j_1} = a_{j_1 j_2 0} \prod\nolimits_{s_1=1}^{m} (\overline{x}_{j_2} - a_{j_1 j_2 s_1}),$$

the flows at $x_{j_1}^* = a_{j_2 j_1 l_1}$ in the x_{j_1}-direction are positive and negative for $\dot{x}_{j_1} > 0$ and $\dot{x}_{j_1} < 0$, respectively.

Thus, the equilibrium of $x_{j_1}^* = a_{j_2 j_1 l_1} (l_1, l_2 \in \{1, 2, \cdots, n\}, l_1 \neq l_2)$ and $\overline{x}_{j_2} \neq a_{j_1 j_2 s_1} (s_1 = 1, 2, \cdots, m)$ on the x_{j_2}-direction has the following properties.

- For $a_{j_1 j_2 0} \prod_{s_1=1}^{m} (\overline{x}_{j_2} - a_{j_1 j_2 s_1}) > 0$ and $a_{j_2 j_1 0} \prod_{l_2=1, l_2 \neq l_1}^{n} (a_{j_2 j_1 l_1} - a_{j_2 j_1 l_2}) > 0$,

$$(\dot{x}_{j_1}, a_{j_2 j_1 l_1}) = \underbrace{(\text{pF, UP})}_{\text{up-parabola flow } (+)}.$$

- For $a_{j_1 j_2 0} \prod_{s_1=1}^{m} (\overline{x}_{j_2} - a_{j_1 j_2 s_1}) < 0$ and $a_{j_2 j_1} \prod_{l_2=1, l_2 \neq l_1}^{n} (a_{j_2 j_1 l_1} - a_{j_2 j_1 l_2}) > 0$,

$$(\dot{x}_{j_1}, a_{j_2 j_1 l_1}) = \underbrace{(\text{nF, DP})}_{\text{down-parabola flow } (-)}.$$

- For $a_{j_1 j_2 0} \prod_{s_1=1}^{m} (\overline{x}_{j_2} - a_{j_1 j_2 s_1}) > 0$ and $a_{j_2 j_1 0} \prod_{l_2=1, l_2 \neq l_1}^{n} (a_{j_2 j_1 l_1} - a_{j_2 j_1 l_2}) < 0$,

$$(\dot{x}_{j_1}, a_{j_2 j_1 l_1}) = \underbrace{(\text{pF, DP})}_{\text{down-parabola flow } (+)}.$$

- For $a_{j_1 j_2 0} \prod_{s_1=1}^{m} (\overline{x}_{j_2} - a_{j_1 j_2 s_1}) < 0$ and $a_{j_2 j_1 0} \prod_{l_2=1, l_2 \neq l_1}^{n} (a_{j_2 j_1 l_1} - a_{j_2 j_1 l_2}) < 0$,

$$\left(\dot{x}_{j_1}, a_{j_2 j_1 l_1}\right) = \underbrace{(\text{nF, UP})}_{\text{up-parabola flow}(-)}.$$

Therefore, from the two cases (I) and (II), the equilibrium of $(x_{j_2}^*, x_{j_1}^*) = (a_{j_1 j_2 s_1}, a_{j_2 j_1 l_1})$ $(s_1, s_2 \in \{1, 2, \cdots, m\},\ s_1 \neq s_2;\ l_1, l_2 \in \{1, 2, \cdots, n\}, l_1 \neq l_2)$ has the following properties as in Eqs. (5.43)–(5.46).

- For $a_{j_1 j_2 0} \prod_{s_2=1, s_2 \neq s_1}^{m} (a_{j_1 j_2 s_1} - a_{j_1 j_2 s_2}) > 0$ and $a_{j_2 j_1 0} \prod_{l_2=1, l_2 \neq l_1}^{n} (a_{j_2 j_1 l_1} - a_{j_2 j_1 l_2}) > 0$,

$$(a_{j_1 j_2 s_1}, a_{j_2 j_1 l_1}) = \underbrace{(\text{UP}_+, \text{UP}_+)}_{\text{positive saddle}}$$

  The equilibrium of $(x_{j_2}^*, x_{j_1}^*) = (a_{j_1 j_2 s_1}, a_{j_2 j_1 l_1})$ is a $(\text{UP}_+, \text{UP}_+)$-positive saddle.
- For $a_{j_1 j_2 0} \prod_{s_2=1, s_2 \neq s_1}^{m} (a_{j_1 j_2 s_1} - a_{j_1 j_2 s_2}) < 0$ and $a_{j_2 j_1 0} \prod_{l_2=1, l_2 \neq l_1}^{n} (a_{j_2 j_1 l_1} - a_{j_2 j_1 l_2}) > 0$,

$$(a_{j_1 j_2 s_1}, a_{j_2 j_1 l_1}) = \underbrace{(\text{DP}_+, \text{DP}_-)}_{\text{CCW center}}.$$

The equilibrium of $(x_{j_2}^*, x_{j_1}^*) = (a_{j_1j_2s_1}, a_{j_2j_1l_1})$ is a $(\mathrm{DP}_+, \mathrm{DP}_-)$-counter-clockwise center.

- For $a_{j_1j_20}\prod_{s_2=1,s_2\neq s_1}^{m}(a_{j_1j_2s_1}-a_{j_1j_2s_2})>0$ and $a_{j_2j_10}\prod_{l_2=1,l_2\neq l_1}^{n}(a_{j_2j_1l_1}-a_{j_2j_1l_2})<0$,

$$(a_{j_1j_2s_1}, a_{j_2j_1l_1}) = \underbrace{(\mathrm{DP}_-, \mathrm{DP}_+)}_{\text{CW center}}.$$

The equilibrium of $(x_{j_2}^*, x_{j_1}^*) = (a_{j_1j_2s_1}, a_{j_2j_1l_1})$ is a $(\mathrm{DP}_-, \mathrm{DP}_+)$-clockwise center.

- For $a_{j_1j_20}\prod_{s_2=1,s_2\neq s_1}^{m}(a_{j_1j_2s_1}-a_{j_1j_2s_2})<0$ and $a_{j_2j_10}\prod_{l_2=1,l_2\neq l_1}^{n}(a_{j_2j_1l_1}-a_{j_2j_1l_2})<0$,

$$(a_{j_1j_2s_1}, a_{j_2j_1l_1}) = \underbrace{(\mathrm{UP}_-, \mathrm{UP}_-)}_{\text{negative saddle}}.$$

The equilibrium of $(x_{j_2}^*, x_{j_1}^*) = (a_{j_1j_2s_1}, a_{j_2j_1l_1})$ is a $(\mathrm{UP}_-, \mathrm{UP}_-)$-negative saddle.

(iv) For $m = 2m_1 + 1$ and $n = 2n_1 + 1$, the bifurcation process is discussed through differential equations as follows. There are three cases (I)–(III).

Case I: Consider a dynamical system having a single equilibrium (A_1) as

$$\dot{x}_{j_1} = a_{j_1j_20}(x_{j_2} - b_{j_1j_21})\prod_{s=1}^{m_1}[(x_{j_2} - a_{j_1j_2}^{(s)})^2 + \Delta_{j_1j_2}^{(s)}],$$
$$\dot{x}_{j_2} = a_{j_2j_10}(x_{j_1} - b_{j_2j_11})\prod_{l=1}^{n_1}[(x_{j_1} - a_{j_2j_1}^{(l)})^2 + \Delta_{j_2j_1}^{(l)}];$$

where $a_{j_1j_21} = b_{j_1j_21}$ and $a_{j_2j_11} = b_{j_2j_11}$. Once

$$\Delta_{j_1j_2}^{(s)} = \Delta_{j_1j_2} = 0\,(s = 1, 2, \cdots, m_1),$$
$$\Delta_{j_2j_1}^{(l)} = \Delta_{j_2j_1} = 0\,(l = 1, 2, \cdots, n_1),$$

if

$$a_{j_1j_21} = b_{j_1j_21} = a_{j_1j_2}^{(s)}\ (s = 1, 2, \cdots, m_1),$$
$$a_{j_2j_11} = b_{j_2j_11} = a_{j_2j_1}^{(l)}\ (l = 1, 2, \cdots, n_1),$$

the foregoing differential equations becomes a dynamical system with a singular equilibrium (B_1) as

$$\dot{x}_{j_1} = a_{j_1j_20}\big(x_{j_2} - a_{j_1j_21}\big)^{2m_1+1},\quad \dot{x}_{j_2} = a_{j_2j_10}\big(x_{j_1} - a_{j_2j_11}\big)^{2n_1+1}.$$

If

$$\begin{aligned}
&\Delta_{j_1j_2}^{(s)} = \Delta_{j_1j_2} + \delta_s \ (s = 1, 2, \cdots, m_1),\\
&\Delta_{j_2j_1}^{(l)} = \Delta_{j_2j_1} + \delta_l \ (l = 1, 2, \cdots, n_1),\\
&\delta_s > 0 \ and \ \delta_l > 0,
\end{aligned}$$

then

$$\begin{aligned}
a_{j_1j_2}^{(s)}, a_{j_1j_2 2}^{(s)} &= a_{j_1j_2 1} \pm \varepsilon_s, (s = 1, 2, \cdots, m_1)\\
a_{j_2j_1 1}^{(l)}, a_{j_2j_1 2}^{(l)} &= a_{j_2j_1 1} \pm \varepsilon_l, (l = 1, 2, \cdots, n_1)\\
\{a_{j_2j_1 1}, a_{j_1j_2 2}, \cdots, a_{j_1j_2(2m_1+1)}\} &= \text{sort}\{b_{j_1j_2 1}, a_{j_1j_2 1}^{(s)}, a_{j_1j_2 2}^{(s)} | s = 1, 2, \cdots, m_1\},\\
\{a_{j_2j_1 1}, a_{j_2j_1 1}, \cdots, a_{j_1j_2(2n_1+1)}\} &= \text{sort}\{b_{j_2j_1 1}, a_{j_2j_1 1}^{(l)}, a_{j_2j_1 2}^{(l)} | l = 1, 2, \cdots, n_1\}.
\end{aligned}$$

Thus, the differential equation becomes a dynamical system with a non-singular equilibrium network (C_1) as

$$\dot{x}_{j_1} = a_{j_1j_2 0} \prod_{s=1}^{2m_1+1} (x_{j_2} - a_{j_1j_2 s}), \ \dot{x}_{j_2} = a_{j_2j_1 0} \prod_{l=1}^{2n_1+1} (x_{j_1} - a_{j_2j_1 l}).$$

From the above bifurcation process analysis, at least, $(m_1 + 1)$-parameter variations in the x_{j_1}-direction and $(n_1 + 1)$-parameter variations in the x_{j_2}-direction are engaged in such a bifurcation from a single equilibrium to the equilibrium network of $(2m_1 + 1) \times (2n_1 + 1)$ through the $(2m_1 + 1)^{\text{th}}$-up-parabola and down-parabola bifurcations in the x_{j_1}-direction and the $(2n_1 + 1)^{\text{th}}$-up-parabola and down-parabola bifurcations in the x_{j_2}-direction. *With both of them, the higher-order positive saddle, counter-clockwise center, clockwise center, and negative saddle bifurcations are developed.*

Thus, the appearing or vanishing bifurcation route is as follows.

$$\underbrace{(a_{j_1j_2 1}, a_{j_2j_1 1})}_{\text{XX}} \rightleftharpoons \underbrace{(a_{j_1j_2 1}, a_{j_2j_1 1})}_{((2m_1+1),(2n_1+1))\text{-XX}} \rightleftharpoons \bigcup_{s=1}^{2m_1+1} \bigcup_{l=1}^{2n_1+1} \underbrace{(a_{j_1j_2 s}, a_{j_2j_1 l})}_{\text{XX}}.$$

Case II: From Case (I), consider a dynamical system having singular equilibriums (A_2) as

$$\begin{aligned}
\dot{x}_{j_1} &= a_{j_1j_2 0} \prod_{r_1=1}^{p_1} (x_{j_2} - b_{j_1j_2 r_1})^{m_{r_1}^{(1)}} \prod_{s=1}^{l_1} [(x_{j_2} - a_{j_1j_2}^{(s)})^2 + \Delta_{j_1j_2}^{(s)}]^{m_s},\\
\dot{x}_{j_2} &= a_{j_2j_1 0} \prod_{s_1=1}^{q_1} (x_{j_1} - b_{j_2j_1 s_1})^{n_{s_1}^{(1)}} \prod_{l=1}^{l_2} [(x_{j_1} - a_{j_2j_1}^{(l)})^2 + \Delta_{j_2j_1}^{(l)}]^{n_l};
\end{aligned}$$

where

$$
\begin{aligned}
&\{a_{j_1j_2r_1}|r_1 = 1, 2, \cdots, p_1\} = \mathrm{sort}\{b_{j_1j_2r_1}|r_1 = 1, 2, \cdots, p_1\},\\
&\{a_{j_2j_1s_1}|s_1 = 1, 2, \cdots, q_1\} = \mathrm{sort}\{b_{j_2j_1s_1}|s_1 = 1, 2, \cdots, q_1\};\\
&2m_1 + 1 - \Sigma_{r_1=1}^{p_1} m_{r_1}^{(1)} = 2\Sigma_{s=1}^{l_1} m_s\\
&\text{with } l_1\text{-quadratic polynomials without real roots}\\
&\text{and } s^{\text{th}}\text{-quadratic polynomial with power } m_s,\\
&2n_1 + 1 - \Sigma_{s_1=1}^{q_1} n_{s_1}^{(1)} = 2\Sigma_{l=1}^{l_2} n_l\\
&\text{with } l_2\text{-quadraitc polynomials without real roots}\\
&\text{and } l^{\text{th}}\text{-quadratic polynomial with power } n_l.
\end{aligned}
$$

Once

$$
\Delta_{j_1j_2}^{(s)} = 0 (s = 1, 2, \cdots, l_1) \text{ and } \Delta_{j_2j_1}^{(l)} = 0 (l = 1, 2, \cdots, l_2),
$$

if

$$
\begin{aligned}
a_{j_1j_21} &\equiv a_{j_1j_2r_1} = a_{j_1j_2}^{(s)} \ (r_1 = 1, 2, \cdots, p_1; s = 1, 2, \cdots, l_1),\\
a_{j_2j_11} &\equiv a_{j_2j_1s_1} = a_{j_2j_1}^{(l)} \ (s_1 = 1, 2, \cdots, q_1; l = 1, 2, \cdots, l_2),
\end{aligned}
$$

the foregoing differential equations becomes a dynamical system with a singular equilibrium (B_2) as

$$
\dot{x}_{j_1} = a_{j_1j_20}\left(x_{j_1} - a_{j_1j_21}\right)^{2m_1+1}, \ \dot{x}_{j_2} = a_{j_2j_10}\left(x_{j_2} - a_{j_2j_11}\right)^{2n_1+1}.
$$

If

$$
\begin{aligned}
&\Delta_{j_1j_2}^{(s)} = \Delta_{j_1j_2} + \delta_s \ (s = 1, 2, \cdots, l_1),\\
&\Delta_{j_2j_1}^{(l)} = \Delta_{j_2j_1} + \delta_l \ (l = 1, 2, \cdots, l_2),\\
&\delta_s > 0 \ \mathit{and} \ \delta_l > 0,
\end{aligned}
$$

then

$$
\begin{aligned}
&a_{j_1j_21}^{(\mathrm{s})}, a_{j_1j_22}^{(\mathrm{s})} = a_{j_1j_21} \pm \varepsilon_s, (s = 1, 2, \cdots, l_1),\\
&a_{j_2j_1}^{(l)}, a_{j_2j_2}^{(l)} = a_{j_2j_1} \pm \varepsilon_l, (l = 1, 2, \cdots, l_2);\\
&\{a_{j_1j_2}, a_{j_1j_22}, \cdots, a_{j_1j_2p_2}\}\\
&= \mathrm{sort}\left\{a_{j_1j_21}, a_{j_1j_22} \cdots, a_{j_1j_2p_1}; a_{j_1j_21}^{(\mathrm{s})}, a_{j_1j_22}^{(\mathrm{s})}|s = 1, 2, \cdots, l_1\right\},\\
&\{a_{j_1j_21}, a_{j_1j_21}, \cdots, a_{j_1j_2q_2}\}
\end{aligned}
$$

$$= \mathrm{sort}\left\{a_{j_1j_21}, a_{j_1j_21}, \cdots, a_{j_1j_2q_1}; a^{(l)}_{j_1j_21}, a^{(l)}_{j_1j_22} \middle| l = 1, 2, \cdots, l_2\right\}.$$

Thus, the differential equation becomes a dynamical system with singular equilibriums (C_2) as

$$\dot{x}_{j_1} = a_{j_1j_10}\prod_{r_2=1}^{p_2}\left(x_{j_2} - a_{j_1j_2r_2}\right)^{m^{(2)}_{r_2}}, \dot{x}_{j_2} = a_{j_2j_20}\prod_{s_2=1}^{q_2}\left(x_{j_1} - a_{j_2j_1s_2}\right)^{n^{(2)}_{s_2}};$$

where

$$\sum_{r_2=1}^{p_2} m^{(2)}_{r_2} = 2m_1 + 1, \sum_{s_2=1}^{q_2} n^{(2)}_{s_2} = 2n_1 + 1.$$

From the above bifurcation process analysis, at least, p_2-parameter variations in the x_{j_1}-direction and q_2-parameter variations in the x_{j_2}-direction are engaged in such a bifurcation from a $p_1 \times q_1$ equilibrium network to a $p_2 \times q_2$ equilibrium network through the $(2m_1 + 1)^{\text{th}}$-up-parabola and down-parabola bifurcations in the x_{j_1}-direction and the $(2n_1 + 1)^{\text{th}}$-up-parabola and down-parabola bifurcations in the x_{j_2}-direction.

Thus, the appearing or vanishing bifurcation route from a $p_1 \times q_1$ to a $p_2 \times q_2$ equilibrium network is expressed as.

$$\bigcup_{r_1=1}^{p_1}\bigcup_{s_1=1}^{q_1}\underbrace{(a_{j_1j_2r_1}, a_{j_2j_1s_1})}_{(m^{(1)}_{r_1}, n^{(1)}_{s_1})\text{-XX}} \rightleftharpoons \underbrace{(a_{j_1j_21}, a_{j_2j_11})}_{((2m_1+1),(2n_1+1))\text{-XX}} \rightleftharpoons \bigcup_{r_2=1}^{p_2}\bigcup_{s_2=1}^{q_2}\underbrace{(a_{j_1j_2r_2}, a_{j_2j_1s_2})}_{(m^{(2)}_{r_2}, n^{(2)}_{s_2})\text{-XX}}.$$

From case (I), the singular equilibriums can be formed through nonsingular equilibrium networks. Thus,

$$\underbrace{(a_{j_1j_2r_1}, a_{j_2j_1s_1})}_{(m^{(1)}_{r_1}, n^{(1)}_{s_1})\text{-XX}} \rightleftharpoons \bigcup_{s=1}^{m^{(1)}_{r_1}}\bigcup_{l=1}^{n^{(1)}_{s_1}}\underbrace{(a_{j_1j_2s}, a_{j_2j_1l})}_{\text{XX}}$$

$$\equiv \left\{\begin{array}{cccc} (a_{j_1j_2m^{(1)}_{r_1}}, a_{j_2j_11}) & (a_{j_1j_2m^{(1)}_{r_1}}, a_{j_2j_12}) & \cdots & (a_{j_1j_2m^{(1)}_{r_1}}, a_{j_2j_1n^{(1)}_{s_1}}) \\ (a_{j_1j_2(m^{(1)}_{r_1}-1)}, a_{j_2j_11}) & (a_{j_1j_2(m^{(1)}_{r_1}-1)}, a_{j_2j_12}) & \cdots & (a_{j_1j_2(m^{(1)}_{r_1}-1)}, a_{j_2j_1n^{(1)}_{s_1}}) \\ \vdots & \vdots & \cdots & \vdots \\ (a_{j_1j_21}, a_{j_2j_11}) & (a_{j_1j_21}, a_{j_2j_12}) & \cdots & (a_{j_1j_21}, a_{j_2j_1n^{(1)}_{s_1}}) \end{array}\right\}_{m^{(1)}_{r_1}\times n^{(1)}_{s_1}},$$

and

$$\underbrace{(a_{j_1j_2r_2},a_{j_1j_1s_2})}_{(m^{(2)}_{r_2},n^{(2)}_{s_2})\text{-XX}} \rightleftharpoons \bigcup_{s=1}^{m^{(2)}_{r_2}}\bigcup_{l=1}^{n^{(2)}_{s_2}}\underbrace{(a_{j_1j_2s},a_{j_2j_1l})}_{\text{XX}}$$

$$\equiv \left\{\begin{array}{cccc} (a_{j_1j_2m^{(2)}_{r_2}},a_{j_2j_1 1}) & (a_{j_1j_2m^{(2)}_{r_2}},a_{j_2j_1 2}) & \cdots & (a_{j_1j_2m^{(2)}_{r_2}},a_{j_2j_1n^{(2)}_{s_2}}) \\ (a_{j_1j_2(m^{(2)}_{r_2}-1)},a_{j_2j_1 1}) & (a_{j_1j_2(m^{(2)}_{r_2}-1)},a_{j_2j_1 2}) & \cdots & (a_{j_1j_2(m^{(2)}_{r_2}-1)},a_{j_2j_1n^{(2)}_{s_2}}) \\ \vdots & \vdots & \cdots & \vdots \\ (a_{j_1j_2 1},a_{j_2j_1 1}) & (a_{j_1j_2 1},a_{j_2j_1 2}) & \cdots & (a_{j_1j_2 1},a_{j_2j_1n^{(2)}_{s_2}}) \end{array}\right\}_{m^{(2)}_{r_2}\times n^{(2)}_{s_2}}.$$

From the above definition, the corner singular equilibriums for $i = 1,2$ are determined by

$$\underbrace{((m^{(i)}_{p_i})^{\text{th}}\text{XX},(n^{(i)}_{q_i})^{\text{th}}\text{XX})}_{(m^{(i)}_{p_i},n^{(i)}_{q_i})\text{-XX}} \rightleftharpoons \bigcup_{s=1}^{m^{(i)}_{p_i}}\bigcup_{l=1}^{n^{(i)}_{q_i}}\underbrace{(a_{j_1j_2s},a_{j_2j_1l})}_{\text{XX}},$$

$$\underbrace{((m^{(i)}_{p_i})^{\text{th}}\text{XX},(n^{(i)}_{1})^{\text{th}}\text{XX})}_{(m^{(i)}_{p_i},n^{(i)}_{1})\text{-XX}} \rightleftharpoons \bigcup_{s=1}^{m^{(i)}_{p_i}}\bigcup_{l=1}^{n^{(i)}_{1}}\underbrace{(a_{j_1j_2s},a_{j_2j_1l})}_{\text{XX}},$$

$$\underbrace{((m^{(i)}_{1})^{\text{th}}\text{XX},(n^{(i)}_{q_i})^{\text{th}}\text{XX})}_{(m^{(i)}_{1},n^{(i)}_{q_i})\text{-XX}} \rightleftharpoons \bigcup_{s=1}^{m^{(i)}_{1}}\bigcup_{l=1}^{n^{(i)}_{q_i}}\underbrace{(a_{j_1j_2s},a_{j_2j_1l})}_{\text{XX}},$$

$$\underbrace{((m^{(i)}_{1})^{\text{th}}\text{XX},(n^{(i)}_{1})^{\text{th}}\text{XX})}_{(m^{(i)}_{1},n^{(i)}_{1})\text{-XX}} \rightleftharpoons \bigcup_{s=1}^{m^{(i)}_{1}}\bigcup_{l=1}^{n^{(i)}_{1}}\underbrace{(a_{j_1j_2s},a_{j_2j_1l})}_{\text{XX}}.$$

Case III: Consider two dynamical systems with the same equilibriums with locations switched (A_3,C_3) for $i = 1,2$ as

$$\dot{x}_{j_1} = a_{j_1j_2 0}\prod_{r_i=1}^{p}(x_{j_2}-a_{j_1j_2r_i})^{m^{(i)}_{r_i}},\ \dot{x}_{j_2} = a_{j_2j_1 0}\prod_{s_i=1}^{q}(x_{j_1}-a_{j_2j_1s_i})^{n^{(i)}_{s_i}};$$

where

$$\sum\nolimits_{r_i=1}^{p} m^{(i)}_{r_i} = 2m_1+1,\ \sum\nolimits_{s_i=1}^{q} n^{(i)}_{s_i} = 2n_1+1.$$

Consider a dynamical system as a singular equilibrium (B_3) as

$$\dot{x}_{j_1} = a_{j_1j_20}(x_{j_2} - a_{j_1j_21})^{2m_1+1},\ \dot{x}_{j_2} = a_{j_2j_10}(x_{j_1} - a_{j_2j_11})^{2n_1+1}.$$

For $r_i = 1, 2, \cdots, p$ and $s_i = 1, 2, \cdots, q\,(i = 1, 2)$, define two functions as

$$\Delta_{j_1j_2}^{(r_1r_2)} = (a_{j_1j_2r_1} - a_{j_1j_2r_2})^2\ (r_1, r_2 \in \{1, 2, \cdots, p\}, r_1 \neq r_2),$$
$$\Delta_{j_2j_1}^{(s_1s_2)} = (a_{j_2j_1s_1} - a_{j_2j_1s_2})^2\ (s_1, s_2 \in \{1, 2, \cdots, q\}, s_1 \neq s_2).$$

For $r_i = 1, 2, \cdots, p$ and $s_i = 1, 2, \cdots, q\,(i = 1, 2)$, if

$$\Delta_{j_1j_2}^{(r_1r_2)} = (a_{j_1j_2r_1} - a_{j_1j_2r_2})^2 = 0,$$
$$\Delta_{j_2j_1}^{(s_1s_2)} = (a_{j_2j_1s_1} - a_{j_2j_1s_2})^2 = 0,$$

two equilibriums of $(a_{j_1j_1r_1}, a_{j_2j_2s_1})$ and $(a_{j_1j_1r_2}, a_{j_2j_2s_2})$ switching at point $(a_{j_1j_11}, a_{j_2j_21})$ with the same order singularity are given through

$$\underbrace{(a_{j_1j_2r_1}, a_{j_2j_1s_1})}_{(m_{r_1}^{(1)},n_{s_1}^{(1)})\text{-XX}} \rightleftharpoons \underbrace{(a_{j_1j_21}, a_{j_2j_11})}_{((2m_1+1),(2n_1+1))\text{-XX}} \rightleftharpoons \underbrace{(a_{j_1j_2r_2}, a_{j_2j_1s_2})}_{(m_{r_2}^{(2)},n_{s_2}^{(2)})\text{-XX}}$$
$$\left\{a_{j_1j_2r_1} \rightleftharpoons a_{j_1j_2r_2}\middle|m_{r_1}^{(1)} = m_{r_2}^{(2)};\ r_1, r_2 \in \{1, 2, \cdots, p\}, r_1 \neq r_2\right\},$$
$$\left\{a_{j_2j_1s_1} \rightleftharpoons a_{j_2j_1s_2}\middle|n_{s_1}^{(1)} = n_{s_2}^{(2)};\ s_1, s_2 \in \{1, 2, \cdots, q\}, s_1 \neq s_2\right\};$$

and

$$\begin{aligned}
&a_{j_1j_2r_1} = a_{j_1j_21} - \varepsilon_{r_1}, a_{j_1j_2r_2} = a_{j_1j_21} + \varepsilon_{r_2},\\
&(a_{j_1j_2r_1} = a_{j_1j_21} + \varepsilon_{r_1}, a_{j_1j_2r_2} = a_{j_1j_21} - \varepsilon_{r_2});\\
&a_{j_2j_1s_1} = a_{j_2j_11} - \varepsilon_{s_1}, a_{j_2j_1s_2} = a_{j_2j_11} + \varepsilon_{s_2},\\
&(a_{j_2j_1s_1} = a_{j_2j_11} + \varepsilon_{s_1}, a_{j_2j_1s_2} = a_{j_2j_11} - \varepsilon_{s_2})\\
&\varepsilon_{r_1}, \varepsilon_{r_2} > 0 \text{ and } \varepsilon_{s_1}, \varepsilon_{s_2} > 0;\\
&r_i = 1, 2, \cdots, p;\ s_i = 1, 2, \cdots, q\,(i = 1, 2).
\end{aligned}$$

From case (I), the singular equilibriums can be formed through nonsingular equilibrium networks as presented in Case (II), and the corner singular equilibriums for $i = 1, 2$ are determined similarly.

In summary, from the cases (I)–(III), the equilibrium of $(x_{j_2}^*, x_{j_1}^*) = (a_{j_1j_21}, a_{j_2j_11})$ for $m = 2m_1+1$ and $n = 2n_1+1$ has the bifurcation properties as stated in (iv$_1$)–(iv$_4$) through Eqs. (5.47)–(5.70).

(v) For $m = 2m_1$ and $n = 2n_1 + 1$, the bifurcation process is discussed through differential equations as similar as in (iv). There are three cases (I)-(III).

Case I: Consider a dynamical system having a 1-diemnsional flow (A_1) as

$$\dot{x}_{j_1} = a_{j_1j_20}\prod_{s=1}^{m_1}[(x_{j_2} - a_{j_1j_2}^{(s)})^2 + \Delta_{j_1j_2}^{(s)}],$$

$$\dot{x}_{j_2} = a_{j_2j_10}(x_{j_1} - b_{j_2j_11})\prod_{l=1}^{n_1}[(x_{j_1} - a_{j_2j_1}^{(l)})^2 + \Delta_{j_2j_1}^{(l)}];$$

where $a_{j_2j_11} = b_{j_2j_11}$. Once

$$\Delta_{j_1j_2}^{(s)} = \Delta_{j_1j_2} = 0\ (s = 1, 2, \cdots, m_1) \text{ and } \Delta_{j_2j_1}^{(l)} = \Delta_{j_2j_1} = 0\ (l = 1, 2, \cdots, n_1),$$

if

$$a_{j_1j_21} = a_{j_1j_2}^{(s)}\ (s = 1, 2, \cdots, m_1)\ and\ a_{j_2j_11} = b_{j_2j_11} = a_{j_2j_1}^{(l)}\ (l = 1, 2, \cdots, n_1),$$

the foregoing differential equations becomes a dynamical system with a singular equilibrium (B_1) as

$$\dot{x}_{j_1} = a_{j_1j_20}\big(x_{j_2} - a_{j_1j_21}\big)^{2m_1},\ \dot{x}_{j_2} = a_{j_2j_10}\big(x_{j_1} - a_{j_2j_11}\big)^{2n_1+1}.$$

If

$$\begin{aligned}&\Delta_{j_1j_2}^{(s)} = \Delta_{j_1j_2} + \delta_s\ (s = 1, 2, \cdots, m_1),\\&\Delta_{j_2j_1}^{(l)} = \Delta_{j_2j_1} + \delta_l\ (l = 1, 2, \cdots, n_1),\\&\delta_s > 0\ and\ \delta_l > 0,\end{aligned}$$

then

$$\begin{aligned}&a_{j_1j_21}^{(s)}, a_{j_1j_22}^{(s)} = a_{j_1j_21} \pm \varepsilon_s,\ (s = 1, 2, \cdots, m_1),\\&a_{j_2j_11}^{(l)}, a_{j_2j_12}^{(l)} = a_{j_2j_11} \pm \varepsilon_l,\ (l = 1, 2, \cdots, n_1);\\&\{a_{j_1j_2}, a_{j_1j_22}, \cdots, a_{j_1j_2(2m_1)}\} = \text{sort}\Big\{a_{j_1j_21}^{(s)}, a_{j_1j_22}^{(s)}\big|s = 1, 2, \cdots, m_1\Big\},\\&\{a_{j_2j_1}, a_{j_2j_1}, \cdots, a_{j_2j_1(2n_1+1)}\} = \text{sort}\Big\{b_{j_2j_11}, a_{j_2j_11}^{(l)}, a_{j_2j_12}^{(l)}\big|l = 1, 2, \cdots, n_1\Big\}.\end{aligned}$$

Thus, the differential equation becomes a dynamical system with a non-singular equilibrium network (C_1) as.

$$\dot{x}_{j_1} = a_{j_1j_20}\prod_{s=1}^{2m_1}\big(x_{j_2} - a_{j_1j_2s}\big),\ \dot{x}_{j_2} = a_{j_2j_10}\prod_{l=1}^{2n_1+1}\big(x_{j_1} - a_{j_2j_1l}\big).$$

From the above bifurcation process analysis, at least, m_1-parameter variations in the x_{j_1}-direction and $(n_1 + 1)$-parameter variations in the x_{j_2}-direction are engaged in such a bifurcation from a parabola flow to the equilibrium network of $(2m_1) \times (2n_1+1)$ through the $(2m_1)^{\text{th}}$-increasing-inflection and decreasing-inflection bifurcations in the x_{j_1}-direction and the $(2n_1+1)^{\text{th}}$-up-parabola and down-parabola bifurcations in the x_{j_2}-direction. *With both of them, the higher-order up- parabola-saddle and down-parabola saddle bifurcations are developed.*

Thus, the appearing or vanishing bifurcation route is as follows.

$$\underbrace{(\dot{x}_{j_1}, a_{j_2j_1 1})}_{\text{XX}} \rightleftharpoons \underbrace{(a_{j_1j_2 1}, a_{j_2j_1 1})}_{((2m_1),(2n_1+1))\text{-XX}} \rightleftharpoons \bigcup_{s=1}^{2m_1} \bigcup_{l=1}^{2n_1+1} \underbrace{(a_{j_1j_2 s}, a_{j_2j_1 l})}_{\text{XX}}.$$

Case II: From Case (I), consider a dynamical system having singular equilibriums (A_2) as

$$\dot{x}_{j_1} = a_{j_1j_20} \prod_{r_1=1}^{p_1} \left(x_{j_2} - b_{j_1j_2r_1}\right)^{m_{r_1}} \prod_{s=1}^{l_1} \left[\left(x_{j_2} - a_{j_1j_2}^{(s)}\right)^2 + \Delta_{j_1j_2}^{(s)}\right]^{m_s},$$
$$\dot{x}_{j_2} = a_{j_2j_10} \prod_{s_1=1}^{q_1} \left(x_{j_1} - b_{j_2j_1s_1}\right)^{n_{s_1}} \prod_{l=1}^{l_2} \left[\left(x_{j_1} - a_{j_2j_1}^{(l)}\right)^2 + \Delta_{j_2j_1}^{(l)}\right]^{n_l};$$

where

$$\begin{array}{l} \left\{a_{j_1j_2r_1} \middle| r_1 = 1, 2, \cdots, p_1\right\} = \text{sort}\left\{b_{j_1j_2r_1} \middle| r_1 = 1, 2, \cdots, p_1\right\}, \\ \left\{a_{j_2j_1s_1} \middle| s_1 = 1, 2, \cdots, q_1\right\} = \text{sort}\left\{b_{j_2j_1s_1} \middle| s_1 = 1, 2, \cdots, q_1\right\}; \\ 2m_1 - \Sigma_{r_1=1}^{p_1} m_{r_1}^{(1)} = 2\Sigma_{s=1}^{l_1} m_s \\ \text{with } l_1\text{-quadratic polynomials without real roots} \\ \text{and } s^{\text{th}}\text{-quadratic polynomial with power } m_s, \\ 2n_1 + 1 - \Sigma_{s_1=1}^{q_1} n_{s_1}^{(1)} = 2\Sigma_{l=1}^{l_2} n_l \\ \text{with } l_2\text{-quadraitc polynomials without real roots} \\ \text{and } l^{\text{th}}\text{-quadratic polynomial with power } n_l. \end{array}$$

Once

$$\Delta_{j_1j_2}^{(s)} = 0\,(s = 1, 2, \cdots, l_1) \text{ and } \Delta_{j_2j_1}^{(l)} = 0\,(l = 1, 2, \cdots, l_2),$$

if

$$a_{j_1j_21} \equiv a_{j_1j_2r_1} = a_{j_1j_2}^{(s)}\,(r_1 = 1, 2, \cdots, p_1; s = 1, 2, \cdots, l_1),$$
$$a_{j_2j_11} \equiv a_{j_2j_1s_1} = a_{j_2j_1}^{(l)}\,(s_1 = 1, 2, \cdots, q_1; l = 1, 2, \cdots, l_2),$$

the foregoing differential equations becomes a dynamical system with a singular equilibrium (B_2) as

$$\dot{x}_{j_1} = a_{j_1 j_2 0}(x_{j_1} - a_{j_1 j_2 1})^{2m_1}, \; \dot{x}_{j_2} = a_{j_2 j_1 0}(x_{j_2} - a_{j_2 j_1 1})^{2n_1+1}.$$

If

$$\begin{aligned}
&\Delta^{(s)}_{j_1 j_2} = \Delta_{j_1 j_2} + \delta_s \; (s = 1, 2, \cdots, l_1),\\
&\Delta^{(l)}_{j_2 j_1} = \Delta_{j_2 j_1} + \delta_l \; (l = 1, 2, \cdots, l_2),\\
&\delta_s > 0 \, and \, \delta_l > 0,
\end{aligned}$$

then

$$\begin{aligned}
&a^{(s)}_{j_1 j_2 1}, a^{(s)}_{j_1 j_2 2} = a_{j_1 j_2 1} \pm \varepsilon_s, \; (s = 1, 2, \cdots, l_1),\\
&a^{(l)}_{j_2 j_1}, a^{(l)}_{j_2 j_2} = a_{j_2 j_1} \pm \varepsilon_l, \; (l = 1, 2, \cdots, l_2);\\
&\left\{a_{j_1 j_2}, a_{j_1 j_2 2}, \cdots, a_{j_1 j_2 p_2}\right\}\\
&= \text{sort}\left\{a_{j_1 j_2 1}, a_{j_1 j_2 2} \cdots, a_{j_1 j_2 p_1}; a^{(s)}_{j_1 j_2 1}, a^{(s)}_{j_1 j_2 2} \middle| s = 1, 2, \cdots, l_1\right\},\\
&\left\{a_{j_1 j_2 1}, a_{j_1 j_2 1}, \cdots, a_{j_1 j_2 q_2}\right\}\\
&= \text{sort}\left\{a_{j_1 j_2 1}, a_{j_1 j_2 1}, \cdots, a_{j_1 j_2 q_1}; a^{(l)}_{j_1 j_2 1}, a^{(l)}_{j_1 j_2 2} \middle| l = 1, 2, \cdots, l_2\right\}.
\end{aligned}$$

Thus, the differential equation becomes a dynamical system with singular equilibriums (C_2) as

$$\dot{x}_{j_1} = a_{j_1 j_1 0} \prod_{r_2=1}^{p_2} (x_{j_2} - a_{j_1 j_2 r_2})^{m^{(2)}_{r_2}}, \; \dot{x}_{j_2} = a_{j_2 j_2 0} \prod_{s_2=1}^{q_2} (x_{j_1} - a_{j_2 j_1 s_2})^{n^{(2)}_{s_2}};$$

where

$$\sum_{r_2=1}^{p_2} m^{(2)}_{r_2} = 2m_1, \; \sum_{s_2=1}^{q_2} n^{(2)}_{s_2} = 2n_1 + 1.$$

From the above bifurcation process analysis, at least, p_2-parameter variations in the x_{j_1}-direction and q_2-parameter variations in the x_{j_2}-direction are engaged in such a bifurcation from a $p_1 \times q_1$ equilibrium network to a $p_2 \times q_2$ equilibrium network through the $(2m_1)^{\text{th}}$-increasing-inflection and decreasing-inflection bifurcations in the x_{j_1}-direction and the $(2n_1 + 1)^{\text{th}}$-up-parabola and down-parabola bifurcations in the x_{j_2}-direction.

Thus, the appearing or vanishing bifurcation route from a $p_1 \times q_1$ to a $p_2 \times q_2$ equilibrium network is expressed as

$$\bigcup_{r_1=1}^{p_1}\bigcup_{s_1=1}^{q_1}\underbrace{(a_{j_1j_2r_1}, a_{j_2j_1s_1})}_{(m_{r_1}^{(1)},\, n_{s_1}^{(1)})\text{-XX}} \rightleftharpoons \underbrace{(a_{j_1j_21}, a_{j_2j_11})}_{((2m_1),(2n_1+1))\text{-XX}} \rightleftharpoons \bigcup_{r_2=1}^{p_2}\bigcup_{s_2=1}^{q_2}\underbrace{(a_{j_1j_2r_2}, a_{j_2j_1s_2})}_{(m_{r_2}^{(2)},\, n_{s_2}^{(2)})\text{-XX}}.$$

From case (I), the singular equilibriums can be formed through nonsingular equilibrium networks. Thus,

$$\underbrace{(a_{j_1j_2r_1}, a_{j_2j_1s_1})}_{(m_{r_1}^{(1)},\, n_{s_1}^{(1)})\text{-XX}} \rightleftharpoons \bigcup_{s=1}^{m_{r_1}^{(1)}}\bigcup_{l=1}^{n_{s_1}^{(1)}}\underbrace{(a_{j_1j_2s}, a_{j_2j_1l})}_{\text{XX}}$$

$$\equiv \left\{\begin{array}{cccc} (a_{j_1j_2m_{r_1}^{(1)}}, a_{j_2j_11}) & (a_{j_1j_2m_{r_1}^{(1)}}, a_{j_2j_12}) & \cdots & (a_{j_1j_2m_{r_1}^{(1)}}, a_{j_2j_1n_{s_1}^{(1)}}) \\ (a_{j_1j_2(m_{r_1}^{(1)}-1)}, a_{j_2j_11}) & (a_{j_1j_2(m_{r_1}^{(1)}-1)}, a_{j_2j_12}) & \cdots & (a_{j_1j_2(m_{r_1}^{(1)}-1)}, a_{j_2j_1n_{s_1}^{(1)}}) \\ \vdots & \vdots & \cdots & \vdots \\ (a_{j_1j_21}, a_{j_2j_11}) & (a_{j_1j_21}, a_{j_2j_12}) & \cdots & (a_{j_1j_21}, a_{j_2j_1n_{s_1}^{(1)}}) \end{array}\right\}_{m_{r_1}^{(1)}\times n_{s_1}^{(1)}},$$

and

$$\underbrace{(a_{j_1j_2r_2}, a_{j_1j_1s_2})}_{(m_{r_2}^{(2)},\, n_{s_2}^{(2)})\text{-XX}} \rightleftharpoons \bigcup_{s=1}^{m_{r_2}^{(2)}}\bigcup_{l=1}^{n_{s_2}^{(2)}}\underbrace{(a_{j_1j_2s}, a_{j_2j_1l})}_{\text{XX}}$$

$$\equiv \left\{\begin{array}{cccc} (a_{j_1j_2m_{r_2}^{(2)}}, a_{j_2j_11}) & (a_{j_1j_2m_{r_2}^{(2)}}, a_{j_2j_12}) & \cdots & (a_{j_1j_2m_{r_2}^{(2)}}, a_{j_2j_1n_{s_2}^{(2)}}) \\ (a_{j_1j_2(m_{r_2}^{(2)}-1)}, a_{j_2j_11}) & (a_{j_1j_2(m_{r_2}^{(2)}-1)}, a_{j_2j_12}) & \cdots & (a_{j_1j_2(m_{r_2}^{(2)}-1)}, a_{j_2j_1n_{s_2}^{(2)}}) \\ \vdots & \vdots & \cdots & \vdots \\ (a_{j_1j_21}, a_{j_2j_11}) & (a_{j_1j_21}, a_{j_2j_12}) & \cdots & (a_{j_1j_21}, a_{j_2j_1n_{s_2}^{(2)}}) \end{array}\right\}_{m_{r_2}^{(2)}\times n_{s_2}^{(2)}}.$$

From the above definition, the corner singular equilibriums for $i = 1, 2$ are determined by

$$\underbrace{((m_{p_i}^{(i)})^{\text{th}}\text{XX},(n_{q_i}^{(i)})^{\text{th}}\text{XX})}_{(m_{p_i}^{(i)},n_{q_i}^{(i)})\text{-XX}} \rightleftharpoons \bigcup_{s=1}^{m_{p_i}^{(i)}}\bigcup_{l=1}^{n_{q_i}^{(i)}}\underbrace{(a_{j_1j_2s}, a_{j_2j_1l})}_{\text{XX}},$$

$$\underbrace{((m_{p_i}^{(i)})^{\text{th}}\text{XX},(n_1^{(i)})^{\text{th}}\text{XX})}_{(m_{p_i}^{(i)},n_1^{(i)})\text{-XX}} \rightleftharpoons \bigcup_{s-1}^{m_{p_i}^{(i)}}\bigcup_{l-1}^{n_1^{(i)}}\underbrace{(a_{j_1j_2s}, a_{j_2j_1l})}_{\text{XX}},$$

$$\underbrace{((m_1^{(i)})^{\text{th}}\text{XX},(n_{q_i}^{(i)})^{\text{th}}\text{XX})}_{(m_1^{(i)},n_{q_i}^{(i)})\text{-XX}} \rightleftharpoons \bigcup_{s=1}^{m_1^{(i)}}\bigcup_{l=1}^{n_{q_i}^{(i)}}\underbrace{(a_{j_1j_2s}, a_{j_2j_1l})}_{\text{XX}},$$

$$\underbrace{((m_1^{(i)})^{\text{th}}\text{XX},(n_1^{(i)})^{\text{th}}\text{XX})}_{(m_1^{(i)},n_1^{(i)})\text{-XX}} \rightleftharpoons \bigcup\nolimits_{s=1}^{m_1^{(i)}} \bigcup\nolimits_{l=1}^{n_1^{(i)}} \underbrace{(a_{j_1j_2s}, a_{j_2j_1l})}_{\text{XX}}.$$

Case III: Consider two dynamical systems with the same equilibriums with locations switched (A_3,C_3) for $i = 1, 2$ as

$$\dot{x}_{j_1} = a_{j_1j_20} \prod_{r_i=1}^{p} (x_{j_2} - a_{j_1j_2r_i})^{m_{r_i}^{(i)}},\ \dot{x}_{j_2} = a_{j_2j_10} \prod_{s_i=1}^{q} (x_{j_1} - a_{j_2j_1s_i})^{n_{s_i}^{(i)}};$$

where

$$\sum_{r_i=1}^{p} m_{r_i}^{(i)} = 2m_1,\ \sum_{s_i=1}^{q} n_{s_i}^{(i)} = 2n_1 + 1.$$

Consider a dynamical system as a singular equilibrium (B_3) as

$$\dot{x}_{j_1} = a_{j_1j_20}(x_{j_2} - a_{j_1j_21})^{2m_1},\ \dot{x}_{j_2} = a_{j_2j_10}(x_{j_1} - a_{j_2j_11})^{2n_1+1}.$$

For $r_i = 1, 2, \cdots, p$ and $s_i = 1, 2, \cdots, q\,(i = 1, 2)$, define two functions as

$$\Delta_{j_1j_2}^{(r_1r_2)} = (a_{j_1j_2r_1} - a_{j_1j_2r_2})^2\ (r_1, r_2 \in \{1, 2, \cdots, p\}, r_1 \neq r_2),$$
$$\Delta_{j_2j_1}^{(s_1s_2)} = (a_{j_2j_1s_1} - a_{j_2j_1s_2})^2\ (s_1, s_2 \in \{1, 2, \cdots, q\}, s_1 \neq s_2).$$

For $r_i = 1, 2, \cdots, p$ and $s_i = 1, 2, \cdots, q\,(i = 1, 2)$, if

$$\Delta_{j_1j_2}^{(r_1r_2)} = (a_{j_1j_2r_1} - a_{j_1j_2r_2})^2 = 0,$$
$$\Delta_{j_2j_1}^{(s_1s_2)} = (a_{j_2j_1s_1} - a_{j_2j_1s_2})^2 = 0,$$

two equilibriums of $(a_{j_1j_1r_1}, a_{j_2j_2s_1})$ and $(a_{j_1j_1r_2}, a_{j_2j_2s_2})$ switching at point $(a_{j_1j_11}, a_{j_2j_21})$ with the same order singularity are given through

$$\underbrace{(a_{j_1j_2r_1}, a_{j_2j_1s_1})}_{(m_{r_1}^{(1)},n_{s_1}^{(1)})\text{-XX}} \rightleftharpoons \underbrace{(a_{j_1j_21}, a_{j_2j_11})}_{((2m_1),(2n_1+1))\text{-XX}} \rightleftharpoons \underbrace{(a_{j_1j_2r_2}, a_{j_2j_1s_2})}_{(m_{r_2}^{(2)},n_{s_2}^{(2)})\text{-XX}}$$
$$\{a_{j_1j_2r_1} \rightleftharpoons a_{j_1j_2r_2} | m_{r_1}^{(1)} = m_{r_2}^{(2)}; r_1, r_2 \in \{1, 2, \cdots, p\}, r_1 \neq r_2\},$$
$$\{a_{j_2j_1s_1} \rightleftharpoons a_{j_2j_1s_2} | n_{s_1}^{(1)} = n_{s_2}^{(2)}; s_1, s_2 \in \{1, 2, \cdots, q\}, s_1 \neq s_2\},$$

and

$$a_{j_1j_2r_1} = a_{j_1j_21} - \varepsilon_{r_1}, a_{j_1j_2r_2} = a_{j_1j_21} + \varepsilon_{r_2},$$
$$(a_{j_1j_2r_1} = a_{j_1j_21} + \varepsilon_{r_1}, a_{j_1j_2r_2} = a_{j_1j_21} - \varepsilon_{r_2});$$

$$\begin{aligned}
&a_{j_2j_1s_1} = a_{j_2j_1 1} - \varepsilon_{s_1}, a_{j_2j_1s_2} = a_{j_2j_1 1} + \varepsilon_{s_2},\\
&(a_{j_2j_1s_1} = a_{j_2j_1 1} + \varepsilon_{s_1}, a_{j_2j_1s_2} = a_{j_2j_1 1} - \varepsilon_{s_2})\\
&\varepsilon_{r_1}, \varepsilon_{r_2} > 0 \text{ and } \varepsilon_{s_1}, \varepsilon_{s_2} > 0;\\
&r_i = 1, 2, \cdots, p; s_i = 1, 2, \cdots, q\,(i = 1, 2).
\end{aligned}$$

From case (I), the singular equilibriums can be formed through nonsingular equilibrium networks as presented in Case (II), and the corner singular equilibriums for $i = 1, 2$ are determined similarly.

In summary, from the cases (I)–(III), the equilibrium of $(x^*_{j_2}, x^*_{j_1}) = (a_{j_1j_2 1}, a_{j_2j_1 1})$ for $m = 2m_1$ and $n = 2n_1 + 1$ has the bifurcation properties as stated in (v_1)-(v_4) through Eqs.(5.71)–(5.94).

(vi) For $m = 2m_1 + 1$ and $n = 2n_1$, this case is quite similar to (v) for $m = 2m_1$ and $n = 2n_1 + 1$, and there are three cases (I)–(III).

Case I: Consider a dynamical system having a 1-dimensional flow (A_1) as

$$\begin{aligned}
\dot{x}_{j_1} &= a_{j_1j_2 0}(x_{j_2} - b_{j_1j_2 1})\prod_{s=1}^{m_1}[(x_{j_2} - a^{(s)}_{j_1j_2})^2 + \Delta^{(s)}_{j_1j_2}],\\
\dot{x}_{j_2} &= a_{j_2j_1 0}\prod_{l=1}^{n_1}[(x_{j_1} - a^{(l)}_{j_2j_1})^2 + \Delta^{(l)}_{j_2j_1}];
\end{aligned}$$

where $a_{j_1j_2 1} = b_{j_1j_2 1}$. Once

$$\Delta^{(s)}_{j_1j_2} = \Delta_{j_1j_2} = 0\ (s = 1, 2, \cdots, m_1) \text{ and } \Delta^{(l)}_{j_2j_1} = \Delta_{j_2j_1} = 0\ (l = 1, 2, \cdots, n_1),$$

if

$$a_{j_1j_2 1} = b_{j_1j_2 1} = a^{(s)}_{j_1j_2}\ (s = 1, 2, \cdots, m_1) \ and\ a_{j_2j_1 1} = a^{(l)}_{j_2j_1}\ (l = 1, 2, \cdots, n_1),$$

the foregoing differential equations becomes a dynamical system with a singular equilibrium (B_1) as

$$\dot{x}_{j_1} = a_{j_1j_2 0}\big(x_{j_2} - a_{j_1j_2 1}\big)^{2m_1+1}, \ \dot{x}_{j_2} = a_{j_2j_1 0}\big(x_{j_1} - a_{j_2j_1 1}\big)^{2n_1}.$$

If

$$\begin{aligned}
&\Delta^{(s)}_{j_1j_2} = \Delta_{j_1j_2} + \delta_s\ (s = 1, 2, \cdots, m_1),\\
&\Delta^{(l)}_{j_2j_1} = \Delta_{j_2j_1} + \delta_l\ (l = 1, 2, \cdots, n_1),\\
&\delta_s > 0 \ and\ \delta_l > 0,
\end{aligned}$$

then

$$a^{(s)}_{j_1j_21}, a^{(s)}_{j_1j_22} = a_{j_1j_21} \pm \varepsilon_s, (s = 1, 2, \cdots, m_1),$$
$$a^{(l)}_{j_2j_11}, a^{(l)}_{j_2j_12} = a_{j_2j_1} \pm \varepsilon_l, (l = 1, 2, \cdots, n_2);$$

$$\{a_{j_1j_21}, a_{j_1j_22} \cdots, a_{j_1j_2(2m_1+1)}\} = \text{sort}\Big\{b^{(s)}_{j_1j_21}, a^{(s)}_{j_1j_22} \big| s = 1, 2, \cdots, m_1\Big\},$$
$$\{a_{j_1j_21}, a_{j_1j_21} \cdots, a_{j_1j_2(2n_1)}\} = \text{sort}\Big\{a^{(l)}_{j_2j_11}, a^{(l)}_{j_2j_12} \big| l = 1, 2, \cdots, n_1\Big\}.$$

Thus, the differential equation becomes a dynamical system with a non-singular equilibrium network (C_1) as

$$\dot{x}_{j_1} = a_{j_1j_20} \prod_{s=1}^{2m_1+1} (x_{j_2} - a_{j_1j_2s}),\ \dot{x}_{j_2} = a_{j_2j_10} \prod_{l=1}^{2n_1} (x_{j_1} - a_{j_2j_1l}).$$

From the above bifurcation process analysis, at least, (m_1+1)-parameter variations in the x_{j_1}-direction and n_1-parameter variations in the x_{j_2}-direction are engaged in such a bifurcation from a parabola flow to the equilibrium network of $(2m_1 + 1) \times (2n_1)$ through the $(2m_1 + 1)^{\text{th}}$-up-parabola and down-parabola bifurcations in the x_{j_1}-direction and the $(2n_1)^{\text{th}}$-increasing-inflection and decreasing-inflection bifurcations in the x_{j_2}-direction. *With both of them, the higher-order up-parabola and down-parabola saddle bifurcations are developed.*

Thus, the appearing or vanishing bifurcation route is as follows.

$$\underbrace{(a_{j_1j_21}, \dot{x}_{j_2})}_{\text{XX}} \rightleftharpoons \underbrace{(a_{j_1j_21}, a_{j_2j_11})}_{((2m_1+1),(2n_1))\text{-XX}} \rightleftharpoons \bigcup_{s=1}^{2m_1+1} \bigcup_{l=1}^{2n_1} \underbrace{(a_{j_1j_2s}, a_{j_2j_1l})}_{\text{XX}}.$$

Case II: From Case (I), consider a dynamical system having singular equilibriums (A_2) as

$$\dot{x}_{j_1} = a_{j_1j_20} \prod_{r_1=1}^{p_1} (x_{j_2} - b_{j_1j_2r_1})^{m^{(1)}_{r_1}} \prod_{s=1}^{l_1} [(x_{j_2} - a^{(s)}_{j_1j_2})^2 + \Delta^{(s)}_{j_1j_2}]^{m_s},$$
$$\dot{x}_{j_2} = a_{j_2j_10} \prod_{s_1=1}^{q_1} (x_{j_1} - b_{j_2j_1s_1})^{n^{(1)}_{s_1}} \prod_{l=1}^{l_2} [(x_{j_1} - a^{(l)}_{j_2j_1})^2 + \Delta^{(l)}_{j_2j_1}]^{n_l};$$

where

$$\{a_{j_1j_2r_1} | r_1 = 1, 2, \cdots, p_1\} = \text{sort}\{b_{j_1j_2r_1} | r_1 = 1, 2, \cdots, p_1\},$$
$$\{a_{j_2j_1s_1} | s_1 = 1, 2, \cdots, q_1\} = \text{sort}\{b_{j_2j_1s_1} | s_1 = 1, 2, \cdots, q_1\};$$

$$2m_1 + 1 - \Sigma_{r_1=1}^{p_1} m_{r_1}^{(1)} = 2\Sigma_{s=1}^{l_1} m_s$$
with l_1-quadratic polynomials without real roots
and s^{th}-quadratic polynomial with power m_s,
$$2n_1 - \Sigma_{s_1=1}^{q_1} n_{s_1}^{(1)} = 2\Sigma_{l=1}^{l_2} n_l$$
with l_2-quadraitc polynomials without real roots
and l^{th}-quadratic polynomial with power n_l.

Once

$$\Delta_{j_1j_2}^{(s)} = 0\ (s = 1, 2, \cdots, l_1) \text{ and } \Delta_{j_2j_1}^{(l)} = 0\ (l = 1, 2, \cdots, l_2),$$

if

$$a_{j_1j_2 1} \equiv a_{j_1j_2r_1} = a_{j_1j_2}^{(s)}\ (r_1 = 1, 2, \cdots, p_1;\ s = 1, 2, \cdots, l_1),$$
$$a_{j_2j_1 1} \equiv a_{j_2j_1s_1} = a_{j_2j_1}^{(l)}\ (s_1 = 1, 2, \cdots, q_1;\ l = 1, 2, \cdots, l_2),$$

the foregoing differential equations becomes a dynamical system with a singular equilibrium (B_2) as

$$\dot{x}_{j_1} = a_{j_1j_20}\big(x_{j_1} - a_{j_1j_21}\big)^{2m_1+1},\ \dot{x}_{j_2} = a_{j_2j_10}\big(x_{j_2} - a_{j_2j_11}\big)^{2n_1}.$$

If

$$\Delta_{j_1j_2}^{(s)} = \Delta_{j_1j_2} + \delta_s\ (s = 1, 2, \cdots, l_1),$$
$$\Delta_{j_2j_1}^{(l)} = \Delta_{j_2j_1} + \delta_l\ (l = 1, 2, \cdots, l_2),$$
$$\delta_s > 0 \textit{ and } \delta_l > 0,$$

then

$$a_{j_1j_21}^{(s)}, a_{j_1j_22}^{(s)} = a_{j_1j_21} \pm \varepsilon_s, (s = 1, 2, \cdots, l_1),$$
$$a_{j_2j_11}^{(l)}, a_{j_2j_12}^{(l)} = a_{j_2j_11} \pm \varepsilon_l, (l = 1, 2, \cdots, l_2);$$
$$\big\{a_{j_1j_21}, a_{j_1j_22}, \cdots, a_{j_1j_2p_2}\big\}$$
$$= \text{sort}\Big\{a_{j_1j_21}^{(s)}, a_{j_1j_22}^{(s)}, \cdots a_{j_1j_2p_1}; a_{j_1j_21}^{(s)}, a_{j_1j_22}^{(s)} \big| s = 1, 2, \cdots, l_1\Big\},$$
$$\big\{a_{j_1j_21}, a_{j_1j_22}, \cdots, a_{j_1j_2q_2}\big\}$$
$$= \text{sort}\Big\{a_{j_1j_21}, a_{j_1j_22}, \cdots, a_{j_1j_2q_1}; a_{j_1j_21}^{(l)}, a_{j_1j_22}^{(l)} \big| l = 1, 2, \cdots, l_2\Big\}.$$

Thus, the differential equation becomes a dynamical system with singular equilibriums (C_2) as

$$\dot{x}_{j_1} = a_{j_1j_10} \prod_{r_2=1}^{p_2} \left(x_{j_2} - a_{j_1j_2r_2}\right)^{m_{r_2}^{(2)}},\ \dot{x}_{j_2} = a_{j_2j_20} \prod_{s_2=1}^{q_2} \left(x_{j_1} - a_{j_2j_1s_2}\right)^{n_{s_2}^{(2)}};$$

where

$$\sum_{r_2=1}^{p_2} m_{r_2}^{(2)} = 2m_1 + 1,\ \sum_{s_2=1}^{q_2} n_{s_2}^{(2)} = 2n_1.$$

From the above bifurcation process analysis, at least, p_2-parameter variations in the x_{j_1}-direction and q_2-parameter variations in the x_{j_2}-direction are engaged in such a bifurcation from a $p_1 \times q_1$ equilibrium network to a $p_2 \times q_2$ equilibrium network through the $(2m_1 + 1)^{\text{th}}$-up-parabola and down-parabola bifurcations in the x_{j_1}-direction and the $(2n_1)^{\text{th}}$-increasing-inflection and decreasing-inflection bifurcations in the x_{j_2}-direction.

Thus, the appearing or vanishing bifurcation route from a $p_1 \times q_1$ to a $p_2 \times q_2$ equilibrium network is expressed as

$$\bigcup_{r_1=1}^{p_1} \bigcup_{s_1=1}^{q_1} \underbrace{\left(a_{j_1j_2r_1}, a_{j_2j_1s_1}\right)}_{\left(m_{r_1}^{(1)}, n_{s_1}^{(1)}\right)\text{-XX}} \rightleftharpoons \underbrace{\left(a_{j_1j_21}, a_{j_2j_11}\right)}_{((2m_1+1),\ (2n_1))\text{-XX}} \rightleftharpoons \bigcup_{r_2=1}^{p_2} \bigcup_{s_2=1}^{q_2} \underbrace{\left(a_{j_1j_2r_2}, a_{j_2j_1s_2}\right)}_{\left(m_{r_2}^{(2)}, n_{s_2}^{(2)}\right)\text{-XX}}.$$

From case (I), the singular equilibriums can be formed through nonsingular equilibrium networks. Thus,

$$\underbrace{(a_{j_1j_2r_1}, a_{j_2j_1s_1})}_{(m_{r_1}^{(1)}, n_{s_1}^{(1)})\text{-XX}} \rightleftharpoons \bigcup_{s=1}^{m_{r_1}^{(1)}} \bigcup_{l=1}^{n_{s_1}^{(1)}} \underbrace{(a_{j_1j_2s}, a_{j_2j_1l})}_{\text{XX}}$$

$$\equiv \left\{\begin{array}{cccc} (a_{j_1j_2m_{r_1}^{(1)}}, a_{j_2j_11}) & (a_{j_1j_2m_{r_1}^{(1)}}, a_{j_2j_12}) & \cdots & (a_{j_1j_2m_{r_1}^{(1)}}, a_{j_2j_1n_{s_1}^{(1)}}) \\ (a_{j_1j_2(m_{r_1}^{(1)}-1)}, a_{j_2j_11}) & (a_{j_1j_2(m_{r_1}^{(1)}-1)}, a_{j_2j_12}) & \cdots & (a_{j_1j_2(m_{r_1}^{(1)}-1)}, a_{j_2j_1n_{s_1}^{(1)}}) \\ \vdots & \vdots & \cdots & \vdots \\ (a_{j_1j_21}, a_{j_2j_11}) & (a_{j_1j_21}, a_{j_2j_12}) & \cdots & (a_{j_1j_21}, a_{j_2j_1n_{s_1}^{(1)}}) \end{array}\right\}_{m_{r_1}^{(1)} \times n_{s_1}^{(1)}},$$

and

$$\underbrace{(a_{j_1j_2r_2}, a_{j_1j_1s_2})}_{(m_{r_2}^{(2)}, n_{s_2}^{(2)})\text{-XX}} \rightleftharpoons \bigcup_{s=1}^{m_{r_2}^{(2)}} \bigcup_{l=1}^{n_{s_2}^{(2)}} \underbrace{(a_{j_1j_2s}, a_{j_2j_1l})}_{\text{XX}}$$

$$\equiv \left\{ \begin{array}{cccc} (a_{j_1j_2m^{(2)}_{r_2}}, a_{j_2j_11}) & (a_{j_1j_2m^{(2)}_{r_2}}, a_{j_2j_12}) & \cdots & (a_{j_1j_2m^{(2)}_{r_2}}, a_{j_2j_1n^{(2)}_{s_2}}) \\ (a_{j_1j_2(m^{(2)}_{r_2}-1)}, a_{j_2j_11}) & (a_{j_1j_2(m^{(2)}_{r_2}-1)}, a_{j_2j_12}) & \cdots & (a_{j_1j_2(m^{(2)}_{r_2}-1)}, a_{j_2j_1n^{(2)}_{s_2}}) \\ \vdots & \vdots & \cdots & \vdots \\ (a_{j_1j_21}, a_{j_2j_11}) & (a_{j_1j_21}, a_{j_2j_12}) & \cdots & (a_{j_1j_21}, a_{j_2j_1n^{(2)}_{s_2}}) \end{array} \right\}_{m^{(2)}_{r_2} \times n^{(2)}_{s_2}}.$$

From the above definition, the corner singular equilibriums for $i = 1, 2$ are determined by

$$\underbrace{((m^{(i)}_{p_i})^{\text{th}}\text{XX},(n^{(i)}_{q_i})^{\text{th}}\text{XX})}_{(m^{(i)}_{p_i},n^{(i)}_{q_i})\text{-XX}} \rightleftharpoons \bigcup_{s=1}^{m^{(i)}_{p_i}} \bigcup_{l=1}^{n^{(i)}_{q_i}} \underbrace{(a_{j_1j_2s}, a_{j_2j_1l})}_{\text{XX}},$$

$$\underbrace{((m^{(i)}_{p_i})^{\text{th}}\text{XX},(n^{(i)}_{1})^{\text{th}}\text{XX})}_{(m^{(i)}_{p_i},n^{(i)}_{1})\text{-XX}} \rightleftharpoons \bigcup_{s=1}^{m^{(i)}_{p_i}} \bigcup_{l=1}^{n^{(i)}_{1}} \underbrace{(a_{j_1j_2s}, a_{j_2j_1l})}_{\text{XX}},$$

$$\underbrace{((m^{(i)}_{1})^{\text{th}}\text{XX},(n^{(i)}_{q_i1})^{\text{th}}\text{XX})}_{(m^{(i)}_{1},n^{(i)}_{q_i})\text{-XX}} \rightleftharpoons \bigcup_{s=1}^{m^{(i)}_{1}} \bigcup_{l=1}^{n^{(i)}_{q_i}} \underbrace{(a_{j_1j_2s}, a_{j_2j_1l})}_{\text{XX}},$$

$$\underbrace{((m^{(i)}_{1})^{\text{th}}\text{XX},(n^{(i)}_{1})^{\text{th}}\text{XX})}_{(m^{(i)}_{1},n^{(i)}_{1})\text{-XX}} \rightleftharpoons \bigcup_{s=1}^{m^{(i)}_{1}} \bigcup_{l=1}^{n^{(i)}_{1}} \underbrace{(a_{j_1j_2s}, a_{j_2j_1l})}_{\text{XX}}.$$

Case III: Consider two dynamical systems with the same equilibriums with locations switched (A_3,C_3) for $i = 1, 2$ as

$$\dot{x}_{j_1} = a_{j_1j_20} \prod_{r_i=1}^{p} (x_{j_2} - a_{j_1j_2r_i})^{m^{(i)}_{r_i}},\ \dot{x}_{j_2} = a_{j_2j_10} \prod_{s_i=1}^{q} (x_{j_1} - a_{j_2j_1s_i})^{n^{(i)}_{s_i}};$$

where

$$\sum_{r_i=1}^{p} m^{(i)}_{r_i} = 2m_1 + 1,\ \sum_{s_i=1}^{q} n^{(i)}_{s_i} = 2n_1.$$

Consider a dynamical system as a singular equilibrium (B_3) as

$$\dot{x}_{j_1} = a_{j_1j_20}(x_{j_2} - a_{j_1j_21})^{2m_1+1},\ \dot{x}_{j_2} = a_{j_2j_10}(x_{j_1} - a_{j_2j_11})^{2n_1}.$$

For $r_i = 1, 2, \cdots, p$ and $s_i = 1, 2, \cdots, q\,(i = 1, 2)$, define two functions as

$$\Delta_{j_1j_2}^{(r_1r_2)} = (a_{j_1j_2r_1} - a_{j_1j_2r_2})^2 \; (r_1, r_2 \in \{1, 2, \cdots, p\}, r_1 \neq r_2),$$
$$\Delta_{j_2j_1}^{(s_1s_2)} = (a_{j_2j_1s_1} - a_{j_2j_1s_2})^2 \; (s_1, s_2 \in \{1, 2, \cdots, q\}, s_1 \neq s_2).$$

For $r_i = 1, 2, \cdots, p$ and $s_i = 1, 2, \cdots, q \, (i = 1, 2)$, if

$$\Delta_{j_1j_2}^{(r_1r_2)} = (a_{j_1j_2r_1} - a_{j_1j_2r_2})^2 = 0,$$
$$\Delta_{j_2j_1}^{(s_1s_2)} = (a_{j_2j_1s_1} - a_{j_2j_1s_2})^2 = 0,$$

two equilibriums of $(a_{j_1j_1r_1}, a_{j_2j_2s_1})$ and $(a_{j_1j_1r_2}, a_{j_2j_2s_2})$ switching at point $(a_{j_1j_1 1}, a_{j_2j_2 1})$ with the same order singularity are given through

$$\underbrace{(a_{j_1j_2r_1}, a_{j_2j_1s_1})}_{(m_{r_1}^{(1)}, n_{s_1}^{(1)})\text{-XX}} \rightleftharpoons \underbrace{(a_{j_1j_2 1}, a_{j_2j_1 1})}_{((2m_1+1),(2n_1))\text{-XX}} \rightleftharpoons \underbrace{(a_{j_1j_2r_2}, a_{j_2j_1s_2})}_{(m_{r_2}^{(2)}, n_{s_2}^{(2)})\text{-XX}},$$
$$\left\{a_{j_1j_2r_1} \rightleftharpoons a_{j_1j_2r_2} \middle| m_{r_1}^{(1)} = m_{r_2}^{(2)}; r_1, r_2 \in \{1, 2, \cdots, p\}, r_1 \neq r_2\right\},$$
$$\left\{a_{j_2j_1s_1} \rightleftharpoons a_{j_2j_1s_2} \middle| n_{s_1}^{(1)} = n_{s_2}^{(2)}; s_1, s_2 \in \{1, 2, \cdots, q\}, s_1 \neq s_2\right\},$$

and

$$\begin{aligned}
a_{j_1j_2r_1} &= a_{j_1j_2 1} - \varepsilon_{r_1}, a_{j_1j_2r_2} = a_{j_1j_2 1} + \varepsilon_{r_2},\\
(a_{j_1j_2r_1} &= a_{j_1j_2 1} + \varepsilon_{r_1}, a_{j_1j_2r_2} = a_{j_1j_2 1} - \varepsilon_{r_2});\\
a_{j_2j_1s_1} &= a_{j_2j_1 1} - \varepsilon_{s_1}, a_{j_2j_1s_2} = a_{j_2j_1 1} + \varepsilon_{s_2},\\
(a_{j_2j_1s_1} &= a_{j_2j_1 1} + \varepsilon_{s_1}, a_{j_2j_1s_2} = a_{j_2j_1 1} - \varepsilon_{s_2})\\
&\varepsilon_{r_1}, \varepsilon_{r_2} > 0 \text{ and } \varepsilon_{s_1}, \varepsilon_{s_2} > 0;\\
&r_i = 1, 2, \cdots, p; s_i = 1, 2, \cdots, q \, (i = 1, 2).
\end{aligned}$$

From case (I), the singular equilibriums can be formed through nonsingular equilibrium networks as presented in Case (II), and the corner singular equilibriums for $i = 1, 2$ are determined similarly.

In summary, from the cases (I)-(III), the equilibrium of $(x_{j_2}^*, x_{j_1}^*) = (a_{j_1j_2 1}, a_{j_2j_1 1})$ for $m = 2m_1 + 1$ and $n = 2n_1$ has the bifurcation properties as stated in (vi_1)-(vi_4) through Eqs. (5.95)–(5.118).

(vii) For $m = 2m_1$ and $n = 2n_1$, the bifurcation process is discussed through differential equations as follows. There are three cases (I)-(III).

Case I: Consider a dynamical system having a two-dimensional flow (A_1) as

$$\dot{x}_{j_1} = a_{j_1j_2 0} \prod_{s=1}^{m_1} [(x_{j_2} - a_{j_1j_2}^{(s)})^2 + \Delta_{j_1j_2}^{(s)}],$$
$$\dot{x}_{j_2} = a_{j_2j_1 0} \prod_{l=1}^{n_1} [(x_{j_1} - a_{j_2j_1}^{(l)})^2 + \Delta_{j_2j_1}^{(l)}].$$

Once

$$\Delta^{(s)}_{j_1j_2} = \Delta_{j_1j_2} = 0\,(s = 1, 2, \cdots, m_1) \text{ and } \Delta^{(l)}_{j_2j_1} = \Delta_{j_2j_1} = 0\,(l = 1, 2, \cdots, n_1),$$

if

$$a_{j_1j_21} = a^{(s)}_{j_1j_2}\,(s = 1, 2, \cdots, m_1)\,\textit{and}\; a_{j_2j_11} = a^{(l)}_{j_2j_1}\,(l = 1, 2, \cdots, n_1),$$

the foregoing differential equations becomes a dynamical system with a singular equilibrium (B_1) as

$$\dot{x}_{j_1} = a_{j_1j_20}(x_{j_2} - a_{j_1j_21})^{2m_1},\; \dot{x}_{j_2} = a_{j_2j_10}(x_{j_1} - a_{j_2j_11})^{2n_1}.$$

If

$$\begin{aligned}
&\Delta^{(s)}_{j_1j_2} = \Delta_{j_1j_2} + \delta_s\,(s = 1, 2, \cdots, m_1),\\
&\Delta^{(l)}_{j_2j_1} = \Delta_{j_2j_1} + \delta_l\,(l = 1, 2, \cdots, n_1),\\
&\delta_s > 0\,\textit{and}\;\delta_l > 0,
\end{aligned}$$

then

$$\begin{aligned}
&a^{(\mathrm{s})}_{j_1j_21}, a^{(\mathrm{s})}_{j_1j_22} = a_{j_1j_21} \pm \varepsilon_s, (s = 1, 2, \cdots, m_1),\\
&a^{(l)}_{j_2j_11}, a^{(l)}_{j_2j_12} = a_{j_2j_11} \pm \varepsilon_l, (l = 1, 2, \cdots, n_1);\\
&\{a_{j_1j_21}, a_{j_1j_22}, \cdots, a_{j_1j_2(2m_1)}\} = \mathrm{sort}\Big\{a^{(\mathrm{s})}_{j_1j_21}, a^{(\mathrm{s})}_{j_1j_22}\,|s = 1, 2, \cdots, m_1\Big\},\\
&\{a_{j_2j_11}, a_{j_2j_11}, \cdots, a_{j_2j_1(2n_1)}\} = \mathrm{sort}\Big\{a^{(l)}_{j_2j_11}, a^{(l)}_{j_2j_12}\,|l = 1, 2, \cdots, n_1\Big\}.
\end{aligned}$$

Thus, the differential equation becomes a dynamical system with a non-singular equilibrium network (C_1) as

$$\dot{x}_{j_1} = a_{j_1j_20}\prod_{s=1}^{2m_1}(x_{j_2} - a_{j_1j_2s}),\; \dot{x}_{j_2} = a_{j_2j_10}\prod_{l=1}^{2n_1}(x_{j_1} - a_{j_2j_1l}).$$

From the above bifurcation process analysis, at least, m_1-parameter variations in the x_{j_1}-direction and n_1-parameter variations in the x_{j_2}-direction are engaged in such a bifurcation from a flow to the equilibrium network of $(2m_1) \times (2n_1)$ through the $(2m_1 + 1)^{\text{th}}$-up-parabola and down-parabola bifurcations in the x_{j_1}-direction and the $(2n_1)^{\text{th}}$-increasing-inflection and decreasing-inflection bifurcations in the x_{j_2}-direction. *With both of them, the higher-order double-inflection saddle bifurcations are developed.*

Thus, the appearing or vanishing bifurcation route is as follows.

$$\underbrace{(\dot{x}_{j_1}, \dot{x}_{j_2})}_{\text{XX}} \rightleftharpoons \underbrace{(a_{j_1 j_2 1}, a_{j_2 j_1 1})}_{((2m_1),(2n_1))\text{-XX}} \rightleftharpoons \bigcup_{s=1}^{2m_1} \bigcup_{l=1}^{2n_1} \underbrace{(a_{j_1 j_2 s}, a_{j_2 j_1 l})}_{\text{XX}}.$$

Case II: From Case (I), consider a dynamical system having singular equilibriums (A_2) as

$$\dot{x}_{j_1} = a_{j_1 j_2 0} \prod_{r_1=1}^{p_1} (x_{j_2} - b_{j_1 j_2 r_1})^{m_{r_1}^{(1)}} \prod_{s=1}^{l_1} [(x_{j_2} - a_{j_1 j_2}^{(s)})^2 + \Delta_{j_1 j_2}^{(s)}]^{m_s},$$
$$\dot{x}_{j_2} = a_{j_2 j_1 0} \prod_{s_1=1}^{q_1} (x_{j_1} - b_{j_2 j_1 s_1})^{n_{s_1}^{(1)}} \prod_{l=1}^{l_2} [(x_{j_1} - a_{j_2 j_1}^{(l)})^2 + \Delta_{j_2 j_1}^{(l)}]^{n_l};$$

where

$$\begin{aligned}
&\{a_{j_1 j_2 r_1} | r_1 = 1, 2, \cdots, p_1\} = \text{sort}\{b_{j_1 j_2 r_1} | r_1 = 1, 2, \cdots, p_1\}, \\
&\{a_{j_2 j_1 s_1} | s_1 = 1, 2, \cdots, q_1\} = \text{sort}\{b_{j_2 j_1 s_1} | s_1 = 1, 2, \cdots, q_1\}; \\
&2m_1 - \Sigma_{r_1=1}^{p_1} m_{r_1}^{(1)} = 2\Sigma_{s=1}^{l_1} m_s \\
&\text{with } l_1\text{-quadratic polynomials without real roots} \\
&\text{and } s^{\text{th}}\text{-quadratic polynomial with power } m_s, \\
&2n_1 - \Sigma_{s_1=1}^{q_1} n_{s_1}^{(1)} = 2\Sigma_{l=1}^{l_2} n_l \\
&\text{with } l_2\text{-quadraitc polynomials without real roots} \\
&\text{and } l^{\text{th}}\text{-quadratic polynomial with power } n_l.
\end{aligned}$$

Once

$$\Delta_{j_1 j_2}^{(s)} = 0\,(s = 1, 2, \cdots, l_1) \text{ and } \Delta_{j_2 j_1}^{(l)} = 0\,(l = 1, 2, \cdots, l_2),$$

if

$$a_{j_1 j_2 1} \equiv a_{j_1 j_2 r_1} = a_{j_1 j_2}^{(s)}\ (r_1 = 1, 2, \cdots, p_1; s = 1, 2, \cdots, l_1),$$
$$a_{j_2 j_1 1} \equiv a_{j_2 j_1 s_1} = a_{j_2 j_1}^{(l)}\ (s_1 = 1, 2, \cdots, q_1; l = 1, 2, \cdots, l_2),$$

the foregoing differential equations becomes a dynamical system with a singular equilibrium (B_2) as

$$\dot{x}_{j_1} = a_{j_1 j_2 0}(x_{j_1} - a_{j_1 j_2 1})^{2m_1},\ \dot{x}_{j_2} = a_{j_2 j_1 0}(x_{j_2} - a_{j_2 j_1 1})^{2n_1}.$$

If

$$\Delta_{j_1 j_2}^{(s)} = \Delta_{j_1 j_2} + \delta_s\ (s = 1, 2, \cdots, l_1),$$

$$\Delta_{j_2 j_1}^{(l)} = \Delta_{j_2 j_1} + \delta_l \ (l = 1, 2, \cdots, l_2),$$
$$\delta_s > 0 \ and \ \delta_l > 0,$$

then

$$\begin{aligned}
& a_{j_1 j_2 1}^{(s)}, a_{j_1 j_2 2}^{(s)} = a_{j_1 j_2 1} \pm \varepsilon_s, (s = 1, 2, \cdots, l_1), \\
& a_{j_2 j_1 1}^{(l)}, a_{j_2 j_1 2}^{(l)} = a_{j_2 j_1 1} \pm \varepsilon_l, (l = 1, 2, \cdots, l_2); \\
& \{a_{j_1 j_2}, a_{j_1 j_2 2}, \cdots, a_{j_1 j_2 p_2}\} \\
& = \mathrm{sort}\left\{a_{j_1 j_2 1}, a_{j_1 j_2 2}, \cdots, a_{j_1 j_2 p_1}; a_{j_1 j_2 1}^{(s)}, a_{j_1 j_2 2}^{(s)} \middle| 1, 2 \cdots, l_1\right\}, \\
& \{a_{j_2 j_1 1}, a_{j_2 j_1 1}, \cdots, a_{j_2 j_1 q_2}\} \\
& = \mathrm{sort}\left\{a_{j_1 j_2 1}, a_{j_1 j_2 2}, \cdots, a_{j_1 j_2 q_1}; a_{j_2 j_1 1}^{(l)}, a_{j_2 j_1 2}^{(l)} \middle| l = 1, 2 \cdots, l_2\right\}.
\end{aligned}$$

Thus, the differential equation becomes a dynamical system with singular equilibriums (C_2) as

$$\dot{x}_{j_1} = a_{j_1 j_1 0} \prod_{r_2=1}^{p_2} \left(x_{j_2} - a_{j_1 j_2 r_2}\right)^{m_{r_2}^{(2)}}, \ \dot{x}_{j_2} = a_{j_2 j_2 0} \prod_{s_2=1}^{q_2} \left(x_{j_1} - a_{j_2 j_1 s_2}\right)^{n_{s_2}^{(2)}};$$

where

$$\sum_{r_2=1}^{p_2} m_{r_2}^{(2)} = 2m_1, \ \sum_{s_2=1}^{q_2} n_{s_2}^{(2)} = 2n_1.$$

From the above bifurcation process analysis, at least, p_2-parameter variations in the x_{j_1}-direction and q_2-parameter variations in the x_{j_2}-direction are engaged in such a bifurcation from a $p_1 \times q_1$ equilibrium network to a $p_2 \times q_2$ equilibrium network through the $(2m_1)^{\text{th}}$-increasing-inflection and decreasing-inflection bifurcations in the x_{j_1}-direction and the $(2n_1)^{\text{th}}$-increasing-inflection and decreasing-inflection bifurcations in the x_{j_2}-direction.

Thus, the appearing or vanishing bifurcation route from a $p_1 \times q_1$ to a $p_2 \times q_2$ equilibrium network is expressed as

$$\bigcup_{r_1=1}^{p_1} \bigcup_{s_1=1}^{q_1} \underbrace{(a_{j_1 j_2 r_1}, a_{j_2 j_1 s_1})}_{(m_{r_1}^{(1)}, n_{s_1}^{(1)})\text{-XX}} \rightleftharpoons \underbrace{(a_{j_1 j_2 1}, a_{j_2 j_1 1})}_{((2m_1),(2n_1))\text{-XX}} \rightleftharpoons \bigcup_{r_2=1}^{p_2} \bigcup_{s_2=1}^{q_2} \underbrace{(a_{j_1 j_2 r_2}, a_{j_2 j_1 s_2})}_{(m_{r_2}^{(2)}, n_{s_2}^{(2)})\text{-XX}}.$$

From case (I), the singular equilibriums can be formed through nonsingular equilibrium networks. Thus,

$$\underbrace{(a_{j_1j_2r_1},a_{j_2j_1s_1})}_{(m_{r_1}^{(1)},n_{s_1}^{(1)})\text{-XX}} \rightleftharpoons \bigcup_{s=1}^{m_{r_1}^{(1)}}\bigcup_{l=1}^{n_{s_1}^{(1)}}\underbrace{(a_{j_1j_2s},a_{j_2j_1l})}_{\text{XX}}$$

$$\equiv\left\{\begin{array}{cccc} (a_{j_1j_2m_{r_1}^{(1)}},a_{j_2j_1 1}) & (a_{j_1j_2m_{r_1}^{(1)}},a_{j_2j_1 2}) & \cdots & (a_{j_1j_2m_{r_1}^{(1)}},a_{j_2j_1n_{s_1}^{(1)}}) \\ (a_{j_1j_2(m_{r_1}^{(1)}-1)},a_{j_2j_1 1}) & (a_{j_1j_2(m_{r_1}^{(1)}-1)},a_{j_2j_1 2}) & \cdots & (a_{j_1j_2(m_{r_1}^{(1)}-1)},a_{j_2j_1n_{s_1}^{(1)}}) \\ \vdots & \vdots & \cdots & \vdots \\ (a_{j_1j_2 1},a_{j_2j_1 1}) & (a_{j_1j_2 1},a_{j_2j_1 2}) & \cdots & (a_{j_1j_2 1},a_{j_2j_1n_{s_1}^{(1)}}) \end{array}\right\}_{m_{r_1}^{(1)}\times n_{s_1}^{(1)}},$$

and

$$\underbrace{(a_{j_1j_2r_2},a_{j_1j_1s_2})}_{(m_{r_2}^{(2)},n_{s_2}^{(2)})\text{-XX}} \rightleftharpoons \bigcup_{s=1}^{m_{r_2}^{(2)}}\bigcup_{l=1}^{n_{s_2}^{(2)}}\underbrace{(a_{j_1j_2s},a_{j_2j_1l})}_{\text{XX}}$$

$$\equiv\left\{\begin{array}{cccc} (a_{j_1j_2m_{r_2}^{(2)}},a_{j_2j_1 1}) & (a_{j_1j_2m_{r_2}^{(2)}},a_{j_2j_1 2}) & \cdots & (a_{j_1j_2m_{r_2}^{(2)}},a_{j_2j_1n_{s_2}^{(2)}}) \\ (a_{j_1j_2(m_{r_2}^{(2)}-1)},a_{j_2j_1 1}) & (a_{j_1j_2(m_{r_2}^{(2)}-1)},a_{j_2j_1 2}) & \cdots & (a_{j_1j_2(m_{r_2}^{(2)}-1)},a_{j_2j_1n_{s_2}^{(2)}}) \\ \vdots & \vdots & \cdots & \vdots \\ (a_{j_1j_2 1},a_{j_2j_1 1}) & (a_{j_1j_2 1},a_{j_2j_1 2}) & \cdots & (a_{j_1j_2 1},a_{j_2j_1n_{s_2}^{(2)}}) \end{array}\right\}_{m_{r_2}^{(2)}\times n_{s_2}^{(2)}}.$$

From the above definition, the corner singular equilibriums for $i = 1, 2$ are determined by

$$\underbrace{((m_{p_i}^{(i)})^{\text{th}}\text{XX},(n_{q_i}^{(i)})^{\text{th}}\text{XX})}_{(m_{p_i}^{(i)},n_{q_i}^{(i)})\text{-XX}} \rightleftharpoons \bigcup_{s=1}^{m_{p_i}^{(i)}}\bigcup_{l=1}^{n_{q_i}^{(i)}}\underbrace{(a_{j_1j_2s},a_{j_2j_1l})}_{\text{XX}},$$

$$\underbrace{((m_{p_i}^{(i)})^{\text{th}}\text{XX},(n_{1}^{(i)})^{\text{th}}\text{XX})}_{(m_{p_i}^{(i)},n_{1}^{(i)})\text{-XX}} \rightleftharpoons \bigcup_{s=1}^{m_{p_i}^{(i)}}\bigcup_{l=1}^{n_{1}^{(i)}}\underbrace{(a_{j_1j_2s},a_{j_2j_1l})}_{\text{XX}},$$

$$\underbrace{((m_{1}^{(i)})^{\text{th}}\text{XX},(n_{q_i1}^{(i)})^{\text{th}}\text{XX})}_{(m_{1}^{(i)},n_{q_i}^{(i)})\text{-XX}} \rightleftharpoons \bigcup_{s=1}^{m_{1}^{(i)}}\bigcup_{l=1}^{n_{q_i}^{(i)}}\underbrace{(a_{j_1j_2s},a_{j_2j_1l})}_{\text{XX}},$$

$$\underbrace{((m_{1}^{(i)})^{\text{th}}\text{XX},(n_{1}^{(i)})^{\text{th}}\text{XX})}_{(m_{1}^{(i)},n_{1}^{(i)})\text{-XX}} \rightleftharpoons \bigcup_{s=1}^{m_{1}^{(i)}}\bigcup_{l=1}^{n_{1}^{(i)}}\underbrace{(a_{j_1j_2s},a_{j_2j_1l})}_{\text{XX}}.$$

Case III: Consider two dynamical systems with the same equilibriums with locations switched (A_3,C_3) for $i = 1, 2$ as

$$\dot{x}_{j_1} = a_{j_1 j_2 0} \prod_{r_i=1}^{p} (x_{j_2} - a_{j_1 j_2 r_i})^{m_{r_i}^{(i)}}, \ \dot{x}_{j_2} = a_{j_2 j_1 0} \prod_{s_i=1}^{q} (x_{j_1} - a_{j_2 j_1 s_i})^{n_{s_i}^{(i)}};$$

where

$$\sum_{r_i=1}^{p} m_{r_i}^{(i)} = 2m_1, \ \sum_{s_i=1}^{q} n_{s_i}^{(i)} = 2n_1.$$

Consider a dynamical system as a singular equilibrium (B_3) as

$$\dot{x}_{j_1} = a_{j_1 j_2 0}(x_{j_2} - a_{j_1 j_2 1})^{2m_1}, \ \dot{x}_{j_2} = a_{j_2 j_1 0}(x_{j_1} - a_{j_2 j_1 1})^{2n_1}.$$

For $r_i = 1, 2, \cdots, p$ and $s_i = 1, 2, \cdots, q\,(i = 1, 2)$, define two functions as

$$\begin{aligned}\Delta_{j_1 j_2}^{(r_1 r_2)} &= (a_{j_1 j_2 r_1} - a_{j_1 j_2 r_2})^2 \ (r_1, r_2 \in \{1, 2, \cdots, p\}, r_1 \neq r_2),\\ \Delta_{j_2 j_1}^{(s_1 s_2)} &= (a_{j_2 j_1 s_1} - a_{j_2 j_1 s_2})^2 \ (s_1, s_2 \in \{1, 2, \cdots, q\}, s_1 \neq s_2).\end{aligned}$$

For $r_i = 1, 2, \cdots, p$ and $s_i = 1, 2, \cdots, q\,(i = 1, 2)$, if

$$\begin{aligned}\Delta_{j_1 j_2}^{(r_1 r_2)} &= (a_{j_1 j_2 r_1} - a_{j_1 j_2 r_2})^2 = 0,\\ \Delta_{j_2 j_1}^{(s_1 s_2)} &= (a_{j_2 j_1 s_1} - a_{j_2 j_1 s_2})^2 = 0,\end{aligned}$$

two equilibriums of $(a_{j_1 j_1 r_1}, a_{j_2 j_2 s_1})$ and $(a_{j_1 j_1 r_2}, a_{j_2 j_2 s_2})$ switching at point $(a_{j_1 j_1 1}, a_{j_2 j_2 1})$ with the same order singularity are given through

$$\begin{aligned}&\underbrace{(a_{j_1 j_2 r_1}, a_{j_2 j_1 s_1})}_{(m_{r_1}^{(1)}, n_{s_1}^{(1)})\text{-XX}} \rightleftharpoons \underbrace{(a_{j_1 j_2 1}, a_{j_2 j_1 1})}_{((2m_1),(2n_1))\text{-XX}} \rightleftharpoons \underbrace{(a_{j_1 j_2 r_2}, a_{j_2 j_1 s_2})}_{(m_{r_2}^{(2)}, n_{s_2}^{(2)})\text{-XX}},\\ &\left\{a_{j_1 j_2 r_1} \rightleftharpoons a_{j_1 j_2 r_2} \middle| m_{r_1}^{(1)} = m_{r_2}^{(2)}; r_1, r_2 \in \{1, 2, \cdots, p\}, r_1 \neq r_2\right\},\\ &\left\{a_{j_2 j_1 s_1} \rightleftharpoons a_{j_2 j_1 s_2} \middle| n_{s_1}^{(1)} = n_{s_2}^{(2)}; s_1, s_2 \in \{1, 2, \cdots, q\}, s_1 \neq s_2\right\},\end{aligned}$$

and

$$\begin{aligned}&a_{j_1 j_2 r_1} = a_{j_1 j_2 1} - \varepsilon_{r_1}, a_{j_1 j_2 r_2} = a_{j_1 j_2 1} + \varepsilon_{r_2},\\ &(a_{j_1 j_2 r_1} = a_{j_1 j_2 1} + \varepsilon_{r_1}, a_{j_1 j_2 r_2} = a_{j_1 j_2 1} - \varepsilon_{r_2});\\ &a_{j_2 j_1 s_1} = a_{j_2 j_1 1} - \varepsilon_{s_1}, a_{j_2 j_1 s_2} = a_{j_2 j_1 1} + \varepsilon_{s_2},\\ &(a_{j_2 j_1 s_1} = a_{j_2 j_1 1} + \varepsilon_{s_1}, a_{j_2 j_1 s_2} = a_{j_2 j_1 1} - \varepsilon_{s_2})\\ &\varepsilon_{r_1}, \varepsilon_{r_2} > 0 \text{ and } \varepsilon_{s_1}, \varepsilon_{s_2} > 0;\\ &r_i = 1, 2, \cdots, p; s_i = 1, 2, \cdots, q\,(i = 1, 2).\end{aligned}$$

From case (I), the singular equilibriums can be formed through nonsingular equilibrium networks as presented in Case (II), and the corner singular equilibriums for $i = 1, 2$ are determined similarly.

In summary, from the cases (I)-(III), the equilibrium of $(x^*_{j_2}, x^*_{j_1}) = (a_{j_1j_21}, a_{j_2j_11})$ for $m = 2m_1$ and $n = 2n_1$ has the bifurcation properties as stated in (vii$_1$)-(vii$_4$) through Eqs. (5.119)–(5.142).

In the end, Theorem 5.1 is proved.

5.3 Bifurcations to Homoclinic Networks with Centers

As in [1], four types of bifurcations are presented for non-singular equilibriums.

5.3.1 $((2m_1 + 1), (2n_1 + 1))$*-Saddle and Center Bifurcations*

Consider a 2-dimensional singular system of the $(2m_1 + 1) \times (2n_1 + 1)$-type as

$$\dot{x}_1 = a_{110}(x_2 - a_{121})^{2m_1+1}, \dot{x}_2 = a_{220}(x_1 - a_{211})^{2n_1+1}. \tag{5.143}$$

The first integral manifold is

$$\begin{aligned}&\frac{1}{2m_1 + 2}\left[(x_2 - a_{121})^{2m_1+2} - (x_{20} - a_{121})^{2m_1+2}\right]\\&= \frac{1}{2n_1 + 2}\frac{a_{210}}{a_{120}}\left[(x_1 - a_{211})^{2n_1+2} - (x_{10} - a_{211})^{2n_1+2}\right].\end{aligned} \tag{5.144}$$

Phase portraits of the singular equilibriums for the 2-dimensional singular system of the $((2m_1+1), (2n_1+1))$-type are presented in Fig. 5.1a–d for $(a_{120} > 0, a_{210} > 0)$, $(a_{120} < 0, a_{210} > 0)$, $(a_{120} > 0, a_{210} < 0)$ and $(a_{120} < 0, a_{210} < 0)$.

$$\left.\begin{aligned}&(a_{121}, a_{211}) = \underbrace{((2m_1+1)^{\text{th}}\text{UP}_+, (2n_1+1)^{\text{th}}\text{UP}_+)}_{((2m_1+1),(2n_1+1))\text{-positive saddle}} \text{ for } a_{120} > 0 \text{ and } a_{210} > 0,\\&(a_{121}, a_{211}) = \underbrace{((2m_1+1)^{\text{th}}\text{DP}_+, (2n_1+1)^{\text{th}}\text{DP}_-)}_{((2m_1+1),(2n_1+1))\text{-CCW center}} \text{ for } a_{120} < 0 \text{ and } a_{210} > 0,\\&(a_{121}, a_{211}) = \underbrace{((2m_1+1)^{\text{th}}\text{DP}_-, (2n_1+1)^{\text{th}}\text{DP}_+)}_{((2m_1+1),(2n_1+1))\text{-CW center}} \text{ for } a_{120} > 0 \text{ and } a_{210} < 0,\\&(a_{121}, a_{211}) = \underbrace{((2m_1+1)^{\text{th}}\text{UP}_-, (2n_1+1)^{\text{th}}\text{UP}_-)}_{((2m_1+1),(2n_1+1))\text{-negative saddle}} \text{ for } a_{120} < 0 \text{ and } a_{210} < 0.\end{aligned}\right\} \tag{5.145}$$

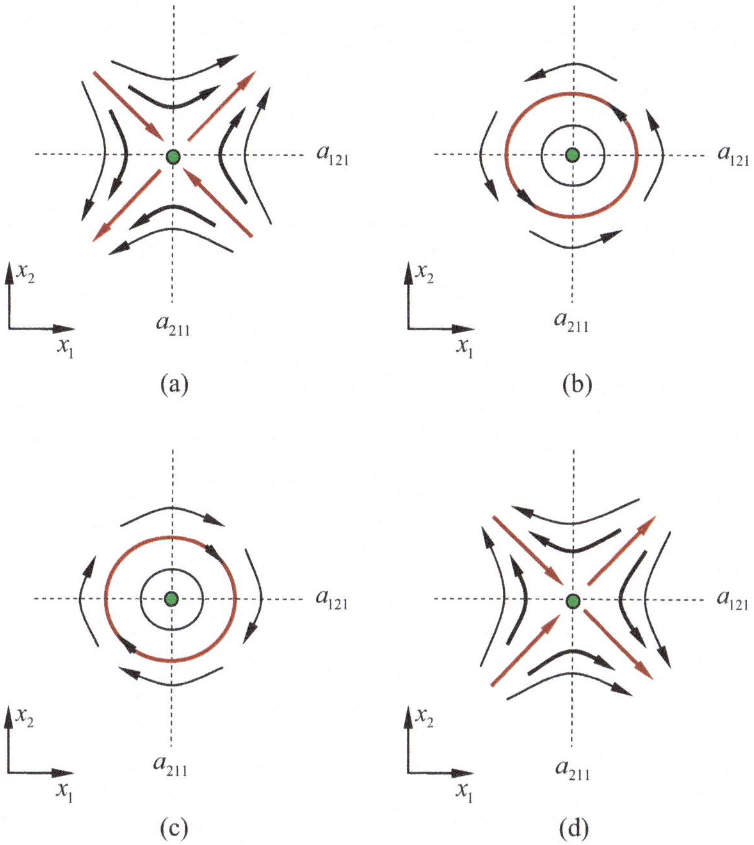

Fig. 5.1 Phase portraits of the $((2m_1+1), (2n_1+1))$-saddles and centers for 2-dimensional systems at $(x_1^*, x_2^*) = (a_{111}, a_{221})$. **a** $(a_{120} > 0, a_{210} > 0)$, **b** $(a_{120} < 0, a_{210} > 0)$, **c** $(a_{120} > 0, a_{210} < 0)$, **d** $(a_{120} < 0, a_{210} < 0)$

From the above bifurcations, a polynomial system with $(2m_1 + 1) \times (2n_1 + 1)$ -non-singular equilibriums exists as

$$\dot{x}_1 = a_{120} \prod\nolimits_{s=1}^{2m_1+1} (x_2 - a_{12s}) \text{ and } \dot{x}_2 = a_{210} \prod\nolimits_{l=1}^{2n_1+1} (x_1 - a_{21l}). \tag{5.146}$$

The first integral manifold is given by

$$\begin{aligned} a_{210}\Big\{ & \frac{1}{2n_1+2}\big[(x_1 - a_{21l_1})^{2n_1+2} - (x_{10} - a_{21l_1})^{2n_1+2}\big] \\ & + \sum_{k=1}^{2n_1} \frac{1}{2n_1 - k + 2} b_{21k}\big[(x_1 - a_{21l_1})^{2n_1-k+2} - (x_{10} - a_{21l_1})^{2n_1-k+2}\big]\Big\} \end{aligned}$$

$$= a_{j_1j_20}\Big\{\frac{1}{2m_1+2}\big[(x_2-a_{12s_1})^{2m_1+2}-(x_{20}-a_{12s_1})^{2m_1+2}\big]$$

$$+\sum_{k=1}^{2m_1}\frac{1}{2m_1-k+2}b_{12k}\big[(x_1-a_{21s_1})^{2m_1-k+2}-(x_{10}-a_{21s_1})^{2m_1-k+2}\big]\Big\} \quad (5.147)$$

where

$$\begin{aligned}
b_{211} &= \sum_{l_2=1,l_2\neq l_1}^{2n_1+1}(a_{21l_1}-a_{21l_2}),\\
b_{212} &= \sum_{\substack{l_2,l_3=1;l_2,l_3\neq l_1\\(l_2<l_3)}}^{2n_1+1}\prod_{r=2}^{3}(a_{21l_1}-a_{21l_r}),\cdots,\\
b_{21k} &= \sum_{\substack{l_2,l_3,\cdots,l_{k+1}=1;\\ l_2,l_3,\cdots,l_{k+1}\neq l_1\\(l_2<l_3<\cdots<l_{k+1})}}^{2n_1+1}\prod_{r=2}^{k+1}(a_{21l_1}-a_{21l_r}),\cdots,\\
b_{21(2n_1)} &= \prod_{l_2=1,l_2\neq l_1}^{2n_1}(a_{2l_1}-a_{j_2j_1l_2});\\
b_{121} &= \sum_{s_2=1,s_2\neq s_1}^{2m_1+1}(a_{12s_1}-a_{12s_2}),\\
b_{122} &= \sum_{\substack{s_2,s_3=1;s_2,s_3\neq s_1\\(s_2<s_3)}}^{2m_1+1}\prod_{r=2}^{3}(a_{12s_1}-a_{12s_r}),\cdots,\\
b_{12k} &= \sum_{\substack{s_2,s_3,\cdots,s_{k+1}=1;\\ s_2,s_3,\cdots,s_{k+1}\neq s_1\\(s_2<s_3<\cdots<s_{k+1})}}^{2m_1+1}\prod_{r=2}^{k+1}(a_{12s_1}-a_{12s_r}),\cdots,\\
b_{12(2m_1)} &= \prod_{s_1=1,s_2\neq s_1}^{2m_1}(a_{12s_1}-a_{12s_2}).
\end{aligned} \quad (5.148)$$

Phase portraits for the 2-dimensional systems near the simple equilibriums of the saddles and centers are presented in Fig. 5.2a–d for ($a_{120}>0, a_{210}>0$), ($a_{120}<0, a_{210}>0$), ($a_{120}>0, a_{210}<0$) and ($a_{120}<0, a_{210}<0$). For all cases, the $(2m_1+1)\times(2n_1+1)$-simple equilibriums are based on the $((2m_1+1),(2n_1+1))$-saddle and center bifurcations.

$$\bigcup_{s=1}^{2m_1+1}\bigcup_{l=1}^{2n_1+1}\underbrace{(a_{j_1j_2s},a_{j_2j_1l})}_{\text{XX}}$$

$$= \left\{ \begin{array}{cccc} \underbrace{(\mathrm{UP}_+,\mathrm{UP}_+)}_{\text{positive saddle}} & \cdots & \underbrace{(\mathrm{DP}_-,\mathrm{DP}_+)}_{\text{CW center}} & \underbrace{(\mathrm{UP}_+,\mathrm{UP}_+)}_{\text{positive saddle}} \\ \underbrace{(\mathrm{DP}_+,\mathrm{DP}_-)}_{\text{CCW center}} & \cdots & \underbrace{(\mathrm{UP}_-,\mathrm{UP}_-)}_{\text{negative saddle}} & \underbrace{(\mathrm{DP}_+,\mathrm{DP}_-)}_{\text{CCW center}} \\ \vdots & \vdots & \vdots & \vdots \\ \underbrace{(\mathrm{UP}_+,\mathrm{UP}_+)}_{\text{positive saddle}} & \cdots & \underbrace{(\mathrm{DP}_-,\mathrm{DP}_+)}_{\text{CW center}} & \underbrace{(\mathrm{UP}_+,\mathrm{UP}_+)}_{\text{positive saddle}} \end{array} \right\}_{(2m_1+1)\times(2n_1+1)}$$

for $a_{120} > 0, a_{210} > 0;$ (5.149)

$$\bigcup_{s=1}^{2m_1+1} \bigcup_{l=1}^{2n_1+1} \underbrace{(a_{j_1j_2s}, a_{j_2j_1l})}_{\mathrm{XX}}$$

$$= \left\{ \begin{array}{cccc} \underbrace{(\mathrm{DP}_+,\mathrm{DP}_-)}_{\text{CCW center}} & \cdots & \underbrace{(\mathrm{UP}_-,\mathrm{UP}_-)}_{\text{negative saddle}} & \underbrace{(\mathrm{DP}_+,\mathrm{DP}_-)}_{\text{CCW center}} \\ \underbrace{(\mathrm{UP}_+,\mathrm{UP}_+)}_{\text{positive saddle}} & \cdots & \underbrace{(\mathrm{DP}_-,\mathrm{DP}_+)}_{\text{CW center}} & \underbrace{(\mathrm{UP}_+,\mathrm{UP}_+)}_{\text{positive saddle}} \\ \vdots & \vdots & \vdots & \vdots \\ \underbrace{(\mathrm{DP}_+,\mathrm{DP}_-)}_{\text{CCW center}} & \cdots & \underbrace{(\mathrm{UP}_-,\mathrm{UP}_-)}_{\text{negative saddle}} & \underbrace{(\mathrm{DP}_+,\mathrm{DP}_-)}_{\text{CCW center}} \end{array} \right\}_{(2m_1+1)\times(2n_1+1)}$$

for $a_{120} < 0, a_{210} > 0;$ (5.150)

$$\bigcup_{s=1}^{2m_1+1} \bigcup_{l=1}^{2n_1+1} \underbrace{(a_{j_1j_2s}, a_{j_2j_1l})}_{\mathrm{XX}}$$

$$= \left\{ \begin{array}{cccc} \underbrace{(\mathrm{DP}_-,\mathrm{DP}_+)}_{\text{CW center}} & \cdots & \underbrace{(\mathrm{UP}_+,\mathrm{UP}_+)}_{\text{positive saddle}} & \underbrace{(\mathrm{DP}_-,\mathrm{DP}_+)}_{\text{CW center}} \\ \underbrace{(\mathrm{UP}_-,\mathrm{UP}_-)}_{\text{negative saddle}} & \cdots & \underbrace{(\mathrm{DP}_+,\mathrm{DP}_-)}_{\text{CCW center}} & \underbrace{(\mathrm{UP}_-,\mathrm{UP}_-)}_{\text{negative saddle}} \\ \vdots & \vdots & \vdots & \vdots \\ \underbrace{(\mathrm{DP}_-,\mathrm{DP}_+)}_{\text{CW center}} & \cdots & \underbrace{(\mathrm{UP}_+,\mathrm{UP}_+)}_{\text{positive saddle}} & \underbrace{(\mathrm{DP}_-,\mathrm{DP}_+)}_{\text{CW center}} \end{array} \right\}_{(2m_1+1)\times(2n_1+1)}$$

for $a_{120} > 0, a_{210} < 0;$ (5.151)

$$\bigcup_{s=1}^{2m_1+1} \bigcup_{l=1}^{2n_1+1} \underbrace{(a_{j_1j_2s}, a_{j_2j_1l})}_{\mathrm{XX}}$$

$$
= \left\{ \begin{array}{cccc}
\underbrace{(\mathrm{UP}_-,\mathrm{UP}_-)}_{\text{negative saddle}} & \cdots & \underbrace{(\mathrm{DP}_+,\mathrm{DP}_-)}_{\text{CCW center}} & \underbrace{(\mathrm{UP}_-,\mathrm{UP}_-)}_{\text{negative saddle}} \\
\underbrace{(\mathrm{DP}_-,\mathrm{DP}_+)}_{\text{CW center}} & \cdots & \underbrace{(\mathrm{UP}_+,\mathrm{UP}_+)}_{\text{positive saddle}} & \underbrace{(\mathrm{DP}_-,\mathrm{DP}_+)}_{\text{CW center}} \\
\vdots & \vdots & \vdots & \vdots \\
\underbrace{(\mathrm{UP}_-,\mathrm{UP}_-)}_{\text{negative saddle}} & \cdots & \underbrace{(\mathrm{DP}_+,\mathrm{DP}_-)}_{\text{CCW center}} & \underbrace{(\mathrm{UP}_-,\mathrm{UP}_-)}_{\text{negative saddle}}
\end{array} \right\}_{(2m_1+1)\times(2n_1+1)}
$$

for $a_{120} < 0, a_{210} < 0$. (5.152)

5.3.2 $((2m_1), (2n_1+1))$-*Parabola-Saddle Bifurcations*

Consider a 2-dimensional singular system of the $(2m_1) \times (2n_1+1)$-type as

$$
\dot{x}_1 = a_{110}(x_2 - a_{121})^{2m_1+1}, \dot{x}_2 = a_{220}(x_1 - a_{211})^{2n_1+1}. \tag{5.153}
$$

The first integral manifold is

$$
\begin{aligned}
&\frac{1}{2m_1+1}\left[(x_2 - a_{121})^{2m_1+1} - (x_{20} - a_{121})^{2m_1+1}\right] \\
&= \frac{1}{2n_1+2}\frac{a_{210}}{a_{120}}\left[(x_1 - a_{211})^{2n_1+2} - (x_{10} - a_{211})^{2n_1+2}\right].
\end{aligned} \tag{5.154}
$$

Phase portraits of the singular equilibriums for the 2-dimensional singular system of the $((2m_1), (2n_1+1))$-type are presented in Fig. 5.3a–d for $(a_{120} > 0, a_{210} > 0)$, $(a_{120} < 0, a_{210} > 0)$, $(a_{120} > 0, a_{210} < 0)$ and $(a_{120} < 0, a_{210} < 0)$.

$$
\begin{aligned}
(a_{121}, a_{211}) &= \underbrace{((2m_1)^{\text{th}}\text{US},(2n_1+1)^{\text{th}}\text{UP})}_{((2m_1),(2n_1+1))\text{-up-parabola upper-saddle}} \quad \text{for } a_{120} > 0 \text{ and } a_{210} > 0, \\
(a_{121}, a_{211}) &= \underbrace{((2m_1)^{\text{th}}\text{LS},(2n_1+1)^{\text{th}}\text{DP})}_{((2m_1),(2n_1+1))\text{-down-parabola lower-saddle}} \quad \text{for } a_{120} < 0 \text{ and } a_{210} > 0, \\
(a_{121}, a_{211}) &= \underbrace{((2m_1)^{\text{th}}\text{US},(2n_1+1)^{\text{th}}\text{DP})}_{((2m_1),(2n_1+1))\text{-down-parabola upper-saddle}} \quad \text{for } a_{120} > 0 \text{ and } a_{210} < 0, \\
(a_{121}, a_{211}) &= \underbrace{((2m_1)^{\text{th}}\text{LS},(2n_1+1)^{\text{th}}\text{UP})}_{((2m_1),(2n_1+1))\text{-down-parabola lower-saddle}} \quad \text{for } a_{120} < 0 \text{ and } a_{210} < 0.
\end{aligned} \tag{5.155}
$$

From the above bifurcations, a polynomial system with $(2m_1) \times (2n_1+1)$- non-singular equilibriums exists as

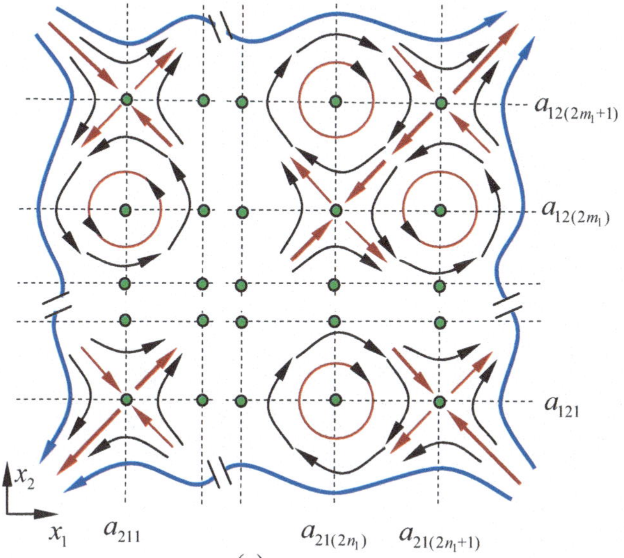

(a)

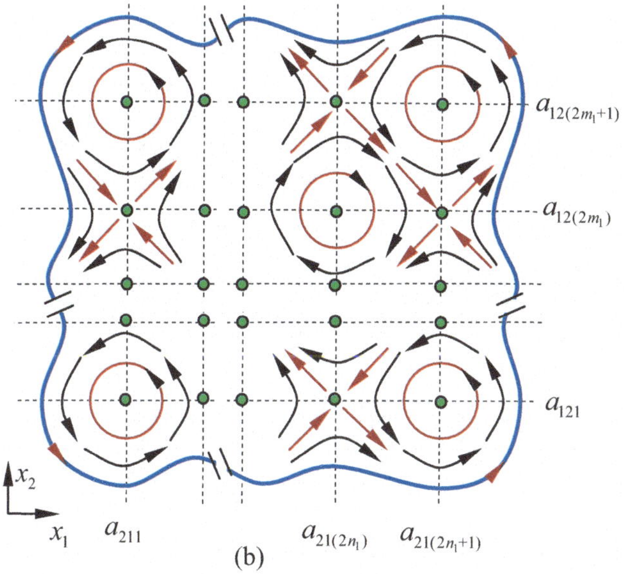

(b)

Fig. 5.2 Phase portraits for 2-dimensional systems with $x_1^* = a_{211}, a_{212}, \cdots, a_{21(2m_1+1)}$ and $x_2^* = a_{211}, a_{212}, \cdots, a_{21(2n_1+1)}$. The four networks of saddles and centers: **a** ($a_{120} > 0, a_{210} > 0$), **b** ($a_{120} < 0, a_{210} > 0$), **c** ($a_{120} > 0, a_{210} < 0$), **d** ($a_{120} < 0, a_{210} < 0$)

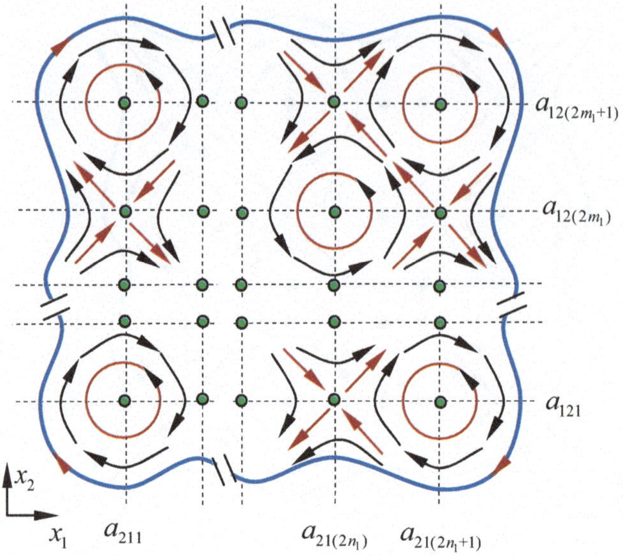

(c)

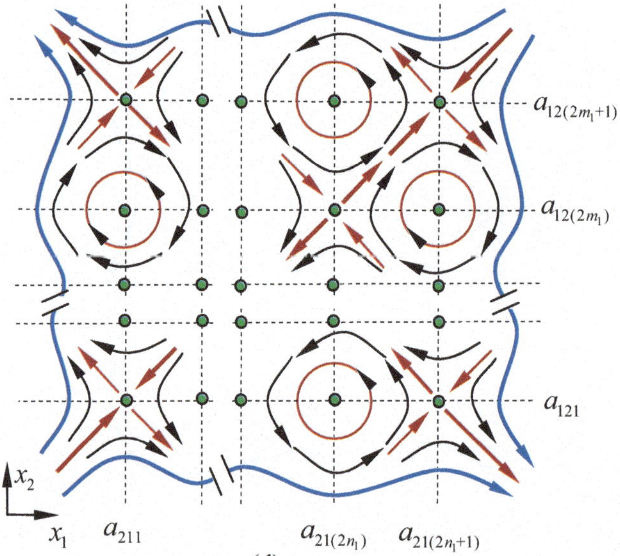

(d)

Fig. 5.2 (continued)

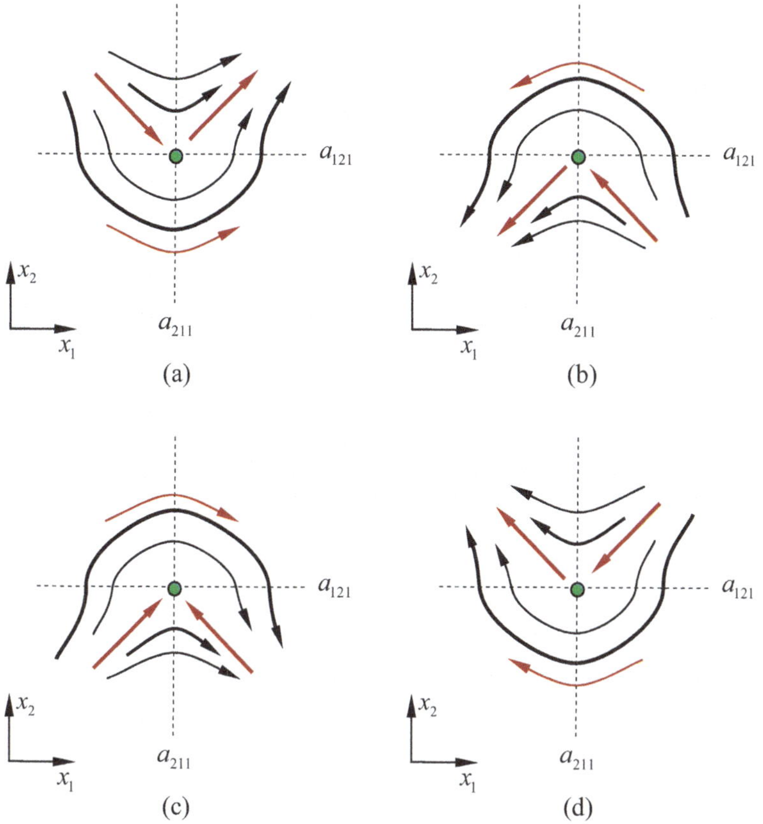

Fig. 5.3 Phase portraits of the $((2m_1), (2n_1+1))$-parabola-saddles for 2-dimensional systems at $(x_1^*, x_2^*) = (a_{121}, a_{211})$. **a** $(a_{120} > 0, a_{210} > 0)$, **b** $(a_{120} < 0, a_{210} > 0)$, **c** $(a_{120} > 0, a_{210} < 0)$, **d** $(a_{120} < 0, a_{210} < 0)$

$$\dot{x}_1 = a_{120} \prod\nolimits_{s=1}^{2m_1} (x_2 - a_{12s}) \text{ and } \dot{x}_2 = a_{210} \prod\nolimits_{l=1}^{2n_1+1} (x_1 - a_{21l}). \tag{5.156}$$

The first integral manifold is given by

$$\begin{aligned} & a_{210}\left\{ \frac{1}{2n_1+2}\left[(x_1 - a_{21l_1})^{2n_1+2} - (x_{10} - a_{21l_1})^{2n_1+2}\right] \right. \\ & \left. + \sum_{k=1}^{2n_1} \frac{1}{2n_1-k+2} b_{21k}\left[(x_1 - a_{21l_1})^{2n_1-k+2} - (x_{10} - a_{21l_1})^{2n_1-k+2}\right]\right\} \\ & = a_{j_1j_20}\left\{ \frac{1}{2m_1+1}\left[(x_2 - a_{12s_1})^{2m_1+1} - (x_{20} - a_{12s_1})^{2m_1+1}\right] \right. \end{aligned}$$

$$+\sum_{k=1}^{2m_1-1}\frac{1}{2m_1-k+1}b_{12k}\left[(x_1-a_{21s_1})^{2m_1-k+1}-(x_{10}-a_{21s_1})^{2m_1-k+1}\right]\Big\} \tag{5.157}$$

where

$$\begin{aligned}
b_{211} &= \sum_{l_2=1,l_2\neq l_1}^{2n_1+1}(a_{21l_1}-a_{21l_2}),\\
b_{212} &= \sum_{\substack{l_2,l_3=1;l_2,l_3\neq l_1\\(l_2<l_3)}}^{2n_1+1}\prod_{r=2}^{3}(a_{21l_1}-a_{21l_r}),\cdots,\\
b_{21k} &= \sum_{\substack{l_2,l_3,\cdots,l_{k+1}=1;\\ l_2,l_3,\cdots,l_{k+1}\neq l_1\\(l_2<l_3<\cdots<l_{k+1})}}^{2n_1+1}\prod\nolimits_{r=2}^{k+1}(a_{21l_1}-a_{21l_r}),\cdots,\\
b_{21(2n_1)} &= \prod_{l_2=1,l_2\neq l_1}^{2n_1}(a_{2l_1}-a_{j_2j_1l_2});\\
b_{121} &= \sum_{s_2=1,s_2\neq s_1}^{2m_1}(a_{12s_1}-a_{12s_2}),\\
b_{122} &= \sum_{\substack{s_2,s_3=1;s_2,s_3\neq s_1\\(s_2<s_3)}}^{2m_1}\prod_{r=2}^{3}(a_{12s_1}-a_{12s_r}),\cdots,\\
b_{12k} &= \sum_{\substack{s_2,s_3,\cdots,s_{k+1}=1;\\ s_2,s_3,\cdots,s_{k+1}\neq s_1\\(s_2<s_3<\cdots<s_{k+1})}}^{2m_1}\prod\nolimits_{r=2}^{k+1}(a_{12s_1}-a_{12s_r}),\cdots,\\
b_{12(2m_1)} &= \prod_{s_1=1,s_2\neq s_1}^{2m_1-1}(a_{12s_1}-a_{12s_2}).
\end{aligned} \tag{5.158}$$

Phase portraits for the 2-dimensional systems near the simple equilibriums of the saddles and centers are presented in Fig. 5.4a–d for $(a_{120}>0, a_{210}>0)$, $(a_{120}<0, a_{210}>0)$, $(a_{120}>0, a_{210}<0)$ and $(a_{120}<0, a_{210}<0)$. For all cases, the $((2m_1),(2n_1+1))$-simple equilibriums are based on the $((2m_1+1),(2n_1))$-parabola-saddle bifurcations.

$$\bigcup_{s=1}^{2m_1}\bigcup_{l=1}^{2n_1+1}\underbrace{(a_{j_1j_2s},a_{j_2j_1l})}_{\text{XX}}$$

$$= \left\{ \begin{matrix} \underbrace{(\mathrm{UP}_+,\mathrm{UP}_+)}_{\text{positive saddle}} & \cdots & \underbrace{(\mathrm{DP}_-,\mathrm{DP}_+)}_{\text{CW center}} & \underbrace{(\mathrm{UP}_+,\mathrm{UP}_+)}_{\text{positive saddle}} \\ \underbrace{(\mathrm{DP}_+,\mathrm{DP}_-)}_{\text{CCW center}} & \cdots & \underbrace{(\mathrm{UP}_-,\mathrm{UP}_-)}_{\text{negative saddle}} & \underbrace{(\mathrm{DP}_+,\mathrm{DP}_-)}_{\text{CCW center}} \\ \vdots & \vdots & \vdots & \vdots \\ \underbrace{(\mathrm{DP}_+,\mathrm{DP}_-)}_{\text{CCW center}} & \cdots & \underbrace{(\mathrm{UP}_-,\mathrm{UP}_-)}_{\text{negative saddle}} & \underbrace{(\mathrm{DP}_+,\mathrm{DP}_-)}_{\text{CCW center}} \end{matrix} \right\}_{(2m_1)\times(2n_1+1)}$$

$$\text{for } a_{120} > 0,\, a_{210} > 0; \tag{5.159}$$

$$\bigcup_{s=1}^{2m_1} \bigcup_{l=1}^{2n_1+1} \underbrace{(a_{j_1 j_2 s},\, a_{j_2 j_1 l})}_{\mathrm{XX}}$$

$$= \left\{ \begin{matrix} \underbrace{(\mathrm{DP}_+,\mathrm{DP}_-)}_{\text{CCW center}} & \cdots & \underbrace{(\mathrm{UP}_-,\mathrm{UP}_-)}_{\text{negative saddle}} & \underbrace{(\mathrm{DP}_+,\mathrm{DP}_-)}_{\text{CCW center}} \\ \underbrace{(\mathrm{UP}_+,\mathrm{UP}_+)}_{\text{positive saddle}} & \cdots & \underbrace{(\mathrm{DP}_-,\mathrm{DP}_+)}_{\text{CW center}} & \underbrace{(\mathrm{UP}_+,\mathrm{UP}_+)}_{\text{positive saddle}} \\ \vdots & \vdots & \vdots & \vdots \\ \underbrace{(\mathrm{UP}_+,\mathrm{UP}_+)}_{\text{positive saddle}} & \cdots & \underbrace{(\mathrm{DP}_-,\mathrm{DP}_+)}_{\text{CW center}} & \underbrace{(\mathrm{UP}_+,\mathrm{UP}_+)}_{\text{positive saddle}} \end{matrix} \right\}_{(2m_1)\times(2n_1+1)}$$

$$\text{for } a_{120} < 0,\, a_{210} > 0; \tag{5.160}$$

$$\bigcup_{s=1}^{2m_1} \bigcup_{l=1}^{2n_1+1} \underbrace{(a_{j_1 j_2 s},\, a_{j_2 j_1 l})}_{\mathrm{XX}}$$

$$= \left\{ \begin{matrix} \underbrace{(\mathrm{DP}_-,\mathrm{DP}_+)}_{\text{CW center}} & \cdots & \underbrace{(\mathrm{UP}_+,\mathrm{UP}_+)}_{\text{positive saddle}} & \underbrace{(\mathrm{DP}_-,\mathrm{DP}_+)}_{\text{CW center}} \\ \underbrace{(\mathrm{UP}_-,\mathrm{UP}_-)}_{\text{negative saddle}} & \cdots & \underbrace{(\mathrm{DP}_+,\mathrm{DP}_-)}_{\text{CCW center}} & \underbrace{(\mathrm{UP}_-,\mathrm{UP}_-)}_{\text{negative saddle}} \\ \vdots & \vdots & \vdots & \vdots \\ \underbrace{(\mathrm{UP}_-,\mathrm{UP}_-)}_{\text{negative saddle}} & \cdots & \underbrace{(\mathrm{DP}_+,\mathrm{DP}_-)}_{\text{CCW center}} & \underbrace{(\mathrm{UP}_-,\mathrm{UP}_-)}_{\text{negative saddle}} \end{matrix} \right\}_{(2m_1)\times(2n_1+1)}$$

$$\text{for } a_{120} > 0,\, a_{210} < 0; \tag{5.161}$$

$$\bigcup_{s=1}^{2m_1} \bigcup_{l=1}^{2n_1+1} \underbrace{(a_{j_1 j_2 s},\, a_{j_2 j_1 l})}_{\mathrm{XX}}$$

$$
= \left\{ \begin{array}{cccc}
\underbrace{(\mathrm{UP}_-,\mathrm{UP}_-)}_{\text{negative saddle}} & \cdots & \underbrace{(\mathrm{DP}_+,\mathrm{DP}_-)}_{\text{CCW center}} & \underbrace{(\mathrm{UP}_-,\mathrm{UP}_-)}_{\text{negative saddle}} \\
\underbrace{(\mathrm{DP}_-,\mathrm{DP}_+)}_{\text{CW center}} & \cdots & \underbrace{(\mathrm{UP}_+,\mathrm{UP}_+)}_{\text{positive saddle}} & \underbrace{(\mathrm{DP}_-,\mathrm{DP}_+)}_{\text{CW center}} \\
\vdots & \vdots & \vdots & \vdots \\
\underbrace{(\mathrm{DP}_-,\mathrm{DP}_+)}_{\text{CW center}} & \cdots & \underbrace{(\mathrm{UP}_+,\mathrm{UP}_+)}_{\text{positive saddle}} & \underbrace{(\mathrm{DP}_-,\mathrm{DP}_+)}_{\text{CW center}}
\end{array} \right\}_{(2m_1)\times(2n_1+1)}
$$

$$
\text{for } a_{120} < 0, a_{210} < 0. \tag{5.162}
$$

5.3.3 $((2m_1 + 1), (2n_1))$-Parabola-Saddle Bifurcations

Consider a 2-dimensional singular system of the $(2m_1 + 1) \times (2n_1)$-type as

$$
\dot{x}_1 = a_{110}(x_2 - a_{121})^{2m_1+1}, \dot{x}_2 = a_{220}(x_1 - a_{211})^{2n_1}. \tag{5.163}
$$

The first integral manifold is

$$
\begin{aligned}
&\frac{1}{2m_1 + 2}\left[(x_2 - a_{121})^{2m_1+2} - (x_{20} - a_{121})^{2m_1+2}\right] \\
&= \frac{1}{2n_1 + 1}\frac{a_{210}}{a_{120}}\left[(x_1 - a_{211})^{2n_1+1} - (x_{10} - a_{211})^{2n_1+1}\right].
\end{aligned} \tag{5.164}
$$

Phase portraits of the singular equilibriums for the 2-dimensional singular system of the $((2m_1 + 1), (2n_1))$-type are presented in Fig. 5.5a–d for $(a_{120} > 0, a_{210} > 0)$, $(a_{120} < 0, a_{210} > 0)$, $(a_{120} > 0, a_{210} < 0)$ and $(a_{120} < 0, a_{210} < 0)$.

$$
\begin{aligned}
(a_{121}, a_{211}) &= \underbrace{((2m_1{+}1)^{\text{th}}\mathrm{UP},(2n_1)^{\text{th}}\mathrm{US})}_{((2m_1+1),(2n_1))\text{-up-parabola upper-saddle}} \quad \text{for } a_{120} > 0 \text{ and } a_{210} > 0, \\
(a_{121}, a_{211}) &= \underbrace{((2m_1{+}1)^{\text{th}}\mathrm{DP},(2n_1)^{\text{th}}\mathrm{US})}_{((2m_1+1),(2n_1)\text{-down-parabola upper-saddle}} \quad \text{for } a_{120} < 0 \text{ and } a_{210} > 0, \\
(a_{121}, a_{211}) &= \underbrace{((2m_1{+}1)^{\text{th}}\mathrm{DP},(2n_1)^{\text{th}}\mathrm{LS})}_{((2m_1+1),(2n_1))\text{-down-parabola lower-saddle}} \quad \text{for } a_{120} > 0 \text{ and } a_{210} < 0, \\
(a_{121}, a_{211}) &= \underbrace{((2m_1{+}1)^{\text{th}}\mathrm{UP},(2n_1)^{\text{th}}\mathrm{LS})}_{((2m_1+1),(2n_1))\text{-up-parabola lower-saddle}} \quad \text{for } a_{120} < 0 \text{ and } a_{210} < 0.
\end{aligned} \tag{5.165}
$$

From the above bifurcations, a non-singular system with $(2m_1 + 1) \times (2n_1)$-equilibriums exists as

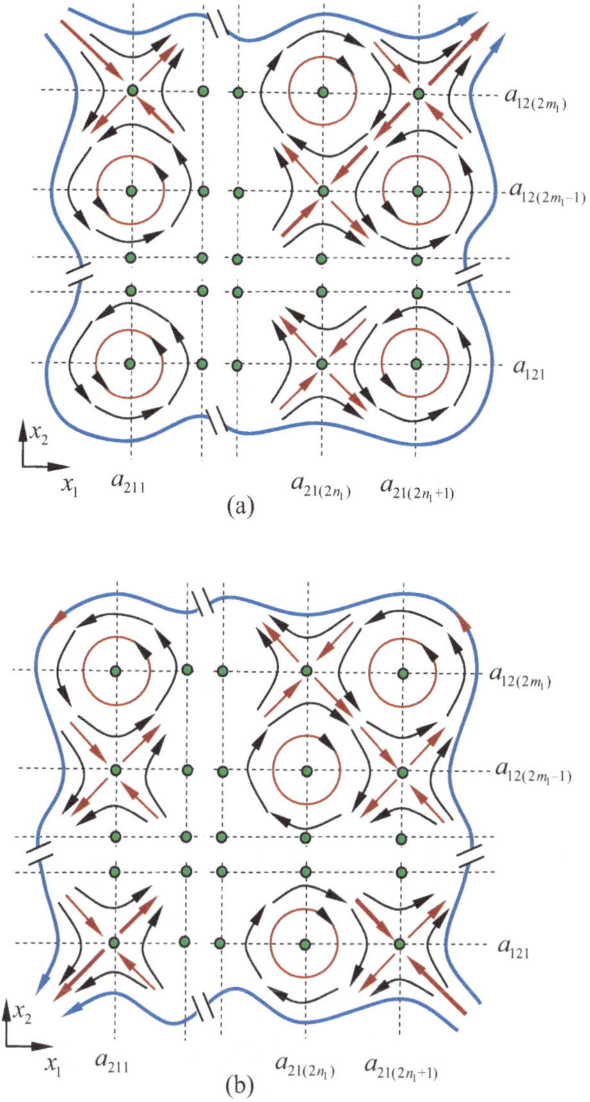

Fig. 5.4 Phase portraits for 2-dimensional systems with $x_1^* = a_{211}, a_{212}, \cdots, a_{21(2n_1+1)}$ and $x_2^* = a_{121}, a_{122}, \cdots, a_{12(2m_1)}$. The four networks of saddles and centers: **a** ($a_{120} > 0, a_{210} > 0$), **b** ($a_{120} < 0, a_{210} > 0$), **c** ($a_{120} > 0, a_{210} < 0$), **d** ($a_{120} < 0, a_{210} < 0$)

$$\dot{x}_1 = a_{120} \prod_{s=1}^{2m_1+1} (x_2 - a_{12s}) \text{ and } \dot{x}_2 = a_{210} \prod_{l=1}^{2n_1} (x_1 - a_{21l}). \tag{5.166}$$

The first integral manifold is given by

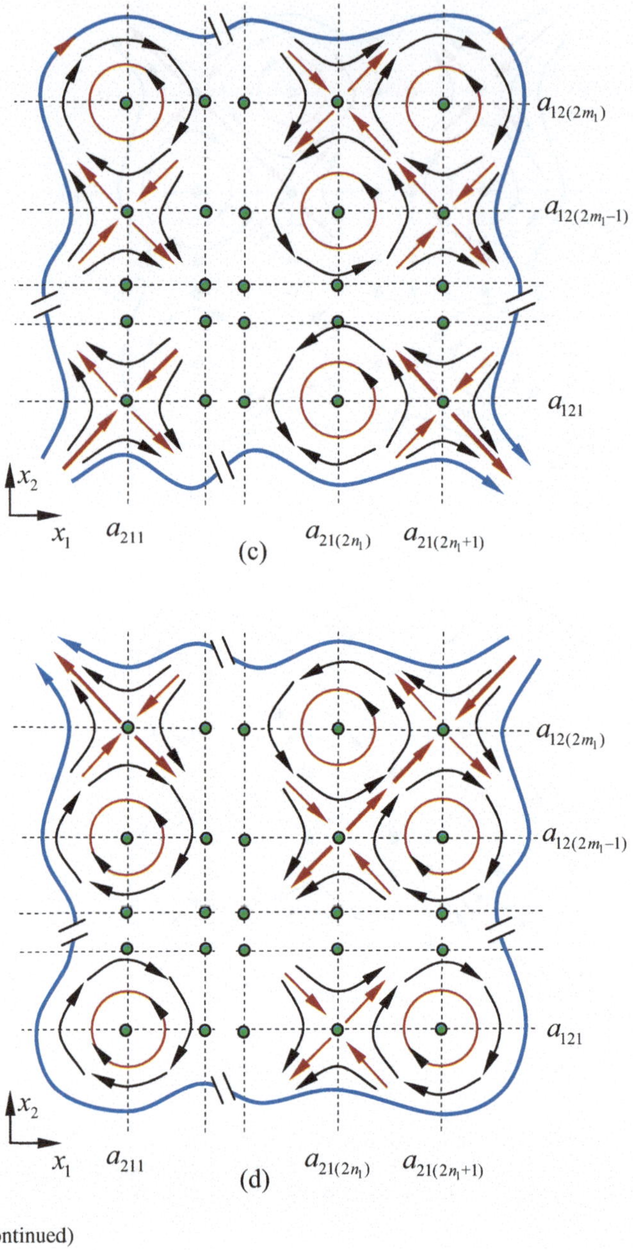

Fig. 5.4 (continued)

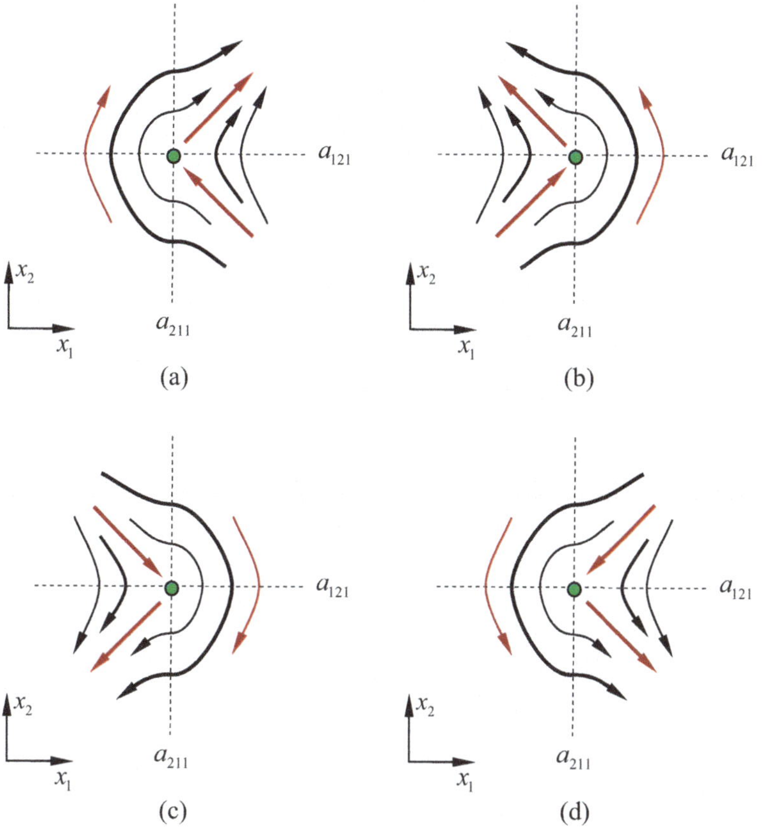

Fig. 5.5 Phase portraits of the $((2m_1+1),(2n_1))$-parabola-saddles for 2-dimensional systems at $(x_1^*, x_2^*) = (a_{111}, a_{221})$. **a** $(a_{120} > 0, a_{210} > 0)$, **b** $(a_{120} < 0, a_{210} > 0)$, **c** $(a_{120} > 0, a_{210} < 0)$, **d** $(a_{120} < 0, a_{210} < 0)$

$$
\begin{aligned}
& a_{210}\left\{\frac{1}{2n_1+1}\left[(x_1-a_{21l_1})^{2n_1+1}-(x_{10}-a_{21l_1})^{2n_1+1}\right]\right. \\
& \quad \left.+\sum_{k=1}^{2n_1-1}\frac{1}{2n_1-k+1}b_{21k}\left[(x_1-a_{21l_1})^{2n_1-k+1}-(x_{10}-a_{21l_1})^{2n_1-k+1}\right]\right\} \\
& = a_{j_1j_20}\left\{\frac{1}{2m_1+2}\left[(x_2-a_{12s_1})^{2m_1+2}-(x_{20}-a_{12s_1})^{2m_1+2}\right]\right. \\
& \quad \left.+\sum_{k=1}^{2m_1}\frac{1}{2m_1-k+2}b_{12k}\left[(x_1-a_{21s_1})^{2m_1-k+2}-(x_{10}-a_{21s_1})^{2m_1-k+2}\right]\right\}
\end{aligned}
\tag{5.167}
$$

where

$$
\begin{aligned}
b_{211} &= \sum_{l_2=1, l_2\neq l_1}^{2n_1} (a_{21l_1} - a_{21l_2}),\\
b_{212} &= \sum_{\substack{l_2,l_3=1; l_2,l_3\neq l_1\\ (l_2<l_3)}}^{2n_1} \prod_{r=2}^{3} (a_{21l_1} - a_{21l_r}), \cdots,\\
b_{21k} &= \sum_{\substack{l_2,l_3,\cdots,l_{k+1}=1;\\ l_2,l_3,\cdots,l_{k+1}\neq l_1\\ (l_2<l_3<\cdots<l_{k+1})}}^{2n_1} \prod_{r=2}^{k+1} (a_{21l_1} - a_{21l_r}), \cdots,\\
b_{21(2n_1)} &= \prod_{l_2=1, l_2\neq l_1}^{2n_1-1} (a_{2l_1} - a_{j_2 j_1 l_2});\\
b_{121} &= \sum_{s_2=1, s_2\neq s_1}^{2m_1+1} (a_{12s_1} - a_{12s_2}),\\
b_{122} &= \sum_{\substack{s_2,s_3=1; s_2,s_3\neq s_1\\ (s_2<s_3)}}^{2m_1+1} \prod_{r=2}^{3} (a_{12s_1} - a_{12s_r}), \cdots,\\
b_{12k} &= \sum_{\substack{s_2,s_3,\cdots,s_{k+1}=1;\\ s_2,s_3,\cdots,s_{k+1}\neq s_1\\ (s_2<s_3<\cdots<s_{k+1})}}^{2m_1+1} \prod_{r=2}^{k+1} (a_{12s_1} - a_{12s_r}), \cdots,\\
b_{12(2m_1)} &= \prod_{s_1=1, s_2\neq s_1}^{2m_1} (a_{12s_1} - a_{12s_2}).
\end{aligned}
\tag{5.168}
$$

Phase portraits for the 2-dimensional systems near the simple equilibriums of the saddles and centers are presented in Fig. 5.6a–d for ($a_{120} > 0, a_{210} > 0$), ($a_{120} < 0, a_{210} > 0$), ($a_{120} > 0, a_{210} < 0$) and ($a_{120} < 0, a_{210} < 0$). For all cases, the $((2m_1 + 1) \times (2n_1))$-simple equilibriums are based on the $((2m_1 + 1), (2n_1))$-parabola-saddle bifurcations.

$$\bigcup_{s=1}^{2m_1+1}\bigcup_{l=1}^{2n_1}\underbrace{(a_{j_1j_2s},a_{j_2j_1l})}_{\text{XX}}$$

$$=\left\{\begin{matrix}\underbrace{(\text{DP}_-,\text{DP}_+)}_{\text{CW center}} & \cdots & \underbrace{(\text{DP}_-,\text{DP}_+)}_{\text{CW center}} & \underbrace{(\text{UP}_+,\text{UP}_+)}_{\text{positive saddle}}\\ \underbrace{(\text{UP}_-,\text{UP}_-)}_{\text{negative saddle}} & \cdots & \underbrace{(\text{UP}_-,\text{UP}_-)}_{\text{negative saddle}} & \underbrace{(\text{DP}_+,\text{DP}_-)}_{\text{CCW center}}\\ \vdots & \vdots & \vdots & \vdots\\ \underbrace{(\text{DP}_-,\text{DP}_+)}_{\text{CW center}} & \cdots & \underbrace{(\text{DP}_-,\text{DP}_+)}_{\text{CW center}} & \underbrace{(\text{UP}_+,\text{UP}_+)}_{\text{positive saddle}}\end{matrix}\right\}_{(2m_1+1)\times(2n_1)}$$

$$\text{for } a_{120}>0,\ a_{210}>0; \tag{5.169}$$

$$\bigcup_{s=1}^{2m_1+1}\bigcup_{l=1}^{2n_1}\underbrace{(a_{j_1j_2s},a_{j_2j_1l})}_{\text{XX}}$$

$$=\left\{\begin{matrix}\underbrace{(\text{UP}_-,\text{UP}_-)}_{\text{negative saddle}} & \cdots & \underbrace{(\text{UP}_-,\text{UP}_-)}_{\text{negative saddle}} & \underbrace{(\text{DP}_+,\text{DP}_-)}_{\text{CCW center}}\\ \underbrace{(\text{DP}_-,\text{DP}_+)}_{\text{CW center}} & \cdots & \underbrace{(\text{DP}_-,\text{DP}_+)}_{\text{CW center}} & \underbrace{(\text{UP}_+,\text{UP}_+)}_{\text{positive saddle}}\\ \vdots & \vdots & \vdots & \vdots\\ \underbrace{(\text{UP}_-,\text{UP}_-)}_{\text{negative saddle}} & \cdots & \underbrace{(\text{UP}_-,\text{UP}_-)}_{\text{negative saddle}} & \underbrace{(\text{DP}_+,\text{DP}_-)}_{\text{CCW center}}\end{matrix}\right\}_{(2m_1+1)\times(2n_1)}$$

$$\text{for } a_{120}<0,\ a_{210}>0; \tag{5.170}$$

$$\bigcup_{s=1}^{2m_1+1}\bigcup_{l=1}^{2n_1}\underbrace{(a_{j_1j_2s},a_{j_2j_1l})}_{\text{XX}}$$

$$=\left\{\begin{matrix}\underbrace{(\text{UP}_+,\text{UP}_+)}_{\text{positive saddle}} & \cdots & \underbrace{(\text{UP}_+,\text{UP}_+)}_{\text{positive saddle}} & \underbrace{(\text{DP}_-,\text{DP}_+)}_{\text{CW center}}\\ \underbrace{(\text{DP}_+,\text{DP}_-)}_{\text{CCW center}} & \cdots & \underbrace{(\text{DP}_+,\text{DP}_-)}_{\text{CCW center}} & \underbrace{(\text{UP}_-,\text{UP}_-)}_{\text{negative saddle}}\\ \vdots & \vdots & \vdots & \vdots\\ \underbrace{(\text{UP}_+,\text{UP}_+)}_{\text{positive saddle}} & \cdots & \underbrace{(\text{UP}_+,\text{UP}_+)}_{\text{positive saddle}} & \underbrace{(\text{DP}_-,\text{DP}_+)}_{\text{CW center}}\end{matrix}\right\}_{(2m_1+1)\times(2n_1)}$$

$$\text{for } a_{120}>0,\ a_{210}<0; \tag{5.171}$$

$$
\bigcup_{s=1}^{2m_1+1}\bigcup_{l=1}^{2n_1}\underbrace{(a_{j_1j_2s},a_{j_2j_1l})}_{\text{XX}}
=\left\{\begin{matrix}
\underbrace{(\text{DP}_+,\text{DP}_-)}_{\text{CCW center}} & \cdots & \underbrace{(\text{DP}_+,\text{DP}_-)}_{\text{CCW center}} & \underbrace{(\text{UP}_-,\text{UP}_-)}_{\text{negative saddle}}\\
\underbrace{(\text{UP}_+,\text{UP}_+)}_{\text{positive saddle}} & \cdots & \underbrace{(\text{UP}_+,\text{UP}_+)}_{\text{positive saddle}} & \underbrace{(\text{DP}_-,\text{DP}_+)}_{\text{CW center}}\\
\vdots & \vdots & \vdots & \vdots\\
\underbrace{(\text{DP}_+,\text{DP}_-)}_{\text{CCW center}} & \cdots & \underbrace{(\text{DP}_+,\text{DP}_-)}_{\text{CCW center}} & \underbrace{(\text{UP}_-,\text{UP}_-)}_{\text{negative saddle}}
\end{matrix}\right\}_{(2m_1+1)\times(2n_1)}
$$
$$
\text{for } a_{120}<0,\ a_{210}<0. \tag{5.172}
$$

5.3.4 $((2m_1), (2n_1))$-*Double-Inflection-Saddle Bifurcations*

Consider a 2-dimensional singular system of the $(2m_1)\times(2n_1)$-type as

$$
\dot{x}_1 = a_{110}(x_2 - a_{121})^{2m_1},\ \dot{x}_2 = a_{220}(x_1 - a_{211})^{2n_1}. \tag{5.173}
$$

The first integral manifold is

$$
\begin{aligned}
&\frac{1}{2m_1+1}\left[(x_2-a_{121})^{2m_1+1}-(x_{20}-a_{121})^{2m_1+1}\right]\\
&=\frac{1}{2n_1+1}\frac{a_{210}}{a_{120}}\left[(x_1-a_{211})^{2n_1+1}-(x_{10}-a_{211})^{2n_1+1}\right].
\end{aligned} \tag{5.174}
$$

Phase portraits of the singular equilibriums for the 2-dimensional singular system of the $((2m_1),(2n_1))$-type are presented in Fig. 5.7a–d for $(a_{120}>0, a_{210}>0)$, $(a_{120}<0, a_{210}>0)$, $(a_{120}>0, a_{210}<0)$ and $(a_{120}<0, a_{210}<0)$.

$$
\begin{aligned}
(a_{121},a_{211}) &= \underbrace{((2m_1)^{\text{th}}\text{II}_+,(2n_1)^{\text{th}}\text{II}_+)}_{((2m_1),(2n_1))\text{-double-inflection saddle}} \quad \text{for } a_{120}>0 \text{ and } a_{210}>0,\\
(a_{121},a_{211}) &= \underbrace{((2m_1)^{\text{th}}\text{DI}_+,(2n_1)^{\text{th}}\text{DI}_-)}_{((2m_1),(2n_1)\text{-double-inflection saddle}} \quad \text{for } a_{120}<0 \text{ and } a_{210}>0,\\
(a_{121},a_{211}) &= \underbrace{((2m_1)^{\text{th}}\text{DI}_-,(2n_1)^{\text{th}}\text{DI}_+)}_{((2m_1),(2n_1))\text{-double-inflection saddle}} \quad \text{for } a_{120}>0 \text{ and } a_{210}<0,\\
(a_{121},a_{211}) &= \underbrace{((2m_1)^{\text{th}}\text{II}_-,(2n_1)^{\text{th}}\text{II}_-)}_{((2m_1),(2n_1))\text{-double-inflection saddle}} \quad \text{for } a_{120}<0 \text{ and } a_{210}<0.
\end{aligned} \tag{5.175}
$$

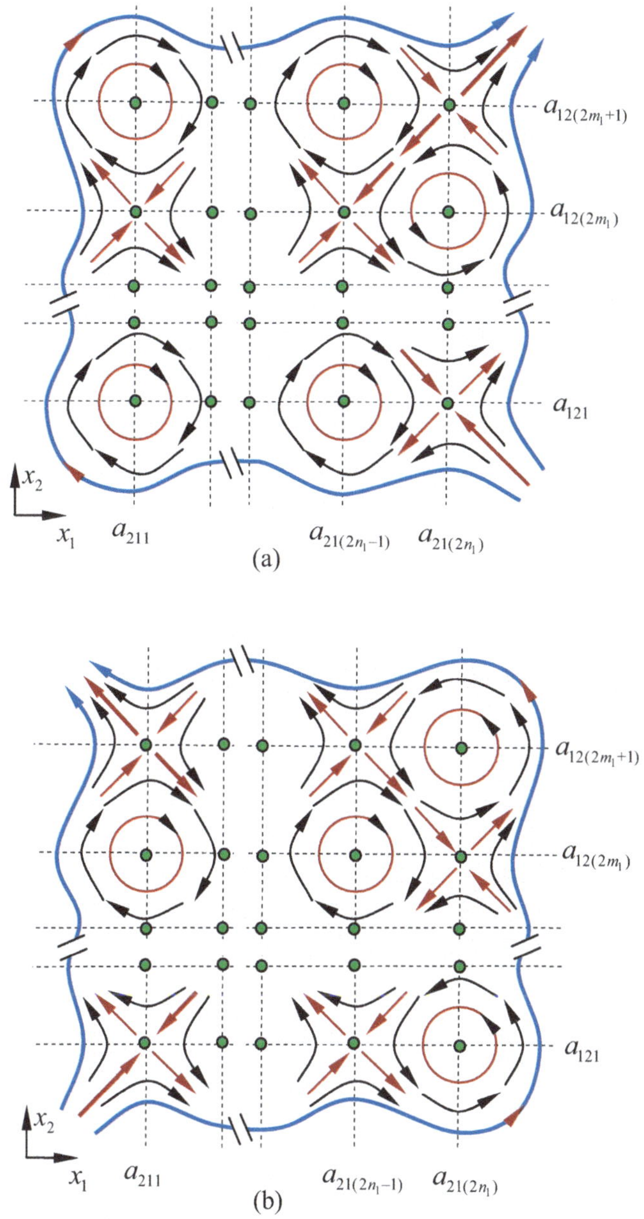

Fig. 5.6 Phase portraits for 2-dimensional systems with $x_1^* = a_{211}, a_{212}, \cdots, a_{21(2m_1+1)}$ and $x_2^* = a_{211}, a_{212}, \cdots, a_{21(2n_1)}$. The four networks of saddles and centers: **a** ($a_{120} > 0, a_{210} > 0$), **b** ($a_{120} < 0, a_{210} > 0$), **c** ($a_{120} > 0, a_{210} < 0$), **d** ($a_{120} < 0, a_{210} < 0$)

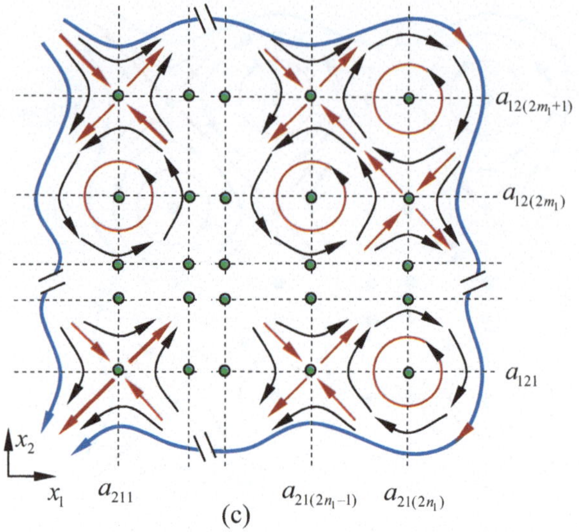

(c)

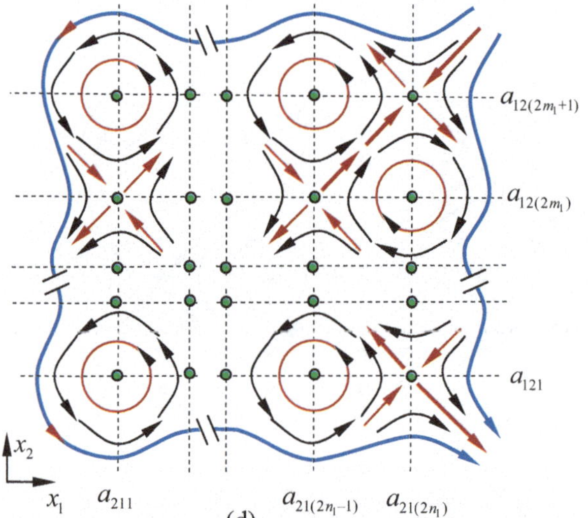

(d)

Fig. 5.6 (continued)

From the above bifurcations, there is a non-singular system with $(2m_1) \times (2n_1)$-equilibriums as

$$\dot{x}_1 = a_{120} \prod_{s=1}^{2m_1} (x_2 - a_{12s}) \text{ and } \dot{x}_2 = a_{210} \prod_{l=1}^{2n_1} (x_1 - a_{21l}). \tag{5.176}$$

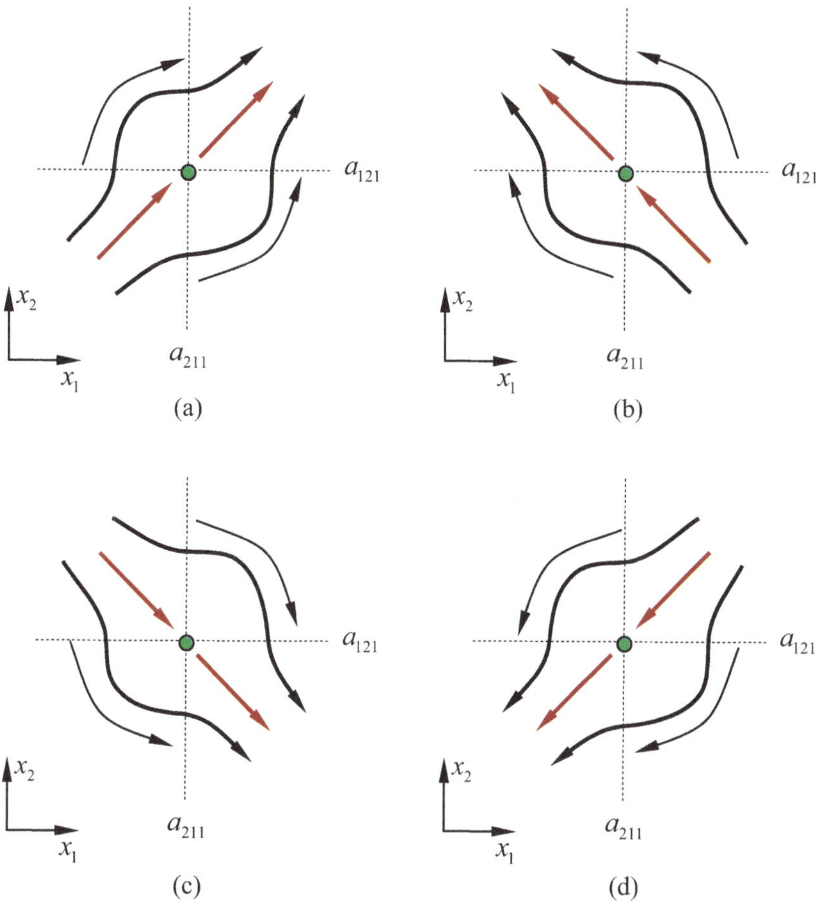

Fig. 5.7 Phase portraits of the $((2m_1), (2n_1))$-double-inflection saddles for 2-dimensional systems at $(x_1^*, x_2^*) = (a_{111}, a_{221})$. **a** $(a_{120} > 0, a_{210} > 0)$, **b** $(a_{120} < 0, a_{210} > 0)$, **c** $(a_{120} > 0, a_{210} < 0)$, **d** $(a_{120} < 0, a_{210} < 0)$

The first integral manifold is given by

$$\begin{aligned} & a_{210}\left\{\frac{1}{2n_1+1}\left[(x_1 - a_{21l_1})^{2n_1+1} - (x_{10} - a_{21l_1})^{2n_1+1}\right]\right. \\ & \left. + \sum_{k=1}^{2n_1-1} \frac{1}{2n_1-k+1} b_{21k}\left[(x_1 - a_{21l_1})^{2n_1-k+1} - (x_{10} - a_{21l_1})^{2n_1-k+1}\right]\right\} \\ & = a_{j_1j_20}\left\{\frac{1}{2m_1+1}\left[(x_2 - a_{12s_1})^{2m_1+1} - (x_{20} - a_{12s_1})^{2m_1+1}\right]\right. \end{aligned}$$

$$+\sum_{k=1}^{2m_1-1}\frac{1}{2m_1-k+1}b_{12k}\big[(x_1-a_{21s_1})^{2m_1-k+1}-(x_{10}-a_{21s_1})^{2m_1-k+1}\big]\Big\} \tag{5.177}$$

where

$$\begin{aligned}
b_{211} &= \sum_{l_2=1,l_2\neq l_1}^{2n_1}(a_{21l_1}-a_{21l_2}),\\
b_{212} &= \sum_{\substack{l_2,l_3=1;l_2,l_3\neq l_1\\(l_2<l_3)}}^{2n_1}\prod_{r=2}^{3}(a_{21l_1}-a_{21l_r}),\cdots,\\
b_{21k} &= \sum_{\substack{l_2,l_3,\cdots,l_{k+1}=1;\\ l_2,l_3,\cdots,l_{k+1}\neq l_1\\(l_2<l_3<\cdots<l_{k+1})}}^{2n_1}\prod_{r=2}^{k+1}(a_{21l_1}-a_{21l_r}),\cdots,\\
b_{21(2n_1)} &= \prod_{l_2=1,l_2\neq l_1}^{2n_1-1}(a_{2l_1}-a_{j_2j_1l_2});\\
b_{121} &= \sum_{s_2=1,s_2\neq s_1}^{2m_1}(a_{12s_1}-a_{12s_2}),\\
b_{122} &= \sum_{\substack{s_2,s_3=1;s_2,s_3\neq s_1\\(s_2<s_3)}}^{2m_1}\prod_{r=2}^{3}(a_{12s_1}-a_{12s_r}),\cdots,\\
b_{12k} &= \sum_{\substack{s_2,s_3,\cdots,s_{k+1}=1;\\ s_2,s_3,\cdots,s_{k+1}\neq s_1\\(s_2<s_3<\cdots<s_{k+1})}}^{2m_1}\prod_{r=2}^{k+1}(a_{12s_1}-a_{12s_r}),\cdots,\\
b_{12(2m_1)} &= \prod_{s_1=1,s_2\neq s_1}^{2m_1-1}(a_{12s_1}-a_{12s_2}).
\end{aligned} \tag{5.178}$$

Phase portraits for the 2-dimensional systems near the simple equilibriums of the saddles and centers are presented in Fig. 5.8a–d for ($a_{120}>0, a_{210}>0$), ($a_{120}<0, a_{210}>0$), ($a_{120}>0, a_{210}<0$) and ($a_{120}<0, a_{210}<0$). For all cases, the $(2m_1)\times(2n_1)$-simple equilibriums are based on the $((2m_1),(2n_1))$-double-inflection saddle bifurcations.

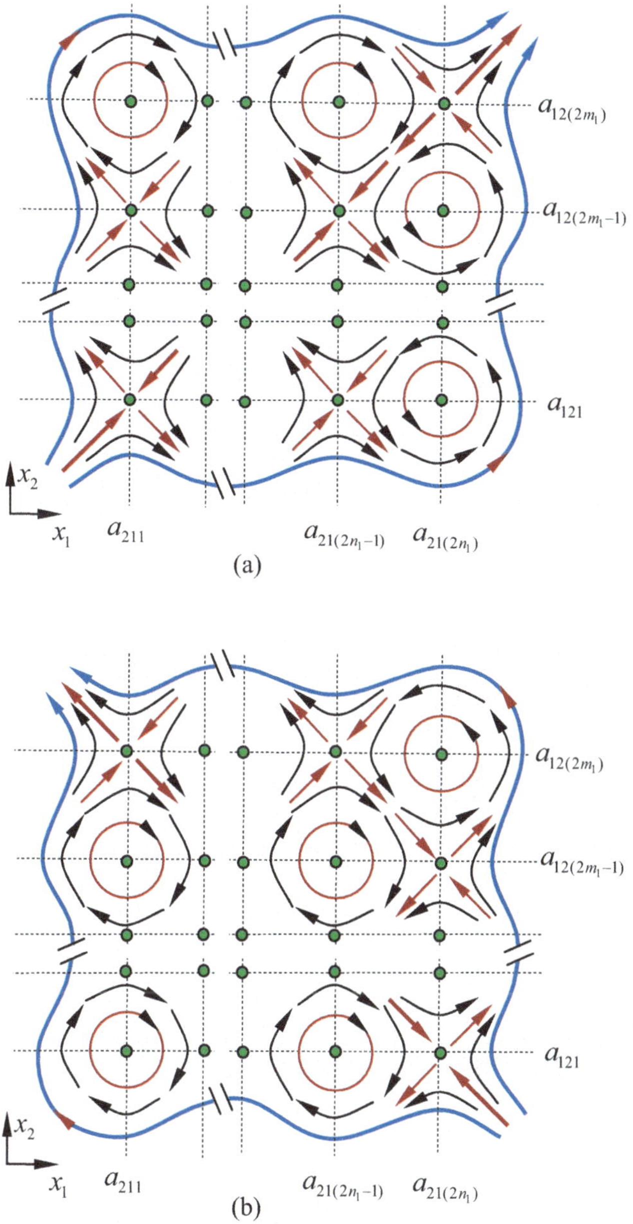

Fig. 5.8 Phase portraits for 2-dimensional systems with $x_1^* = a_{211}, a_{212}, \cdots, a_{21(2m_1)}$ and $x_2^* = a_{211}, a_{212}, \cdots, a_{21(2n_1)}$. The four networks of saddles and centers: **a** ($a_{120} > 0, a_{210} > 0$), **b** ($a_{120} < 0, a_{210} > 0$), **c** ($a_{120} > 0, a_{210} < 0$), **d** ($a_{120} < 0, a_{210} < 0$)

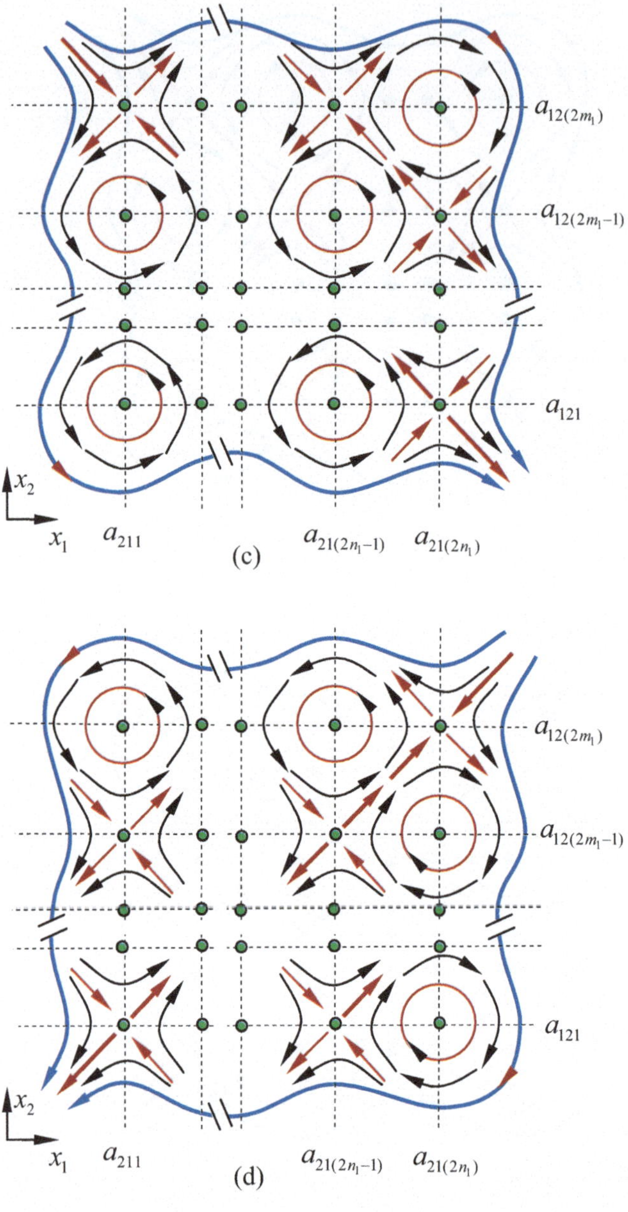

Fig. 5.8 (continued)

$$\bigcup_{s=1}^{2m_1}\bigcup_{l=1}^{2n_1}\underbrace{(a_{j_1j_2s},a_{j_2j_1l})}_{\text{XX}}=\left\{\begin{matrix}\underbrace{(\text{DP}_-,\text{DP}_+)}_{\text{CW center}} & \cdots & \underbrace{(\text{DP}_-,\text{DP}_+)}_{\text{CW center}} & \underbrace{(\text{UP}_+,\text{UP}_+)}_{\text{positive saddle}}\\ \underbrace{(\text{UP}_-,\text{UP}_-)}_{\text{negative saddle}} & \cdots & \underbrace{(\text{UP}_-,\text{UP}_-)}_{\text{negative saddle}} & \underbrace{(\text{DP}_+,\text{DP}_-)}_{\text{CCW center}}\\ \vdots & \vdots & \vdots & \vdots\\ \underbrace{(\text{UP}_-,\text{UP}_-)}_{\text{negative saddle}} & \cdots & \underbrace{(\text{UP}_-,\text{UP}_-)}_{\text{negative saddle}} & \underbrace{(\text{DP}_+,\text{DP}_-)}_{\text{CCW center}}\end{matrix}\right\}_{(2m_1)\times(2n_1)}$$

for $a_{120}>0,\ a_{210}>0;$ (5.179)

$$\bigcup_{s=1}^{2m_1}\bigcup_{l=1}^{2n_1}\underbrace{(a_{j_1j_2s},a_{j_2j_1l})}_{\text{XX}}=\left\{\begin{matrix}\underbrace{(\text{UP}_-,\text{UP}_-)}_{\text{negative saddle}} & \cdots & \underbrace{(\text{UP}_-,\text{UP}_-)}_{\text{negative saddle}} & \underbrace{(\text{DP}_+,\text{DP}_-)}_{\text{CCW center}}\\ \underbrace{(\text{DP}_-,\text{DP}_+)}_{\text{CW center}} & \cdots & \underbrace{(\text{DP}_-,\text{DP}_+)}_{\text{CW center}} & \underbrace{(\text{UP}_+,\text{UP}_+)}_{\text{positive saddle}}\\ \vdots & \vdots & \vdots & \vdots\\ \underbrace{(\text{DP}_-,\text{DP}_+)}_{\text{CW center}} & \cdots & \underbrace{(\text{DP}_-,\text{DP}_+)}_{\text{CW center}} & \underbrace{(\text{UP}_+,\text{UP}_+)}_{\text{positive saddle}}\end{matrix}\right\}_{(2m_1+1)\times(2n_1)}$$

for $a_{120}<0,\ a_{210}>0;$ (5.180)

$$\bigcup_{s=1}^{2m_1}\bigcup_{l=1}^{2n_1}\underbrace{(a_{j_1j_2s},a_{j_2j_1l})}_{\text{XX}}=\left\{\begin{matrix}\underbrace{(\text{UP}_+,\text{UP}_+)}_{\text{positive saddle}} & \cdots & \underbrace{(\text{UP}_+,\text{UP}_+)}_{\text{positive saddle}} & \underbrace{(\text{DP}_-,\text{DP}_+)}_{\text{CW center}}\\ \underbrace{(\text{DP}_+,\text{DP}_-)}_{\text{CCW center}} & \cdots & \underbrace{(\text{DP}_+,\text{DP}_-)}_{\text{CCW center}} & \underbrace{(\text{UP}_-,\text{UP}_-)}_{\text{negative saddle}}\\ \vdots & \vdots & \vdots & \vdots\\ \underbrace{(\text{DP}_+,\text{DP}_-)}_{\text{CCW center}} & \cdots & \underbrace{(\text{DP}_+,\text{DP}_-)}_{\text{CCW center}} & \underbrace{(\text{UP}_-,\text{UP}_-)}_{\text{negative saddle}}\end{matrix}\right\}_{(2m_1)\times(2n_1)}$$

for $a_{120}>0,\ a_{210}<0;$ (5.181)

$$\bigcup_{s=1}^{2m_1}\bigcup_{l=1}^{2n_1}\underbrace{(a_{j_1j_2s},a_{j_2j_1l})}_{\text{XX}}=\left\{\begin{matrix}\underbrace{(\text{DP}_+,\text{DP}_-)}_{\text{CCW center}} & \cdots & \underbrace{(\text{DP}_+,\text{DP}_-)}_{\text{CCW center}} & \underbrace{(\text{UP}_-,\text{UP}_-)}_{\text{negative saddle}}\\ \underbrace{(\text{UP}_+,\text{UP}_+)}_{\text{positive saddle}} & \cdots & \underbrace{(\text{UP}_+,\text{UP}_+)}_{\text{positive saddle}} & \underbrace{(\text{DP}_-,\text{DP}_+)}_{\text{CW center}}\\ \vdots & \vdots & \vdots & \vdots\\ \underbrace{(\text{UP}_+,\text{UP}_+)}_{\text{positive saddle}} & \cdots & \underbrace{(\text{UP}_+,\text{UP}_+)}_{\text{positive saddle}} & \underbrace{(\text{DP}_-,\text{DP}_+)}_{\text{CW center}}\end{matrix}\right\}_{(2m_1)\times(2n_1)}$$

for $a_{120}<0,\ a_{210}<0.$ (5.182)

Reference

1. Luo, A. C. J. (2024). Bifurcations for homoclinic networks in polynomial systems. *International Journal of Bifurcation and Chaos*, 34(3), 2430006 (153 page).

Subject Index

A. C. J. Luo, *Limit Cycles and Homoclinic Networks in Two-Dimensional Polynomial Systems*, https://doi.org/10.1007/978-981-97-2617-2

D

S

U

The manufacturer's authorised representative in the EU is Springer Nature Customer Service Centre GmbH, Europaplatz 3, 69115 Heidelberg, Germany. If you have any concerns regarding our products, please contact ProductSafety@springernature.com

Printed and bound by CPI Group (UK) Ltd, Croydon, CR0 4YY
01/10/2026
02225928-0007